third edition

*Fundamentals
of Algebra
and Trigonometry*

Other Texts by the Same Author

Calculus with Analytic Geometry

Fundamentals of College Algebra, Third Edition

Fundamentals of Trigonometry, Third Edition

Elementary Functions with Coordinate Geometry

Precalculus Mathematics, A Functional Approach

A Precalculus Course in Algebra and Trigonometry
5 Modules co-authored by Roy A. Dobyns, David T. Brown and Gail L. Carns.

Cover photograph by George Sheng. Text design by Deborah Schneider in collaboration with Michael Michaud and the PWS Production staff. Composed in Linofilm Palatino and Optima by Progressive Typographers. Printed and bound by Quinn & Boden, Co., Inc.

third edition

*Fundamentals
of Algebra
and Trigonometry*

EARL W. SWOKOWSKI
MARQUETTE UNIVERSITY

 Prindle, Weber & Schmidt, Incorporated
Boston, Massachusetts

© Copyright 1975 by Prindle, Weber & Schmidt, Incorporated
20 Newbury Street, Boston, Massachusetts 02116

Printed in the United States of America

Library of Congress Cataloging in Publication Data

Swokowski, Earl William
   Fundamentals of algebra and trigonometry.

   Includes index.
   1. Algebra. 2. Trigonometry, Plane. I. Title.
QA152.2.S95 1975   512'.13   74-31120
ISBN 0-87150-191-0

Second Printing: December, 1975

A Programmed Guide by Roy Dobyns is available for use with this text. The Guide is designed to serve as a tutor, aiding in the diagnosis of weaknesses and providing reinforcement for what is being read in the text. Page references to the text are given when further study is needed. The text and the Guide have the same table of contents and use the same notation.

Preface

One of my main objectives for this edition was to make the subject matter easier to understand. To achieve this goal, each sentence of the second edition was scrutinized to determine whether it could be improved, and discussions of concepts were simplified wherever possible. In order to help pinpoint troublesome material, the publisher conducted a survey of many users of the text, and special attention was given to sections that were noted as causing any difficulties.

The number of solved examples has increased significantly in this edition, and certain techniques have been replaced by methods that are easier to apply. Most of the exercise sets were expanded by adding drill-type problems together with some challenging ones for highly motivated students. This edition contains approximately 3000 exercises — almost 800 more than appeared previously. Several sections were added in response to the recommendations of numerous teachers. These are indicated in the partial description of the contents that follows.

As the title suggests, Chapter 1 covers topics fundamental for the study of algebra. Adopters of earlier editions of this text have treated Chapter 1 in different ways, depending on the background of the students. Some have omitted it entirely, regarding it as a review of previous courses. Others have devoted considerable time to a thorough study of this material.

Chapter 2 presents solutions of equations and inequalities. The rather formal technique used for nonlinear inequalities in the previous edition has been replaced by a graphical method that is easier to apply. The interval notation for solutions is introduced early, since it is so prevalent in calculus courses.

A discussion of functions and graphs follows in Chapter 3. This extensively rewritten chapter places additional emphasis on polynomial functions. In particular, Section 6, where quadratic functions and their graphs are considered in some detail, is new. In Section 7 the notions of composite and inverse functions are introduced in a manner that makes them readily applicable to future work in trigonometry or calculus. Section 8 is new, having been written to meet the needs of many teachers who wanted a discussion of conic sections to appear early in the text.

The introduction of more practical illustrations of exponential and logarithmic functions changes the flavor of Chapter 4 somewhat. As before, computational aspects of logarithms appear in the latter half of the chapter, where they may be omitted without interrupting the continuity of the book.

In Chapter 5 the introduction to trigonometric functions using a unit circle has been made more palatable by shortening the discussion and including more examples. Although this approach to

v

trigonometry requires slightly more work than the classical ratio approach, subsequent developments make it well worth the effort. Angles are introduced later in the chapter and the two approaches are then united by means of a theorem. In the previous edition graphs of trigonometric functions were discussed in Chapter 5, however, other graphical techniques were postponed until later. All of this material is now included in Chapter 5. Moreover, additional solved examples should make it easier for students to learn to sketch graphs rapidly.

Most of Chapter 6 consists of work with trigonometric identities and equations. Inverse trigonometric functions are introduced in Section 7. Numerical solutions of triangles appear at the end of the chapter. Although logarithmic solutions are included, they may be easily replaced by arithmetic processes.

Chapter 7 opens with a discussion of the method of substitution for finding solutions of arbitrary systems of equations. The method is then applied to linear systems as a special case. This leads in a natural way to solutions by row operations on matrices or by determinants. Next, systems of inequalities are considered and used in a new section on linear programming. The chapter concludes with a discussion of the fundamentals of matrix algebra.

Complex numbers are defined in Chapter 8, using the symbolic $a + bi$ approach. Many instructors favor this procedure over the ordered pair definition, and most students find it easier to understand. Of course, the only essential difference in the two definitions is the notation, provided that at the outset the plus sign in $a + bi$ is not interpreted as the operation of addition. A brief introduction to vectors appears at the end of the chapter.

Chapter 9 consists of results on polynomial functions that are deeper than those considered earlier in the text.

Finally, Chapter 10 contains sections on mathematical induction, sequences, the Binomial Theorem, permutations and combinations.

The material included in this text is more than can be covered in a typical one-semester course. This was done intentionally to allow instructors to choose the concepts they wish to stress. These choices will vary, depending on the objectives of the course together with the background and abilities of students.

There is a review section at the end of each chapter consisting of a list of important topics and pertinent exercises. The review exercises are similar in scope to those which appear throughout the chapter and may be used to prepare for examinations. Answers to odd-numbered exercises are given at the end of the text. An answer booklet for the even-numbered exercises may be obtained from the publisher.

A *Programmed Guide,* by Roy Dobyns, is again available for use with this book. The Guide, designed to serve as a tutor, aids in the diagnosis of weaknesses and provides reinforcement for material in the text. The text and the Guide have the same table of contents and use the same notation.

This edition has benefited from comments and suggestions of users of the first two editions. The publisher also solicited critical reviews of the book from a number of teachers. To these, and to the many others who have assisted in the formation of this text, I wish to express my sincere appreciation.

Finally, I owe a great deal to my wife Shirley, for her patience, understanding, and moral support over long periods of writing. In recognition of many sacrifices she has made, I dedicate this book to her.

Earl W. Swokowski

Contents

4
Exponential and Logarithmic Functions

5
The Trigonometric Functions

6
Analytic Trigonometry

7

Systems of Equations and Inequalities

8

Complex Numbers

Contents

1

Fundamental Concepts of Algebra

The material in this chapter is basic for the study of algebra. After introducing the terminology of sets we briefly discuss properties of the real number system. Included in our discussion are the important concepts of inequalities and absolute values. We then turn our attention to exponents and radicals, and how they may be used to simplify complicated algebraic expressions.

1 SETS

Throughout mathematics and other areas the concept of **set** is used extensively. A set may be thought of as a collection of objects. For example, we could refer to the set of books in a library, the set of giraffes in a zoo, the set of natural numbers 1, 2, 3, $\cdots$, the set of points in a plane, and so on. Each object in a set is called an **element** of the set. We assume that every set is **well defined** in the sense that there is some rule or property that can be used to determine whether a given object is or is not an element of the set.

Capital letters, A, B, C, R, S, $\cdots$ will often be used to denote sets. Lower-case letters a, b, x, y, $\cdots$ will represent elements of sets. If S represents a set, then the symbol $a \in S$ denotes the fact that a is an element of S. Similarly, a, $b \in S$ means that a and b are elements of S. The notation $a \notin S$ signifies that a is not an element of S.

If every element of a set S is also an element of a set T, then S is called a **subset** of T and we write $S \subseteq T$, or $T \supseteq S$, which may be read "S is contained in T" or "T contains S." For example, if T is the set of letters in the English alphabet and if S is the set of vowels, then $S \subseteq T$.

If S is any set, then $S \subseteq S$, since every element of S is an element of S. The symbol $S \nsubseteq T$ means that S is not a subset of T. In this case there is at least one element of S which is not an element of T.

Two sets S and T are said to be **equal,** written $S = T$, if S and T contain precisely the same elements. If S and T are not equal, we write $S \neq T$. If $S \subseteq T$ and $S \neq T$, then S is called a **proper subset** of T. In this case there is at least one element of T which is not an element of S.

The notation $a = b$, translated "a equals b," means that a and b are symbols which represent the same element of a set. For example, in arithmetic the symbol $2 + 3$ represents the same number as the symbol $4 + 1$ and we write $2 + 3 = 4 + 1$. Similarly, $a = b = c$ means that a, b, and c all represent the same element. Of course, $a \neq b$ means that a and b represent different elements. We assume that equality of elements of a set S satisfies the following three properties:

(1) $a = a$ for every element a,
(2) if $a = b$, then $b = a$,
(3) if $a = b$ and $b = c$, then $a = c$.

There are various ways of describing sets. One method, especially useful for sets containing only a few elements, is to list all the elements within braces. For example, if S consists of the first five letters of the alphabet, we write $S = \{a, b, c, d, e\}$. When sets are described in this way, the order in which the elements are listed is irrelevant. We could also write $S = \{a, c, b, e, d\}$, or $S = \{d, c, b, e, a\}$, and so on. This notation is also useful for representing larger sets if there is some definite pattern for the elements. As an illustration, we might specify the set N of **natural numbers** by

(1.1) $N = \{1, 2, 3, 4, \cdots\}$,

where the dots may be read "and so on."

There is another useful notation for describing sets. If $S \subseteq T$ and each element of S has a certain property, then we write

$$S = \{x \in T : \text{-----}\},$$

where the property which describes x is stated in the space designated by the dashes. If the set T from which the elements are chosen is evident then we shall only use the symbol x in front of the colon. As a specific example, let

$$S = \{x \in N : x + 2 = 7\}.$$

This can be read "S is the set of elements x in N such that $x + 2 = 7$." Hence S contains only one element, the number 5, and we could write $S = \{5\}$. As another illustration, if $E = \{x \in N : x \text{ is even}\}$, then E consists of the collection of all even natural numbers, that is, $E = \{2, 4, 6, 8, \cdots\}$. Another way of describing E is to write $E = \{2n : n \in N\}$.

The **empty** (or **null**) **set** $\emptyset$ may be defined by $\emptyset = \{x : x \neq x\}$. The empty set $\emptyset$ differs from all other sets because it contains no elements. It is mainly a notational device we find convenient to use in certain instances. For example, if $S = \{x \in \mathbf{N} : x + 2 = 1\}$, then $S = \emptyset$, since $x + 2$ is never 1 if $x \in \mathbf{N}$. It is customary to assume that $\emptyset$ is a subset of every set S.

EXAMPLE 1   List the subsets of the set $S = \{a, b, c\}$.

Solution

There are 8 subsets in all. They are

$$\{a\}, \{b\}, \{c\}, \{a, c\}, \{a, b\}, \{b, c\}, \{a, b, c\}, \text{ and } \emptyset.$$

When working with several sets, we always assume they are subsets of some larger set W, called a **universal set,** however, W will not always be stated explicitly. If elements $x, y, z, \cdots$ are employed, without specifying any particular set, we assume they belong to some universal set W. With these remarks in mind, we state the following definition.

(1.2)   Definition

If A and B are sets, their **union** $A \cup B$ and **intersection** $A \cap B$ are given by

$$A \cup B = \{x : x \in A \quad \text{or} \quad x \in B\}$$
$$A \cap B = \{x : x \in A \quad \text{and} \quad x \in B\}.$$

The word "or" in the definition of union means that either $x \in A$ or $x \in B$, or possibly that x is in *both* A and B. The intersection of two sets consists of the elements which are *common* to both sets. If $A \cap B = \emptyset$, that is, if A and B have no elements in common, then A and B are said to be **disjoint sets.**

EXAMPLE 2   If $A = \{a, b, c, d\}$, $B = \{b, c, e, f\}$, and $C = \{a, d\}$, find $A \cup B, A \cap B, A \cup C, A \cap C$, and $B \cap C$.

Solution

By (1.2) we have $A \cup B = \{a, b, c, d, e, f\}$, $A \cap B = \{b, c\}$, $A \cup C = A$, $A \cap C = C$, and $B \cap C = \emptyset$.

Sets are often pictured by drawing circles, squares, or other simple closed curves in a plane, where it is understood that the points within these figures represent the elements of the sets. For example, we might indicate that A and B are subsets of a set W by the sketch in Fig. 1.1. Unions and intersections can then be represented by shading

3

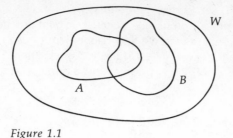

Figure 1.1

appropriate parts of the figure. This is illustrated in Fig. 1.2, where the universal set W is not shown.

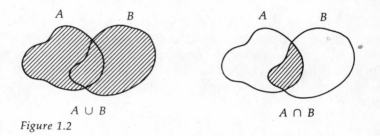

$A \cup B$ $A \cap B$

Figure 1.2

Such geometric representations of sets are often referred to as **Venn diagrams.** The reader should also sketch $A \cup B$ and $A \cap B$ if $A \subseteq B$, $B \subseteq A$, or $A \cap B = \emptyset$. It is important to remember that Venn diagrams are used merely to help motivate and visualize notions concerning sets and are not used to prove or serve as steps in proofs of any theorems.

Unions and intersections of more than two sets can also be formed. For example, if A, B, and C are sets, we may consider the union of $A \cup B$ and C as follows.

$$(A \cup B) \cup C = \{x : x \in A \cup B \text{ or } x \in C\}$$
$$= \{x : x \in A \text{ or } x \in B \text{ or } x \in C\}.$$

Thus $(A \cup B) \cup C$ is the set of all elements x which appear in *at least one* of the sets A, B, or C. We could also first form $B \cup C$ and then consider $A \cup (B \cup C)$. Evidently $(A \cup B) \cup C = A \cup (B \cup C)$. Similarly, it can be shown that $(A \cap B) \cap C = A \cap (B \cap C)$, where this set is the collection of elements which are common to the three sets A, B, and C.

EXAMPLE 3 If A, B, and C are sets, use Venn diagrams to represent

$(A \cap B) \cap C$ and $(A \cup B) \cap C.$

Solution

See Figure 1.3.

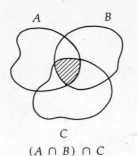

$(A \cap B) \cap C$

Figure 1.3

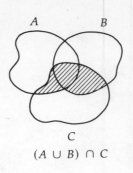

$(A \cup B) \cap C$

EXERCISES

1   Use a notation for sets to designate each of the following:

a   the set consisting of the last six letters of the alphabet

b   the set consisting of the letters of the alphabet which occur after the letter z

c   the natural numbers which are divisible by 10

2   Describe each of the following sets in words.

a $S = \{x \in \mathbf{N} : x \text{ is odd}\}$

b $T = \{x \in \mathbf{N} : x = 5\}$

c $V = \{10n : n \in \mathbf{N}\}$

3   Characterize each of the following as true or false and give reasons for your answers.

a $\{1, 3\} \subseteq \{3, 2, 1\}$ T

b $\{1, 2\} \subseteq \{1, 3\}$ F

c $\{1, 2\} = \{2, 1\}$ T

d $\{1, 2\} \subseteq \{1, 2\}$ T

e $\{2\} \in \{1, 2\}$ F

f $2 \in \{1, 2\}$ T

g $2 \subseteq \{1, 2\}$ F

h $\{2\} \subseteq 2$ F

4   If $A = \{1, 2, 3\}$, determine whether the following are true or false and give reasons for your answers.

a $3 \in A$ b $3 \subseteq A$ c $\{3\} \subseteq A$

d $A \subseteq A$ e $\varnothing \in A$ f $\varnothing \subseteq A$

5   Find all the subsets of $\{a, b, c, d\}$.

6   Find all the subsets of $\{a, b\}$.

7   Find $A \cup B$ and $A \cap B$.

a $A = \{2, 5, 3\}$, $B = \{3, 1, 6\}$

b $A = \{a, b, c, d\}$, $B = \{d, e, a\}$

c $A = \{a, b, c\}$, $B = \{d, e, f\}$

d $A = \{1, 2\}$, $B = \{1, 2, 3\}$

[Handwritten annotations:]

$A \cup B$
$\{1, 2, 3, 5, 6\}$
$\{a, b, c, d, e\}$
$\{a, b, c, d, e, f\}$
$\{1, 2, 3\}$

$A \cap B$
$\{3\}$
$\{a, d\}$
$\varnothing$
$\{1, 2\}$

5

8 If $R = \{4, 7, 1, 5, 2\}$, $S = \{3, 5, 2\}$, $V = \{7, 2, 5\}$, and $P = \{3, 6\}$, find each of the following.

a $R \cup S$

b $R \cup V$

c $R \cap S$

d $R \cap V$

e $R \cap P$

f $R \cup P$

g $(S \cup V) \cap P$

h $S \cup (V \cap P)$

i $(S \cup P) \cap (R \cup V)$

9 If $R = \{1, 5, 6\}$, $S = \{2, 3, 4\}$, and $T = \{6, 2\}$ find each of the following.

a $R \cap (S \cup T) = \{6\}$

b $T \cap (R \cup S) = \{6, 2\}$

c $R \cup (S \cap T) = \{1, 2, 5, 6\}$

d $(R \cup S) \cap (R \cup T) = \{1, 2, 5, 6\}$

10 If A and B are sets, find $A \cap B$ and $A \cup B$ in case

a $A \subseteq B$

b $B \subseteq A$

c A and B are disjoint

d $A = B$

11 Let S and T be sets. Under what conditions will the following be true?

a $S \cup T = S$ $T \subseteq S$

b $S \cap T = S$ $S \subseteq T$

c $S \cap \emptyset = S$ $S = \emptyset$

d $S \cup \emptyset = S$ Any S

e $S \cap T = \emptyset$ $S \& T$ disjoint

f $S \cup T = \emptyset$ $S = T = \emptyset$

g $S \cap T = T \cap S$ any S and T

h $S \cup T = T \cup S$ any S and T

12 If A and B are represented geometrically as in Fig. 1.1, shade the part of W which corresponds to each of the following sets.

a $\{x \in W : x \in A \text{ and } x \notin B\}$

b $\{x \in W : x \notin A \cup B\}$

c $\{x \in W : x \in A \cup B \text{ and } x \notin A \cap B\}$

If A, B, and C are sets, use Venn diagrams to represent each of the following.

13 $A \cap (B \cap C)$

14 $A \cup (B \cup C)$

15 $A \cup (B \cap C)$

16 $(A \cap B) \cup C$

17 $A \cap (B \cup C)$

18 $(A \cup C) \cap B$

19 $(A \cap B) \cup (A \cap C)$

20 $(A \cup B) \cap (A \cup C)$

2 REAL NUMBERS

One of the most important sets in mathematics is the set **R** of real numbers. We refer to **R**, together with the various properties possessed by its elements, as the **real number system.** The reader is undoubtedly well acquainted with symbols such as 2, $-\frac{3}{5}$, $\sqrt{3}$, 0, -8.614, .3333 · · · , 467.2, and so on, which are used to denote real numbers.

In this section we shall list some properties of **R** and review the notation and terminology associated with real numbers.

The system **R** is **closed** relative to operations of addition (denoted by "+") and multiplication (denoted by "·"). This means that for every pair a, $b \in$ **R**, there corresponds a unique real number $a + b$ called the **sum** of a and b and a unique real number $a \cdot b$ (also written ab) called the **product** of a and b. These operations have the following properties, where all lower case letters denote arbitrary real numbers.

(1.3) **Commutative Laws**

$$a + b = b + a, \quad ab = ba.$$

(1.4) **Associative Laws**

$$a + (b + c) = (a + b) + c, \quad a(bc) = (ab)c.$$

(1.5) **Distributive Laws**

$$a(b + c) = ab + ac, \quad (a + b)c = ac + bc.$$

(1.6) **Identity Elements**

There exist special real numbers, denoted by 0 and 1, with the following properties:

$$a + 0 = a = 0 + a, \quad a \cdot 1 = a = 1 \cdot a.$$

(1.7) **Inverse Elements**

For every real number a, there is a real number denoted by $-a$ such that $a + (-a) = 0 = (-a) + a$. For every real number $a \neq 0$, there is a real number denoted by $1/a$ such that $a(1/a) = 1 = (1/a)a$.

A set which satisfies the above properties is referred to as a **field.** For this reason, (1.3)–(1.7) are sometimes called the **field properties** of the real number system.

Since $a + (b + c) = (a + b) + c$, we may, without ambiguity, use the symbol $a + b + c$ to denote the real number they represent. Similarly, the notation abc is used to represent either $a(bc)$ or $(ab)c$. An analogous situation exists if four real numbers a, b, c, and d are added. For example, we could consider

$$(a + b) + (c + d), \, a + [(b + c) + d], \, [(a + b) + c] + d.$$

It can be shown that regardless of how the four numbers are grouped, the same result is always obtained, and consequently it is customary to write $a + b + c + d$ for any of these expressions. Furthermore, it follows from the commutative law (1.3) that the numbers in such a sum can be interchanged in any way. For example, $a + b + c + d = a + d + c + b$, etc. We shall justify a manipulation of this type by referring to "rearrangement properties of real numbers" instead of pointing out the specific field properties which were used. A similar situation exists for multiplication, where the expression $abcd$ is used to denote the product of four real numbers.

The special real numbers 0 and 1 are referred to as **zero** and **one** respectively. They are the only real numbers which satisfy the conditions stated in (1.6). We call $-a$ the **additive inverse** of a (or **minus** a); and we call $1/a$ the **multiplicative inverse** of a (or the **reciprocal** of a) if $a \neq 0$. The symbol a^{-1} is often used in place of $1/a$.

Many facts about real numbers can be derived from (1.3)–(1.7). Two very important results are the following **cancellation laws for addition and multiplication:**

If $a + c = b + c$, then $a = b$.

If $ac = bc$ and $c \neq 0$, then $a = b$.

The field properties can be used to prove that

$a \cdot 0 = 0$ for every real number a.

It can also be shown that

if $ab = 0$, then either $a = 0$ or $b = 0$.

The following can also be proved:

$$-(-a) = a; \quad (-1)a = -a$$
$$(-a)b = -(ab) = a(-b)$$
$$(-a)(-b) = ab.$$

We shall assume that the reader is familiar with the above, and other basic rules pertaining to **R.**

If $a = c$ and $b = d$, then since a and c are merely different names for the same real numbers, and likewise for b and d, it follows that $a + b = c + d$ and $a \cdot b = c \cdot d$. This is often called the **substitution principle,** since we may think of replacing a by c and b by d in the expressions $a + b$ and $a \cdot b$. As a special case, if $a = c$, we can use the fact that $b = b$ to obtain $a + b = c + b$ and $a \cdot b = c \cdot b$. For convenience we sometimes refer to the latter manipulations by the statement "add b to both sides of the equality $a = c$" or "multiply both sides of $a = c$ by b."

The operation of **subtraction** (denoted by "$-$") is defined by $a - b = a + (-b)$. If $b \neq 0$, then **division** (denoted by "$\div$") is defined by

$a \div b = a(1/b) = ab^{-1}$. The symbol a/b is often used in place of $a \div b$, and we refer to it as the **quotient of** a **by** b or the **fraction** a **over** b. The numbers a and b are called the **numerator** and **denominator,** respectively, of the fraction. It is important to note that a/b is not defined if $b = 0$; that is, **division by zero is not permissible.** The following rules for quotients may be established, where all denominators are nonzero real numbers:

$$a/b = c/d \text{ if and only if } ad = bc,$$
$$(ad)/(bd) = a/b,$$
$$(a/b) + (c/d) = (ad + bc)/(bd),$$
$$(a/b)(c/d) = (ac)/(bd).$$

If we begin with the real number 1 and successively add it to itself, we obtain the set of **positive integers.** We shall identify this subset of **R** with the set **N** of natural numbers given in (1.1). The negatives -1, $-2, -3, -4, \cdots$ of the positive integers are referred to as **negative integers.** The set **Z** of **integers** is the totality of positive and negative integers together with the real number 0, that is,

$$\mathbf{Z} = \{\cdots, -3, -2, -1, 0, 1, 2, 3, \cdots\}.$$

If $a, b, c \in \mathbf{Z}$ and $c = ab$, then a and b are called **factors,** or **divisors,** of c. For example, the integer 6 may be written as $2 \cdot 3$, $(-2) \cdot (-3)$, $(-1) \cdot (-6)$, and $1 \cdot 6$. Hence $1, -1, 2, -2, 3, -3, 6$, and -6 are factors of 6. A positive integer p different from 1 is **prime** if its only positive factors are 1 and p. The first few primes are 2, 3, 5, 7, 11, 13, 17, and 19. One of the reasons for the importance of prime numbers is that every positive integer a different from 1 can be expressed in one and only one way (except for order of factors) as a product of primes. The proof of this result will not be given in this book. As examples, we have

$$12 = 2 \cdot 2 \cdot 3, \quad 126 = 2 \cdot 3 \cdot 3 \cdot 7, \quad 540 = 2 \cdot 2 \cdot 3 \cdot 3 \cdot 3 \cdot 5.$$

A real number is called a **rational number** if it can be written in the form a/b, where a and b are integers and $b \neq 0$. Real numbers that are not rational are called **irrational.** The ratio of the circumference of a circle to its diameter is irrational. This real number is denoted by π and is often approximated by the decimal 3.1416 or by the rational number $\frac{22}{7}$. We use the notation $\pi \approx 3.1416$ to indicate that π is *approximately equal* to 3.1416. To cite another example, the positive real number a such that $a^2 = 2$, where a^2 denotes $a \cdot a$, is not rational. This irrational number is denoted by the symbol $\sqrt{2}$.

Decimal representations for rational numbers either terminate or are nonterminating and repeating. For example, it can be shown by long division that a decimal representation for 7434/2310 is $3.2181818 \cdots$, where the dots indicate that the digits 1 and 8 repeat indefinitely. The rational number $\frac{5}{4}$ has the terminating decimal representation 1.25. Decimal representations for irrational numbers may also be obtained; however, they are always nonterminating and nonre-peating. The process of finding decimal representations for irrational numbers is usually difficult. Often some method of successive approximation is employed. For example, the device learned in arithmetic for extracting square roots can be used to find a decimal representation for $\sqrt{2}$. Using this technique we successively obtain the approximations 1, 1.4, 1.41, 1.414, 1.4142, and so on.

An important subset of the real numbers is the collection of **positive real numbers,** which are characterized by the following two properties:

(1.8)   If a and b are positive real numbers, then the sum $a + b$ and product ab are also positive.

(1.9)   For every real number a, one and only one of the following is true: $a = 0$, a is positive, or $-a$ is positive.

The positive integers are examples of positive real numbers, as is every rational number a/b when a and b are both positive or both negative. The nonzero real numbers which are not positive are called **negative real numbers.** The negative integers, or rational numbers such as $(-2)/3$ and $13/(-18)$, are examples of negative real numbers. The real number 0 is considered neither positive nor negative.

It follows from (1.9) that if a is positive, then $-a$ is negative. Similarly, if $-a$ is positive, then $-(-a) = a$ is negative. A common error among beginning students is to always regard $-a$ as a negative number; however, this is not always the case. For example, if $a = -3$, then $-a = -(-3) = 3$, which is positive.

If a and b are real numbers, then by (1.9) precisely one of the following is true: $a - b = 0$, $a - b$ is positive, or $-(a - b)$ is positive. If $a - b$ is positive we say that a **is greater than** b (or b **is less than** a) and write $a > b$ (or $b < a$). The symbols ">" and "<" are called **inequality signs** and expressions such as $a > b$ or $b < a$ are called **inequalities.**

(1.10)   Definition

If a and b are real numbers, then $a > b$, or $b < a$, means that $a - b$ is positive.

From the previous discussion we see that if $a, b \in \mathbf{R}$, then *one and only one of the following is true:*

$$a = b, \quad a > b, \quad \text{or} \quad a < b.$$

As illustrations of (1.10), $5 > 3$, since $5 - 3 = 2$, which is positive. Similarly, $-6 < -2$, since $-2 - (-6) = 4$, which is positive. Similarly, $1 > 0$, $2 > 1$, $3 > 2$, and so on. In general, the following **ordering**

$$\cdots < -4 < -3 < -2 < -1 < 0 < 1 < 2 < 3 < 4 < \cdots.$$

Since $a - 0 = a$, it follows that $a > 0$ if and only if a is positive. Similarly, $a < 0$ if and only if a is negative. The rules stated in the following theorem are very important.

(1.11) Theorem on Inequalities

Let a, b, $c \in \mathbf{R}$.

(1)   If $a > b$ and $b > c$, then $a > c$.

(2)   If $a > b$, then $a + c > b + c$.

(3)   If $a > b$ and $c > 0$, then $ac > bc$.

(4)   If $a > b$ and $c < 0$, then $ac < bc$.

Proof

(1)   If $a > b$ and $b > c$, then from (1.10), $a - b$ and $b - c$ are both positive. Consequently, by (1.8), the sum $(a - b) + (b - c)$ is positive. Since the sum reduces to $a - c$, we see that $a - c$ is positive. According to (1.10) this means that $a > c$.

(2)   If $a > b$, then by (1.10), $a - b$ is positive. Since $(a + c) - (b + c) = a - b$, it follows that $(a + c) - (b + c)$ is positive; that is, $a + c > b + c$.

(3)   If $a > b$ and $c > 0$, then $a - b$ and c are both positive. Hence, by (1.8), the product $(a - b)c$, that is, $ac - bc$, is positive. Hence by (1.10), $ac > bc$.

The proof of (4) is left as an exercise.

Similar results hold for the symbol "$<$." Specifically, the following results can be proved, where all letters represent real numbers:

(1.12)   (1)   If $a < b$ and $b < c$, then $a < c$.

(2)   If $a < b$, then $a + c < b + c$.

(3)   If $a < b$ and $c > 0$, then $ac < bc$.

(4)   If $a < b$ and $c < 0$, then $ac > bc$.

Note that by (3) of (1.11) or (1.12) it is permissible to multiply both sides of an inequality by a positive real number; however, as indicated by (4), multiplying both sides by a negative real number reverses the inequality sign. For example, if we multiply both sides of the inequality $5 > 3$ by -2 we obtain $-10 < -6$. By rule (2) any real number can be added to both sides of an inequality. For example, if we add -2 to both sides of $5 > 3$ we obtain $3 > 1$.

EXAMPLE Use (1.11) to change the inequality $-3a + 4 > 11$ to the form $a < k$ for some real number k.

Solution

Adding -4 to both sides of the given inequality we obtain

$$(-3a + 4) + (-4) > 11 + (-4),$$

which simplifies to $-3a > 7$. Multiplying both sides of the latter inequality by $(-\frac{1}{3})$ and using (4) of (1.11) gives us

$$(-\tfrac{1}{3})(-3a) < (-\tfrac{1}{3})(7).$$

Simplifying, we obtain the desired form: $a < -\frac{7}{3}$.

The symbol $a \geq b$, which is read "a is greater than or equal to b" means that either $a > b$ or $a = b$ (but not both). The symbol $a \leq b$ is defined in like manner. The expression $a < b < c$ means that *both* $a < b$ and $b < c$, in which case we say that b **is between** a **and** c. This may also be expressed by writing $c > b > a$. For instance,

$$1 < 5 < \tfrac{11}{2}, \quad -4 < \tfrac{2}{3} < \sqrt{2}, \quad 3 > -6 > -10.$$

An *incorrect* use of this notation would be to write $5 < a < 2$, since it is impossible for a real number a to satisfy *both* of the conditions $5 < a$ and $a < 2$.

Other variations of the inequality notation are used. For example, $a < b \leq c$ means both $a < b$ and $b \leq c$. Similarly, $a \leq b < c$ means both $a \leq b$ and $b < c$. As another illustration, $a \leq b \leq c$ means both $a \leq b$ and $b \leq c$.

Let us conclude this section by proving what are sometimes called the **Laws of Signs** for real numbers. We know from (1.8) that the product of two positive real numbers is positive. The next theorem covers the cases where one or both numbers is negative.

(1.13) **Theorem**

    (1)   The product of two negative real numbers is positive.

    (2)   The product of a positive and a negative real number is negative.

Proof

    (1)   If b and c are negative real numbers, we may write $0 > b$ and $c < 0$. From (4) of (1.11), with $a = 0$, we obtain $0 \cdot c < bc$, or $0 < bc$. Thus the product of two negative real numbers is positive.

    (2)   If b is negative and c positive, then $0 > b$ and $c > 0$. Using (3) of (1.11), with $a = 0$, we have $0 \cdot c > bc$ or $bc < 0$; that is, the product is negative.

In Exercises 1–10 justify each equality by stating only *one* of the field properties (1.3)–(1.7).

1 $(2 + 3) + 2 = 2 + (2 + 3)$ *commut.*   **2** $(2 + 3) \cdot 2 = 2 \cdot (2 + 3)$

3 $(2 + 3) + 2 = 2 + (3 + 2)$ *assoc.*   **4** $(2 + 3) \cdot 2 = 2 \cdot 2 + 3 \cdot 2$

5 $(2 + 0) \cdot 3 = 2 \cdot 3$ *Identity*   **6** $0 \cdot 1 = 0$

7 $2(\frac{1}{2}) = 1$ *inverse*   **8** $(-5) + 5 = 0$

9 $(-1)(1) = -1$ *Identity*   **10** $2 \cdot (2 + 2) = (2 + 2) \cdot 2$

11 In each of the following, replace the comma between the given pair of real numbers with the appropriate symbol $<$, $>$, or $=$.

    a $-2, -5$ **b** $-2, 5$ **c** $6 - 1, 2 + 3$

    d $\frac{2}{3}, 0.66$ **e** $\frac{1}{2}, 0.50$ **f** $\pi, \frac{22}{7}$

12 Same as Exercise 11 for each of the following:

    a $-3, 0$ **b** $-8, -3$ **c** $8, -3$

    d $\frac{3}{4} - \frac{2}{3}, \frac{1}{15}$ **e** $\sqrt{2}, 1.4$ **f** $\frac{4053}{1110}, 3.6513$

In Exercises 13–20 express the given statement in terms of inequalities.

13 a is positive $a > 0$ **14** a is negative

15 a is between 4 and -3 $-3 < a < 4$ **16** a is between 0.1 and 0.01

17 a is greater than -6 $-6 < a$ **18** a is less than or equal to 9

19 a is nonnegative $a \geq 0$ **20** a is not greater than 2

In Exercises 21–26 justify each inequality by stating only one of the properties in (1.11) or (1.12).

21 If $a > 3$, then $4a > 12$ **22** If $4 < 2y$, then $2 < y$.

23 If $-2a < 10$, then $a > -5$ **24** If $4 > -8b$, then $-\frac{1}{2} < b$

25 If $3a + 2 < 7$, then $3a < 5$ **26** If $5 - a < 0$, then $5 < a$

27 Show, by means of examples, that the operation of subtraction on **R** is neither commutative nor associative.

28 Show that the operation of division, as applied to nonzero real numbers, is neither commutative nor associative.

29 Show that if $a, b \in$ **R**, then $a > b$ if and only if $-a < -b$.

30 Prove that $a = -a$ if and only if $a = 0$.

31 Prove the cancellation laws for addition and multiplication.

32 Prove that $a \cdot 0 = 0$. (HINT: Write $a \cdot 0 = a \cdot (0 + 0) = a \cdot 0 + a \cdot 0$ and add $-(a \cdot 0)$ to both sides.)

33 Prove that if $a + c < b + c$, then $a < b$. What can be said if $ac < bc$?

34 If b and d are positive, prove that $a/b < c/d$ if and only if $ad < bc$. Is the result true if b or d is negative?

35 If a and b are positive, prove that $a < b$ if and only if $1/a > 1/b$. Is this true if negative numbers are allowed?

36 If a and b are positive, prove that $a > b$ if and only if $a^2 > b^2$. Is this true if a or b is negative?

37 Prove (4) of (1.11).   **38** Prove (1.12).

3 COORDINATE LINES AND ABSOLUTE VALUES

It is possible to associate the set of real numbers with the points on a straight line l in such a way that for each real number a there corresponds one and only one point and conversely, to each point P on l there corresponds precisely one real number. Such an association between two sets is referred to as a **one-to-one correspondence.** We begin by choosing an arbitrary point O on l, called the **origin,** and associate with it the real number 0. We then select any other point P_1 and associate with it the real number 1. The line segment OP_1 from O to P_1 is said to have **unit length.** Proceeding from P_1, we lay off successive segments of unit length along l, obtaining points $P_2, P_3, \cdots$, which are associated with the real numbers 2, 3, $\cdots$, respectively. This process corresponds to the algebraic operation of adding 1 to itself many times. Similarly, working on the opposite side of O from P_1, we locate points $P_{-1}, P_{-2}, \cdots$ corresponding to the negative integers -1, $-2, \cdots$. In this manner we set up a correspondence between the set of integers and certain equispaced points on l (see Fig. 1.4).

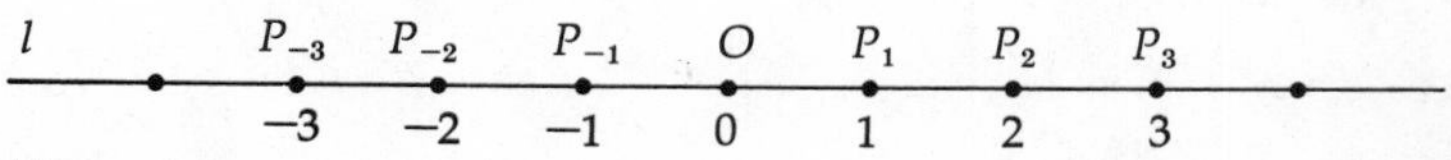

Figure 1.4

For any positive integer n, a line segment can be subdivided into n equal parts. If the segment OP_1 is subdivided in this way, the endpoint of the first such subdivision is associated with the rational number $1/n$. By counting off m such segments, we can associate a point on l with any positive rational number $m/n = m(1/n)$. Thus the point corresponding to $\frac{13}{5}$ is $\frac{13}{5}$ units from the origin, or $\frac{3}{5}$ of the way from P_2 to P_3. The negative rational numbers are handled in like manner.

Certain points on l associated with irrational numbers can be determined. For example, the point corresponding to $\sqrt{2}$ can be found by striking off a circular arc as indicated in Fig. 1.5. Not all points corresponding to irrational numbers can be constructed in this way. The number $\pi = 3.14159 \cdots$ is irrational, but no such construction is possible. However, the point corresponding to π on l can be approximated

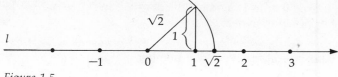

Figure 1.5

with any degree of accuracy. Thus we could successively locate the points corresponding to 3, 3.1, 3.14, 3.141, 3.1415, · · · , and so on. It can be shown that to every irrational number there corresponds a unique point on *l* and, conversely, every point that is not associated with a rational number corresponds to an irrational number.

The number *x* that is associated with a point *X* on *l* is called the **coordinate** of *X*. An assignment of coordinates to points on *l* is called a **coordinate system** for *l*, and *l* is called a **coordinate line.** A coordinate line *l* is often pictured as a horizontal line, the positive real numbers being taken as coordinates of points to the right of *O* and the negative real numbers as coordinates for points to the left of *O*. A direction can be assigned to *l* by taking the **positive direction** along *l* as that in which coordinates increase and the **negative direction** as that in which they decrease. The positive direction is noted by placing an arrowhead on *l* as shown in Fig. 1.5. If coordinates are assigned in this manner, it is evident that if the real numbers *a* and *b* are coordinates of points *A* and *B* respectively, then $a < b$ if and only if *B* lies to the right of *A*.

We shall now introduce another important concept for real numbers.

(1.14)   Definition

The **absolute value** $|a|$ of a real number *a* is defined as follows:

$$|a| = \begin{cases} a & \text{if} \quad a \geq 0 \\ -a & \text{if} \quad a < 0 \end{cases}.$$

From this definition we see that the *absolute value of every nonzero real number is positive,* for, on the one hand, if $a > 0$, then $|a| = a$, which is positive. On the other hand, if $a < 0$, then by (1.14) we have $|a| = -a > 0$; that is, $|a|$ is positive.

EXAMPLE 1   Find $|3|$, $|-3|$, $|0|$, $|\sqrt{2} - 2|$, and $|2 - \sqrt{2}|$.

Solution

Since 3, $2 - \sqrt{2}$, and 0 are nonnegative, we have by (1.14),

$$|3| = 3, \quad |2 - \sqrt{2}| = 2 - \sqrt{2}, \quad \text{and} \quad |0| = 0.$$

Since -3 and $\sqrt{2} - 2$ are negative, we use the formula $|a| = -a$ of (1.14) to obtain

$$|-3| = -(-3) = 3 \quad \text{and} \quad |\sqrt{2} - 2| = -(\sqrt{2} - 2) = 2 - \sqrt{2}.$$

15

Note that in Example 1, $|-3| = |3|$ and $|2 - \sqrt{2}| = |\sqrt{2} - 2|$. It can be shown in general that

$$|a| = |-a| \text{ for every real number } a.$$

In the next definition absolute values are used to assign numerical values to segments of a coordinate line.

(1.15) Definition

Let a and b be the coordinates of two points A and B respectively, on a coordinate line l, and let AB denote the line segment from A to B. The **length** $d(A, B)$ of AB is defined by $d(A, B) = |b - a|$.

The nonnegative number $d(A, B)$ is also called the **distance between A and B**. Note that since $|b - a| = |a - b|$, we have $d(A, B) = d(B, A)$. The distance between the origin O and any point A is

$$d(O, A) = |a - 0| = |a|.$$

In geometric terms this states that the absolute value of a real number a equals the distance between the origin and the point corresponding to a.

EXAMPLE 2 Let A, B, C, and D have coordinates $-5, -3, 1$, and 6, respectively on a coordinate line l (see Fig. 1.6). Find $d(A, B)$, $d(C, B), d(O, A)$, and $d(C, D)$.

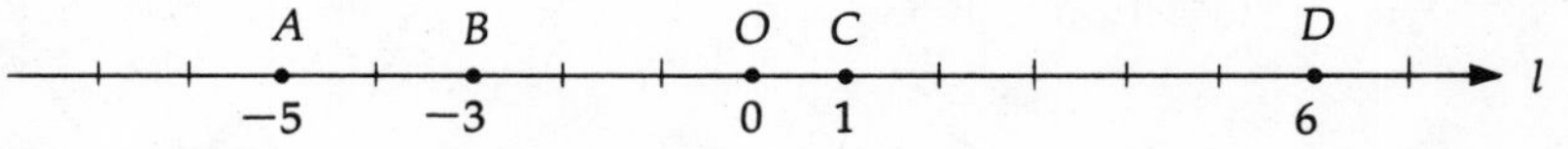

Figure 1.6

Solution

By (1.15),

$$d(A, B) = |-3 - (-5)| = |-3 + 5| = |2| = 2,$$
$$d(C, B) = |-3 - 1| = |-4| = 4,$$
$$d(O, A) = |-5 - 0| = |-5| = 5,$$
$$d(C, D) = |6 - 1| = |5| = 5.$$

The student should check these answers visually by referring to Fig. 1.6.

In some cases we wish to work with distances along l which take into account the direction of l, as in the next definition.

(1.10)   Definition

Let a and b be the coordinates of two points A and B respectively, on a coordinate line l. The **directed distance** $\overline{AB}$ from A to B is defined by $\overline{AB} = b - a$.

Since $\overline{BA} = a - b$, we have $\overline{AB} = -\overline{BA}$. Evidently, the point B is to the right of A if and only if $\overline{AB} > 0$, and B is to the left of A if and only if $\overline{AB} < 0$. Accordingly, the directed distance indicates the *direction* from A to B with respect to the direction assigned to l.

EXAMPLE 3   If A, B, C, and D are the points in Example 2, find the directed distances $\overline{AB}$, $\overline{BA}$, $\overline{CB}$, and $\overline{OA}$.

Solution

By (1.16),

$$\overline{AB} = -3 - (-5) = 2,$$
$$\overline{BA} = -5 - (-3) = -2,$$
$$\overline{CB} = -3 - 1 = -4,$$
$$\overline{OA} = -5 - 0 = -5.$$

The fact that $\overline{AB} = 2$ in Example 3 means that in order for a point to move along l from A to B it must travel two units in the *positive* direction. Since $\overline{BA} = -2$, a point would have to travel two units in the *negative* direction to proceed from B to A. This may be checked by referring to Fig. 1.6.

If M is the midpoint of the segment AB, then $\overline{AM} = \overline{MB}$ whether $\overline{AB} > 0$ or $\overline{AB} < 0$ (see Fig. 1.7).

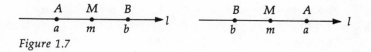

Figure 1.7

Hence, if m is the coordinate of M, then $m - a = b - m$. Adding $m + a$ to both sides and simplifying, we get $2m = a + b$ and therefore, $m = (a + b)/2$. This establishes the following formula.

(1.17)   Midpoint Formula

If a and b are the coordinates of two points A and B respectively, on a coordinate line l, then the coordinate of the midpoint of the line segment AB is $(a + b)/2$.

17

EXAMPLE 4 Let A, B, C, and D be as in Example 2 (see Fig. 1.6). Find the coordinates of the midpoints of the segments AB, BC, and CD.

Solution

By (1.17) the coordinate of the midpoint of AB is

$$\frac{-5 + (-3)}{2} = \frac{-8}{2} = -4.$$

The midpoint of BC has coordinate

$$\frac{-3 + 1}{2} = \frac{-2}{2} = -1.$$

Finally, that of CD is

$$\frac{1 + 6}{2} = \frac{7}{2}.$$

It is easy to check these answers by referring to Fig. 1.6.

In determining the midpoint of a segment AB, a common error is to *subtract* the coordinates of A and B and divide by 2. Note that (1.17) states that the coordinates are to be *added* and the *sum* divided by 2.

We shall conclude this section with some additional remarks about absolute values. If we consider an inequality such as $|a| < 5$, then a could represent real numbers such as $\frac{1}{2}$, $\sqrt{2}$, 2.7, π, 4.75, 4.999, or for that matter, *any* number between 0 and 5. Moreover, since $|a| = |-a|$, a could also denote $-\frac{1}{2}$, $-\sqrt{2}$, -2.7, $-\pi$, -4.75, -4.999, or any number between -5 and 0. Thus it appears that if $|a| < 5$, then $-5 < a < 5$. Let us show that if b is *any* positive real number, then

(1.18) $|a| < b$ if and only if $-b < a < b$.

A general rule such as (1.18) can be proved by examining all cases which may occur, that is, the cases in which a is either positive, zero, or negative. Let us begin by assuming that $|a| < b$.

Case 1 If a is nonnegative, then $|a| = a$ and hence $a < b$. Moreover, since $-b < 0$ and $0 \le a$, we also have $-b < a$. Consequently, $-b < a < b$.

Case 2 If a is negative, then $|a| = -a$ and since, by assumption, $|a| < b$, we have $-a < b$. Multiplying both sides of the latter inequality by -1 leads to $-b < a$. Since b is positive it follows that $-b < a < b$.

This proves that in all cases, if $|a| < b$, then $-b < a < b$.

Let us now prove conversely, that if $-b < a < b$, then $|a| < b$.

Case 1 If a is nonnegative, then $|a| = a$. Since by assumption $a < b$, we have $|a| < b$.

Case 2 If a is negative, then $|a| = -a$. However, by hypothesis, $a > -b$ and hence $-a < b$. (Why?) Consequently $|a| < b$. This completes the proof of (1.18).

It is worth noting that (1.18) implies that all points on a coordinate line corresponding to real numbers a such that $|a| < b$ lie between the points corresponding to $-b$ and b.

EXAMPLE 5 If x is a real number and $|x - 2| < 5$, show that x is between -3 and 7.

Solution

If $|x - 2| < 5$, then applying (1.18) with $a = x - 2$ and $b = 5$ we obtain

$$-5 < x - 2 < 5.$$

This tells us that $-5 < x - 2$ and also $x - 2 < 5$. Adding 2 to both sides of the latter inequalities leads to $-3 < x$ and $x < 7$, that is, $-3 < x < 7$. As a partial check on the answer, the student may wish to substitute some specific real numbers between -3 and 7 for x in the expression $|x - 2|$.

It can also be proved that if $b \geq 0$, then

(1.19) $|a| = b$ if and only if $a = b$ or $a = -b$.

(1.20) $|a| > b$ if and only if $a > b$ or $a < -b$,

EXAMPLE 6 If x is a real number such that $|x + 3| > 2$, show that either $x > -1$ or $x < -5$.

Solution

If $|x + 3| > 2$, then from (1.19) with $a = x + 3$,

$$x + 3 > 2 \quad \text{or} \quad x + 3 < -2.$$

If we add -3 to both sides of these inequalities we obtain

$$x > -1 \quad \text{or} \quad x < -5,$$

which is what we wish to show. The student may find it enlightening to replace x in $|x + 3|$ by several real numbers greater than -1 or less than -5.

In Chapter 2 we shall have more to say about inequalities which involve absolute values.

EXERCISES

1   Rewrite each of the following numbers without using symbols for absolute values:

   a $|2 - 5|$ $= 3$ **b** $|-5| + |-2|$ 7

   c $|5| + |-2|$ 7 **d** $|-5| - |-2|$ 3

   e $|\pi - \frac{22}{7}|$ $\frac{22}{7} - \pi$ **f** $(-2)/|-2|$ -1

   g $|\frac{1}{2} - 0.5|$ 0 **h** $|(-3)^2|$ 9

2   Work Exercise 1 for the following numbers:

   a $|4 - 8|$ **b** $|3 - \pi|$

   c $|-4| - |-8|$ **d** $|-4 + 8|$

   e $|-3|^2$ **f** $|2 - \sqrt{4}|$

   g $|-0.67|$ **h** $-|-3|$

3   Let A, B, and C be points on a coordinate line with coordinates -5, -1, and 7, respectively. Find:

   a $d(A, B)$ 4 **b** $d(B, C)$ 8 **c** $d(C, B)$ 8 **d** $d(A, C)$ 12

4   Work Exercise 3 if A, B, and C have coordinates 2, -8, and -3, respectively.

5   If A, B, and C are the points in Exercise 3, find:

   a $\overline{AB}$ $-1 + 5 = 4$ **b** $\overline{BC}$ $7 - 1 = 8$ **c** $\overline{CB}$ $-1 + 7 = -8$ **d** $\overline{AC}$ $7 - 5 = 12$

6   Work Exercise 5 for the points A, B, C given in Exercise 4.

7   If A, B, and C are as in Exercise 3, find the coordinates of the midpoint of the following segments:

   a AB **b** BC **c** AC

8   Work Exercise 7 for the points of Exercise 4.

9   If P, Q, and R are any three points on a coordinate line l, prove that $\overline{PQ} + \overline{QR} = \overline{PR}$.

10   Show, by means of an example, that if P, Q, and R are points on a coordinate line l, then it is not always true that $d(P, Q) + d(Q, R) = d(P, R)$.

11   If $|x + 4| < 2$, prove that $-6 < x < -2$.

12   If $|x - 5| < 3$, prove that $2 < x < 8$.

13   If $|2x - 3| < 7$, prove that $-2 < x < 5$.

14   If $|3x + 1| < 9$, prove that $-\frac{10}{3} < x < \frac{8}{3}$.

15   If $|x - 1| > 3$, prove that either $x > 4$ or $x < -2$.

16   If $|x + 2| > 6$, prove that either $x > 4$ or $x < -8$.

In Exercises 17–20 change the given inequality to the form $k < a < l$ for some real numbers k and l.

17 $|2a| < 6$ $-3 < a < 3$

18 $|-a| < 5$

19 $|a + 4| < 9$ $-9 < a+4 < 9$
$-9+ -4 < a < 9$
$-13 < a < 9$

20 $|a - 6| < 2$

21 Prove (1.19)

22 Prove (1.20)

4 INTEGRAL EXPONENTS

Throughout the remainder of this chapter, all symbols for elements will denote real numbers. We have had occasion to use the notation x^2 for the real number $x \cdot x$. Similarly, x^3 is used to denote $x \cdot x \cdot x$. In general, if n is any positive integer,

$$(1.21) \quad x^n = \underbrace{x \cdot x \cdots x}_{n \text{ times}}$$

where n factors, all equal to x, appear on the right-hand side of the expression. The positive integer n is called the **exponent** of x in the expression x^n, and x^n is read "x to the nth power." As illustrations we have

$$(\tfrac{1}{2})^5 = \tfrac{1}{2} \cdot \tfrac{1}{2} \cdot \tfrac{1}{2} \cdot \tfrac{1}{2} \cdot \tfrac{1}{2} = \tfrac{1}{32},$$
$$(-3)^3 = (-3)(-3)(-3) = -27,$$
$$(\sqrt{2})^4 = \sqrt{2}\,\sqrt{2}\,\sqrt{2}\,\sqrt{2} = (\sqrt{2})^2(\sqrt{2})^2 = 2 \cdot 2 = 4.$$

It is important to remember that if n is a positive integer, then ax^n means $a(x^n)$ and *not* $(ax)^n$. The real number a is called the **coefficient** of x^n in the expression ax^n. Similarly, $-ax^n$ means $-(ax^n)$, not $(-ax)^n$. For example, we have

$$3 \cdot 2^4 = 3 \cdot 16 = 48 \quad \text{and} \quad -3 \cdot 2^4 = -3 \cdot 16 = -48.$$

The basic laws of exponents are stated in the following theorem.

(1.22) Laws of Exponents

If x, y are real numbers and m, n are positive integers, then

(1) $x^m x^n = x^{m+n}$,

(2) $(x^m)^n = x^{mn}$,

(3) $(xy)^n = x^n y^n$,

(4) if $y \neq 0$, then $\left(\dfrac{x}{y}\right)^n = \dfrac{x^n}{y^n}$.

(5)　if $x \neq 0$, then

$$\frac{x^m}{x^n} = x^{m-n} \text{ for } m > n,$$

$$\frac{x^m}{x^n} = \frac{1}{x^{n-m}} \text{ for } n > m,$$

$$\frac{x^m}{x^n} = 1 \text{ for } m = n.$$

A complete proof of (1.22) requires the method of mathematical induction discussed in Chapter Ten. However, if we are allowed to count an arbitrary number of factors, then it is easy to supply arguments which establish the laws. To prove (1) we have

$$x^m x^n = \underbrace{x \cdot x \cdot \cdot \cdot x}_{m \text{ times}} \cdot \underbrace{x \cdot x \cdot \cdot \cdot x}_{n \text{ times}}.$$

Since the total number of factors x on the right is $m + n$, this expression is equal to x^{m+n}.

Similarly, we could write

$$(x^m)^n = \underbrace{x^m \cdot x^m \cdot \cdot \cdot x^m}_{n \text{ times}}$$

and count the number of times x appears as a factor on the right-hand side. Since $x^m = x \cdot x \cdot \cdot \cdot x$, where x occurs as a factor m times, and since the number of such groups of m factors is n, the total number of factors is $m \cdot n$. This gives us (2) of (1.22).

Laws (3) and (4) of (1.22) can be obtained in similar fashion and are left to the reader. Law (5) is clear if $m = n$. For the case $m > n$, the integer $m - n$ is positive and we may write

$$\frac{x^m}{x^n} = \frac{x^n x^{m-n}}{x^n} = \frac{x^n}{x^n} \cdot x^{m-n} = 1 \cdot x^{m-n} = x^{m-n}.$$

A similar argument can be used if $n > m$.

The following are some specific examples of (1.22).

$$x^3 x^4 = x^{3+4} = x^7; \qquad (x^3)^4 = x^{3 \cdot 4} = x^{12};$$

$$(xy)^3 = x^3 y^3; \qquad \left(\frac{x}{y}\right)^3 = \frac{x^3}{y^3};$$

$$\frac{x^5}{x^2} = x^{5-2} = x^3; \qquad \frac{x^2}{x^5} = \frac{1}{x^{5-2}} = \frac{1}{x^3}.$$

The Laws of Exponents can be extended to rules such as $x^m x^n x^p = x^{m+n+p}$, $(xyz)^n = x^n y^n z^n$, and so on. For convenience we shall also refer to such generalizations as (1) and (3) of (1.22).

EXAMPLE 1 Simplify each of the following:

(a) $(3x^3y^4)(4xy^5)$,   (b) $(2x^2y^3z)^4$,   (c) $\left(\dfrac{2x^3}{y}\right)^2\left(\dfrac{y}{x^3}\right)^3$.

Solutions

We shall justify each step by referring to an appropriate property of real numbers.

(a) $(3x^3y^4)(4xy^5) = (3)(4)x^3xy^4y^5$ (rearrangement properties)

$= 12x^4y^9$ (1) of (1.22)

(b) $(2x^2y^3z)^4 = 2^4(x^2)^4(y^3)^4z^4$ (3) of (1.22)

$= 16x^8y^{12}z^4$ (2) of (1.22).

(c) $\left(\dfrac{2x^3}{y}\right)^2\left(\dfrac{y}{x^3}\right)^3 = \left(\dfrac{2^2x^6}{y^2}\right)\left(\dfrac{y^3}{x^9}\right)$ (4), (3), and (2) of (1.22)

$= 2^2\left(\dfrac{x^6}{x^9}\right)\left(\dfrac{y^3}{y^2}\right)$ (properties of quotients)

$= 4\left(\dfrac{1}{x^3}\right)(y)$ (5) of (1.22)

$= \dfrac{4y}{x^3}$ (properties of quotients).

It is possible to extend our work to exponents which are negative integers or 0. If we want (1) of (1.22) to be true when $n = 0$, then $x^m \cdot x^0 = x^{m+0} = x^m$ and, if $x \neq 0$, multiplication by $1/(x^m)$ leads to $x^0 = 1$. Thus, in order to be consistent with our previous development we introduce the following definition.

(1.23) Definition

If x is any nonzero real number, then

$$x^0 = 1.$$

The symbol 0^0 will be left undefined.

Let us next turn our attention to negative exponents. In Section 2 we used the notation $x^{-1} = 1/x$ if $x \neq 0$. If (1) of (1.22) is to be true in this situation, then we must have

$x^{-2} = x^{-1} \cdot x^{-1} = (1/x)(1/x) = 1/x^2$,
$x^{-3} = x^{-2} \cdot x^{-1} = (1/x^2)(1/x) = 1/x^3$,

and so on. It is natural, therefore, to define negative integral exponents as follows.

(1.24)   Definition

If x is a nonzero real number and n is a positive integer, then

$$x^{-n} = \frac{1}{x^n}.$$

It is possible to show that the Laws of Exponents are valid for all integers m and n, whether positive, negative, or zero; however, we shall not discuss the proofs. If negative exponents are allowed, then (5) of (1.22) may be abbreviated to read

$$\frac{x^m}{x^n} = x^{m-n}, \text{ for } all \text{ integers } m \text{ and } n.$$

In the future it will be assumed that (1.22) is true for all integral exponents and we shall justify steps in proofs by referring to (1.22) even though negative exponents are involved.

To simplify statements, we shall assume that in all problems involving exponents, symbols which appear in denominators represent *nonzero* real numbers.

EXAMPLE 2   Eliminate negative exponents and simplify:

(a) $(x^{-2}y^3)^{-3}$,   (b) $\dfrac{8x^3y^{-5}}{4x^{-1}y^2}$,   (c) $\dfrac{x^{-2} + y^{-2}}{(xy)^{-1}}$.

Solutions

(a) $(x^{-2}y^3)^{-3} = (x^{-2})^{-3}(y^3)^{-3}$ (3) of (1.22)

$\qquad\qquad\quad = x^6 y^{-9}$ (2) of (1.22)

$\qquad\qquad\quad = x^6 (1/y^9)$ (1.24)

$\qquad\qquad\quad = x^6/y^9$ (properties of quotients).

(b) $\dfrac{8x^3y^{-5}}{4x^{-1}y^2} = \dfrac{8}{4} \dfrac{x^3}{x^{-1}} \dfrac{y^{-5}}{y^2}$ (properties of quotients)

$\qquad\qquad\quad = 2x^{3-(-1)}y^{-5-2}$ (5) of (1.22)

$\qquad\qquad\quad = 2x^4 y^{-7}$ (simplifying)

$\qquad\qquad\quad = \dfrac{2x^4}{y^7}$ (1.24).

Another method for simplifying (b) is to multiply numerator and denominator of the given fraction by $y^5 x$, thereby eliminating negative exponents. Thus

$$\frac{8x^3y^{-5}}{4x^{-1}y^2} = \frac{8x^3y^{-5}}{4x^{-1}y^2} \cdot \frac{y^5x}{y^5x} \qquad \text{(identity element)}$$

$$= \frac{8}{4}\frac{x^4y^0}{x^0y^7} \qquad \text{(Why?)}$$

$$= 2\frac{x^4(1)}{(1)y^7} \qquad (1.23)$$

$$= \frac{2x^4}{y^7} \qquad \text{(identity element).}$$

(c) $$\frac{x^{-2}+y^{-2}}{(xy)^{-1}} = \frac{(1/x^2)+(1/y^2)}{1/xy} \qquad (1.24)$$

$$= \frac{\left(\dfrac{y^2+x^2}{x^2y^2}\right)}{1/xy} \qquad \text{(properties of quotients)}$$

$$= \frac{(y^2+x^2)(xy)}{x^2y^2} \qquad \text{(properties of quotients)}$$

$$= \frac{y^2+x^2}{xy} \qquad \text{(5) of (1.22).}$$

Another method of attacking (c) is to multiply numerator and denominator of the original expression by x^2y^2 and simplify.

In the following exercises, the word "simplify" means to replace the given expression by one in which letters representing real numbers appear only once, and no negative exponents occur.

EXERCISES

Express the numbers in Exercises 1–10 in the form a/b, where a and b are integers.

1 $(-\frac{2}{3})^4$ $\frac{16}{81}$ 2 $(-3)^3$ -27 3 $\dfrac{2^{-3}}{3^{-2}}$ 4 $\dfrac{2^0+0^2}{2+0}$

5 $(-2)^3+3^{-2}$ $-8+\frac{1}{9}=-7\frac{8}{9}$ 6 $(-\frac{3}{2})^4-2^{-4}$ $\frac{81}{16}-\frac{1}{16}=\frac{80}{16}$

7 $\dfrac{2^{-2}-3^{-3}}{(-2)^2+(-3)^3}$ $\frac{\frac{1}{4}-\frac{1}{27}}{4+-27}$ 8 $2^{-2}+(-2)^5$

9 4^0+0^4 $1+0=1$ 10 $(1,000,000)^0$

For Exercises 11–50 eliminate negative exponents and simplify.

11 $(\frac{1}{2}x^4)(16x^5)$ $8x^9$ 12 $(-3x^{-2})(4x^4)$

25

13 $\dfrac{(2x^3)(3x^2)}{(x^2)^3}$

14 $\dfrac{(2x^2)^3}{4x^4}$

15 $(\tfrac{1}{6}a^5)(-3a^2)(4a^7)$

16 $(-4b^3)(\tfrac{1}{6}b^2)(-9b^4)$

17 $\dfrac{(6x^3)^2}{(2x^2)^3}$

18 $\dfrac{(3y^3)(2y^2)^2}{(y^4)^3}$

19 $(3u^7v^3)(4u^4v^{-5})$

20 $(x^2yz^3)(-2xz^2)(x^3y^{-2})$

21 $(8x^4y^{-3})(\tfrac{1}{2}x^{-5}y^2)$

22 $\left(\dfrac{4a^2b}{a^3b^2}\right)\left(\dfrac{5a^2b}{2b^4}\right)$

23 $(\tfrac{1}{3}x^4y^{-3})^{-2}$

24 $(-2xy^2)^5\left(\dfrac{x^7}{8y^3}\right)$

25 $(3y^3)^4(4y^2)^{-3}$

26 $(-3a^2b^{-5})^3$

27 $(-2r^4s^{-3})^{-2}$

28 $(2x^2y^{-5})(6x^{-3}y)(\tfrac{1}{3}x^{-1}y^3)$

29 $(5x^2y^{-3})(4x^{-5}y^4)$

30 $(-2r^2s)^5(3r^{-1}s^3)^2$

31 $\left(\dfrac{3x^5y^4}{x^0y^{-3}}\right)^2$

32 $(4a^2b)^4\left(\dfrac{-a^3}{2b}\right)^2$

33 $(-2a^3b^{-4}c^0)^3(\tfrac{1}{2}a^{-5}b^3c^2)$

34 $\dfrac{(3x^{-3}y)^{-2}}{(x^2y^{-2})^3}$

35 $\left(\dfrac{5a^{-1}b^2}{2c^{-3}}\right)^{-1}\left(\dfrac{ab^{-1}}{c^2}\right)^4$

36 $\left(\dfrac{2u^{-2}}{v^3}\right)^{-1}\left(\dfrac{4u^{-1}}{v^2}\right)^3$

37 $\left(\dfrac{3x^3y^{-2}}{7x^{-5}y^8}\right)^0$

38 $(x+y)^{10}(x+y)^{-10}$

39 $((ab^2)^{-2})^{-2}$

40 $a^0+(b^0+c)^0$

41 $\dfrac{x^{-1}}{y^{-1}}-\left(\dfrac{x}{y}\right)^{-1}$

42 $\left(\dfrac{x^{-3}}{y^{-3}}\right)^2\left(\dfrac{x^2}{y^2}\right)^3$

43 $\dfrac{a^{-1}-b^{-1}}{(ab)^{-1}}$

44 $\dfrac{(a^{-1}-b^{-1})^{-1}}{ab}$

45 $\dfrac{a^{-1}+b^{-1}}{(a+b)^{-1}}$

46 $\dfrac{a^{-1}}{b^{-1}}+\dfrac{a}{b}$

47 $\dfrac{x^{-2}-y^{-2}}{x^2-y^2}$

48 $\dfrac{x^0-y^0}{x^0+y^0}$

49 $\dfrac{x^{2n-3}}{x^{3n+1}}\cdot\dfrac{x^{n+5}}{x^{n-2}}$

50 $\dfrac{(2x^{n+1})^2}{x^{2(n+1)}}\cdot\dfrac{x^{3-n}}{(x^n)^2}$

5 RADICALS

It can be shown that if $a > 0$ and n is a positive integer, then there is precisely one real number $b > 0$ such that $b^n = a$. The number b is called the **principal nth root** of a and is denoted by $\sqrt[n]{a}$. It can also be shown that if $a < 0$ and n is an *odd* positive integer, then there is one and only one real number $b < 0$ such that $b^n = a$. In this case we again write $b = \sqrt[n]{a}$ and call b the principal nth root of a. Finally, we let $\sqrt[n]{0} = 0$ for all positive integers n. These remarks lead to the following definition.

(1.25) Definition

If a and b are nonnegative real numbers and n is a positive integer, or if a and b are both negative and n is an odd positive integer, then

$$\sqrt[n]{a} = b \text{ means that } b^n = a.$$

If $n = 2$ and a is nonnegative it is customary to write $\sqrt{a}$ instead of $\sqrt[2]{a}$, and to call $\sqrt{a}$ the (principal) **square root** of a. The number $\sqrt[3]{a}$ is referred to as the **cube root** of a.

EXAMPLE 1 Find $\sqrt[5]{\frac{1}{32}}$, $\sqrt[4]{81}$, $\sqrt[3]{-8}$, and $\sqrt{16}$.

Solution

By (1.25),

$$\sqrt[5]{\tfrac{1}{32}} = \tfrac{1}{2} \text{ since } (\tfrac{1}{2})^5 = \tfrac{1}{32},$$

$$\sqrt[4]{81} = 3 \text{ since } 3^4 = 81,$$

$$\sqrt[3]{-8} = -2 \text{ since } (-2)^3 = -8,$$

$$\sqrt{16} = 4 \text{ since } 4^2 = 16.$$

Note that we have not defined $\sqrt[n]{a}$ if $a < 0$ and n is an *even* positive integer. The reason for this is that if n is even, then $b^n \geq 0$ for every real number b. We shall use the terminology "$\sqrt[n]{a}$ exists" if there is a real number b such that $b^n = a$.

It is important to observe that if $\sqrt[n]{a}$ exists it is a *unique* real number. More generally, if $b^n = a$ for a positive integer n, then b is called an nth root of a. For example, both 4 and -4 are square roots of 16, since $4^2 = 16$ and also $(-4)^2 = 16$. However, as in Example 1, the *principal* square root of 16 is 4 and we write $\sqrt{16} = 4$. In elementary arithmetic the expression $\sqrt{16} = \pm 4$ is sometimes used to denote *all* square roots of 16. It should be emphasized that this is *not* done in advanced mathematics.

To complete our terminology, the symbol $\sqrt[n]{a}$ is called a **radical,** the number a is called the **radicand,** and n is the **index** of the radical. The symbol $\sqrt{}$ is called a **radical sign.**

Several important properties of radicals are stated in the following theorem.

(1.26)   Laws of Radicals

If n is a positive integer and if x and y are real numbers such that $\sqrt[n]{x}$ and $\sqrt[n]{y}$ exist, then

(1) $(\sqrt[n]{x})^n = x$,

(2) $\sqrt[n]{x}\,\sqrt[n]{y} = \sqrt[n]{xy}$,

(3) $\dfrac{\sqrt[n]{x}}{\sqrt[n]{y}} = \sqrt[n]{\dfrac{x}{y}}$ if $y \neq 0$,

(4) $\sqrt[n]{x^n} = x$ if $x > 0$ or if $x < 0$ and n is odd,

(5) $\sqrt[m]{\sqrt[n]{x}} = \sqrt[mn]{x}$ if m is a positive integer and the indicated roots exist.

Partial Proof

If we let $u = \sqrt[n]{x}$, then $u^n = x$ by (1.25). Substituting for u, we obtain $(\sqrt[n]{x})^n = x$. This proves (1). To prove (2), let $u = \sqrt[n]{x}$ and $v = \sqrt[n]{y}$. We then have the following

$u^n = x$ and $v^n = y$ (1.25)

$u^n v^n = xy$ (substitution principle)

$(uv)^n = xy$ (3) of (1.22)

$\sqrt[n]{xy} = uv$ (1.25)

$\sqrt[n]{xy} = \sqrt[n]{x}\,\sqrt[n]{y}$ (substitution principle).

Laws (3), (4), and (5) can be proved in like manner.

We may generalize these laws to $\sqrt[n]{xyz} = \sqrt[n]{x}\,\sqrt[n]{y}\,\sqrt[n]{z}$, and so on. In the future we shall also refer to this generalized version as (2) of (1.26). The law stated in the following theorem is a special case of (4).

(1.27)   Theorem

If x is any real number, then $\sqrt{x^2} = |x|$.

Proof

The theorem is clearly true if $x = 0$. If $x > 0$, then $\sqrt{x^2} = x = |x|$. Finally, if $x < 0$, then $-x > 0$, and since $\sqrt{a^2} = a$ for all $a > 0$, we have $\sqrt{(-x)^2} = -x$. However, $|x| = -x$ if $x < 0$. This gives us $\sqrt{x^2} = \sqrt{(-x)^2} = -x = |x|$, and the theorem is proved.

As an illustration of (1.27), we have $\sqrt{(-4)^2} = |-4| = 4$. Of course, this equality could also be obtained by writing $\sqrt{(-4)^2} = \sqrt{16} = 4$. Incidentally, we have shown that it is not always true that $\sqrt{x^2} = x$. Indeed, if x is negative, then $\sqrt{x^2} = -x$.

Theorem (1.27) may be used to prove that for all real numbers a and b,

(1.28) $|ab| = |a|\,|b|$.

To establish this formula we let $x = ab$ in (1.27) and proceed as follows:

$$|ab| = \sqrt{(ab)^2} = \sqrt{a^2b^2} = \sqrt{a^2}\,\sqrt{b^2} = |a|\,|b|.$$

It follows in similar fashion that if $b \neq 0$, then

(1.29) $\left|\dfrac{a}{b}\right| = \dfrac{|a|}{|b|}$.

If c is a real number and c^n occurs as a factor in a radical of index n, then c can be removed from the radicand provided the sign of c is taken into account. For example, by (2) and (4) of (1.26),

$$\sqrt[n]{c^n d} = \sqrt[n]{c^n}\,\sqrt[n]{d} = c\sqrt[n]{d},$$

where we have assumed that the sign of c is such that (4) of (1.26) is valid. If, in the preceding equalities, c is negative and $n = 2$, then by (1.27)

$$\sqrt{c^2 d} = \sqrt{c^2}\,\sqrt{d} = |c|\sqrt{d}.$$

A similar situation exists if n is any even integer. In order to avoid considering positive and negative cases separately in the examples and exercises to follow, we shall assume that all letters represent positive real numbers.

The technique discussed above is particularly useful in simplifying radicals of the form $\sqrt[n]{a}$, where a is an integer. In this case we first obtain the prime factorization of a. If a positive prime power p^n appears as a factor, then p may be taken out from under the radical sign. The following are illustrations of this procedure.

$$\sqrt[3]{135} = \sqrt[3]{3^3 \cdot 5} = \sqrt[3]{3^3}\,\sqrt[3]{5} = 3\sqrt[3]{5},$$

$$\sqrt{1400} = \sqrt{2^3 5^2 7} = \sqrt{(2 \cdot 5)^2 \cdot 2 \cdot 7} = \sqrt{(2 \cdot 5)^2}\,\sqrt{14} = 10\sqrt{14}.$$

A similar technique may be used if the radicand contains symbols for unspecified real numbers. For example,

$$\sqrt{x^7} = \sqrt{x^6 x} = \sqrt{(x^3)^2 x} = \sqrt{(x^3)^2}\,\sqrt{x} = x^3\sqrt{x}.$$

$$\sqrt[4]{x^9 y^6} = \sqrt[4]{x^8 y^4 xy^2} = \sqrt[4]{(x^2 y)^4 xy^2}$$

$$= \sqrt[4]{(x^2 y)^4}\,\sqrt[4]{xy^2} = x^2 y\sqrt[4]{xy^2}.$$

If we use the term "simplify" when referring to a radical, we mean to proceed as above until the radicand contains no factors whose exponent is greater than or equal to the index of the radical. Moreover, no fractions should appear under the final radical sign and denominators should be free of radicals. The index n should also be as low as possible.

EXAMPLE 2 Simplify the following.

(a) $\sqrt[3]{16x^3y^8z^4}$, (b) $\sqrt{3a^2b^3}\sqrt{6a^5b}$, (c) $\sqrt{\dfrac{27x^3}{8y^5}}$.

Solutions

(a) $\sqrt[3]{16x^3y^8z^4} = \sqrt[3]{(2^3x^3y^6z^3)(2y^2z)}$ (1) of (1.22)

$\qquad = \sqrt[3]{(2xy^2z)^3(2y^2z)}$ (2) and (3) of (1.22)

$\qquad = \sqrt[3]{(2xy^2z)^3}\sqrt[3]{2y^2z}$ (2) of (1.26)

$\qquad = 2xy^2z\sqrt[3]{2y^2z}$ (4) of (1.26).

(b) $\sqrt{3a^2b^3}\sqrt{6a^5b} = \sqrt{18a^7b^4}$ (2) of (1.26)

$\qquad = \sqrt{9a^6b^4(2a)}$ Why?

$\qquad = \sqrt{(3a^3b^2)^2(2a)}$ (2) and (3) of (1.22)

$\qquad = \sqrt{(3a^3b^2)^2}\sqrt{2a}$ (2) of (1.26)

$\qquad = 3a^3b^2\sqrt{2a}$ (4) of (1.26).

(c) $\sqrt{\dfrac{27x^3}{8y^5}} = \sqrt{\dfrac{3^3x^3}{8y^5}\cdot\dfrac{2y}{2y}}$ (identity element)

$\qquad = \sqrt{\dfrac{3^2x^2(3x)2y}{16y^6}}$ (Why?)

$\qquad = \sqrt{\left(\dfrac{3x}{4y^3}\right)^2\cdot 6xy}$ (2) and (4) of (1.22)

$\qquad = \dfrac{3x}{4y^3}\sqrt{6xy}$ (2) and (4) of (1.26).

The process used in (c) is sometimes called **rationalizing the denominator.** When it is used properly, no radicals appear in the denominators of the final result. Some fractions contain denominators of the form $a + \sqrt{b}$ or $\sqrt{a} + \sqrt{b}$. These can be put in simplest form by multiplying numerator and denominator by $a - \sqrt{b}$ or $\sqrt{a} - \sqrt{b}$, respectively.

EXAMPLE 3 Simplify $\dfrac{1}{\sqrt{x} - \sqrt{y}}$.

Solution

$$\dfrac{1}{\sqrt{x} - \sqrt{y}} = \dfrac{1}{\sqrt{x} - \sqrt{y}}\dfrac{\sqrt{x} + \sqrt{y}}{\sqrt{x} + \sqrt{y}}$$

$$= \dfrac{\sqrt{x} + \sqrt{y}}{(\sqrt{x})^2 - \sqrt{y}\sqrt{x} + \sqrt{x}\sqrt{y} - (\sqrt{y})^2}$$

$$= \dfrac{\sqrt{x} + \sqrt{y}}{x - y}.$$

Radicals with the same index can sometimes be combined by using the distributive laws. This is illustrated in the next example.

EXAMPLE 4 Simplify $\sqrt{12} - \sqrt{27} + \sqrt{4x^2 y} + \sqrt{y^3}$.

Solution

$$\sqrt{12} - \sqrt{27} + \sqrt{4x^2 y} + \sqrt{y^3}$$
$$= \sqrt{2^2 \cdot 3} - \sqrt{3^3} + \sqrt{(2x)^2 y} + \sqrt{y^2 \cdot y}$$
$$= 2\sqrt{3} - 3\sqrt{3} + 2x\sqrt{y} + y\sqrt{y}$$
$$= -\sqrt{3} + (2x + y)\sqrt{y}.$$

EXERCISES

Simplify the following. All letters denote positive real numbers.

1 $\sqrt{25}$

2 $\sqrt[5]{-32}$

3 $\sqrt[3]{-64}$ —4

4 $\sqrt[4]{625}$

5 $\sqrt[8]{256}$ 2

6 $\sqrt[3]{10,000}$

7 $\sqrt[3]{\frac{1}{3}}$ $\frac{1}{3}\sqrt[3]{9}$

8 $\sqrt{\frac{1}{5}} \cdot \frac{5}{5} = \sqrt{\frac{5}{5^2}} = \frac{\sqrt{5}}{5}$

9 $\sqrt{4a^4 b^{-8}}$ $2a^2/b^4$

10 $\sqrt[3]{-8a^3 b^{-9}}$

11 $\sqrt[3]{8x^3 y^{-6}}$

12 $\sqrt[4]{\frac{16v^9}{u^{12}}}$

13 $\sqrt[3]{x^{-6} y^9}$

14 $\sqrt[5]{-32y^5 x^{-10}}$

15 $\sqrt{\frac{50x^6 y^{10}}{z^2}}$

16 $\sqrt[3]{\frac{-27x^7}{y^9}}$

17 $\sqrt{\frac{1}{2xy^3}}$

18 $\sqrt{\frac{3c^2}{d}} \sqrt{\frac{6d^3}{c^2}}$

19 $\sqrt{3u^5 v^3} \sqrt{4u^{-1}v}$

20 $\sqrt[5]{16x^3 y^4} \sqrt[5]{-2x^4 y}$

21 $\sqrt[4]{(3a^5 b^{-1}c^3)^4}$

22 $\sqrt[3]{16x} \sqrt{16x^4}$

23 $(\sqrt[3]{-2x^2 y})^3$

24 $\sqrt[10]{(a+b)^{10}}$

25 $\frac{\sqrt[3]{-81x^5 y^{-4}}}{\sqrt[3]{3x^2 y}}$

26 $\sqrt[3]{\frac{1}{x}}$

27 $\frac{1}{\sqrt{a}}$

28 $(\sqrt{x})^6$

29 $\sqrt[3]{\frac{16a^4}{3b}}$

30 $\sqrt{\frac{27a^8 b}{c^3 d^4}}$

31 $\sqrt[3]{(x^7 y^{-2} z)^3}$

32 $(\sqrt[3]{-2x^3 y})^4$

33 $\sqrt{2x^3 y} \sqrt{6x^2 y^3}$

34 $\sqrt[4]{8u^2 v^3} \sqrt[4]{4u^3 v^2}$

35 $\dfrac{\sqrt[3]{-250x^{-4}y}}{\sqrt[3]{x^2 y^{-3}}}$

36 $\sqrt{\dfrac{x}{y}} \div \sqrt{\dfrac{y}{x}}$

37 $\sqrt[3]{\sqrt{x^6 y^9}}$

38 $\sqrt[3]{4x^2}\,\sqrt[3]{8x^6}$

39 $\dfrac{1}{\sqrt{x^3}}$

40 $\dfrac{1}{\sqrt{x+1}}$

41 $\dfrac{1}{a + \sqrt{b}}$

42 $\dfrac{\sqrt{a} - \sqrt{b}}{\sqrt{a} + \sqrt{b}}$

43 $\dfrac{\sqrt{x}}{\sqrt{x} + 1}$

44 $\dfrac{\sqrt{x} + \sqrt{x^3}}{\sqrt{x}}$

45 $\sqrt{(x-y)^2}$

46 $\sqrt[3]{(x-y)^3}$

47 $2\sqrt{20} - \sqrt{125} + \sqrt{45}$

48 $\sqrt[3]{-32} + 2\sqrt[3]{108} - \sqrt[3]{\tfrac{1}{16}}$

49 $\sqrt{4x^5} - \sqrt{x^3} + \sqrt{16x}$

50 $\sqrt[3]{8u^4 v^5} - \sqrt[3]{-27u^{10}v^{17}}$

6 RATIONAL EXPONENTS

Radicals can be used to introduce rational exponents. Let us begin by defining $x^{1/n}$, where n is a positive integer and $x > 0$. This must be done in a manner consistent with our previous work. In particular, if (2) of (1.22) is to be true, then $(x^{1/n})^n = x^{(1/n)n} = x$. According to (1.25), with $b = x^{1/n}$ and $a = x$, this implies that $x^{1/n} = \sqrt[n]{x}$. We are led, therefore, to the following definition.

(1.30) Definition

If x is a real number and n is a positive integer, and if $\sqrt[n]{x}$ exists, then

$$x^{1/n} = \sqrt[n]{x}$$

Whenever we use the symbol $x^{1/n}$ in the future, it will be assumed that x and n are chosen so that $\sqrt[n]{x}$ exists.

Let us now consider $x^{m/n}$, where m and n are integers with $n > 0$, and where m and n have no common prime factor. It is always possible to express a rational number m/n in this way by factoring the numerator and denominator into primes and then eliminating the common factors. For example, instead of $x^{-(12/18)}$ we write $x^{(-2)/3}$, where the common factor $3 \cdot 2$ has been removed from the numerator and denominator.

Note that if (2) of (1.22) is true for rational exponents, then necessarily $(x^{1/n})^m = (x^m)^{1/n}$, since each of these would equal $x^{m/n}$. We can prove that these expressions are equal as follows. If $m > 0$, then

$$(x^{1/n})^m = (x^{1/n})(x^{1/n}) \cdots (x^{1/n}),$$

where there are m factors $x^{1/n}$ on the right. Hence

$$(x^{1/n})^m = \sqrt[n]{x}\ \sqrt[n]{x}\ \cdots\ \sqrt[n]{x} \qquad (1.30)$$
$$= \sqrt[n]{x \cdot x \cdots x} \qquad (2)\ \text{of}\ (1.26)$$
$$= \sqrt[n]{x^m} \qquad (1.21)$$
$$= (x^m)^{1/n} \qquad (1.30)$$

The cases $m \le 0$ are left to the reader. Since $(\sqrt[n]{x})^m = \sqrt[n]{x^m}$, the following definition is meaningful.

(1.31)   Definition

Let m/n be a rational number, where n is positive and the integers m and n have no common prime factor. If x is a real number such that $\sqrt[n]{x}$ exists, then

$$x^{m/n} = (\sqrt[n]{x})^m = \sqrt[n]{x^m}.$$

We may also write (1.31) as

$$x^{m/n} = (x^{1/n})^m = (x^m)^{1/n}.$$

It may be shown that the Laws of Exponents are true for all rational exponents. Henceforth we shall assume that (1.22) is valid for *rational* as well as integral exponents.

EXAMPLE 1   Simplify:

    (a) $(-27)^{2/3}(4)^{-5/2}$.

    (b) $(4a^{1/3})(2a^{1/2})$.

    (c) $(r^2 s^6)^{1/3}$.

    (d) $\left(\dfrac{2x^{2/3}}{y^{1/2}}\right)^2 \left(\dfrac{3x^{-5/6}}{y^{1/3}}\right)$.

Solutions

The reader should supply reasons for each step.

(a) $(-27)^{2/3}(4)^{-5/2} = (\sqrt[3]{-27})^2(\sqrt{4})^{-5}$
                   $= (-3)^2(2)^{-5} = \frac{9}{32}$.

(b) $(4a^{1/3})(2a^{1/2}) = 8a^{1/3+1/2} = 8a^{5/6}$.

(c) $(r^2 s^6)^{1/3} = (r^2)^{1/3}(s^6)^{1/3} = r^{2/3}s^2$.

(d) $\left(\dfrac{2x^{2/3}}{y^{1/2}}\right)^2 \left(\dfrac{3x^{-5/6}}{y^{1/3}}\right) = \left(\dfrac{4x^{4/3}}{y}\right)\left(\dfrac{3x^{-5/6}}{y^{1/3}}\right)$

                            $= \dfrac{12x^{1/2}}{y^{4/3}}$.

Rational exponents are often useful for simplifying expressions which involve radicals. The technique is to use (1.31) to transform radicals into expressions with rational exponents, then simplify, and finally change back to radical form. We shall illustrate the procedure in the following example. The reader should supply the reason for each equality.

EXAMPLE 2   Simplify:

(a) $\sqrt[3]{4}/\sqrt{2}$.   (b) $\sqrt[4]{a^3}/\sqrt[3]{a^2}$.   (c) $\sqrt{xy}\sqrt[3]{x^2y}$.

Solutions

(a) $\dfrac{\sqrt[3]{4}}{\sqrt{2}}=\dfrac{4^{1/3}}{2^{1/2}}=\dfrac{4^{2/6}}{2^{3/6}}=\left(\dfrac{4^2}{2^3}\right)^{1/6}=2^{1/6}=\sqrt[6]{2}$.

(b) $\dfrac{\sqrt[4]{a^3}}{\sqrt[3]{a^2}}=\dfrac{a^{3/4}}{a^{2/3}}=a^{3/4-2/3}=a^{1/12}=\sqrt[12]{a}$.

(c) $\sqrt{xy}\ \sqrt[3]{x^2y}=(xy)^{1/2}(x^2y)^{1/3}$
$=x^{1/2}y^{1/2}x^{2/3}y^{1/3}$
$=x^{7/6}y^{5/6}$
$=(x^7y^5)^{1/6}$
$=\sqrt[6]{x^7y^5}$
$=\sqrt[6]{x^6xy^5}$
$=x\ \sqrt[6]{xy^5}$.

The concept of exponent can be generalized to irrational exponents such as $x^{\sqrt{2}}$, x^{π}, etc. We shall discuss this in Chapter Four.

In scientific areas it is not unusual to work with numbers that are very large or very small. To simplify matters, it is customary to write such numbers in the form $a\cdot 10^n$, where a is expressed as a decimal between 1 and 10 and n is an integer. This is referred to as the **scientific form** for real numbers. As an illustration, the distance a ray of light travels in one year is approximately 5,900,000,000,000 miles. This may be written in scientific form as $(5.9)10^{12}$. The positive exponent 12 indicates that the decimal point should be moved 12 places to the *right*. The notation works equally well for small numbers. To illustrate, it is estimated that the weight of an oxygen molecule is 0.000000000000000000000053 grams or, in scientific form, $(5.3)10^{-23}$ grams. The negative exponent indicates that the decimal point should be moved 23 places to the *left*. There are a number of advantages to this notation. It is compact and it enables the reader to quickly see (without counting zeros) the relative magnitudes of large or small quantities. It may also be used to simplify calculations which involve such quantities.

EXAMPLE 3   Calculate $\dfrac{(1{,}100{,}000)^2\ \sqrt{0.00000004}}{(8{,}000{,}000{,}000)^{2/3}}$.

Solution

Writing each number in scientific form and using laws of exponents, we have

$$\frac{[(1.1)10^6]^2[(4)10^{-8}]^{1/2}}{[(8)10^9]^{2/3}} = \frac{(1.1)^2 10^{12}(4)^{1/2}10^{-4}}{(8)^{2/3}10^6}$$

$$= \frac{(1.21)(2)10^8}{(4)10^6}$$

$$= (0.605)10^2 = 60.5.$$

EXERCISES

Rewrite the expressions in Exercises 1–6 by using fractional exponents instead of radicals.

1 $x\sqrt[3]{x}$

2 $x^2\sqrt[5]{x^{-3}}$

3 $\sqrt{(a+1)^3}$

4 $\sqrt{\sqrt{x}+\sqrt{y}}$

5 $\sqrt[3]{x^2+y^2}$

6 $\sqrt[3]{(x+y)^2}$

Rewrite the expressions in Exercises 7 and 8 by using radicals instead of fractional exponents.

7 a $8a^{2/3}$ **b** $(8a)^{2/3}$ **c** $8+a^{2/3}$

8 a $(a^{1/2}+b^{1/2})^{-1/2}$ **b** $(a^{-1/2}+b^{-1/2})^{1/2}$

Simplify the expressions in Exercises 9–34.

9 $16^{-3/4}$

10 $9^{5/2}$

11 $(-0.008)^{2/3}$

12 $(0.008)^{-2/3}$

13 $(4a^{3/2})(2a^{1/2})$

14 $(-6x^{7/5})(2x^{8/5})$

15 $(3x^{5/6})(8x^{2/3})$

16 $(8r)^{1/3}(2r^{1/2})$

17 $(27a^6)^{-2/3}$

18 $(25z^4)^{-3/2}$

19 $(8x^{-2/3})x^{1/6}$

20 $(3x^{1/2})(-2x^{5/2})$

21 $\left(\frac{-8x^3}{y^{-6}}\right)^{2/3}$

22 $\left(\frac{-y^{3/2}}{y^{-1/3}}\right)^3$

23 $\left(\frac{x^6}{9y^{-4}}\right)^{-1/2}$

24 $\left(\frac{c^{-4}}{16d^8}\right)^{3/4}$

25 $\frac{(x^6y^3)^{-1/3}}{(x^4y^2)^{-1/2}}$

26 $a^{4/3}a^{-3/2}a^{1/6}$

27 $\left(\frac{a^{2/3}}{b^{-2}}\right)^{-1}\left(\frac{a^{1/2}}{b^{1/3}}\right)^3$

28 $\left(\frac{x^{-1/4}y^{3/2}}{x^{3/4}y^{-2/3}}\right)^6$

29 $x^{1/2}x^{1/3}x^{1/6}$

30 $((2a^{1/2}b^{-3})^{1/2})^{-4}$

31 $((-8u^6b^{-3})^{1/3})^2$

32 $(a^{1/3}-a^{-5/3})a^{2/3}$

33 $x^{1/2}(x^{3/2}+x^{1/2}+x^{-1/2})$

34 $(x^{1/2}+y^{1/2})(x^{1/2}-y^{1/2})$

35

Express each of Exercises 35–42 as a single radical with the least possible index.

35 $\sqrt{3}\ \sqrt[3]{2}$

36 $\sqrt[3]{3}\ \sqrt{2}$

37 $\sqrt[4]{xy}\ \sqrt[3]{x^2y}$

38 $\sqrt[3]{x^2y^4}\ \sqrt[6]{x^5y}\ = x y \sqrt[]{xy}$

39 $\dfrac{\sqrt[6]{a^2b^3}}{\sqrt[3]{a^4b}}$

40 $\dfrac{\sqrt[3]{32}}{\sqrt[6]{4}}$

41 $\sqrt{x}\ \sqrt{x}$

42 $\sqrt[3]{a^4}\ \sqrt{a^3}$

43 The mass of a hydrogen atom is approximately

0.00000000000000000000000017 grams.

Express this number in scientific form.

44 The mass of an electron is approximately 9.1×10^{-31} kilograms. Express this number in decimal form.

Express the numbers in Exercises 45 and 46 in scientific form.

45 **a** 64,300   **b** 0.00000000012   **c** 34,200,000

46 **a** 420,000   **b** .0000067   **c** 3,400,000

Express the numbers in Exercises 47 and 48 in decimal form.

47 **a** $(4.4)10^4$ **b** $(2.7)10^{-20}$ **c** $(7.68)10^{10}$

48 **a** $(6.9)10^9$ **b** $(1.2)10^{-8}$ **c** $(9.5)10^{13}$

Approximate the quantities in Exercises 49–52.

49 $\dfrac{\sqrt{90,000}\ (2,000,000)^5}{(0.000000008)^{2/3}}$

50 $\dfrac{(8,000,000)^{2/3}}{(200,000)^4\ \sqrt{0.00000004}}$

51 $\dfrac{(0.00000009)^{3/2}}{(3,000,000)^2\ \sqrt[5]{(0.00243)^2}}$

52 $\dfrac{(0.000025)\ \sqrt{16,000,000}}{(1,100,000)^2}$

53 If the speed of light is 186,000 miles per second, approximate the distance light travels in one day; in 365 days.

54 Approximately how long does it take light to travel one mile? One foot?

7 ALGEBRAIC EXPRESSIONS

We frequently use symbols to denote arbitrary elements of a set. For example, the notation $x \in \mathbf{R}$ means that x is a real number, although no *particular* real number is specified. A letter which is used to represent an arbitrary element of a given set is called a **variable.** Letters near the end of the alphabet, such as $x, y, z, w, \cdots$, are often employed for variables. In this section we shall assume that all symbols for variables represent real numbers.

We often wish to restrict a variable to some subset of **R**. For example, if we want $\sqrt{x}$ to be real, then x must be nonnegative. In general, the subset of **R** whose elements are represented by a variable is called the **domain** of the variable. The domain of a variable x is also referred to as the set of **permissible** or **allowable** values for x.

If we begin with any collection of variables and real numbers, then an **algebraic expression** is the result obtained by applying a finite number of additions, subtractions, multiplications, divisions, or the taking of roots. The following are examples of algebraic expressions:

$$x^3 - 2x + \frac{3^{1/9}}{\sqrt{2x}}, \quad \frac{2xy + 3x}{y - 1}, \quad \frac{3^{1/7}yz^{-2} + \left(\dfrac{-\pi}{x + w}\right)^2}{\sqrt[3]{y^2 + \sqrt{5z}}},$$

where x, y, z, and w are variables. If specific numbers are substituted for the variables in an algebraic expression, the resulting real number is called the **value** of the expression for these numbers. For example, the value of the second expression above when $x = -2$ and $y = 3$ is

$$\frac{2(-2)(3) + 3(-2)}{3 - 1} = \frac{-12 - 6}{2} = -9.$$

When we work with algebraic expressions, it will be assumed that the domains are chosen so that variables do not represent numbers which make the expressions meaningless. Thus it is assumed that denominators do not vanish, roots always exist, and so on. For example, if we wish to restrict ourselves to the real number system when working with the first expression above, then it is necessary to have $x > 0$. In the second expression we do not allow $y = 1$. To simplify our work, we shall not always state the domains of variables. If meaningless expressions occur when certain numbers are substituted for the variables, then these numbers are *not* in the domains of the variables.

Let us introduce some terms which refer to special types of algebraic expressions. A **monomial** in certain variables is an algebraic expression which can be written as a product of a real number and *nonnegative integral powers* of the variables. For example, a monomial in the variable x is an algebraic expression of the form ax^n, where a is a real number and n is a nonnegative integer. A monomial in the variables x and y is any expression of the form ax^ny^m, where n and m are nonnegative integers. Similarly, a monomial in x, y, and z is an expression of the form $ax^ny^mz^p$, where n, m, and p are nonnegative integers. In all of these, a is called the **coefficient** of the monomial. A real number may be thought of as a monomial in *any* variable, since we can always write $a = ax^0 = ax^0y^0$, and so on. The **zero monomial** is defined to be the number 0.

The **degree** of a nonzero monomial *in one of its variables* is the exponent of that variable which appears in the monomial. The **degree**

of a nonzero monomial is the sum of the exponents of all the variables which occur. The monomial $2x^4yz^3$ is of degree 4 in x, degree 1 in y, degree 3 in z, and the degree of the monomial is 8. Observe that the degree of a monomial consisting of a nonzero real number a is zero. We shall not assign degree to the real number 0. It is the only monomial which has no degree.

A sum of two monomials, such as $2x^2y + 3yz$, is called a **binomial.** Since any difference $a - b$ can also be written in the form $a + (-b)$, we can always interpret a difference as a sum, where the minus sign indicates that the coefficient which follows is negative. With this agreement, $2x^2 - 3yz$ is a binomial, as are $-2x^2y + 3yz$ and $-2x^2y - 3yz$.

A **polynomial** is any finite sum of monomials. Again, instead of using negative coefficients, we use minus signs before appropriate terms. Some examples of polynomials are

$$(1.32) \quad x^4 - x^3 - 2x^2 + 5, \quad 3x^3y + y^2z - 4z, \quad -u^1v^5 + 7v^3u^4.$$

Expressions such as $(1/x^2) + 3x$, $(x + 1)/(y - 1)$, $x + \sqrt{y}$, are not polynomials since they cannot be written as sums in which each term is a monomial.

The *degree* of a polynomial in a given variable is the exponent of the highest power of that variable which appears. The first polynomial displayed above is of degree 4 in x. The second polynomial is of degree 3 in x, 2 in y, and 1 in z. The third polynomial has degree 4 in u and degree 5 in v. We define the **degree of a polynomial** as the degree of a term which has highest degree. The degrees of the above polynomials are 4, 4, and 7 respectively. We shall study the theory of polynomials in more detail in Chapter 4. At that time a number of important theorems will be discussed. For the present our main objective is to become proficient in formal manipulations with polynomials.

Since polynomials, and the monomials which make up polynomials, are symbols representing real numbers, all of the rules given in Section 2 can be applied. The principal difference is that symbols such as a, b, and c used in Section 2 will now be replaced by polynomials and other algebraic expressions. If additions, multiplications, and subtractions are carried out with polynomials, we ordinarily simplify the result by using various properties of real numbers.

EXAMPLE 1 Find the sum of the polynomials $x^3 + 2x^2 - 5x + 7$ and $4x^3 - 5x^2 + 3$.

Solution

$$(x^3 + 2x^2 - 5x + 7) + (4x^3 - 5x^2 + 3)$$
$$= x^3 + 4x^3 + 2x^2 - 5x^2 - 5x + 3 + 7$$

(rearrangement properties)

$$= (1 + 4)x^3 + (2 - 5)x^2 + (-5)x + (3 + 7)$$

(distributive law)

$$= 5x^3 - 3x^2 - 5x + 10$$

(simplifying).

Although Example 1 is a special case of the sum of two polynomials, the method used is perfectly general. It can be seen that the sum of any two polynomials in x can be obtained by adding coefficients of like powers of x. Similarly, the difference of two polynomials is found by subtracting coefficients of like powers, as in the next example.

EXAMPLE 2 Subtract $4x^3 - 5x^2 + 3$ from $x^3 + 2x^2 - 5x + 7$.

Solution

$$
\begin{aligned}
(x^3 + 2x^2 - 5x + 7) &- (4x^3 - 5x^2 + 3) \\
&= x^3 + 2x^2 - 5x + 7 - 4x^3 + 5x^2 - 3 \qquad \text{(Why?)} \\
&= x^3 - 4x^3 + 2x^2 + 5x^2 - 5x + 7 - 3 \\
&\qquad\qquad\qquad\qquad \text{(rearrangement properties)} \\
&= (1 - 4)x^3 + (2 + 5)x^2 - 5x + (7 - 3) \\
&\qquad\qquad\qquad\qquad \text{(distributive law)} \\
&= -3x^3 + 7x^2 - 5x + 4 \qquad\qquad \text{(simplifying).}
\end{aligned}
$$

The intermediate steps in the solution above were used for completeness. After the student becomes proficient with such manipulations, these may be omitted. Similar techniques are used for polynomials in more than one variable. In order to multiply two polynomials, we merely use the distributive law together with laws of exponents and combine like terms.

EXAMPLE 3 Find the product of $2x^3 + 3x - 1$ and $x^2 - x + 4$.

Solution

$$
\begin{aligned}
(2x^3 + 3x - 1)&(x^2 - x + 4) \\
&= (2x^3 + 3x - 1)x^2 + (2x^3 + 3x - 1)(-x) \\
&\quad + (2x^3 + 3x - 1)4 \qquad\qquad \text{(distributive law)} \\
&= 2x^5 + 3x^3 - x^2 - 2x^4 - 3x^2 + x \\
&\quad + 8x^3 + 12x - 4 \qquad\qquad\qquad \text{(Why?)} \\
&= 2x^5 - 2x^4 + (3 + 8)x^3 + (-1 - 3)x^2 \\
&\quad + (1 + 12)x - 4 \qquad\qquad\qquad \text{(Why?)} \\
&= 2x^5 - 2x^4 + 11x^3 - 4x^2 + 13x - 4 \qquad \text{(simplifying).}
\end{aligned}
$$

For convenience, the work above may be arranged as follows:

$$
\begin{array}{l}
2x^3 + 3x - 1 \\
\underline{x^2 - x + 4} \\
2x^5 \qquad\quad + 3x^3 - x^2 \qquad\qquad = (2x^3 + 3x - 1)x^2 \\
\quad - 2x^4 \qquad\qquad - 3x^2 + x \qquad = (2x^3 + 3x - 1)(-x) \\
\underline{\qquad\qquad\quad 8x^3 \qquad\quad + 12x - 4} = (2x^3 + 3x - 1)4 \\
2x^5 - 2x^4 + 11x^3 - 4x^2 + 13x - 4 = \text{sum of the above.}
\end{array}
$$

EXAMPLE 4 Multiply $x^2 + xy + y^2$ by $x - y$.

Solution

$$\begin{array}{l}
x^2 + xy\ + y^2 \\
x\ -\ y \\
\hline
x^3 + x^2y + xy^2 \qquad = x(x^2 + xy + y^2) \\
\quad - x^2y - xy^2 - y^3 = (-y)(x^2 + xy + y^2) \\
\hline
x^3 \qquad\qquad\quad\ -\ y^3 = \text{sum of the above.}
\end{array}$$

Division by a monomial is relatively easy, as seen in the next example.

EXAMPLE 5 Divide $6x^2y^3 + 4x^3y^2 - 10xy$ by $2xy$.

Solution

$$\frac{6x^2y^3 + 4x^3y^4 - 10xy}{2xy} = \frac{6x^2y^3}{2xy} + \frac{4x^3y^2}{2xy} - \frac{10xy}{2xy}$$

$$\text{(properties of quotients)}$$

$$= 3xy^2 + 2x^2y - 5$$

$$\text{(laws of exponents).}$$

Certain products occur so frequently in algebra that they deserve special attention. We list some of these in (1.33), where the letters represent real numbers. The reader should check the validity of each formula by actually carrying out the multiplications.

(1.33) **Product Formulas**

(1) $(x + y)(x - y) = x^2 - y^2$.

(2) $(ax + b)(cx + d) = acx^2 + (ad + bc)x + bd$.

(3) $(x + y)^2 = x^2 + 2xy + y^2$.

(4) $(x - y)^2 = x^2 - 2xy + y^2$.

(5) $(x + y)^3 = x^3 + 3x^2y + 3xy^2 + y^3$.

(6) $(x - y)^3 = x^3 - 3x^2y + 3xy^2 - y^3$.

Since the symbols x and y used in (1.33) represent real numbers, they may be replaced by algebraic expressions. This is illustrated in the next example.

EXAMPLE 6 Find the products

(a) $(2r^2 - \sqrt{s})(2r^2 + \sqrt{s})$,

(b) $\left(\sqrt{a} + \dfrac{1}{\sqrt{a}}\right)^2$,

(c) $(2a - 5b)^3$.

Solutions

(a) Using (1) of (1.33) together with laws of exponents and radicals,

$$(2r^2 - \sqrt{s})(2r^2 + \sqrt{s}) = (2r^2)^2 - (\sqrt{s})^2$$
$$= 4r^4 - s.$$

(b) Using (3) of (1.33) and simplifying,

$$\left(\sqrt{a} + \frac{1}{\sqrt{a}}\right)^2 = (\sqrt{a})^2 + 2\sqrt{a} \cdot \frac{1}{\sqrt{a}} + \left(\frac{1}{\sqrt{a}}\right)^2$$

$$= a + 2 + \frac{1}{a}.$$

(c) Applying (6) of (1.33),

$$(2a - 5b)^3 = (2a)^3 - 3(2a)^2(5b) + 3(2a)(5b)^2 - (5b)^3$$
$$= 8a^3 - 60a^2b + 150ab^2 - 125b^3.$$

EXERCISES

In Exercises 1–22 perform the indicated operations and find the degree of the resulting polynomial.

1 $(x^4 + 6x^3 - 4x + 5) + (2x^3 + 2x - 7)$

2 $(3x^2 - 2x - 3) + (x^5 + x^2 + 4x - 8)$

3 $(2x^3 + 6x - 4 + 3x^2) - (3x^3 + 3x^2 + 5x + 4)$

4 $(3y + 4 - 3y^4) - (5y^2 - 1 + 7y)$

5 $(x^3 - 2x^2 + 3x + 5) + (x^4 - 3x^3 + 5x - 2)$

6 $(x^5 - 3x + 1) + (x^2 + 3x - x^4 - 1)$

7 $(x^3 + 7 - 5x + 2x^2) - (x^3 - 3x^2 + 7 + 5x)$

8 $(a^2 + 1 + 3a^4) - (1 - a^3 + 4a^4)$

9 $(2z^2 + z - 5)(4z^3 + z^2 - 3z + 2)$

10 $(2a^2 + 1)(a - 3)(a^3 + 4a + 5)$

11 $(3c^2 - cd + d^2)(2c - d)$

12 $(u^2v - uv + uv^2)(u + v)$

13 $(2y^3 - y + 5)(3y^2 + 2y - 4)$

14 $(x^2 + 1)(x^3 + 1)(x^4 + 1)$

15 $(a + b)(a^2 - ab + b^2)$

16 $(x - y)(x^2 + xy + y^2)$

17 $\dfrac{9rs^2 + 6r^2}{3r}$

18 $\dfrac{4y^2z + 6yz^3}{6yz}$

19 $\dfrac{15x^2y - 12xy^3}{3xy}$

20 $\dfrac{x^2y^3 + 3x^4y^2 + 2x^3y^5}{x^2y^2}$

21 $\dfrac{20r^4s^2 - 3r^5s^4 + 6r^3s^3}{4r^3s^2}$

22 $\dfrac{4a^2bc^4 - 6ab^2c^3}{3abc}$

In Exercises 23–48 use (1.33) to find the indicated products.

23 $(x - 7)(x + 4)$

24 $(2x + 4y)(6x - 7y)$

25 $(3x + \frac{1}{2}y)(2x - \frac{1}{3}y)$

26 $(a^2 + 4)(a^2 - 3)$

27 $(2x - 3y)^2$

28 $(3y - 2x)^2$

29 $(2x + 7)(3x - 5)$

30 $(\frac{1}{2} - 4x)(\frac{1}{4} - 2x)$

31 $(8x - 3y)(10x + y)$

32 $(a^2 + 3b^2)(2a^2 - 5b^2)$

33 $(6x - 11y)^2$

34 $(4u + 7v)^2$

35 $\left(\dfrac{2}{x^2} + \dfrac{x^2}{2}\right)^2$

36 $(\sqrt{a} + 2\sqrt{b})^2$

37 $(3a - 2b^2)(3a + 2b^2)$

38 $(5x^2 - 4y^2)(5x^2 + 4y^2)$

39 $(\sqrt{u} + \sqrt{v})(\sqrt{u} - \sqrt{v})$

40 $(x - y - z)(x - y + z)$

41 $(2r + s)^3$

42 $(u^2 + v^2)^3$

43 $\left(x^2 - \dfrac{2}{x}\right)^3$

44 $(4a - 5b)^3$

45 $(2\sqrt{x} + 4y^2)(2\sqrt{x} - 4y^2)$

46 $(a + b + c)(a + b - c)$

47 $(\sqrt{x} + x + x^{3/2})^2$

48 $(x + 1)^4$

8 FACTORING

If a polynomial is written as a product of other polynomials, then each polynomial in the product is called a **factor** of the original polynomial. The process of expressing a polynomial as a product is called **factoring.** For example since $x^2 - 9 = (x + 3)(x - 3)$, we see that $x + 3$ and $x - 3$ are factors of $x^2 - 9$. Any polynomial has, as a factor, *every* nonzero real number c. For example, given $3x^2y - 5yz^2$ and any nonzero real number c, we can write

$$3x^2y - 5yz^2 = c\left(\frac{3}{c}x^2y - \frac{5}{c}yz^2\right).$$

A factor c of this type is called a **trivial factor.** We shall be primarily interested in **nontrivial factors** of polynomials, that is, factors which contain polynomials of degree greater than zero in certain variables. An exception to this rule is that if the coefficients are restricted to *integers,* then it is customary to remove a common integral factor from each term of the polynomial. This can be done by means of the distributive law, as in this factorization:

$$4x^2y + 8z^3 = 4(x^2y + 2z^3).$$

Before carrying out factorizations of polynomials it is necessary to specify the system from which the coefficients of the factors are to be chosen. In this chapter we shall use the rule that *if a polynomial with integer coefficients is given, then the factors should be polynomials with integer coefficients.* For example,

$$x^2 + x - 6 = (x + 3)(x - 2),$$
$$4x^2 - 9y^2 = (2x - 3y)(2x + 3y).$$

An integer $a > 1$ is prime if it cannot be written as a product of two positive integers greater than 1. A polynomial is said to be **prime** or **irreducible** if it cannot be written as a product of two polynomials of positive degree. When factoring a polynomial the objective is to express it as a product of prime polynomials or powers of prime polynomials. According to the rule given above, $x^2 - 2$ is prime since it cannot be factored as a product of two polynomials of positive degree which have *integral* coefficients. If we allow the factors to have *real* coefficients, then $x^2 - 2$ is not prime, since

$$x^2 - 2 = (x + \sqrt{2})(x - \sqrt{2}).$$

As we have mentioned, however, we shall not allow irrational coefficients in the factorizations made in this chapter. Later, in our work with solutions of equations, this rule will be rescinded.

In general, it is very difficult to factor polynomials of high degree. In simple cases some of the product formulas given in Section 7 are useful. One of the most important formulas is (1) of (1.33) for **the difference of two squares:**

(1.34) $x^2 - y^2 = (x + y)(x - y).$

The next example illustrates several applications of this formula.

EXAMPLE 1   Factor each of the following.

(a) $25r^2 - 49s^2$.   (b) $81x^4 - y^4$.   (c) $16x^2 - (y - 2z)^2$.

Solutions

(a)   Applying (1.34) with $x = 5r$ and $y = 7s$ gives us

$$25r^2 - 49s^2 = (5r)^2 - (7s)^2 = (5r - 7s)(5r + 7s).$$

(b) We make two applications of (1.34) as follows:

$$81x^4 - y^4 = (9x^2)^2 - (y^2)^2$$
$$= (9x^2 - y^2)(9x^2 + y^2)$$
$$= (3x - y)(3x + y)(9x^2 + y^2).$$

(c) Since the given expression is the difference of two squares we have, by (1.34)

$$16x^4 - (y - 2z)^2 = (4x^2)^2 - (y - 2z)^2$$
$$= [(4x^2) + (y - 2z)][4x^2 - (y - 2z)]$$
$$= (4x^2 + y - 2z)(4x^2 - y + 2z).$$

If a polynomial $ax^2 + bx + c$, where a, b, and c are integers, is factorable, then the factorization must be of the form $(dx + e)(fx + h)$, where d, f, e, and h are integers. In this case we have $df = a$, $eh = c$, and $dh + ef = b$. If a, b, and c are written as products of primes, then evidently there are only a limited number of choices for d, f, e, and h which satisfy the conditions above. If none of these choices works, then $ax^2 + bx + c$ is prime. This method is also applicable to polynomials of the form $ax^2 + bxy + cy^2$.

EXAMPLE 2 Factor

(a) $6x^2 - 7x - 3$.

(b) $4x^2 - 12xy + 9y^2$.

(c) $4x^4y - 11x^3y^2 + 6x^2y^3$.

Solutions

(a) If we write

$$6x^2 - 7x - 3 = (ax + b)(cx + d),$$

then the product of a and c is 6, and the product of b and d is -3. Trying various possibilities, we arrive at the factorization

$$6x^2 - 7x - 3 = (2x - 3)(3x + 1).$$

(b) If a factorization as a product of two first-degree polynomials exists, then it must be of the form

$$(ax + by)(cx + dy).$$

By trial we obtain

$$4x^2 - 12xy + 9y^2 = (2x - 3y)(2x - 3y) = (2x - 3y)^2.$$

(c) Since each term has x^2y as a factor, we begin by writing

$$4x^4y - 11x^3y^2 + 6x^2y^3 = x^2y(4x^2 - 11xy + 6y^2).$$

We then obtain the prime factorization

$$4x^4y - 11x^3y^2 + 6x^2y^3 = x^2y(4x - 3y)(x - 2y).$$

Sometimes, if terms in a sum are grouped in suitable fashion, then a factorization can be found by means of the distributive laws (1.5). This is illustrated in the next example.

EXAMPLE 3 Factor

(a) $a(x^2 - y) + 2b(x^2 - y).$ (b) $3x^3 + 2x^2 - 12x - 8.$

Solutions

(a) The given expression is of the form $ac + 2bc$ with $c = x^2 - y^2$. Since $ac + 2bc = (a + 2b)c$, we have

$$a(x^2 - y) + 2b(x^2 - y) = (a + 2b)(x^2 - y).$$

(b) We begin by grouping the first two terms and the last two terms of the given expression. Thus

$$3x^3 + 2x^2 - 12x - 8 = x^2(3x + 2) - 4(3x + 2).$$

The right-hand side has the form $x^2c - 4c$ where $c = 3x + 2$. Since $x^2c - 4c = (x^2 - 4)c$, we have

$$3x^3 + 2x^2 - 12x - 8 = (x^2 - 4)(3x + 2).$$

Finally, using (1.33) we obtain this prime factorization:

$$3x^3 + 2x^2 - 12x - 8 = (x + 4)(x - 4)(3x + 2).$$

Each of the next two formulas may be verified by multiplying the two factors on the right-hand side of the equation.

(1.35) $$x^3 + y^3 = (x + y)(x^2 - xy + y^2)$$
$$x^3 - y^3 = (x - y)(x^2 + xy + y^2)$$

The expressions on the left in (1.35) are referred to as the **sum of two cubes** and the **difference of two cubes** respectively. The student should take extreme care to note where the plus or minus signs occur.

EXAMPLE 4 Factor

(a) $a^3 + 64b^3$.   (b) $8x^6 - 27y^9$.

Solution

(a) By (1.35),

$$a^3 + 64b^3 = a^3 + (4b)^3$$
$$= (a + 4b)[a^2 - a(4b) + (4b)^2]$$
$$= (a + 4b)(a^2 - 4ab + 16b^2).$$

(b) By (1.35),

$$8x^6 - 27y^9 = (2x^2)^3 - (3y^3)^3$$
$$= (2x^2 - 3y^3)[(2x^2)^2 + (2x^2)(3y^3) + (3y^3)^2]$$
$$= (2x^2 - 3y^3)(4x^4 + 6x^2y^3 + 9y^6).$$

The next example illustrates an interesting technique for factoring certain types of expressions.

EXAMPLE 5 Factor $9a^4 + 8a^2b^2 + 4b^4$.

Solution

If the middle term were $12a^2b^2$ instead of $8a^2b^2$, then the polynomial could be written $(3a^2 + 2b^2)^2$. This suggests the device of *adding and subtracting* $4a^2b^2$ to obtain a factorization as follows:

$$9a^4 + 8a^2b^2 + 4b^4 = 9a^4 + 12a^2b^2 + 4b^4 - 4a^2b^2$$
$$= (3a^2 + 2b^2)^2 - (2ab)^2$$
$$= (3a^2 + 2b^2 + 2ab)(3a^2 + 2b^2 - 2ab).$$

The method used in Example 5 is sometimes referred to as the method of **completing the square.**

EXERCISES

Factor the expressions in Exercises 1–50.

1 $2ab + ac$

2 $3x^2 - 6yx$

3 $2xyz + xy + 5yz$

4 $4a^2b^2 - 3a^3b^2 + a^2b$

5 $3x^2 + 10x - 8$

6 $10y^2 + 29y - 21$

7 $12x^2 + 24x - 15$

8 $12x^2 + xy - 20y^2$

9 $49m^2 - 36n^2$

10 $u^3 - 125v^3$

11 $xy + xz$

12 $-6ax + 2ya$

13 $ux + yu + 3uz$

14 $2bz - 4bw + 6zb$

15 $2x^2 - 9x - 5$

16 $3y^2 + 7y - 6$

17 $12x^2 + 32x + 5$

18 $16a^8 - 7a^6 + 6a^7$

19 $9 - 25y^2$

20 $\frac{1}{36} y^2 - \frac{1}{49} x^6$

21 $16r^4 - 81s^4$

22 $49a^6 - 36b^2$

23 $8a^3 + 27b^3$

24 $r^5s^2 - 9r^3s^4$

25 $9x^2y^3 - 6x^3y^2 + x^4y$

26 $a^8 - 1$

27 $9z^5 + 24z^3 + 16z$

28 $(2x + y)^2 + (2x + y)$

29 $2ac - 2bc + 3ad - 3bd$

30 $6xu + 2vx - 3uy - yv$

31 $3x^3 + 2x^2 + 15x + 10$

32 $4ac + 2bc - 2ad - bd$

33 $a^6b^2 - b^6a^2$

34 $x^{64} - 1$

35 $y^3 - 8x^3$

36 $27y^3 - 64$

37 $4x^3y^2 - 12x^2y^2 + 9xy^2$

38 $-4x^2 - 10yx^2 + 6(xy)^2$

39 $x^2 + 16$

40 $x^2 + x + 1$

41 $(z + 1)^2 + 8(z + 1) + 15$

42 $(4u + 3v)^2 + (4u + 3v)(4u - 3v)$

43 $2k(y - m) - 3n(m - y)$

44 $2x^3 - x^2 - 2x + 1$

45 $r^6 - s^6$

46 $(a + b)^3 - 8$

47 $ac + bd - ad - bc$

48 $xy^3 + 2y^2 - xy - 2$

49 $x^4 + 16$

50 $(x^2 + 4)^2$

Factor each of the following by completing the square.

51 $a^4 + 2a^2b^2 + 9b^4$

52 $a^4 + 3a^2b^2 + 4b^4$

53 $x^4 - 4x^2y^2 + 36y^4$

54 $9x^4 + 16y^4 - x^2y^2$

9 RATIONAL EXPRESSIONS

A quotient of two polynomials is called a **rational expression.** Some examples are

$$\frac{x^2 - 5x + 1}{x^3 + 7}, \quad \frac{z^2x^4 - 3yz}{5z}, \quad \text{and} \quad \frac{1}{4xy}.$$

Many problems in mathematics involve combining rational expressions and then simplifying the result. Since rational expressions are quotients containing symbols which represent real numbers, the rules for quotients stated in Section 2 may be used. Of course, the letters a, b, c, d, etc., will now be replaced by polynomials. Of particular importance in simplification problems is the formula

$$\frac{ad}{bd} = \frac{a}{b},$$

which is obtained by dividing the numerator and denominator of the given fraction by d. This rule is sometimes expressed thus: "a common factor in the numerator and denominator may be canceled from the quotient." To use this technique in problems, we factor both the numerator and denominator of the given rational expression into prime factors. We then divide numerator and denominator by these common factors.

EXAMPLE 1 Simplify $\dfrac{3x^2 - 5x - 2}{x^2 - 4}$.

Solution

$$\frac{3x^2 - 5x - 2}{x^2 - 4} = \frac{(3x + 1)(x - 2)}{(x - 2)(x + 2)} = \frac{3x + 1}{x + 2}.$$

In the preceding example we divided numerator and denominator by $x - 2$. This simplification is valid only if $x - 2 \neq 0$, that is, $x \neq 2$. However, 2 is not in the domain of x since it leads to a zero denominator when substituted in the original expression. Hence our manipulations are valid. We shall always assume such restrictions when simplifying rational expressions.

EXAMPLE 2 Simplify $\dfrac{2 - x - 3x^2}{6x^2 - x - 2}$.

Solution

$$\frac{2 - x - 3x^2}{6x^2 - x - 2} = \frac{(1 + x)(2 - 3x)}{(2x + 1)(3x - 2)} = \frac{-(1 + x)}{2x + 1}.$$

The fact that $(2 - 3x) = -(3x - 2)$ accounts for the minus sign in the final answer. Another method of attack is to change the form of the numerator as follows:

$$\frac{2 - x - 3x^2}{6x^2 - x - 2} = \frac{-(3x^2 + x - 2)}{6x^2 - x - 2}$$

$$= -\frac{(3x - 2)(x + 1)}{(3x - 2)(2x + 1)} = -\frac{x + 1}{2x + 1}.$$

Multiplication and division are performed using rules for quotients and then simplifying, as illustrated in the next example.

EXAMPLE 3 Perform the indicated operations and simplify:

(a) $\dfrac{x^2 - 6x + 9}{x^2 - 1} \cdot \dfrac{2x - 2}{x - 3}.$ (b) $\dfrac{x + 2}{2x - 3} \div \dfrac{x^2 - 4}{2x^2 - 3x}.$

Solutions

(a) $\dfrac{x^2 - 6x + 9}{x^2 - 1} \cdot \dfrac{2x - 2}{x - 3} = \dfrac{(x - 3)^2 \cdot 2(x - 1)}{(x - 1)(x + 1) \cdot (x - 3)}$

$$= \dfrac{2(x - 3)}{x + 1}.$$

(b) $\dfrac{x + 2}{2x - 3} \div \dfrac{x^2 - 4}{2x^2 - 3x} = \dfrac{x + 2}{2x - 3} \cdot \dfrac{2x^2 - 3x}{x^2 - 4}$

$$= \dfrac{(x + 2) \cdot x(2x - 3)}{(2x - 3)(x + 2)(x - 2)}$$

$$= \dfrac{x}{x - 2}.$$

When adding or subtracting two rational expressions, it is customary to find a common denominator and use the rule $a/d \pm c/d = (a \pm c)/d$. A common denominator may be introduced by multiplying numerator and denominator of each of the given expressions by suitable polynomials. It is usually desirable to use the **least common denominator (l.c.d.)** of the two fractions. The l.c.d. can be found by obtaining the prime factorization for each denominator and then forming the product of the different prime factors, using the *highest* exponent which appears with each prime factor. The method is illustrated in the next example.

EXAMPLE 4 Simplify $\dfrac{2x + 5}{x^2 + 6x + 9} + \dfrac{x}{x^2 - 9} + \dfrac{1}{x - 3}.$

Solution

The factored forms of the denominators are $(x + 3)^2$, $(x + 3)(x - 3)$, and $(x - 3)$. Hence the l.c.d. is $(x + 3)^2(x - 3)$. We then change each rational expression to an equal one having denominator $(x + 3)^2(x - 3)$ and proceed as follows:

$$\dfrac{2x + 5}{(x + 3)^2} + \dfrac{x}{(x + 3)(x - 3)} + \dfrac{1}{x - 3}$$

$$= \dfrac{(2x + 5)}{(x + 3)^2} \cdot \dfrac{(x - 3)}{(x - 3)} + \dfrac{x}{(x + 3)(x - 3)} \cdot \dfrac{(x + 3)}{(x + 3)}$$

$$+ \dfrac{1}{(x - 3)} \cdot \dfrac{(x + 3)^2}{(x + 3)^2}$$

$$= \dfrac{(2x^2 - x - 15) + (x^2 + 3x) + (x^2 + 6x + 9)}{(x + 3)^2(x - 3)}$$

$$= \dfrac{4x^2 + 8x - 6}{(x + 3)^2(x - 3)} = \dfrac{2(2x^2 + 4x - 3)}{(x + 3)^2(x - 3)}.$$

It is sometimes necessary to simplify quotients in which the numerator and denominator are not polynomials, as illustrated in the next example.

EXAMPLE 5 Simplify

$$\frac{1 - \dfrac{2}{x+1}}{\dfrac{1}{x} - x}.$$

Solution

$$\frac{1 - \dfrac{2}{x+1}}{\dfrac{1}{x} - x} = \frac{\dfrac{(x+1)-2}{x+1}}{\dfrac{1-x^2}{x}}$$

$$= \frac{x-1}{x+1} \cdot \frac{x}{1-x^2}$$

$$= \frac{(x-1) \cdot x}{(x+1)(1-x)(1+x)}$$

$$= \frac{-x}{(x+1)^2}.$$

EXERCISES

Simplify

1. $\dfrac{2x^2 - x - 6}{3x^2 - 5x - 2}$

2. $\dfrac{10a^2 - a - 2}{4 - 11a + 6a^2}$

3. $\dfrac{x^3 - y^3}{x^2 - y^2}$

4. $\dfrac{2a^2 - 5ab - 3b^2}{a^2 - ab - 6b^2}$

5. $\dfrac{6x^2 + 5x - 4}{4x^2 - 4x + 1}$

6. $\dfrac{2 - 9y + 4y^2}{4 - y^2}$

7. $\dfrac{a^4 + 6a^2 + 9}{9 - a^4}$

8. $(a^2 + 2a + 1)\left(\dfrac{a-5}{a^3+1}\right)$

9. $\dfrac{4}{2x-1} - \dfrac{5}{x+3}$

10. $\dfrac{r}{r-t} - \dfrac{t}{r+t}$

11. $\dfrac{2a^2 - 4a - 6}{9a^2 - 16} \cdot \dfrac{6a - 8}{a + 1}$

12. $\dfrac{2x^2 - 7x + 6}{4x^2 + 27x - 7} \cdot \dfrac{2x^2 - 3x}{6x^2 - 21x + 18}$

13. $\dfrac{3}{3x+1} - \dfrac{5}{7x-2}$

14. $\dfrac{x}{2x-y} - \dfrac{y}{x+2y}$

15. $\dfrac{3y^2 + 14y - 5}{4y^2 - 9} \div \dfrac{3y - 1}{2y + 3}$

16. $\dfrac{4 - 6y}{2y^2 + 3y - 5} \cdot \dfrac{y^3 - 1}{3y - 2}$

17 $\dfrac{x}{2x+6} + \dfrac{5}{x^2-9} - \dfrac{4x}{3-x}$

18 $\dfrac{r^2 + rs + s^2}{\dfrac{r^2}{s} - \dfrac{s^2}{r}}$

19 $\left(\dfrac{1}{3x + 3h - 5} - \dfrac{1}{3x - 5}\right) \div h$

20 $\left(\dfrac{1}{(x+h)^2 + 9} - \dfrac{1}{x^2 + 9}\right) \div h$

21 $\dfrac{2}{6x - 2} + \dfrac{x}{9x^2 - 1} - \dfrac{2x + 1}{1 - 3x}$

22 $\dfrac{5x}{2x + 4} - \dfrac{3}{x} + 6$

23 $\dfrac{1 - \dfrac{2}{x+1}}{\dfrac{1}{x} - x}$

24 $\dfrac{3}{4a - 8} - \dfrac{a + 2}{a^2 - 2a}$

25 $\dfrac{x^3 + y^3}{x^3 - y^3} \cdot \dfrac{x^2 - y^2}{x^2 + 2xy + y^2}$

26 $\dfrac{a^2 - b^2}{b^2 - c^2} \div \dfrac{ac + bc}{ab - ac}$

27 $\dfrac{a}{9a^2 - 1} + \dfrac{2}{6a - 2} - \dfrac{2a + 1}{1 - 3a}$

28 $\dfrac{x^2/y^2 - y^2/x^2}{x/y + y/x}$

29 $\dfrac{\sqrt{x^2 + y^2} - \dfrac{x^2}{\sqrt{x^2 + y^2}}}{\sqrt{x^2 + y^2}}$

30 $\sqrt{a^2 + b^2} - \dfrac{b^2}{\sqrt{a^2 + b^2}} + \dfrac{a}{\sqrt{1 + \dfrac{b^2}{a^2}}}$

10 REVIEW EXERCISES

Oral

Define or explain each of the following concepts.

1 The union and intersection of two sets.

2 Subset of a set

3 Disjoint sets

4 The commutative and associative laws for real numbers

5 The distributive laws

6 Rational and irrational numbers

7 Prime number

8 A number a is greater than a number b

9 A number a is less than a number b

10 Properties of inequalities

11 The absolute value of a real number

12 Coordinate line

13 The distance between points on a coordinate line

14 Directed distance

15 The midpoint formula   **16** The Laws of Exponents

17 Negative exponents   **18** The zero exponent

19 Principal nth root   **20** The radical notation

21 Rational exponents   **22** Scientific form for real numbers

23 Variable   **24** Algebraic expression

25 Monomial   **26** Degree of a monomial

27 Polynomial   **28** Degree of a polynomial

29 Prime polynomial   **30** Rational expression

Written

In Exercises 1 and 2 find $A \cup B$ and $A \cap B$.

1 $A = \{x, z, w, y\}, B = \{u, v, y, z\}$

2 $A = \{3, 1, 4\}, B = \{1, 2, 3\}$

3 List all the subsets of $\{u, v, w\}$

4 If A, B, C and D are sets, use Venn diagrams to represent $(A \cap C) \cup (A \cap B)$.

Rewrite Exercises 5–8 without using the absolute value symbol.

5 $|-7| - |-3|$ **6** $|-4| - 4$

7 $|\sqrt{9} - 3|$ **8** $|3 - \sqrt{10}|$

Express the statements in Exercises 9–12 in terms of inequalities.

9 a is not positive   **10** a is between -8 and -10

11 a is not less than 1   **12** a is not between 2 and 5

Use properties of inequalities to change each of Exercises 13–16 to one of the forms $a > k$ or $a < k$, where k is a real number.

13 $3a > 8$ **14** $-3a > 8$

15 $5 - a < 3$ **16** $2a + 3 < a$

Change each of Exercises 17–20 to the form $k < a < l$, where k and l are real numbers.

17 $|-a| < 6$ **18** $|a - 3| < 6$

19 $|2a - 1| < 5$ **20** $|1 - a| < 2$

21 If $|x - 3| > 5$, prove that either $x > 8$ or $x < -2$.

22 If $|x + 3| > 1$, prove that either $x < -4$ or $x > -2$.

23 If A and B are points on a coordinate line with coordinates -8 and -16, respectively, find:

a $d(A, B)$ **b** $\overline{AB}$ **c** $\overline{BA}$

d the coordinate of the midpoint of AB

24 Work Exercise 23 if A and B have coordinates -3 and 8, respectively.

Simplify the expressions in each of Exercises 25–42.

25 $(4z^3 x)(2zx^2)^2$ **26** $\dfrac{5x^2 y^2}{2x^4 y^{-3}}$

27 $\dfrac{(2^{-1} a^{-4} b^2)^{-3}}{a^{-4} b^2}$ **28** $\left(\dfrac{x^{1/3} y^{3/2}}{x^{1/2} y^{2/3}}\right)^6$

29 $(u^{1/3} + v^{1/3})(u^{1/3} - v^{1/3})$ **30** $\left(\dfrac{2x^{-2}}{y^3 z^2}\right)^{-2} \left(\dfrac{xz}{2y^2}\right)^3$

31 $\dfrac{(ab)^{-1}}{a^{-1} + b^{-1}}$ **32** $\left(\dfrac{27a^6}{-8b^3}\right)^{2/3}$

33 $((a^2 b^{-1})^{1/2})^{-4}$ **34** 2^{-0}

35 $\sqrt{\dfrac{x^3 y^2}{z}}$ **36** $\sqrt[3]{4x^2 y^4} \ \sqrt[3]{2xy}$

37 $\dfrac{\sqrt{2a^5 b}}{\sqrt{8a^2 b^3}}$ **38** $\sqrt[5]{(x - y + z)^5}$

39 $\sqrt{4a^3} - \sqrt{a}$ **40** $\sqrt[3]{-27} + \sqrt[4]{16} - \sqrt[5]{-1}$

41 $1/\sqrt{5}$ **42** $1/(\sqrt{w} + 2\sqrt{v})$

Perform the operations indicated in Exercises 43–48.

43 $(2x^3 - x^2 + 5x - 7) - (3x^3 + 2x^2 - x + 2)$

44 $(x^4 + 3x^2 - 7) + (x^3 - 2x^2 + x + 1)$

45 $(y^2 - 3y + 4)(y - 5)$ **46** $(z^2 - 1)(z^2 + 3z + 2)$

47 $(x + 1)^2 - (3x - 1)^2$ **48** $(4a^3 b^2 - 6a^2 b^3)/a^2 b$

Find the products in Exercises 49–54.

49 $(5a - 2b)^2$ **50** $(6r - s)(2r + 3s)$

51 $(2x - y)^3$ **52** $(a^{-1} + a)^3$

53 $(\sqrt{p} + \sqrt{q})(\sqrt{p} - \sqrt{q})$ **54** $(a - b - c)(a + b - c)$

Factor the expressions in Exercises 55–60.

55 $8x^2 + 10x - 3$ **56** $9m^4 + 12m^2 + 4$

57 $x^4 y - x^2 y^3$ **58** $ac + 2bc + 2bd + ad$

53

59 $x^3 - x^2 + x - 1$ **60** $64r^3 + 8s^6$

Simplify the expressions in Exercises 61–66.

61 $\dfrac{4x^2 + 7x - 2}{2x^2 + x - 6}$ **62** $\dfrac{x^2 - 4}{x^2 + 4x + 4}$

63 $\dfrac{a^2b^{-1} - b^2a^{-1}}{a^2 + ab + b^2}$ **64** $\left(\dfrac{1}{(x + h)^3 + 1} - \dfrac{1}{x^3 + 1} \right) \div h$

65 $\dfrac{a^{1/2} - a^{-1/2}}{a^{1/2} + a^{-1/2}}$ **66** $\dfrac{2}{2 - x} + \dfrac{4x}{x^2 - 4} + \dfrac{3}{x + 2}$

2

Equations and Inequalities

For hundreds of years one of the main concerns in algebra has been the solutions of equations. More recently the study of inequalities has reached the same level of importance. Both of these topics are used extensively in applications of mathematics. In this chapter we shall discuss the rudiments of solving equations and inequalities. All variables will represent real numbers. Later in our work we shall study equations that involve complex numbers.

1 ELEMENTARY EQUATIONS

If x is a variable, then expressions such as

$$x + 3 = 0,$$
$$x^2 - 5 = 4x,$$
$$(x^2 - 9)\sqrt[3]{x + 1} = 0,$$

are called **equations** in the variable x. If certain numbers are substituted for x in these equations, true statements are obtained, whereas other numbers produce false statements. For example, the equation $x + 3 = 0$ leads to a false statement for every value of x except -3. If 2 is substituted for x in the second equation above, we obtain $4 - 5 = 8$, or $-1 = 8$, a false statement. On the other hand, if we let $x = 5$, then we obtain $(5)^2 - 5 = 4 \cdot 5$, or $20 = 20$, which is true.

If all expressions involved are algebraic, then an equation in x is called an **algebraic equation** in x. If a true statement is obtained when x is replaced by some real number a from the domain of x, then a is called a **solution** or a **root** of the equation. We also say that a **satisfies** the equation. The **solution set** of an equation is the set of all solutions. To **solve** an equation means to find its solution set. The solution sets of the

55

three equations on page 55 are $\{-3\}$, $\{5, -1\}$, and $\{3, -3, -1\}$, respectively. Upon substitution it can be seen that the numbers in these sets are indeed solutions for their respective equations. It will follow from our later work that they are the *only* solutions.

If the solution set of an equation is the entire domain of x, then the equation is called an **identity.** For example

$$\sqrt{1 - x^2} = \sqrt{(1 - x)(1 + x)}$$

is an identity since it is true for every number in the domain of x. If there are numbers in the domain of x which are not solutions, then the equation is called a **conditional equation.**

The solution set of an equation depends on the system of numbers we decide to allow for solutions. For example, if we demand that solutions be rational numbers, then the solution set of the equation $x^2 = 2$ is empty, since there is no rational number whose square is 2. However, if we allow real numbers, then the solution set is $\{-\sqrt{2}, \sqrt{2}\}$. Similarly, the equation $x^2 = -1$ has no real solutions; however we shall see later that this equation has solutions if *complex* numbers are allowed.

Two equations are said to be **equivalent** if they have the same solution set. For example, the equations

$$x - 1 = 2, \quad x = 3, \quad 5x = 15, \quad \text{and} \quad 2x + 1 = 7$$

are equivalent since they all have the solution set $\{3\}$.

One method for solving an equation is to replace it by a chain of equivalent equations, each one in some sense simpler than the preceding one and terminating in an equation for which the solution set is obvious. This is accomplished by using various properties of real numbers. For example, since x represents a real number, we may add the same expression in x to both sides of an equation without changing the solution set. Similarly, subtraction from both sides leads to an equivalent equation. We can also multiply or divide both sides of an equation by an expression which represents a nonzero real number. The following example illustrates these remarks. The reader should supply reasons for each step in the solution.

EXAMPLE 1 Solve $2x - 5 = 3$.

Solution

The following is a chain of equations, each of which is equivalent to the preceding one:

$$2x - 5 = 3$$
$$(2x - 5) + 5 = 3 + 5$$
$$2x = 8$$
$$\tfrac{1}{2}(2x) = \tfrac{1}{2}(8)$$
$$x = 4.$$

Since the solution set of the last equation is obviously $\{4\}$, then $\{4\}$ is also the solution set of $2x - 5 = 3$.

In order to guard against errors in manipulations and simplifications, answers should be checked by substitution in the original equation. If we apply a check in Example 1 we obtain $2(4) - 5 = 3$, which reduces to $3 = 3$, a true statement.

If we inadvertently multiply both sides of an equation by an expression which equals zero for some value of x, then an equivalent equation might not be obtained. The following example illustrates a manner in which this happens.

EXAMPLE 2 Solve $\dfrac{3x}{x-2} = 1 + \dfrac{6}{x-2}$.

Solution

Multiplying both sides by $x - 2$ and simplifying leads to

$$\left(\frac{3x}{x-2}\right)(x-2) = 1(x-2) + \left(\frac{6}{x-2}\right)(x-2)$$
$$3x = (x-2) + 6.$$
$$2x = 4$$
$$x = 2.$$

Let us check to see whether 2 is a solution of the given equation. Substituting 2 for x we obtain

$$\frac{3(2)}{2-2} = 1 + \frac{6}{2-2}, \quad \text{or} \quad \frac{6}{0} = 1 + \frac{6}{0}.$$

Since division by 0 is not permissible, 2 is not a solution. Actually, the given equation has no solutions; that is, the solution set is empty.

The preceding solution indicates that *it is essential to check answers* after multiplying both sides of an equation by an expression containing variables. Checks are unnecessary if manipulations consist only of additions or subtractions involving expressions which contain variables, or of multiplications by specific real numbers.

In the exercises we shall restrict our efforts to equations which are equivalent to equations of the form

$$ax + b = 0,$$

where a and b are real numbers and $a \neq 0$. An equation of this type is called a **linear equation** in x. Evidently $ax + b = 0$ is equivalent to $ax = -b$ and, therefore, to $x = -b/a$. Hence there is precisely one solution, $-b/a$. In Section 3 we shall consider equations having more than one solution.

EXAMPLE 3 Solve $\dfrac{3}{2x-4} - \dfrac{5}{x+3} = \dfrac{2}{x-2}$.

Solution

Since the equation may be rewritten as

$$\frac{3}{2(x-2)} - \frac{5}{x+3} = \frac{2}{x-2},$$

we see that the l.c.d. of the three fractions is $2(x-2)(x+3)$. Multiplying both sides by this l.c.d. leads to

$$3(x+3) - 5 \cdot 2(x-2) = 2 \cdot 2(x+3).$$

Simplifying, we obtain

$$3x + 9 - 10x + 20 = 4x + 12$$
$$-11x = -17$$
$$x = 17/11.$$

We now check this result in the given equation. Substituting for x we have

$$\frac{3}{2(17/11) - 4} - \frac{5}{(17/11) + 3} = \frac{2}{(17/11) - 2}.$$

This reduces to $-22/5 = -22/5$, a true statement. Hence the solution set is $\{17/11\}$.

   Students sometimes have difficulty in determining whether an equation is conditional or an identity. An identity will usually be indicated when, after applying methods which lead to equivalent equations, an equation of the form $p = p$ is obtained, where p is an algebraic expression in x. Of course, to show that an equation is *not* an identity, we need only find *one* real number in the domain of x that fails to satisfy the original equation.

EXERCISES

Find the solution sets of the equations in Exercises 1–28.

1 $2x + 3 = 5x - 4$

2 $4x - 3 = 2x + 8$

3 $\frac{1}{3}x - 7 = 3 + \frac{1}{2}x$

4 $\frac{1}{2}x - 3 = \frac{3}{4}x + 1$

5 $7(2x + 5) = 2(3x - 1)$

6 $0.5 - 2.4w = 0.3(4 - 2w)$

7 $0.2(6x - 5) + 0.3x = 1.4$

8 $2(3x - 1) - 4x = -3(x + 2) + 2$

9 $\dfrac{3 - 4y}{2y + 9} = \dfrac{5}{2}$

10 $\dfrac{3x - 10}{2} = 2 - \dfrac{1 - 2x}{3}$

11 $\dfrac{3}{u} - 5 = \dfrac{7}{u} - 1$

12 $(4x - 5)^2 = (2x + 3)(8x - 1)$

13 $(y - 5)^2 - (y + 5)^2 = 5^2$

14 $\dfrac{1}{w} + \dfrac{2}{w} + \dfrac{3}{w} + \dfrac{4}{w} = 5$

15 $\dfrac{5 - 3x}{2x - 7} = \dfrac{6x + 5}{1 - 4x}$

16 $\dfrac{4}{3 - z} + \dfrac{7}{3z - 9} = \dfrac{5}{3}$

17 $\dfrac{1}{2} - \dfrac{2}{3x + 2} = \dfrac{3}{6x + 4}$

18 $\dfrac{2x + 2}{3x - 5} - \dfrac{4x + 1}{6x - 7} = 0$

19 $\dfrac{2}{x^2 - 9} - \dfrac{5}{x + 3} = \dfrac{7}{x - 3}$

20 $\dfrac{x - 8}{x^2 - x - 2} + \dfrac{2}{x - 2} = \dfrac{3}{x + 1}$

21 $\dfrac{5}{2w - 3} - \dfrac{3}{2} = \dfrac{1}{4w - 6}$

22 $\dfrac{2}{v + 3} = \dfrac{v - 4}{v^2 - 9} + \dfrac{5}{v - 3}$

23 $\dfrac{5x}{2x - 3} = \dfrac{x + 6}{2x - 3}$

24 $\dfrac{3}{4 + 6x} = \dfrac{5}{2 + 3x}$

25 $\dfrac{x + 1}{x} = 1 + \dfrac{1}{x}$

26 $\dfrac{2}{x + 1} = 0$

27 $(t + 1)^3 = (t - 1)^3 + 6t(t + 2)$

28 $\dfrac{8 + 2z}{2z} = 1 + \dfrac{4}{z}$

29 For what value of a is $\{-5\}$ the solution set of the equation
$3x + 2 + 2a = 5a - x + 9$?

30 For what value of a is $\{-3\}$ the solution set of the equation
$2x - a = 7x - 5 + 3a$?

31 Determine values for a and b such that the equation $ax + b = 0$ has solution set $\{\frac{3}{2}\}$. Are these the only possible values for a and b? For what values of a and b will the solution set be $\varnothing$? For what values of a and b will the solution set be **R**?

32 Which of the following pairs of equations are not equivalent? Explain.
(a) $x = 3$ and $x^2 = 9$. (b) $x = 3$ and $x = \sqrt{9}$. (c) $x = 3$ and $x^2 = 3x$. (d) $x = 3$ and $3x = 9$.

33 **a** Find an equation in the following chain which is *not* equivalent to the preceding equation:

$$x^2 - 1 = x^2 - x$$
$$(x + 1)(x - 1) = x(x - 1)$$
$$x + 1 = x$$
$$1 = 0.$$

b Find the solution set of the first equation in (a).

34 Find an equation in the following chain which is *not* equivalent to the preceding equation:

$$x^2 - x - 6 = x^2 - 2x - 3$$
$$(x - 3)(x + 2) = (x - 3)(x + 1)$$
$$x + 2 = x + 1$$
$$2 = 1.$$

2 APPLICATIONS

Formulas or equations involving variables are often used in areas such as physics, chemistry, engineering, biology, psychology, sociology, and finance. For certain applications it is necessary to solve for a particular variable in terms of the remaining variables which appear in the formula. This is done by treating the equation as if the desired variable were the only one present and transforming the original equation to an equivalent equation in which that variable is isolated on one side.

EXAMPLE 1 The formula $C = \frac{5}{9}(F - 32)$ gives the relation between Fahrenheit temperature, F, and centigrade temperature, C. Solve for F in terms of C.

Solution

The following equations are equivalent. (Why?):

$$C = \tfrac{5}{9}(F - 32)$$
$$\tfrac{9}{5}C = F - 32$$
$$\tfrac{9}{5}C + 32 = F$$
$$F = \tfrac{9}{5}C + 32.$$

EXAMPLE 2 The formula $R = \dfrac{R_1 R_2}{R_1 + R_2}$ occurs in electrical theory. Solve for R_1 in terms of R and R_2.

Solution

The following equations are equivalent to the given equation.

$$(R_1 + R_2)R = (R_1 + R_2)\left(\frac{R_1 R_2}{R_1 + R_2}\right)$$

$$R_1 R + R_2 R = R_1 R_2$$

$$R_1 R - R_1 R_2 = -R_2 R$$

$$R_1(R - R_2) = -R_2 R$$

$$R_1 = \frac{-R_2 R}{R - R_2}, \ (R \neq R_2)$$

$$R_1 = \frac{R_2 R}{R_2 - R}, \ (R \neq R_2).$$

In many applications of mathematics, problems are stated in words. If variables are introduced, it may be possible to formulate a specific problem by means of an equation, from which the solution can be obtained. This technique is illustrated below.

EXAMPLE 3 A student has test scores of 64 and 78. What score on a third test will give the student an average of 80?

Solution

If we let x denote the score on the third test, then the average for the three tests is

$$\frac{64 + 78 + x}{3}.$$

In order for this average to equal 80, we must have

$$\frac{64 + 78 + x}{3} = 80.$$

The preceding equation is equivalent to each of the following:

$$64 + 78 + x = 240$$
$$142 + x = 240$$
$$x = 240 - 142$$
$$x = 98.$$

Hence a score of 98 must be obtained on the third test.

Check: If three test scores are 64, 78, and 98, then the average is $(64 + 78 + 98)/3 = 240/3 = 80$.

As indicated in Example 3, when we wish to solve verbal problems involving one unknown quantity, we usually introduce a symbol such as x to denote the quantity which is to be found. Next, by carefully analyzing the problem, we form an equation which describes precisely what is stated in words. Finally, we solve for x and check the answer. The check should be made by referring to the original statement and *not* to the equations which were formulated.

EXAMPLE 4 A store holding a clearance sale advertises that all prices have been discounted 20%. If a certain article is on sale for $28, what was its price before the sale?

Solution

We begin by letting x denote the unknown quantity, that is, the presale price. It is convenient to arrange our work as follows, where the quantities are measured in dollars.

$$x = \text{presale price}$$
$$.20x = \text{discount}$$
$$x - .20x = \text{sale price.}$$

Since the sale price is \$28, we have

$$x - .20x = 28.$$

This equation may be solved as follows:

$$x - \tfrac{1}{5}x = 28$$
$$\tfrac{4}{5}x = 28$$
$$x = (\tfrac{5}{4})28 = 35.$$

Hence the price before the sale was \$35.

Check: If a \$35 article is discounted 20%, then the discount (in dollars) is $(.20)(35) = 7$, and the selling price is $35 - 7$, or \$28.

An important type of problem which may be solved by using linear equations involves mixing two substances to obtain a prescribed mixture. Typical illustrations are given in the next two examples.

EXAMPLE 5   A chemist has 10 ounces of a solution which contains a 30% concentration of a certain chemical. How many ounces of pure chemical must be added in order to increase the concentration to 50%?

Solution

Since 30% of the given 10-ounce solution consists of the chemical, the number of ounces of chemical present is $.3(10)$, or 3. Let x denote the number of ounces of pure chemical to be added. We then have the following:

$$x = \text{ounces of chemical to be added,}$$
$$3 + x = \text{ounces of chemical in the new solution,}$$
$$10 + x = \textit{total ounces in the new solution}$$

Since the new solution is to contain 50% of the chemical, we may write

$$.5(10 + x) = \text{ounces of chemical in the new solution.}$$

We now have two different ways of expressing the amount of chemical in the new solution, namely $3 + x$ and $.5(10 + x)$. This leads to the following chain of equations.

$$3 + x = .5(10 + x).$$
$$3 + x = 5 + .5x$$
$$.5x = 2$$
$$\tfrac{1}{2}x = 2$$
$$x = 4.$$

Hence 4 ounces of the chemical should be added.

Check: If 4 ounces of chemical are added, then the new solution contains 14 ounces, 7 of which are chemical. This is the desired 50% concentration.

EXAMPLE 6 A radiator contains 8 quarts of a mixture of water and antifreeze. If 40% of the mixture is antifreeze, how much of the mixture should be drained and replaced by pure antifreeze in order that the resultant mixture will contain 60% antifreeze?

Solution

The amount of antifreeze in the original mixture is .4(8), or 3.2 quarts. If we let x denote the number of quarts of mixture to be drained, then x is also the additional amount of pure antifreeze that will be added. If each quart of the original mixture contains 40% antifreeze, then in draining x quarts we lose .4x quarts of antifreeze. Hence *the amount of antifreeze in the final mixture* is $(3.2 + x) - .4x$ quarts. Since we want the final mixture of 8 quarts to contain 60% antifreeze, we must have

$$3.2 + x - .4x = .6(8)$$
$$3.2 + .6x = 4.8$$
$$.6x = 1.6$$
$$x = \tfrac{8}{3} \text{ quarts.}$$

Check: In draining $\tfrac{8}{3}$ quarts of the original mixture, we lose $.4(\tfrac{8}{3})$ quarts of antifreeze and hence there remain $3.2 - .4(\tfrac{8}{3})$ quarts of antifreeze. If we then add $\tfrac{8}{3}$ quarts of pure antifreeze, the amount of antifreeze in the final mixture is $3.2 - .4(\tfrac{8}{3}) + \tfrac{8}{3}$ quarts. This reduces to 4.8, which is 60% of 8.

Many applied problems have to do with objects that move at a constant, or uniform, rate of speed. If an object travels at a uniform (or average) rate r, then the distance d traversed in time t is given by $d = rt$. Of course we assume that the units are properly chosen; that is, if r is in feet per second, then t is in seconds, and so on.

EXAMPLE 7 Two cities A and B are connected by means of a highway 150 miles long. An automobile leaves A at 1:00 P.M. and travels at a uniform rate of 40 miles per hour toward B. Thirty minutes later, another automobile leaves A and travels toward B at a uniform rate of 55 miles per hour. At what time will the second car overtake the first car?

Solution

Let t denote the time, in hours, *after* 1:30 P.M. At 1:30 P.M. the first automobile has already traveled 20 miles. Hence at time t *after* 1:30 P.M. the distance it has traveled is $20 + 40t$ miles. Since the second automobile starts the trip at 1:30 P.M., the distance it has traveled at time t is $55t$ miles. We wish to find the time t at which the distances traveled by the two automobiles are equal. This will be true when

$$55t = 20 + 40t.$$

Solving for t, we obtain

$$15t = 20, \quad \text{or} \quad t = \tfrac{4}{3}.$$

Consequently, $t = 1\tfrac{1}{3}$ hours, that is, 1 hour and 20 minutes. Since this is the amount of time after 1:30 P.M., it follows that the second car overtakes the first at 2:50 P.M.

Check: At 2:50 P.M. the first car has traveled for $1\tfrac{5}{6}$ hours and its distance from A is $40(11/6) = 220/3$ miles. At 2:50 P.M. the second car has traveled for $1\tfrac{1}{3}$ hours and is $55(4/3) = 220/3$ miles from A. Hence they are together at 2:50 P.M.

EXERCISES

The formulas in Exercises 1–12 occur in mathematics and its applications. Solve for the indicated variable in terms of the remaining variables.

1 $s = \tfrac{1}{2}gt^2$ for g

2 $R = \dfrac{E}{I}$ for I

3 $A = \tfrac{1}{2}(b_1 + b_2)h$ for b_2

4 $A = 4\pi r^2 + 2\pi rh$ for h

5 $\dfrac{1}{R} = \dfrac{1}{R_1} + \dfrac{1}{R_2} + \dfrac{1}{R_3}$ for R_2

6 $\dfrac{1}{p} = \dfrac{1}{f} + \dfrac{1}{g}$ for f

7 $I = \dfrac{nE}{R + nr}$ for r; for n

8 $mv = Ft + mv_0$ for m

9 $S = a + (k - 1)d$ for k

10 $R = \dfrac{nE - rI}{nI}$ for I

11 $A = p + prt$ for p; for t

12 $S = \tfrac{1}{2}(a + I)$ for a

13 Find three consecutive even integers whose sum is 234.

14 Find three consecutive integers whose sum is 762.

15 The sum of the digits of a two-digit number is 8. If the digits are reversed, the resulting number is 18 less than the given number. Find the given number.

16 Find two consecutive integers such that the difference of their squares is 237.

17 A boy has 25 coins consisting of nickels and dimes. If the total amount is $1.65, how many coins of each type does he have?

18 A boy has 40 coins consisting of pennies and nickels. If the total amount is 72 cents, how many coins of each type does he have?

19 An airplane flew with the wind for 1 hour and returned the same distance against the wind in $1\frac{1}{2}$ hours. If the cruising speed of the plane is 350 miles per hour, find the velocity of the wind.

20 A man can row 1 mile upstream in the same amount of time it takes him to row 2 miles downstream. If the rate of the current is 3 miles per hour, how fast can he row in still water?

21 The distance around an oval track is $2\frac{1}{2}$ miles. If two racing cars A and B start from the same point and the average speed of A is 130 miles per hour while the average speed of B is 126 miles per hour, how long will it take A to be one lap ahead of B? After A has traveled 200 miles, how far ahead is A?

22 For a certain event 600 tickets were sold for a total of $1176.00. If some tickets sold for $2.25 and others for $1.50, how many of each kind were sold?

23 A motorist traveled from city A to city B at an average rate of 40 miles per hour. On the return trip he drove at an average rate of 50 miles per hour and made the trip in 45 minutes less time. Find the distance from A to B.

24 An airplane flew with the wind for 50 minutes and returned the same distance against the wind in 1 hour. If the cruising speed of the plane is 330 miles per hour, find the velocity of the wind.

25 A projectile is fired horizontally at a target and the sound of its impact is heard 1.5 seconds later. If the speed of the projectile is 3300 ft/sec and the speed of sound is 1100 ft/sec, how far away is the target?

26 A merchant wishes to mix peanuts costing 70¢ per pound with cashews costing $1.50 per pound, obtaining 50 pounds of a mixture which costs $1.20 per pound. How many pounds of each type should be mixed?

27 An 80-pound solution of salt water is 5% salt. How much water must be evaporated in order that the resulting mixture will be 10% salt?

28 A chemist has two acid solutions, the first containing 10% acid and the second 30% acid. How many ounces of each should be mixed in order to obtain 10 ounces of a solution containing 25% acid?

29 Generalize Exercise 27 as follows: Let a, b, and d be positive real numbers with $a < b < 100$. Given a solution of d pounds of salt water that is $a\%$ salt, how much water must be evaporated in order that the resulting mixture will be $b\%$ salt? Express the answer in terms of a, b, and d.

30 Let a, b, c, and d be positive real numbers where $a < c < b < 100$. Generalize Exercise 28 to the case where the given solutions contain $a\%$ and $b\%$ acid respectively, and where d ounces of a solution containing $c\%$ acid is desired. Express the answer in terms of a, b, c, and d.

31 A chemist has 12 liters of acid solution which is 25% acid. How much should be removed and replaced with a 50% solution in order to obtain a solution which is 30% acid?

32 If man *A* can do a certain job alone in 6 days, and if man *B* can do the job alone in 10 days, in how many days can the job be completed if *A* and *B* work together?

3 EQUATIONS OF DEGREE GREATER THAN 1

In the previous two sections, all equations were equivalent to linear equations. We shall now turn to equations which are not of this type. Of fundamental importance is the fact that if *a* and *b* are real numbers such that $ab = 0$, then either $a = 0$ or $b = 0$. This result is useful for solving equations which can be written in the form $pq = 0$, where *p* and *q* are algebraic expressions in a variable *x*.

If *p* and *q* are algebraic expressions in *x*, then p' and q' will denote the values of *p* and *q* respectively, when $x = a$. For example, if $p = x^2 + 3$, then $p' = a^2 + 3$. It follows that the value of the expression pq is $p'q'$ when $x = a$. Of course we use only the values of *x* for which both *p* and *q* have meaning.

If *a* is a solution of either the equation $p = 0$ or the equation $q = 0$, then by definition $p' = 0$ or $q' = 0$. This implies that $p'q' = 0$; that is, *a* is a solution of the equation $pq = 0$. Conversely, if *a* is a solution of $pq = 0$, then $p'q' = 0$. This implies that either $p' = 0$ or $q' = 0$; that is, *a* is a solution of either $p = 0$ or $q = 0$. We have proved the following result.

(2.1)   Theorem

If *S* and *T* are the solution sets of the equations $p = 0$ and $q = 0$, respectively, then the solution set of the equation $pq = 0$ is their union, $S \cup T$.

Theorem (2.1) can be extended to cases where more than two factors are involved. For example, if *p*, *q*, and *r* are algebraic expressions in *x*, then the solution set of $pqr = 0$ is the union of the solution sets of $p = 0$, $q = 0$, and $r = 0$. We shall refer to this method of solving equations as the **method of solution by factoring.**

EXAMPLE 1   Solve $x^3 + 2x^2 - x - 2 = 0$ by factoring.

Solution

The following equations are all equivalent:

$$x^3 + 2x^2 - x - 2 = 0$$
$$x^2(x + 2) - (x + 2) = 0$$
$$(x^2 - 1)(x + 2) = 0$$
$$(x + 1)(x - 1)(x + 2) = 0.$$

According to (2.1), the solution set of $x^3 + 2x^2 - x - 2 = 0$ consists of the union of the solution sets of $x + 1 = 0$, $x - 1 = 0$, and $x + 2 = 0$, and therefore is $\{-1, 1, -2\}$.

EXAMPLE 2 Solve $x^2 - 8x + 16 = 0$ by factoring.

Solution

The equation is equivalent to $(x - 4)^2 = 0$. Setting each factor equal to zero, we have $x - 4 = 0$ and $x - 4 = 0$. Hence $\{4\}$ is the solution set of $x^2 - 8x + 16 = 0$.

Since $x - 4$ appears as a factor twice in the previous solution, the real number 4 is called a **double root,** or **root of multiplicity two,** of the equation $x^2 - 8x + 16 = 0$.

If we are given an equation of the form $x^2 = d$, where $d \geq 0$, then $x^2 - d = 0$ is an equivalent equation. The latter equation may be written

$$(x + \sqrt{d})(x - \sqrt{d}) = 0.$$

Setting each factor equal to zero, we obtain the solution set $\{-\sqrt{d}, \sqrt{d}\}$. We sometimes use $\{\pm\sqrt{d}\}$ as an abbreviation for a solution set of this type. The same argument may be used to prove that if p is any algebraic expression in x, then the solution set of the equation $p^2 = d$, where $d \geq 0$, is the union of the solution sets of the equations $p = \sqrt{d}$ and $p = -\sqrt{d}$.

EXAMPLE 3 Solve $(x - 3)^2 = 5$.

Solution

As in the preceding paragraph, we write

$$x - 3 = \sqrt{5} \quad \text{or} \quad x - 3 = -\sqrt{5}.$$

Hence the solution set of $(x - 3)^2 = 5$ is $\{3 + \sqrt{5}, 3 - \sqrt{5}\}$ or, more compactly, $\{3 \pm \sqrt{5}\}$.

The method used in the solution of Example 3 may be employed to obtain a formula for solving any equation of the form

(2.2) $ax^2 + bx + c = 0,$

where a, b, and c are real numbers and $a \neq 0$. An equation of this type is called a **quadratic equation in** x.

To obtain solutions of (2.2), consider the following chain of equivalent equations (supply the reasons for each step):

$$ax^2 + bx + c = 0$$
$$ax^2 + bx = -c$$
$$x^2 + \frac{b}{a}x = -\frac{c}{a}$$
$$x^2 + \frac{b}{a}x + \frac{b^2}{4a^2} = -\frac{c}{a} + \frac{b^2}{4a^2}$$
$$\left(x + \frac{b}{2a}\right)^2 = \frac{b^2 - 4ac}{4a^2}.$$

The addition of $b^2/4a^2$ to both sides in the fourth step is referred to as **completing the square** in x, since it enables us to write the left-hand side as the square of the binomial $x + (b/2a)$. If the right-hand side of the last equation is nonnegative, then as in the solution of Example 3,

$$x + \frac{b}{2a} = \sqrt{\frac{b^2 - 4ac}{4a^2}} \quad \text{or} \quad x + \frac{b}{2a} = -\sqrt{\frac{b^2 - 4ac}{4a^2}}.$$

For simplicity let us combine these equations into one expression by writing

$$x = -\frac{b}{2a} \pm \sqrt{\frac{b^2 - 4ac}{4a^2}}.$$

Note that since $4a^2 > 0$, the radicand $(b^2 - 4ac)/4a^2$ is nonnegative if and only if $b^2 - 4ac \geq 0$. From our work with radicals we may change the radical on the right-hand side of the last equation as follows:

$$\pm\sqrt{\frac{b^2 - 4ac}{4a^2}} = \pm\frac{\sqrt{b^2 - 4ac}}{\sqrt{(2a)^2}} = \pm\frac{\sqrt{b^2 - 4ac}}{|2a|}.$$

If $a > 0$, then $|2a| = 2a$ and the radical equals $\pm\dfrac{\sqrt{b^2 - 4ac}}{2a}$. If $a < 0$, then $|2a| = -2a$ and we obtain $\mp\dfrac{\sqrt{b^2 - 4ac}}{2a}$. Consequently, whether a is positive or negative, we may write

$$x = -\frac{b}{2a} \pm \frac{\sqrt{b^2 - 4ac}}{2a}.$$

It follows that the solution set of $ax^2 + bx + c = 0$ is

$$\left\{\frac{-b + \sqrt{b^2 - 4ac}}{2a}, \quad \frac{-b - \sqrt{b^2 - 4ac}}{2a}\right\}.$$

The expression

(2.3) $\qquad x = \dfrac{-b \pm \sqrt{b^2 - 4ac}}{2a}$

is called **the quadratic formula.** Given any quadratic equation of the form (2.2), the solutions can be obtained by substituting the coefficients for a, b, c in (2.3).

EXAMPLE 4   Solve $4x^2 + x - 3 = 0$.

Solution

Letting $a = 4$, $b = 1$, and $c = -3$ in (2.3), we obtain

$$x = \frac{-1 \pm \sqrt{1 - 4(4)(-3)}}{2(4)}$$

$$= \frac{-1 \pm \sqrt{49}}{8}$$

$$= \frac{-1 \pm 7}{8}.$$

Hence the solution set is

$$\left\{ \frac{-1 + 7}{8}, \ \frac{-1 - 7}{8} \right\}, \quad \text{or} \quad \left\{ \frac{3}{4}, -1 \right\}.$$

Note that Example 4 could also have been solved by factoring. If we write $(4x - 3)(x + 1) = 0$ and set each factor equal to zero, we obtain the solution set $\{\frac{3}{4}, -1\}$.

The number $b^2 - 4ac$ which appears under the radical sign in the quadratic formula is called the **discriminant** of the quadratic equation. It can be used to determine the nature of the roots of the equation. Specifically,

(1)   if $b^2 - 4ac = 0$, the equation has a double root;

(2)   if $b^2 - 4ac > 0$, the equation has two real and unequal solutions;

(3)   if $b^2 - 4ac < 0$, the equation has no real solutions.

Case (3) will be discussed further in Chapter Eight. If a, b, and c are rational and $b^2 - 4ac$ is the square of a rational number, then the quadratic formula indicates that the equation has rational roots. This was the case in Example 4. The next example illustrates the case of irrational roots.

EXAMPLE 5   Solve $2x^2 - 6x + 3 = 0$.

Solution

Letting $a = 2$, $b = -6$, and $c = 3$ in the quadratic formula, we obtain

$$x = \frac{6 \pm \sqrt{(-6)^2 - 4(2)(3)}}{2(2)}$$

$$= \frac{6 \pm \sqrt{12}}{4}$$

$$= \frac{3 \pm \sqrt{3}}{2}.$$

Therefore the solution set is $\{3/2 + \sqrt{3}/2, 3/2 - \sqrt{3}/2\}$.

The following example illustrates the case of a double root.

EXAMPLE 6 Solve the equation $9x^2 - 30x + 25 = 0$.

Solution

Letting $a = 9$, $b = -30$, and $c = 25$ in the quadratic formula gives us

$$x = \frac{30 \pm \sqrt{(-30)^2 - 4(9)(25)}}{2(9)}$$

$$= \frac{30 \pm \sqrt{900 - 900}}{18}$$

$$= \frac{30 \pm 0}{18} = \frac{5}{3}.$$

Consequently the solution set is $\{5/3\}$.

There are many applications of quadratic equations. One is illustrated in the following example.

EXAMPLE 7 A box with a square base and no top is to be made from a square piece of tin by cutting out 3-inch squares from each corner and folding up the sides. If the box is to hold 48 cubic inches, what size piece of tin should be used?

Solution

If x denotes the length of the side of the piece of tin, then the length of the base of the box is $x - 6$ (see Fig. 2.1). Since the area of the base is $(x - 6)^2$ and the height is 3, the volume of the box is $3(x - 6)^2$. The required condition gives us the equation

$$3(x - 6)^2 = 48.$$

Hence

$$(x - 6)^2 = 16 \quad \text{and} \quad x - 6 = \pm 4.$$

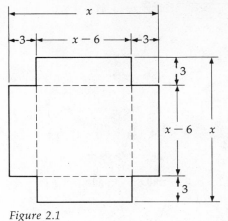

Figure 2.1

The solution set of the latter equation is $\{10, 2\}$.

Check: It is clear that 2 is not acceptable since no box is possible in this case. However, if we begin with a 10-inch square of tin, cut out 3-inch corners and fold, we obtain a box having dimensions 4 inches, 4 inches, and 3 inches. The box has the desired volume of 48 cubic inches. Thus 10 inches is the answer to the problem.

As illustrated in Example 7, even though an equation is formulated correctly, it is possible, owing to the physical nature of a given problem, to arrive at meaningless solutions. These solutions should be discarded. For example, we would not accept the answer -7 years for the age of an individual nor $\sqrt{50}$ for the number of automobiles in a parking lot.

EXERCISES

Solve the equations in Exercises 1–10 by factoring.

1 $3x^2 + x - 10 = 0$

2 $2x^2 - 9x - 5 = 0$

3 $25x^2 = 20x - 4$

4 $16y^2 + 1 = 8y$

5 $2y^3 = y^2 + 3y$

6 $x^3 = 9x$

7 $w^4 = 5w^2 - 4$

8 $u^4 - 25u^2 + 144 = 0$

9 $2x^3 + 3x^2 - 50x - 75 = 0$

10 $7z^3 - 28z = 4z^2 - 16$

Solve the equations in Exercises 11–26 by means of the quadratic formula.

11 $4x^2 - 11x - 3 = 0$

12 $25x^2 + 10x + 1 = 0$

13 $9x^2 = 12x - 4$

14 $10y^2 + y - 21 = 0$

71

15 $2z^2 + z - 5 = 0$

16 $2x^2 + 4x - 3 = 0$

17 $\dfrac{2x + 1}{x - 3} = \dfrac{x - 1}{x + 2}$

18 $9x^2 + 13x = 0$

19 $\dfrac{3}{y^2} - \dfrac{5}{y} + 2 = 0$

20 $0.3x^2 + 1.4x + 0.7 = 0$

21 $2x^2 - 7x - 15 = 0 \cdot$

22 $4x^2 + 28x + 49 = 0$

23 $t^2 = \frac{1}{2}t - \frac{1}{16}$

24 $\frac{1}{2}w^2 + \frac{1}{5}w = \frac{3}{10}$

25 $3x^2 + 4x = 0$

26 $4x^2 - 9 = 0$

27 Find two different values for b such that the equation $x^2 + bx - 5 + 3b = 0$ has a double root.

28 Find all values of b such that the equation $3x^2 + bx - 4b = 0$ has no real roots.

29 Find two consecutive even integers whose product is 224.

30 Find two consecutive integers the sum of whose squares is 181.

31 The hypotenuse of a right triangle is 10 inches long. Find the lengths of the two legs if their sum is 12 inches.

32 The diameter of a circle is 16 inches. What change in the radius will decrease the area by 48π square inches?

33 A projectile is fired straight upward with an initial velocity of 640 ft/sec. The distance s (in feet) above the ground after t seconds is given by $s = -16t^2 + 640t$. When will the projectile be 1600 feet above the ground? When will it hit the ground?

34 If the projectile in Exercise 33 is fired upward from a height of s_0 feet above the ground with an initial velocity of v_0 feet per second, then its height above the ground at time t is given by $s = -16t^2 + v_0 t + s_0$. Solve for t in terms of s, v_0, and s_0. When will the projectile hit the ground? Express your answer in terms of v_0 and s_0.

35 If one solution of the equation $kx^2 + 3x + k = 0$ is -2, find the other solution.

36 If a, b, and c are rational numbers, can the equation $ax^2 + bx + c = 0$ have one rational and one irrational root? Explain.

37 If r_1 and r_2 are the two roots of $ax^2 + bx + c = 0$, show that $r_1 + r_2 = -b/a$ and $r_1 r_2 = c/a$.

38 Prove that the roots of $ax^2 + bx + c = 0$ are numerically equal but opposite in sign to the roots of $ax^2 - bx + c = 0$.

39 Given the equation $2x^2 - 2xy + 4 - y^2 = 0$, use the quadratic formula to solve for (a) x in terms of y, (b) y in terms of x.

40 Same as Exercise 33 for the equation $3x^2 + xy + 4y^2 = 9$.

4 MISCELLANEOUS EQUATIONS

Many equations which are not of the specific types discussed in the preceding sections can often be put into one of those forms by a suitable manipulation. If equations involve radicals or fractional exponents, the method of raising both sides of the equation to a positive integral power is often useful. This method is based on the following theorem.

(2.4)   Theorem

If p and q are expressions in a variable x, and if n is any positive integer, then the solution set of the equation $p = q$ is a subset of the solution set of $p^n = q^n$.

To state this theorem in another way we could say that *every solution of $p = q$ is also a solution of $p^n = q^n$*. The proof is relatively easy and will be omitted.

It may happen that the solution set of $p = q$ is a *proper* subset of the solution set of $p^n = q^n$. For example, if we start with the equation $x = 2$ and square both sides we obtain $x^2 = 4$. Note that the solution set $\{2\}$ of the first equation is contained in the solution set $\{2, -2\}$ of the second equation. However the latter set also contains -2, which is *not* a solution of the given equation, $x = 2$. Any solution of $p^n = q^n$ which is not a solution of $p = q$ is called an **extraneous solution** of $p = q$. Since extraneous solutions may be introduced, it is *absolutely essential* to check all answers obtained by using (2.4).

EXAMPLE 1   Solve $\sqrt[3]{x^2 - 1} = 2$.

Solution

If we cube both sides, then by (2.4) the solutions of the equation are included among the solutions of

$$(\sqrt[3]{x^2 - 1})^3 = 2^3,$$

and hence among those of

$$x^2 - 1 = 8,$$

or

$$x^2 = 9.$$

Thus the solution set of $\sqrt[3]{x^2 - 1} = 2$ is a subset of $\{3, -3\}$.

Check: Substituting 3 for x in the given equation, we obtain $\sqrt[3]{3^2 - 1} = 2$, or $\sqrt[3]{8} = 2$, which is a true statement. Similarly, -3 is a solution. Hence the solution set of the given equation is $\{3, -3\}$.

73

EXAMPLE 2 Solve $3 + \sqrt{3x+1} = x$.

Solution

The equation $\sqrt{3x+1} = x - 3$ is equivalent to the given one. Squaring both sides gives us

$$3x + 1 = x^2 - 6x + 9,$$

which is equivalent to

$$x^2 - 9x + 8 = 0,$$

or

$$(x - 8)(x - 1) = 0.$$

Since the last equation has solutions 1 and 8, it follows from (2.4) that the solution set of $3 + \sqrt{3x+1} = x$ is a subset of $\{1, 8\}$.

Check: Letting $x = 1$ in the original equation, we obtain $3 + \sqrt{4} = 1$, or $5 = 1$, which is false. Consequently, 1 is not a solution of the given equation. Letting $x = 8$ in the original equation we obtain $3 + \sqrt{25} = 8$, or $8 = 8$, which is true. Hence the solution set of $3 + \sqrt{3x+1} = x$ is $\{8\}$.

Sometimes it is necessary to use (2.4) several times in the same problem, as illustrated in the next example.

EXAMPLE 3 Solve $\sqrt{2x-3} - \sqrt{x+7} + 2 = 0$.

Solution

Let us begin by writing $\sqrt{2x-3} = \sqrt{x+7} - 2$. Sqaring both sides we obtain

$$2x - 3 = (x + 7) - 4\sqrt{x+7} + 4,$$

or

$$x - 14 = -4\sqrt{x+7}.$$

Squaring both sides of the last equation gives us

$$x^2 - 28x + 196 = 16(x + 7),$$

which can be written

$$x^2 - 44x + 84 = 0,$$

or as

$$(x - 42)(x - 2) = 0.$$

Hence the solution set of the original equation is a subset of $\{42, 2\}$.

Check: When $x = 42$, the given equation yields $\sqrt{84 - 3} - \sqrt{42 + 7} + 2 = 0$, or $9 - 7 + 2 = 0$, which is false. When $x = 2$, we obtain $\sqrt{4 - 3} - \sqrt{2 + 7} + 2 = 0$, or $1 - 3 + 2 = 0$, a true statement. Hence the solution set of the given equation is $\{2\}$.

An equation in a variable x is said to be of **quadratic type** if it can be written in the form

(2.5) $au^2 + bu + c = 0, \quad a \neq 0,$

where u is an expression in x. If we find the solution set of (2.5), then the solutions of the original equation can be obtained by referring to the expression u. The method is illustrated in the following example.

EXAMPLE 4   Solve $x^{2/3} + x^{1/3} - 6 = 0$.

Solution

If we let $u = x^{1/3}$, then the equation can be written

$$u^2 + u - 6 = 0,$$

or

$$(u + 3)(u - 2) = 0.$$

The solutions of this equation are $u = -3$ and $u = 2$. Since $u = x^{1/3}$ we have $x^{1/3} = -3$ or $x^{1/3} = 2$. Cubing we obtain $x = -27$ or $x = 8$.

Check: Letting $x = -27$ in the given equation we obtain $[(-27)^{1/3}]^2 + (-27)^{1/3} - 6 = 9 - 3 - 6 = 0$. Thus -27 is a solution. Similarly it can be shown that 8 is a solution. Hence the solution set of the equation is $\{-27, 8\}$.

EXAMPLE 5   Solve $x^4 - 3x^2 + 1 = 0$.

Solution

Letting $x^2 = u$ gives us $u^2 - 3u + 1 = 0$. Using the quadratic formula, we obtain

$$u = \frac{3 \pm \sqrt{9 - 4}}{2}.$$

Since $u = x^2$ we have

$$x^2 = \frac{3 \pm \sqrt{5}}{2},$$

or

$$x = \pm \sqrt{\frac{3 \pm \sqrt{5}}{2}}.$$

Thus there are four possible solutions:

$$\sqrt{\frac{3 + \sqrt{5}}{2}}, \quad -\sqrt{\frac{3 + \sqrt{5}}{2}}, \quad \sqrt{\frac{3 - \sqrt{5}}{2}}, \quad -\sqrt{\frac{3 - \sqrt{5}}{2}}.$$

By checking these in the original equation, it can be shown that each is a solution.

EXAMPLE 6   Solve $|2x + 7| = 3$.

Solution

By (1.19) either

$$2x + 7 = 3 \quad \text{or} \quad 2x + 7 = -3.$$

Since these equations are equivalent to $2x = -4$ or $2x = -10$, the solution set is $\{-2, -5\}$.

EXERCISES

Find the solution sets of the following equations.

1   $\sqrt{4x + 9} = 7$ 2 $\sqrt[4]{4 - 11x} - 3 = 0$

3   $5 - \sqrt[3]{3x^2 + 2} = 0$ 4 $\sqrt{5x - 4} + 2 = x$

5   $2x - \sqrt{2x - 3} = 3$ 6 $\sqrt{7x + 1} + 8 = 0$

7   $\sqrt{5y + 6} = 3 + \sqrt{y + 3}$ 8 $\sqrt{3x + 4} - \sqrt{x - 6} = \sqrt{2x + 2}$

9   $\sqrt{9x + 10} - \sqrt{x + 1} = \sqrt{4x + 5}$ 10 $\sqrt{3\sqrt{x - 1}} = \sqrt{x - 1}$

11   $4z^4 - 39z^2 + 27 = 0$ 12 $2u^{2/3} - u^{1/3} - 6 = 0$

13   $y^{1/3} + y^{1/6} - 2 = 0$ 14 $10x^{-2} - 19x^{-1} - 15 = 0$

15   $(\sqrt{y} - 1)^2 - 6(\sqrt{y} - 1) + 9 = 0$ 16 $12x - 67\sqrt{x} + 90 = 0$

17   $6z^{1/4} - 2z^{-1/4} - 1 = 0$ 18 $2x^4 - 11x^2 + 12 = 0$

19 $\dfrac{2}{(w+1)^2} - \dfrac{5}{w+1} + 2 = 0$

20 $\dfrac{12}{(2x+3)^2} + \dfrac{1}{2x+3} - 1 = 0$

21 $\dfrac{x^4}{(x-1)^2} - 13\,\dfrac{x^2}{x-1} + 36 = 0$

22 $(z^2-1)^{-2} = 5(z^2-1)^{-1} - 4$

23 $|3 + 4x| = 9$

24 $\sqrt{3x+4} = -8$

25 $|6 - 5x| = 4$

26 $|7 - 2x| = 1$

27 $|5x - 2| = |2x + 13|$

28 $|(x-3)^2| = 4$

29 $|x + 1| = -1$

30 $|4x - 1| = |3 - 8x|$

5 ELEMENTARY INEQUALITIES

In Chapter One we defined the inequalities $a < b$ and $c > d$, where a, b, c, and d are real numbers. Now, in a manner similar to our work with equations, we shall consider inequalities which contain variables. The domains of the variables will be subsets of **R**.

As a first illustration, let us consider the inequality

$$x^2 - 3 < 2x + 4.$$

If certain numbers such as 4 or 5 are substituted for x, we obtain the false statements $13 < 12$ or $22 < 14$, respectively. Other numbers such as 1 or 2 produce the true statements $-2 < 6$ or $1 < 8$, respectively. The object of the following work will be to determine the totality of real numbers which produce true statements when substituted for the variable in inequalities of this type.

Let us begin with several definitions which are similar to those given for equations in Section 1. If we are given an inequality in x and if a true statement is obtained when x is replaced by a real number a, then a is called a **solution** of the inequality. The **solution set** of an inequality is the collection of all solutions. To **solve** an inequality means to find its solution set. We say that two inequalities are **equivalent** if they have the same solution set.

As with equations, a standard method for solving an inequality is to replace it with a chain of equivalent inequalities, terminating in one for which the solution set is obvious. This is often done by using the rules for inequalities established in Chapter One. For example, if x is a variable, then it follows from (2) of (1.11) or (1.12) that adding the same expression in x to both sides of an inequality does not change the solution set. Similarly, by (3) of (1.11) or (1.12), we may multiply both sides of an inequality by an expression containing x *if we are certain that the expression is positive* for all values of x under consideration. For example, multiplication by $x^4 + 3x^2 + 5$ will not change the solution set since this expression is always positive. If we multiply both sides of an inequality by an expression that is always negative, such as $-7 - x^2$, then the inequality sign is reversed by (4) of (1.11) or (1.12).

EXAMPLE 1   Solve the inequality $4x - 3 < 2x + 5$.

Solution

The following is a chain of equivalent inequalities:

$$
\begin{array}{ll}
4x - 3 < 2x + 5 & \text{(given)} \\
(4x - 3) + 3 < (2x + 5) + 3 & \text{(2) of (1.12)} \\
4x < 2x + 8 & \text{(simplifying)} \\
4x + (-2x) < (2x + 8) + (-2x) & \text{(2) of (1.12)} \\
2x < 8 & \text{(simplifying)} \\
x < 4 & \text{(multiplication by } \tfrac{1}{2}\text{)}.
\end{array}
$$

The solution set S of the final inequality is obviously the set of all real numbers which are less than 4, that is, $S = \{x : x < 4\}$.

It is convenient to give graphical interpretations for solution sets of inequalities. By the **graph** of a set S of real numbers we mean the collection of all points on a coordinate line l which correspond to the numbers in S. To **sketch a graph** we darken an appropriate portion of l. The graph of the solution set S of Example 1 consists of all points to the left of the point corresponding to 4 and is sketched in Fig. 2.2. The parenthesis in the figure indicates that the point corresponding to 4 is not part of the graph. If we wish to include such an **endpoint** of a graph we shall employ a bracket instead of a parenthesis. This will be illustrated in Example 2.

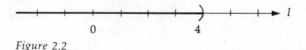

0   4

Figure 2.2

If $a < b$, the symbol (a, b) will sometimes be used for the set of all real numbers between a and b; that is,

$$(a, b) = \{x : a < x < b\}.$$

The set (a, b) is called an **open interval.** The graph of (a, b) consists of all points on a coordinate line which lie between the points corresponding to a and b. In Fig. 2.3 we have sketched the graph of a general open interval (a, b) and also the special open intervals $(-1, 3)$ and $(2, 4)$. The parentheses in the figure indicate that the endpoints of the intervals are not to be included.

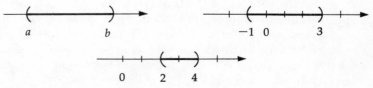

a b -1 0 3

0   2   4

Figure 2.3

If we wish to include the endpoints a bracket is used instead of a parenthesis. **Closed intervals,** denoted by $[a, b]$, and **half-open intervals,**
denoted by $[a, b)$ or $(a, b]$, are defined as follows:

$$[a, b] = \{x : a \leq x \leq b\}$$
$$[a, b) = \{x : a \leq x < b\}$$
$$(a, b] = \{x : a < x \leq b\}.$$

Typical graphs are sketched in Fig. 2.4. For convenience we shall use
the terms *interval* and *graph of an interval* interchangeably.

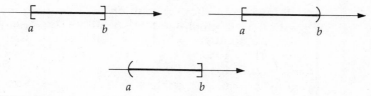

Figure 2.4

The solution set of an inequality may often be described in
terms of intervals. An illustration is given in the next example.

EXAMPLE 2 Solve the inequality $-6 \leq 2x - 4 < 2$ and sketch the
graph of the solution set.

Solution

A real number x is a solution of the given inequality if and
only if it is a solution of *both* of the inequalities

$$-6 \leq 2x - 4 \quad \text{and} \quad 2x - 4 < 2.$$

The first inequality is equivalent to each of the following:

$$-6 + 4 \leq (2x - 4) + 4$$
$$-2 \leq 2x$$
$$-1 \leq x.$$

Similarly, the second inequality is equivalent to each of the
following:

$$2x - 4 < 2$$
$$2x < 6$$
$$x < 3.$$

Thus x is a solution of the given inequality if and only if
both

$$-1 \leq x \quad \text{and} \quad x < 3,$$

79

that is,

$$-1 \le x < 3.$$

Hence the solution set is the half-open interval $[-1, 3)$. The graph is sketched in Fig. 2.5.

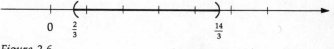

−1   0   3

Figure 2.5

An alternate (and shorter) solution may be given by working with both inequalities at the same time, as follows:

$$-6 \le 2x - 4 < 2$$
$$-6 + 4 \le 2x < 2 + 4$$
$$-2 \le 2x < 6$$
$$-1 \le x < 3.$$

The last inequality agrees with the first solution.

EXAMPLE 3 Solve the inequality $-5 < \dfrac{4 - 3x}{2} < 1$ and sketch the graph of the solution set.

Solution

A number x is a solution if and only if it satisfies both of the inequalities

$$-5 < \frac{4 - 3x}{2} \quad \text{and} \quad \frac{4 - 3x}{2} < 1.$$

We could work with each of these separately or simultaneously, as in the alternate solution of Example 2. Let us employ the latter technique as follows:

$$-5 < \frac{4 - 3x}{2} < 1 \qquad \text{(given)}$$
$$-10 < 4 - 3x < 2 \qquad \text{(3) of (1.12)}$$
$$-14 < -3x < -2 \qquad \text{(2) of (1.12)}$$
$$\tfrac{14}{3} > x > \tfrac{2}{3} \qquad \text{(4) of (1.12)}$$
$$\tfrac{2}{3} < x < \tfrac{14}{3} \qquad \text{(definition of >)}$$

Hence the solution set is the open interval $(\tfrac{2}{3}, \tfrac{14}{3})$. The graph is sketched in Fig. 2.6.

0     $\frac{2}{3}$ $\frac{14}{3}$

Figure 2.6

In order to describe solution sets of certain inequalities we must employ **infinite intervals.** In particular, if a is a real number we define

$$(-\infty, a) = \{x : x < a\}.$$

The symbol ∞ (**infinity**) is merely a notational device and is not to be interpreted as representing a real number. To illustrate, the solution set in Example 1 is the infinite interval $(-\infty, 4)$. The graph is sketched in Fig. 2.2. If we wish to include the point corresponding to a we write

$$(-\infty, a] = \{x : x \le a\}.$$

Other types of infinite intervals are defined by

$$(a, \infty) = \{x : x > a\}$$
$$[a, \infty) = \{x : x \ge a\}.$$

EXAMPLE 4 Solve $1/(x - 2) > 0$ and sketch the graph of the solution set.

Solution

Since the numerator is positive, the given fraction is positive if and only if $x - 2 > 0$ or equivalently, $x > 2$. Hence the solution set is the infinite interval $(2, \infty)$. The graph is sketched in Fig. 2.7.

Figure 2.7

Among the most important inequalities which occur in advanced mathematics are those involving absolute values. Formulas (1.18) and (1.20) are vital for solving such inequalities. For our present work the real number a in those formulas is replaced by an expression in one variable. The following theorem provides the forms of (1.18) and (1.20) which are needed for exercises in this section.

(2.6) Theorem

If d is a positive real number and if p is an algebraic expression, then

    (1)   the inequality $|p| < d$ is equivalent to $-d < p < d$;

    (2)   the solution set of $|p| > d$ is the union of the solution sets of $p > d$ and $p < -d$.

Since the theorem follows directly from our work in Chapter One, we shall omit the proof.

EXAMPLE 5 Solve $|x - 3| < 0.1$.

Solution

By (2.6) the given inequality is equivalent to

$$-0.1 < x - 3 < 0.1,$$

and hence to

$$-0.1 + 3 < (x - 3) + 3 < 0.1 + 3,$$

or

$$2.9 < x < 3.1.$$

Consequently the solution set is the open interval $(2.9, 3.1)$.

EXAMPLE 6 Solve $|2x + 3| > 9$ and sketch the graph of the solution set.

Solution

By (2.6) the solution set of the given inequality is the union of the solution sets of

$$2x + 3 > 9 \quad \text{and} \quad 2x + 3 < -9.$$

The first inequality is equivalent to $2x > 6$, or $x > 3$, and hence has solution set $(3, \infty)$. The second inequality is equivalent to $2x < -12$, or $x < -6$, and therefore has solution set $(-\infty, -6)$. Consequently the solution set of $|2x + 3| > 9$ is the following union of two infinite intervals:

$$(-\infty, -6) \cup (3, \infty).$$

The graph is sketched in Fig. 2.8.

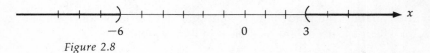

Figure 2.8

EXERCISES

Solve the following inequalities and sketch the graph of each solution set.

1 $3x + 2 < 5x - 8$.

2 $2 + 7x < 3x - 10$

3 $\frac{1}{4} - \frac{1}{3}x > \frac{1}{6} + \frac{1}{2}x$.

4 $4.2x + 0.3 \geq 2.7x - 7.2$

5 $5x + 1 \geq 2x - 6$.

6 $\frac{1}{2}x - \frac{3}{4} < 2 - \frac{1}{3}x$

7 $(x-3)(x-4) < (x+1)(x+3)$ **8** $(x-2)^2 > x(x+5)$

9 $12 \ge 5x - 3 > -7$ **10** $5 > 2 - 9x > -4$

11 $-1 < \dfrac{3 - 7x}{4} \le 6$ **12** $0 \le 4x - 1 \le 2$

13 $-2 \le 3x + 1 \le 7$ **14** $1 < 4 - 5x < 3$

15 $\dfrac{5}{7 - 2x} < 0$ **16** $\dfrac{3}{9 - 5x} > 0$

17 $\dfrac{4}{x^2 + 9} > 0$ **18** $\dfrac{10}{x^2 + 16} < 0$

19 $\dfrac{1}{2x + 3} > 0$ **20** $\dfrac{1}{2x + 3} < 0$

21 $|x - 10| < 0.3$ **22** $\left|\dfrac{2x + 3}{5}\right| < 2$

23 $\left|\dfrac{7 - 3x}{2}\right| \le 1$ **24** $|\tfrac{1}{4}x - \tfrac{2}{3}| \le \tfrac{1}{2}$

25 $|25x - 8| > 7$ **26** $|3 - 11x| \ge 41$

27 $|35x - 12| < 0$ **28** $|2x + 1| > 0$

29 $|2 - 3x| \ge 8$ **30** $|x - 5| < 0.001$

31 $2x + 3 < 5x + 6 < -3x + 7$ **32** $4x - 2 < x + 8 < 9x + 1$

33 $3 + 5x \ge 2x + 1 > 4x - 6$ **34** $1 - x > 4 - 3x > 7 + 2x$

35 $|x|(2x + 5) < 0$ **36** $|-x|(3x - 4) > 0$

6 MORE ON INEQUALITIES

Most of the inequalities considered in the preceding section were equivalent to inequalities containing only first degree polynomials. In this section we shall investigate inequalities involving polynomials of degree greater than 1.

EXAMPLE 1 Solve $2x^2 < x + 3$.

Solution

The following inequalities are equivalent.

$$2x^2 < x + 3$$
$$2x^2 - x - 3 < 0$$
$$(2x - 3)(x + 1) < 0$$

Consequently, x is a solution of the given inequality if and only if the product $(2x - 3)(x + 1)$ is negative. In order for the latter to occur, the factors $2x - 3$ and $x + 1$ must have opposite signs. Let us therefore examine the signs of each factor. Since $2x - 3 > 0$ if and only if $x > \frac{3}{2}$, the factor $2x - 3$ is positive whenever x is in the interval $(\frac{3}{2}, \infty)$. In like manner $2x - 3$ is negative if and only if x is in the interval $(-\infty, \frac{3}{2})$. Similarly, $x + 1 > 0$ if and only if $x > -1$, and hence $x + 1$ is positive for all x in $(-1, \infty)$. On the other hand $x + 1$ is negative whenever x is in $(-\infty, -1)$. It is convenient to use the graphical technique shown in Fig. 2.9 to display the intervals in which the factors $2x - 3$ and $x + 1$ are positive or negative. Referring to the figure we see that these factors have opposite signs, and hence the product is negative, whenever x is in the open interval $(-1, \frac{3}{2})$. This is the desired solution.

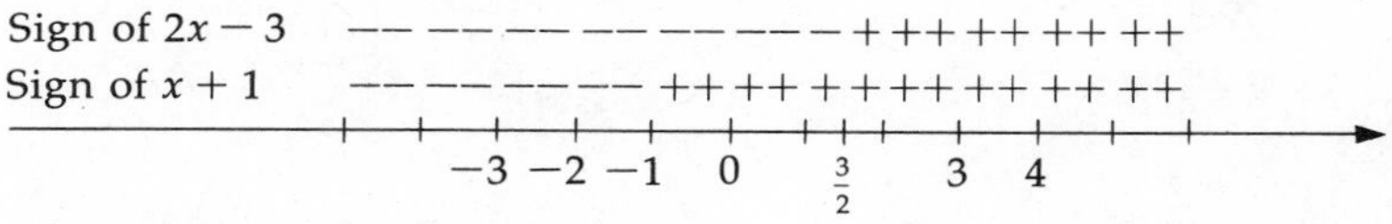

Sign of $2x - 3$ $\quad$ $- - \ - - \ - - \ - - \ - - \ - - \ - + + + + + + + + + +$
Sign of $x + 1$ $\quad$ $- - \ - - \ - - - - + + + + + + + + + + + + + + +$

$$-3 \ -2 \ -1 \quad 0 \qquad \tfrac{3}{2} \quad 3 \quad 4$$

Figure 2.9

EXAMPLE 2 Solve the inequality $x^2 - 7x + 10 > 0$ and sketch the graph of the solution set.

Solution

Since the given inequality may be written

$$(x - 5)(x - 2) > 0,$$

it follows that x is a solution if and only if both factors $x - 5$ and $x - 2$ are positive, or both are negative. The diagram in Fig. 2.10 indicates the signs of these factors for various real numbers. Evidently, both factors are positive if x is in the interval $(5, \infty)$, and both are negative if x is in $(-\infty, 2)$. Hence the solution set is $(-\infty, 2) \cup (5, \infty)$. The graph is sketched in Fig. 2.10.

Sign of $x - 5$ $\quad$ $- - \ - - \ - - \ - - \ - - + + + + + + + +$
Sign of $x - 2$ $\quad$ $- - \ - - \ -- \ + + + + + + + + + + + + + +$

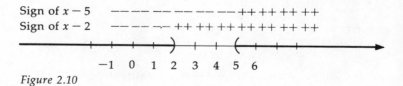

$$-1 \quad 0 \quad 1 \quad 2 \quad 3 \quad 4 \quad 5 \quad 6$$

Figure 2.10

EXAMPLE 3 Solve $\dfrac{x + 1}{x + 2} \le 3$ and sketch the graph of the solution set.

Solution

The following inequalities are equivalent.

$$\frac{x+1}{x+2} \le 3$$

$$\frac{x+1}{x+2} - 3 \le 0$$

$$\frac{x+1-3x-6}{x+2} \le 0$$

$$\frac{-2x-5}{x+2} \le 0$$

$$\frac{2x+5}{x+2} \ge 0.$$

The fraction $\frac{2x+5}{x+2}$ equals zero if $2x+5=0$; that is, if $x=-\frac{5}{2}$. Hence $-\frac{5}{2}$ is in the solution set. It remains to determine the solutions of the inequality

$$\frac{2x+5}{x+2} > 0.$$

Since a quotient is positive if and only if the numerator and denominator have the same signs, we must determine where both $2x+5$ and $x+2$ are positive or where both are negative. The diagram in Fig. 2.11 shows that this occurs in

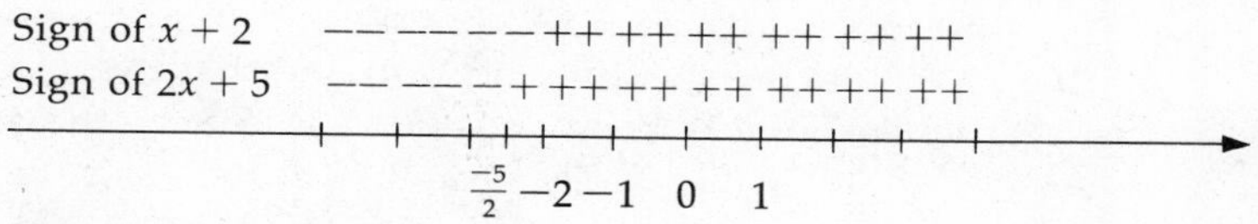

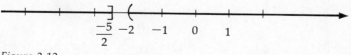

Sign of $x+2$ — — — — — ++ ++ ++ ++ ++ ++

Sign of $2x+5$ — — — — + ++ ++ ++ ++ ++ ++

$$\frac{-5}{2} \quad -2 \; -1 \quad 0 \quad 1$$

Figure 2.11

either of the intervals $(-\infty, -\frac{5}{2})$ or $(-2, \infty)$. Since $-\frac{5}{2}$ is also a solution, it follows that the solution set of the given inequality is the union

$$(-\infty, -\tfrac{5}{2}] \cup (-2, \infty).$$

The graph is sketched in Fig. 2.12.

$$\frac{-5}{2} \quad -2 \quad -1 \quad 0 \quad 1$$

Figure 2.12

If we agree that -2 is not in the domain of x, then another method of solution is to multiply both sides of the given inequality by $(x+2)^2$. This is permissible since $(x+2)^2$ is always positive if $x \neq -2$. This leads to the following chain of equivalent inequalities (if $x \neq -2$).

$$\left(\frac{x+1}{x+2}\right)(x+2)^2 \leq 3(x+2)^2$$

$$(x+1)(x+2) \leq 3x^2 + 12x + 12$$
$$0 \leq 2x^2 + 9x + 10$$
$$0 \leq (2x+5)(x+2).$$

We could now proceed as before. It is important to observe that it is *not* permissible to multiply both sides of the inequality by $x + 2$. The student may find it instructive to determine what would happen if this were done.

EXAMPLE 4 Solve the inequality $(x + 2)(x - 1)(x - 5) > 0$ and sketch the graph of the solution set.

Solution

If a number is substituted for x in $(x + 2)(x - 1)(x - 5)$, the result will be positive provided all three factors are positive or provided two of the factors are negative and one is positive. As in the previous sections we find that $x + 2$ is negative in $(-\infty, -2)$ and positive in $(-2, \infty)$. Similarly, the factor $x - 1$ is negative in $(-\infty, 1)$ and positive in $(1, \infty)$. Finally, $x - 5$ is negative in $(-\infty, 5)$ and positive in $(5, \infty)$. These facts are displayed in Fig. 2.13. We see from this diagram that all factors are positive in $(5, \infty)$ and that two factors are negative and one is positive in $(-2, 1)$. Hence the solution set is $(-2, 1) \cup (5, \infty)$. The graph is sketched in Fig. 2.14.

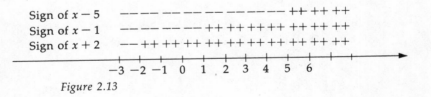

Sign of $x - 5$
Sign of $x - 1$
Sign of $x + 2$

Figure 2.13

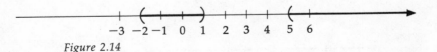

Figure 2.14

EXAMPLE 5 Solve $|x^2 - 5| < 4$ and sketch the graph of the solution set.

Solution

The following inequalities are equivalent by (1.18) and (2) of (1.12).

$$|x^2 - 5| < 4$$
$$-4 < x^2 - 5 < 4,$$
$$1 < x^2 < 9.$$

Hence the solution set is the intersection of the solution sets of

$$x^2 > 1 \quad \text{and} \quad x^2 < 9.$$

Taking square roots we obtain

$$\sqrt{x^2} > 1 \quad \text{and} \quad \sqrt{x^2} < 3,$$

which, by (1.27) are equivalent to

$$|x| > 1 \quad \text{and} \quad |x| < 3$$

respectively. Applying (1.18) and (1.20), the solution sets of the latter inequalities are

$$\{x : x > 1 \text{ or } x < -1\} \quad \text{and} \quad \{x : -3 < x < 3\},$$

respectively. The graph of each of these is sketched in Fig. 2.15. We see from this diagram that the intersection of the solution sets is $(-3, -1) \cup (1, 3)$, as shown in Fig. 2.16.

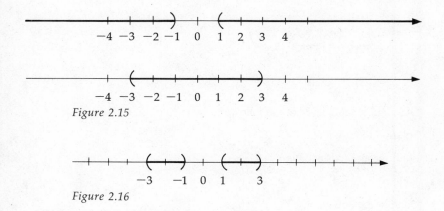

Figure 2.15

Figure 2.16

EXERCISES

Solve the following inequalities and sketch the graph of each solution set.

1. $3x^2 + 5x - 2 < 0$ 2. $2x^2 - 9x + 7 < 0$

3. $x^2 \leq 13x$ 4. $x^2 - 10x \leq 200$

5. $2x^2 + 9x + 4 > 0$ 6. $x^2 > 13x - 42$

7 $x^2 > 10x - 25$

8 $5 + \sqrt{x} < 1$

9 $x^2 > x$

10 $x^3 > x^2$

11 $\dfrac{3x + 2}{2x - 7} \leq 0$

12 $\dfrac{3}{x - 9} > \dfrac{2}{x + 2}$

13 $\dfrac{-x}{x^2 - x - 12} \geq 0$

14 $x^2 + 9 > 6x$

15 $\dfrac{x + 2}{2x - 1} < 5$

16 $\dfrac{2x - 7}{3x + 2} \geq 1$

17 $(x + 1)^2 \geq 0$

18 $\dfrac{(x + 1)^2}{(2x - 3)^3} \leq 0$

19 $x^3 - x^2 \geq 2x$

20 $(x + 4)^2 > x^2 + 16$

21 $\dfrac{1}{|x^2 - 5|} > 1$

22 $(x + 2)^2(3x - 4) < 0$

23 $(x - 9)^2(2x + 15)^3 < 0$

24 $x^3 + 7x^2 + 6x < 0$

25 $x^4 + 2x^2 + 3 \leq 0$

26 $x^4 - 13x^2 + 36 \geq 0$

27 $|x^2 - 4| < 3$

28 $|x^2 - 4| < 5$

29 $|x^2 - 2| > 1$

30 $|x^2 - 1| > 2$

31 $\dfrac{x(x - 1)(x - 2)}{(x + 1)(x + 2)} < 0$

32 $\dfrac{(x - 5)(x + 4)(x - 6)}{(x + 2)(x - 3)} \geq 0$

7 REVIEW EXERCISES

Oral

Define or discuss each of the following.

1 Solution of an equation

2 Solution set of an equation

3 Identity

4 Conditional equation

5 Equivalent equations

6 Linear equation

7 Quadratic equation

8 The quadratic formula

9 Intervals

10 Solution set of an inequality

Find the solution sets in Exercises 1–34.

1 $(3x + 1)(2x + 5) = (1 - 6x)(4 - x)$

2 $\dfrac{4}{2x - 4} + \dfrac{3x - 14}{x^2 - 4} = \dfrac{5}{x + 2}$

3 $\dfrac{3}{3x - 4} + \dfrac{5}{9x^2 - 16} = \dfrac{4}{3x + 4}$ **4** $\dfrac{6y}{3y - 1} = \dfrac{2}{3y - 1}$

5 $4x^3 = 2x - 7x^2$ **6** $z^4 = 13z^2 - 36$

7 $\dfrac{x + 3}{x + 2} = \dfrac{x + 2}{2x - 1}$ **8** $(x - 3)(x + 1) = 1$

9 $\sqrt{2x - 3} - \sqrt{x + 2} = 1$ **10** $\sqrt[3]{x + 1} = \sqrt[6]{3x + 7}$

11 $x^{2/3} + 2x^{1/3} - 3 = 0$ **12** $\sqrt{(2x - 2)^2} = 2|x - 1|$

13 $4 - 3x > 7 + 2x$ **14** $\dfrac{7}{2} > \dfrac{1 - 4x}{5} > \dfrac{3}{2}$

15 $|2x - 7| \le 0.01$ **16** $|6x - 7| > 1$

17 $2x^2 < 5x - 3$ **18** $\dfrac{2x^2 - 3x - 20}{x + 3} < 0$

19 $\dfrac{1}{3x - 1} < \dfrac{2}{x + 5}$ **20** $x^2 + 4 \ge 4x$

21 $\tfrac{2}{3}y - 5 = 4 + \tfrac{1}{2}y$ **22** $\dfrac{1}{h^2} + \dfrac{2}{h} = 3$

23 $1 - 6a - 2a^2 = 0$ **24** $2y^3 - y^2 - 10y + 5 = 0$

25 $\left|\dfrac{4x + 1}{8}\right| < \dfrac{1}{2}$ **26** $|x - \sqrt{3}| < 0$

27 $\dfrac{2}{x^3} < \dfrac{1}{x^2}$ **28** $15x^{-2} + 7x^{-1} - 4 = 0$

29 $\dfrac{2}{(x - 1)^2} + \dfrac{5}{(x - 1)} - 3 = 0$ **30** $\sqrt{3x + 2} - 5 = 0$

31 $x - \sqrt{2x - 5} = 4$ **32** $|7 - 3x| = 2$

33 $|2x + 1| = |3 - x|$ **34** $|x^2 - 2| < 4$

35 A motorist averaged 50 miles per hour driving outside the city limits and 30 miles per hour within the city limits. If a 70-mile trip took him 2 hours, how much of the time was he outside the city limits?

36 The sum of the digits of a two-digit number is 16. If the digits are reversed, the resulting number is 18 more than the given number. Find the given number.

37 The sum of two positive integers is 47 and their product is 532. Find the numbers.

38   An airplane flying south at the rate of 300 miles per hour passed over a point on the ground at 1:00 P.M. Three minutes later a second plane flying west at the rate of 300 miles per hour passed over the same point at the same altitude. At what time after 1:00 P.M. will the planes be 25 miles apart?

39   Sixty pounds of a solution of salt water is 10% salt. How much water must be evaporated in order that the resulting mixture will be 15% salt?

40   A chemist has two acid solutions, the first containing 15% acid and the second 40% acid. How many ounces of each should be mixed in order to obtain 16 ounces of solution containing 20% acid?

41   A projectile is fired straight upward from a height of 200 feet above the ground with an initial velocity of 800 feet per second. The distance s (in feet) above the ground after t seconds is given by $s = -16t^2 + 800t + 200$. When will the projectile be 2000 feet above the ground? When will it hit the ground?

42   If one solution of $x^2 + kx + 3 = 0$ is $\sqrt{2}$, find the other solution.

In Exercises 43–46 solve for the indicated variable in terms of the remaining variables:

43 $V = \pi R^2 h - \pi r^2 h$ for h $\qquad$ **44** $S = \dfrac{a - rl}{1 - r}$ for r

45 $s = \frac{1}{2}gt^2 + v_0 t$ for t $\qquad$ **46** $\dfrac{1}{R} = \dfrac{1}{R_1} + \dfrac{1}{R_2}$ for R_1

3

Functions and Graphs

One of the most useful concepts in mathematics is that of function. Indeed, it is safe to say that without the notion of function, little progress could be made in mathematics or in any area of science. In the first two sections of this chapter we shall consider several concepts which are useful in the study of functions. The remainder of the chapter contains a discussion of functions and graphs.

1 COORDINATE SYSTEMS IN TWO DIMENSIONS

In Section 3 of Chapter One we indicated how coordinates may be assigned to points on a straight line. Coordinate systems can also be introduced in planes by using the notion of **ordered pair.** As mentioned earlier, if a and b are elements of a set, then there is no difference between $\{a, b\}$ and $\{b, a\}$. However, an ordered pair is not merely a set. It consists of two elements a and b in which one of the elements is designated the "first" element and the other the "second" element. The symbol (a, b) is used to denote the ordered pair consisting of the elements a and b with first element a and second element b. We consider two ordered pairs (a, b) and (c, d) equal, and write $(a, b) = (c, d)$, if and only if $a = c$ and $b = d$. In particular this implies that $(a, b) \neq (b, a)$ whenever $a \neq b$.

There are many uses for ordered pairs. They were used in Chapter 2 to denote open intervals. In this chapter we shall use ordered pairs to represent points in a plane. Although ordered pairs are employed in different situations, there is little chance for confusion, since it should always be clear from the discussion whether the symbol (a, b) represents an interval, a point, or some other mathematical object.

A **rectangular,** or **Cartesian,** * **coordinate system** may be introduced in a plane by considering two perpendicular coordinate lines in the plane which intersect in the origin O on each line. Unless specified otherwise, the same unit of length is chosen on each line. Usually one of the lines is horizontal with positive direction to the right, and the other line is vertical with positive direction upward, as indicated by the arrowheads in Fig. 3.1. The two lines are called **coordinate axes** and the point O is called the **origin.** The horizontal line is often referred to as the **x-axis** and the vertical line as the **y-axis,** and they are labeled x and y, respectively. The plane is then called a **coordinate plane** or, with the preceding notation for coordinate axes, the **xy-plane.** Although the symbols x and y are used to denote lines as well as numbers, there should be no misunderstanding as to what these letters represent when they appear alongside of coordinate lines as in Fig. 3.1. In certain applications different labels such as d, t, etc., are used for the coordinate lines. The coordinate axes divide the plane into four parts called the **first, second, third,** and **fourth quadrants** and labeled I, II, III, and IV, respectively, as in Fig. 3.1.

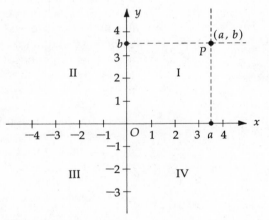

Figure 3.1

Each point P in the xy-plane may be assigned a unique ordered pair of real numbers. If vertical and horizontal lines through P intersect the x and y-axes at points with coordinates a and b respectively (see Fig. 3.1), then P is assigned the ordered pair (a, b). The number a is called the **x-coordinate** (or **abscissa**), of P, and the number b is called the **y-coordinate** (or **ordinate**), of P. We sometimes say that P has *coordinates* (a, b). Conversely, every ordered pair (a, b) of real numbers determines

---

* The term "Cartesian" is used in honor of the French mathematician and philosopher René Descartes (1596–1650), who was one of the first to employ such coordinate systems.

a point P in the xy-plane with coordinates a and b. Specifically, P is the point of intersection of lines perpendicular to the x-axis and y-axis at the points having coordinates a and b respectively. This establishes a one-to-one correspondence between points in the xy-plane and the set of all ordered pairs of real numbers. It is sometimes convenient to refer to the *point* (a, b) meaning the point with abscissa a and ordinate b. The symbol $P(a, b)$ will denote the point P with coordinates (a, b). To *plot* a point $P(a, b)$ means to locate, in a coordinate plane, the point P with coordinates (a, b). This point is represented by a dot in the appropriate position, as illustrated in Fig. 3.2.

The coordinates a and b of a point $P(a, b)$ are *directed* distances. For example if a is positive, then we know that P lies to the right of the y-axis; if a is negative, then P is to the left of the y-axis. Note that abscissas are positive for points in quadrants I or IV and negative for points in quadrants II or III. Ordinates are positive for points in quadrants I or II and negative for points in quadrants III or IV. Some typical points in a coordinate plane are illustrated in Fig. 3.2.

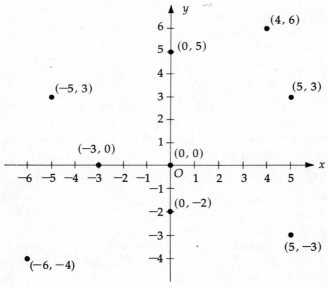

Figure 3.2

We shall now derive a formula for finding the distance between any two points in a coordinate plane. The distance between two points P and Q will be denoted by $d(P, Q)$. If $P = Q$, then we agree that $d(P, Q) = 0$, whereas if $P \neq Q$ the distance is considered positive. Thus the number $d(P, Q)$ is *not* a directed distance and so we may write $d(P, Q) = d(Q, P)$.

Let us consider any two points $P_1(x_1, y_1)$ and $P_2(x_2, y_2)$. If the points lie on the same horizontal line then $y_1 = y_2$, and we may denote the points by $P_1(x_1, y_1)$ and $P_2(x_2, y_1)$. If lines through P_1 and P_2 parallel to the y-axis intersect the x-axis at $A_1(x_1, 0)$ and $A_2(x_2, 0)$, as shown in (i) of Fig. 3.3, then $d(P_1, P_2) = d(A_1, A_2)$. However, by (1.15), $d(A_1, A_2) = |x_2 - x_1|$ and hence

(3.1) $d(P_1, P_2) = |x_2 - x_1|.$

Since $|x_2 - x_1| = |x_1 - x_2|$, formula (3.1) is valid whether P_1 lies to the left of P_2 or to the right of P_2. Moreover, the formula is independent of the quadrants in which the points lie.

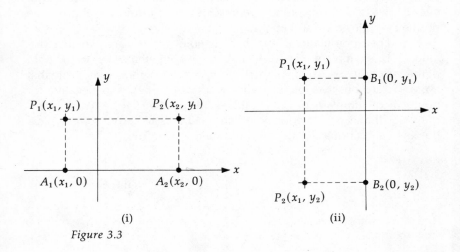

(i)   (ii)

Figure 3.3

In similar fashion, if P_1 and P_2 are on the same vertical line, then $x_1 = x_2$ and we may denote the points by $P_1(x_1, y_1)$ and $P_2(x_1, y_2)$. If we consider the points $B_1(0, y_1)$ and $B_2(0, y_2)$ on the y-axis as shown in (ii) of Fig. 3.3, then

(3.2) $d(P_1, P_2) = d(B_1, B_2) = |y_2 - y_1|.$

Finally, let us consider the general case, in which the points $P_1(x_1, y_1)$ and $P_2(x_2, y_2)$ do not lie on the same horizontal or vertical line. The line through $P_1(x_1, y_1)$ parallel to the x-axis and the line through $P_2(x_2, y_2)$ parallel to the y-axis intersect at some point P_3. Since P_3 has the same y-coordinate as P_1 and the same x-coordinate as P_2, we can denote it by $P_3(x_2, y_1)$ (see Fig. 3.4). From the previous discussion $d(P_1, P_3) = |x_2 - x_1|$ and $d(P_3, P_2) = |y_2 - y_1|$. Since P_1, P_2, and P_3

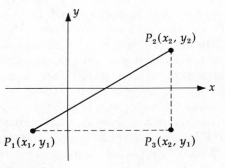

Figure 3.4

form a right triangle with hypotenuse from P_1 to P_2 we have, by the Pythagorean Theorem,

$$[d(P_1, P_2)]^2 = [d(P_1, P_3)]^2 + [d(P_3, P_2)]^2.$$

Applying (3.1) and (3.2) gives us

$$[d(P_1, P_2)]^2 = |x_2 - x_1|^2 + |y_2 - y_1|^2.$$

Using the fact that $d(P_1, P_2)$ is nonnegative and that $|a|^2 = a^2$ for every real number a, we obtain the following formula.

(3.3) Distance Formula

If $P_1(x_1, y_1)$ and $P_2(x_2, y_2)$ are in a coordinate plane, then the **distance between P_1 and P_2** is given by

$$d(P_1, P_2) = \sqrt{(x_2 - x_1)^2 + (y_2 - y_1)^2}.$$

Although we referred to the special case indicated in Fig. 3.4, the argument used in the proof of the distance formula is independent of the positions of P_1 and P_2.

EXAMPLE 1 Prove that the triangle with vertices $A(-1, -3)$, $B(6, 1)$, and $C(2, -5)$ is a right triangle.

Solution

By the distance formula,

$$d(A, B) = \sqrt{(-1 - 6)^2 + (-3 - 1)^2}$$
$$= \sqrt{49 + 16} = \sqrt{65},$$

$$d(B, C) = \sqrt{(6 - 2)^2 + (1 + 5)^2}$$
$$= \sqrt{16 + 36} = \sqrt{52},$$

$$d(A, C) = \sqrt{(-1 - 2)^2 + (-3 + 5)^2}$$
$$= \sqrt{9 + 4} = \sqrt{13}.$$

Since $[d(A, B)]^2 = [d(B, C)]^2 + [d(A, C)]^2$, the triangle is a right triangle with hypotenuse joining A to B.

It is easy to obtain a formula for the midpoint of a line segment. Let $P_1(x_1, y_1)$ and $P_2(x_2, y_2)$ be two points in a coordinate plane and let M be the midpoint of the segment from P_1 to P_2. The lines through P_1 and P_2 parallel to the y-axis intersect the x-axis at $A_1(x_1, 0)$ and $A_2(x_2, 0)$, respectively (see Fig. 3.5). By (1.17) the midpoint M_1 of the segment from A_1 to A_2 is $((x_1 + x_2)/2, 0)$. It follows from plane geometry that the abscissa of M is $(x_1 + x_2)/2$. It can be shown in similar fashion that the ordinate of M is $(y_1 + y_2)/2$. This establishes the following formula:

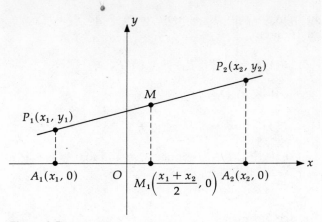

Figure 3.5

(3.4)   Midpoint Formula

The midpoint of the line segment from $P_1(x_1, y_1)$ to $P_2(x_2, y_2)$ is

$$\left(\frac{x_1 + x_2}{2}, \frac{y_1 + y_2}{2}\right).$$

EXAMPLE 2   Find the midpoint of the segment from $P_1(-2, 3)$ to $P_2(4, -2)$.

Solution

By (3.4) the coordinates of the midpoint are

$$\left(\frac{-2 + 4}{2}, \frac{3 + (-2)}{2}\right), \quad \text{or} \quad \left(1, \frac{1}{2}\right).$$

(Check by plotting points.)

EXERCISES

1   Plot the following points on a rectangular coordinate system: $A(4, -1)$, $B(-4, 1)$, $C(-4, -1)$, $D(-3, 0)$, $E(0, -3)$, $F(-\sqrt{3}, \pi)$.

2   Plot the following points on a rectangular coordinate system: $A(-2, 3)$, $B(2, -3)$, $C(-2, -3)$, $D(0, -5)$, $E(-3/2, 0)$, $F(\pi, -\sqrt{2})$.

3   Plot $A(0, 0)$, $B(1, 1)$, $C(4, 4)$, $D(-3, -3)$. Describe the set of points $\{(x, x) : x \in \mathbf{R}\}$.

4   Plot: $A(0, 0)$, $B(1, -1)$, $C(3, -3)$, $D(-\frac{5}{2}, \frac{5}{2})$. Describe the set of points $\{(x, -x) : x \in \mathbf{R}\}$.

5 Describe the set of all points $P(x, y)$ in a coordinate plane such that:

    a $x = 5$ **b** $y = -2$ **c** $xy < 0$ **d** $xy = 0$

6 Describe the set of all points $P(x, y)$ in a coordinate plane such that

    a $x = 0$ **b** $y = 100$ **c** $\dfrac{x}{y} > 0$ **d** $xy \neq 0$

In Exercises 7–12, find (a) the distance $d(A, B)$ between the given points A and B; (b) the midpoint of the segment AB.

7 $A(6, -2)$, $B(2, 1)$ **8** $A(-4, -1)$, $B(2, 3)$

9 $A(0, -7)$, $B(-1, -2)$ **10** $A(4, 5)$, $B(4, -4)$

11 $A(-3, -2)$, $B(-8, -2)$ **12** $A(11, -7)$, $B(-9, 0)$

In Exercises 13 and 14 prove that the triangle with the indicated vertices is a right triangle and find its area.

13 $A(-3, 4)$, $B(2, -1)$, $C(9, 6)$

14 $A(7, 2)$, $B(-4, 0)$, $C(4, 6)$

15 Prove that the following are vertices of a square: $A(-2, -2)$, $B(11, -9)$, $C(8, 1)$, $D(1, -12)$.

16 Prove that the following points are vertices of a parallelogram: $A(-9, -5)$, $B(2, 4)$, $C(-3, -3)$, $D(-4, 2)$.

17 Given $A(4, -9)$, find the coordinates of the point B such that $M(-3, 19)$ is the midpoint of AB.

18 Given $A(2, -5)$ and $B(-8, 11)$, find the point on AB that is one-fourth of the way from A to B.

19 Given $A(6, 4)$ and $B(-2, 2)$, prove that $P(5, -9)$ is on the perpendicular bisector of AB.

20 If A and B are as in Exercise 19, find a formula which expresses the fact that $P(x, y)$ is on the perpendicular bisector of AB.

21 Find a formula which expresses the fact that $P(x, y)$ is 8 units from the origin. Describe the totality of all such points.

22 If r is a positive real number, find a formula that states that $P(x, y)$ is r units from a fixed point (h, k). Describe the totality of all such points.

23 Find all points on the x-axis that are 5 units from the point $(2, 3)$.

24 Find all points on the y-axis that are 8 units from $(3, 5)$.

25 Let S be the set of points $\{(x, x) : x \in \mathbf{R}\}$. Find the points in S that are 5 units from $(2, 4)$.

26 Let S be the set of points $\{(x, 2x) : x \in \mathbf{R}\}$. Find the point in S that lies in the first quadrant and is 8 units from $(1, 2)$.

27 For what values of a is the distance between $(a, 3)$ and $(5, 2a)$ less than $\sqrt{26}$?

28 Given the points $A(-1, 0)$ and $B(1, 0)$, find a formula not containing radicals that expresses the fact that the sum of the distances from $P(x, y)$ to A and to B, respectively, is 4.

29 Prove that the midpoint of the hypotenuse of any right triangle is equidistant from the vertices. (*Hint:* Label the vertices of the triangle $O(0, 0)$, $A(a, 0)$, and $B(0, b)$.)

30 Prove that the diagonals of any parallelogram bisect each other. (*Hint:* Label three of the vertices of the parallelogram $O(0, 0)$, $A(a, b)$, and $C(0, c)$.)

2 RELATIONS AND GRAPHS

If S and T are sets then the totality of ordered pairs (s, t) obtained by letting s range through all elements of S and letting t range through all elements of T is called the **product set** of S and T, and is denoted by $S \times T$. Thus, by definition,

$$S \times T = \{(s, t) : s \in S, t \in T\}.$$

EXAMPLE 1 Given $A = \{a, b, c\}$ and $B = \{c, d\}$, find $A \times B$, $B \times A$, and $B \times B$.

Solution

Using the definition of product set we obtain

$A \times B = \{(a, c), (b, c), (c, c), (a, d), (b, d), (c, d)\}$,
$B \times A = \{(c, a), (c, b), (c, c), (d, a), (d, b), (d, c)\}$,
$B \times B = \{(c, c), (c, d), (d, c), (d, d)\}$.

Note that in Example 1, $A \times B \neq B \times A$. Also, as illustrated by the third part of Example 1, the sets in the product may be identical. We shall often work with subsets of product sets. Such subsets are given a special name in the next definition.

(3.5)   Definition

A **relation** between two sets S and T is any subset of $S \times T$.

If W is a relation between S and T, that is, if $W \subseteq S \times T$, and if an ordered pair (s, t) is in W we say that **s is related to t.** As an illustration let S denote the set of all points in a plane, T the set of all lines, and let W denote the subset of $S \times T$ defined by

$$W = \{(s, t) : s \text{ lies on } t\}.$$

Thus the point s is related to the line t if and only if s lies on t.

If $S = T$ in (3.5), it is customary to use the phrase *relation on S* in place of *relation between S and S*. For example, if W is the subset of $\mathbf{R} \times \mathbf{R}$ defined by

$$W = \{(a, b) : a < b\},$$

then W is a relation on $\mathbf{R}$ and a real number a is related to a real number b if and only if a is less than b. Thus 2 is related to 5, -3 is related to 2, $\sqrt{2}$ is related to 1.5 and so on. We sometimes refer to W as the **less than relation** on $\mathbf{R}$.

The following discussion concerns relations on $\mathbf{R}$, that is, subsets of $\mathbf{R} \times \mathbf{R}$. Since a relation W on $\mathbf{R}$ is a set of ordered pairs of real numbers, we may speak of the point $P(x, y)$ in a coordinate plane that corresponds to the ordered pair (x, y) in W. This leads to the next definition.

(3.6) Definition

If W is a relation on $\mathbf{R}$, then the **graph** of W is the set of all points in a coordinate plane that correspond to the ordered pairs in W.

In order to simplify statements we shall not always include the phrase *relation on* $\mathbf{R}$ but shall merely specify a relation by defining a certain subset of $\mathbf{R} \times \mathbf{R}$. This is illustrated in the next example.

EXAMPLE 2 Describe the graph of

$$W = \{(x, y) : -1 < x < 4, 2 < y < 3\}.$$

Solution

A point $P(x, y)$ is on the graph of W if and only if the abscissa x is between -1 and 4 and the ordinate y is between 2 and 3. Hence the graph of W consists of all points inside the shaded rectangle in Fig. 3.6.

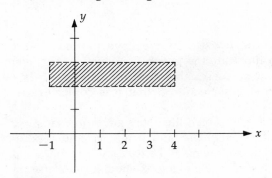

Figure 3.6

The phrase *sketch the graph of a relation* means to illustrate the significant features of the graph geometrically on a coordinate plane. This was done in Example 2 by shading a rectangular region. Graphs of relations sometimes consist of lines, circles, or other curves. An illustration of this is given in the next example.

EXAMPLE 3   Sketch the graph of $W = \{(x, y) : y = 2x - 1\}$.

Solution

We begin by finding points with coordinates of the form (x, y) where $(x, y) \in W$. It is convenient to list these coordinates in tabular form as shown below, where for each real number x the corresponding value for y is $2x - 1$.

x	-2	-1	0	1	2	3
y	-5	-3	-1	1	3	5

After plotting, it appears that the points with these coordinates all lie on a line and so we sketch the graph accordingly (see Fig. 3.7).

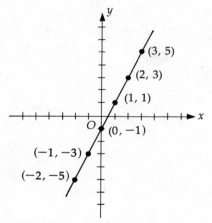

Figure 3.7

Ordinarily the few points we have plotted would not be enough to illustrate the graph; however, in this elementary case we can be reasonably sure that the graph is a line. It will be proved in Section 5 that this conjecture is correct.

It is impossible to sketch the entire graph in Example 3 since x may be assigned values which are numerically as large as desired. Nevertheless, we often call a drawing of the type given in Fig. 3.7 *the graph of the relation* or the *sketch of the graph*, where it is understood that the drawing is only a device for visualizing the actual graph and the line does not terminate as shown in the figure. In general, when sketching a graph one should illustrate enough of the graph so that the remaining parts are evident.

The relation in Example 3 is determined by the equation $y = 2x - 1$ in the sense that for any real number x, the equation can be used to find a number y such that $(x, y) \in W$. Given an equation involving x and y, we say that an ordered pair (a, b) is a **solution** of the equation if equality is obtained when a is substituted for x and b for y. For example, $(2, 3)$ is a solution of $y = 2x - 1$ since substitution of 2 for x and 3 for y leads to $3 = 4 - 1$, or $3 = 3$. The **solution set** of such an equation is the set of all solutions and hence is a relation on **R**. Two equations in x and y are said to be **equivalent** if they have the same solution sets.

The **graph of an equation** in x and y is defined as the graph of its solution set and the phrase *sketch the graph of an equation* means to sketch the graph of its solution set. Notice that the solutions of the equation $y = 2x - 1$ are the pairs (a, b) such that $b = 2a - 1$, and hence the solution set is identical with the relation W given in Example 3. Consequently the graph of the equation $y = 2x - 1$ is the same as the graph of W.

Given a geometric figure in a coordinate plane, it is sometimes possible to find its equation in the sense that the graph of the equation is the figure. We shall demonstrate how this can be accomplished for circles.

If $C(h, k)$ is a point in a coordinate plane, then a circle in the plane with center C and radius r may be defined as the collection of all points in the plane that are r units from C. If $P(x, y)$ is an arbitrary point in the plane, then as illustrated in Fig. 3.8, P is on the circle if and only if $d(C, P) = r$.

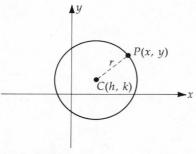

Figure 3.8

Using the distance formula (3.3) gives us the equation

$$\sqrt{(x - h)^2 + (y - k)^2} = r.$$

The equivalent equation

(3.7) $(x - h)^2 + (y - k)^2 = r^2, \quad r > 0$

is called the **standard equation of a circle of radius _r_ and center** (h, k).
If $h = 0$ and $k = 0$, then (3.7) reduces to

(3.8) $x^2 + y^2 = r^2$,

which is an equation of a circle of radius _r_ with center at the origin (see
Fig. 3.9). If $r = 1$, the graph of (3.8) is a **unit circle** with center at the origin. Note that a point $P(x, y)$ is on this unit circle if and only if
$x^2 + y^2 = 1$.

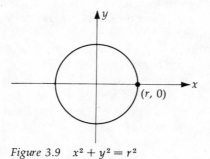

Figure 3.9 $x^2 + y^2 = r^2$

EXAMPLE 4   Find an equation of the circle with center $C(-2, 3)$ and
containing the point $D(4, 5)$.

Solution

The circle is illustrated in Fig. 3.10. Since D is on the circle,

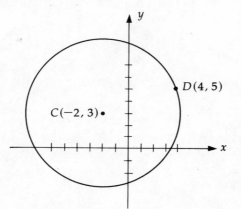

Figure 3.10

the radius r is $d(C, D)$. By the distance formula (3.3) we
have

$$r = d(C,D) = \sqrt{(-2 - 4)^2 + (3 - 5)^2}$$
$$= \sqrt{36 + 4} = \sqrt{40}.$$

Applying (3.7) with $h = -2$ and $k = 3$ we obtain

$$(x + 2)^2 + (y - 3)^2 = 40.$$

Squaring the indicated terms and simplifying gives us the equivalent equation

$$x^2 + y^2 + 4x - 6y - 27 = 0.$$

If we square the binomials in (3.7) and simplify, we obtain an equation of the form

(3.9) $x^2 + y^2 + ax + by + c = 0,$

where $a, b, c \in \mathbf{R}$. Conversely, if we *begin* with an equation of the form (3.9) it is always possible, by completing the squares in x and y, to obtain an equation of the form

$$(x - h)^2 + (y - k)^2 = s$$

where $h, k, s \in \mathbf{R}$. The method is illustrated below in Example 5. If $s > 0$, then the graph is a circle with center (h, k) and radius $r = \sqrt{s}$. If $s = 0$, then since $(x - h)^2 \geq 0$ and $(y - k)^2 \geq 0$, the only solution of the equation is (h, k) and hence the graph consists of only one point. Finally, if $s < 0$ the solution set of the equation is empty and there is no graph. This proves that *the graph of* (3.9) *is either a circle, a point, or the empty set.*

EXAMPLE 5   Find the center and radius of the circle with equation

$$x^2 + y^2 - 4x + 6y - 3 = 0.$$

Solution

We begin by arranging the given equation in the form

$$(x^2 - 4x) + (y^2 + 6y) = 3.$$

Next we complete the squares by adding appropriate numbers within the parentheses. Of course, to obtain an equivalent equation we must add the numbers to *both* sides of the equation. In order to complete the square for an expression of the form $x^2 + ax$ we *add the square of half the coefficient of* x, that is, $(a/2)^2$, to both sides of the equation. Similarly, for $y^2 + by$ we add $(b/2)^2$ to both sides. This leads to

$$(x^2 - 4x + 4) + (y^2 + 6y + 9) = 3 + 4 + 9,$$

or

$$(x - 2)^2 + (y + 3)^2 = 16.$$

By (3.7) the center is $(2, -3)$ and the radius is 4.

EXERCISES

1   If $S = \{1, 2, 3\}$ and $T = \{x, y\}$, list the elements of each of the following:
    a $S \times T$ **b** $T \times S$ **c** $S \times S$ **d** $T \times T$

2   **a** If S contains three elements and T contains two elements, how many elements are in the product set $S \times T$?

    b How many elements are in $S \times T$ if S contains four elements and T contains three elements?

    c Give a reason for calling $S \times T$ a *product* set.

In each of Exercises 3–10 sketch the graph of the given relation W on **R**.

3   $W = \{(x, y) : x = -5\}$ 4 $W = \{(x, y) : y = -x\}$

5   $W = \{(x, y) : y = x\}$ 6 $W = \{(x, y) : y < x\}$

7   $W = \{(x, y) : xy > 0\}$ 8 $W = \{(x, y) : |x| \le 1, |y| \ge 1\}$

9   $W = \{(x, y) : |x| < 1, |y| < 1\}$ 10 $W = \{(x, y) : y \le 0\}$

Sketch the graph of the equation in each of Exercises 11–18 after plotting a sufficient number of points.

11   $y = 3x$ 12 $y = -2x$

13   $y = 3x + 2$ 14 $y = -2x + 3$

15   $y = x^2$ 16 $y = -x^2$

17   $y = x^2 + 1$ 18 $y = -x^2 + 1$

Find an equation of a circle satisfying the stated condition in each of Exercises 19–26.

19   Center $C(-2, 4)$, radius 3

20   Center $C(-3, -1)$, radius $\sqrt{5}$

21   Center $C(\frac{3}{2}, -\frac{1}{2})$; radius 2

22   Center at the origin, passing through $P(-2, -3)$

23   Center $C(4, -5)$, passing through $P(1, 1)$

24   Center $C(-8, 5)$, tangent to the x-axis

25   Endpoints of diameter $A(2, 5)$, $B(6, -4)$

26   Circumscribed about the right triangle with vertices $A(2, 1)$, $B(-1, 4)$, and $C(-2, -3)$ (*Hint:* The center of the circle is the midpoint of the hypotenuse.)

In each of Exercises 27–32 sketch the graph of the circle which has the given equation.

27   $x^2 + y^2 = 4$ 28 $x^2 = 16 - y^2$

29 $(x + 5)^2 + (y - 2)^2 = 9$

30 $(x - 4)^2 + y^2 = 16$

31 $(x + 1)^2 + (y + 2)^2 = 2$

32 $4x^2 + 4y^2 = 1$

Sketch the graphs of the relations given in Exercises 33–36.

33 $\{(x, y) : x^2 + (y - 4)^2 \leq 25\}$

34 $\{(x, y) : x^2 + y^2 \geq 16\}$

35 $\{(x, y) : (x - 2)^2 + (y + 3)^2 > 1\}$

36 $\{(x, y) : (x - 5)^2 + (y + 1)^2 < 1\}$

In each of the following exercises, find the center and radius of the circle with the given equation.

37 $x^2 + y^2 + 4x - 6y + 4 = 0$

38 $x^2 + y^2 - 10x + 2y + 22 = 0$

39 $x^2 + y^2 + 6x = 0$

40 $x^2 + y^2 + x + y - 1 = 0$

41 $2x^2 + 2y^2 - x + y - 3 = 0$

42 $x^2 + y^2 + 8x - 12y + 52 = 0$

43 $9x^2 + 9y^2 - 6x + 12y - 31 = 0$

44 $x^2 + y^2 + 3y = 0$

3 FUNCTIONS

The notion of **correspondence** is encountered frequently in everyday life. For example, to each book in a library there corresponds the number of pages in the book. As another example, to each human being there corresponds a birth date. To cite a third example, if the temperature of the air is recorded throughout a day, then at each instant of time there is a corresponding temperature.

The examples of correspondences we have given involve two sets X and Y. In our first example, X denotes the set of books in a library and Y the set of positive integers. For each book x in X there corresponds a positive integer y, namely the number of pages in the book. In the second example, if we let X denote the set of all human beings and Y the set of all possible dates, then to each person x in X there corresponds a birth date y.

We sometimes represent correspondences pictorially by diagrams of the type shown in Fig. 3.11. The curved arrow indicates that the element y of Y corresponds to the element x of X. We have pictured X and Y as disjoint sets. However, X and Y may have elements in common. As a matter of fact, we often have $X = Y$.

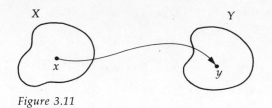

Figure 3.11

Our examples indicate that to each $x \in X$ there corresponds *one and only one* $y \in Y$, that is, *y is unique* for a given *x*. However, the same element of *Y* may correspond to different elements of *X*. For example, two different books may have the same number of pages, two different people may have the same birthday, and so on.

In most of our work *X* and *Y* will be sets of numbers. To illustrate, let *X* and *Y* both denote the set **R** of real numbers, and to each real number *x* let us assign its square x^2. Thus to 3 we assign 9, to -5 we assign 25, to $\sqrt{2}$ the number 2, and so on. This gives us a correspondence from **R** to **R**.

All the examples of correspondences we have given are *functions,* as defined in (3.10) below. It is customary (but not a definite requirement) to use letters near the middle of the alphabet such as *f*, *g*, and *h* to denote functions. Sometimes capital letters, *F*, *G*, *H*, and so on are also employed.

(3.10) Definition

A **function** *f* from a set *X* to a set *Y* is a correspondence that assigns to each element *x* of *X* a unique element *y* of *Y*. The element *y* is called the **image** of *x* under *f* and is denoted by $f(x)$. The set *X* is called the **domain** of the function. The **range** of the function is the set of all images of elements of *X*.

In (3.10) we introduced the notation $f(x)$ for the element which corresponds to *x*. This is usually read "*f* of *x*." We also call $f(x)$ the **value** of *f* at *x*. In terms of the pictorial representation given earlier, we may now sketch a diagram as in Fig. 3.12. The curved arrows indicate

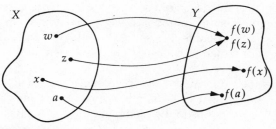

Figure 3.12

that the elements $f(x)$, $f(w)$, $f(z)$, and $f(a)$ of Y correspond to the elements x, w, z, and a of X. Let us repeat the important fact that to each element x of X there is assigned precisely one image $f(x)$ in Y; however, different elements of X such as w and z in Fig. 3.12 may have the same image in Y.

Beginning students are sometimes confused by the symbols f and $f(x)$. Remember that f is used to represent the function. It is neither an element of X nor an element of Y. However, $f(x)$ is an element of Y, namely the element which f assigns to x.

EXAMPLE 1 Let f be the function with domain $\mathbf{R}$ such that $f(x) = x^2$ for x in $\mathbf{R}$. Find $f(-6)$, $f(\sqrt{3})$, and $f(a)$, where $a \in \mathbf{R}$. What is the range of f?

Solution:

Values of f (or images under f) may be found by substituting for x in the equation $f(x) = x^2$. Thus

$$f(-6) = (-6)^2 = 36, \quad f(\sqrt{3}) = (\sqrt{3})^2 = 3, \quad \text{and} \quad f(a) = a^2.$$

If S denotes the range of f, then by definition

$$S = \{f(a) : a \in \mathbf{R}\} = \{a^2 : a \in \mathbf{R}\}.$$

Evidently the range of f is the set of all nonnegative real numbers.

If a function is defined as in the preceding example, the symbol used for the variable is immaterial; that is, expressions such as $f(x) = x^2$, $f(s) = s^2$, $f(t) = t^2$ and so on, all define the same function f. This is true because if a is any number in the domain of f, then the same image a^2 is obtained no matter which of those expressions is employed.

EXAMPLE 2 Let X denote the set of nonnegative real numbers and let f be the function from X to $\mathbf{R}$ defined by $f(x) = \sqrt{x} + 1$ for all x in X. Find $f(1)$, $f(5/2)$, $f(\sqrt[3]{2})$, and $f(b+c)$, where $b, c \in X$. What is the range of f?

Solution

As in Example 1, finding images under f is simply a matter of substituting the appropriate number for x in the expression for $f(x)$. Thus

$$
\begin{aligned}
f(1) &= \sqrt{1} + 1 = 2, \\
f(\tfrac{5}{2}) &= \sqrt{\tfrac{5}{2}} + 1 = \tfrac{1}{2}\sqrt{10} + 1, \\
f(\sqrt[3]{2}) &= \sqrt{\sqrt[3]{2}} + 1 = \sqrt[6]{2} + 1, \\
f(b+c) &= \sqrt{b+c} + 1.
\end{aligned}
$$

The range of f is

$$\{f(a) : a \geq 0\} = \{\sqrt{a} + 1 : a \geq 0\}.$$

It follows that the range of f is the set of all real numbers y such that $y \geq 1$.

In certain branches of mathematics such as calculus, it is important to carry out manipulations of the type given in the next example.

EXAMPLE 3   Suppose $f(x) = x^2 + 3x - 2$ for all real numbers x. If a, $t \in \mathbf{R}$, where $t \neq 0$, find

$$\frac{f(a+t) - f(a)}{t}.$$

Solution

We have

$$f(a + t) = (a + t)^2 + 3(a + t) - 2$$

and

$$f(a) = a^2 + 3a - 2.$$

Hence

$$\frac{f(a+t) - f(a)}{t}$$

$$= \frac{(a^2 + 2at + t^2) + (3a + 3t) - 2 - (a^2 + 3a - 2)}{t}$$

$$= \frac{2at + t^2 + 3t}{t}$$

$$= 2a + t + 3.$$

Occasionally one of the notations

$$X \xrightarrow{f} Y, \quad f : X \to Y, \quad \text{or} \quad f : x \to f(x)$$

is used to signify that f is a function from X to Y. It is not unusual in this event to say f *maps X into Y* or f *maps x into f(x)*. If f is the function in Example 1, then f maps x into x^2 and we may write $f : x \to x^2$.

Two functions f and g from X to Y are said to be *equal*, written $f = g$, provided $f(x) = g(x)$ for all $x \in X$. For example, if $g(x) = (\frac{1}{2})(2x^2 - 6) + 3$ and if $f(x) = x^2$, for all $x \in \mathbf{R}$, then $g = f$ (Why?).

Many formulas which occur in mathematics and the sciences determine functions. As an illustration, the formula $A = \pi r^2$ for the area

A of a circle of radius *r* assigns to each positive real number *r* a unique value of *A*. This determines a function *f*, where $f(r) = \pi r^2$, and we may write $A = f(r)$. The letter *r*, which represents an arbitrary number from the domain of *f*, is often called an **independent variable.** The letter *A*, which represents a number from the range of *f*, is called a **dependent variable,** since its value depends on the number assigned to *r*. When two variables *r* and *A* are related in this manner, it is customary to use the phrase *A is a function of r*. To cite another example, if an automobile travels at a uniform rate of 50 miles per hour, then the distance *d* (miles) traveled in time *t* (hours) is given by $d = 50t$ and hence the distance *d* is a function of time *t*.

We have seen that different elements in the domain of a function may have the same image. If a function *f* from *X* to *Y* has the property that whenever $a \neq b$ in *X*, then $f(a) \neq f(b)$ in *Y*, the function is called a **one-to-one function** (or **one-to-one correspondence**). In this case each element of the range is the image of precisely one element of *X*. The association between real numbers and points on a coordinate line is an example of a one-to-one correspondence. The function illustrated in Fig. 3.12 is *not* one-to-one since two different elements *w* and *z* of *X* have the same image in *Y*.

EXAMPLE 4 (a) If $f(x) = 3x + 2$, where $x \in \mathbf{R}$, prove that *f* is one-to-one. (b) If $g(x) = x^2 + 5$, where $x \in \mathbf{R}$, prove that *g* is not one-to-one.

Solution

(a) If $a \neq b$, then $3a \neq 3b$ and hence $3a + 2 \neq 3b + 2$, or $f(a) \neq f(b)$. Therefore *f* is one-to-one.

(b) The function *g* is not one-to-one since different numbers in the domain may have the same image. For example, although $-1 \neq 1$, $g(-1)$ and $g(1)$ are both equal to 6.

(3.11) Definition

The **identity function** *f* on a set *X* is defined by $f(x) = x$ for every $x \in X$.

Note that if *f* is the identity function on *X*, then every element *x* of *X* is mapped into itself.

(3.12) Definition

A function *f* from *X* to *Y* is a **constant function** if there is some (fixed) element *c* in *Y* such that $f(x) = c$ for every $x \in X$.

The diagram in Fig. 3.13 illustrates the fact that if *f* is a constant function, then every arrow from *X* terminates at the same element in *Y*.

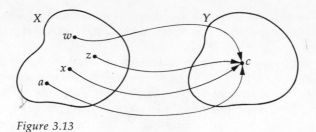

Figure 3.13

The concept of relation can be used to obtain an alternate approach to functions. We first observe that a function f from X to Y determines the following relation W between X and Y:

$$W = \{(x, f(x)) : x \in X\}$$

Thus W is the totality of ordered pairs for which the first element is in X and the second element is the image of the first element. In Example 2, W consists of all pairs of the form $(x, \sqrt{x} + 1)$, where x is a nonnegative real number. It is important to note that for each x there is exactly one pair (x, y) in W having x as its first element.

Conversely, if we begin with a relation W between X and Y such that each element of X appears *exactly once* as a first element of an ordered pair, then W determines a function from X to Y. Specifically, for any x in X there is a unique pair (x, y) in W, and by letting y correspond to x, we obtain a function from X to Y.

It follows from the preceding discussion that statement (3.13) below could also be used as a definition of function. We prefer, however, to think of it as an alternate approach to this concept and shall refer to it only sparingly in this book.

(3.13)   Alternate Definition of a Function

A **function** from a set X to a set Y is a relation W between X and Y such that for each x in X, there is exactly one ordered pair (x, y) in W having x as its first element.

In terms of (3.13), the function f of Example 1 is the relation $W = \{(x, x^2) : x \in \mathbf{R}\}$. Similarly, the relation $W = \{(x, x^2 + 3x - 2) : x \in \mathbf{R}\}$ determines the function of Example 3.

In the remainder of our work, unless specified otherwise, the phrase f *is a function* will mean that the domain and range are sets of real numbers. If a function is defined by means of some expression as in Examples 1–3, and the domain X is not stated explicitly, then X is considered to be the totality of real numbers for which the given expression is meaningful. To illustrate, if $f(x) = \sqrt{x}/(x - 1)$, then the domain is assumed to be the set of nonnegative real numbers different from 1. If x is in the domain we sometimes say that f **is defined at** x, or that $f(x)$ **exists.** If an interval I is contained in the domain we often say that f **is defined** on *I*. The terminology *f* **is undefined** at *x* means that *x* is not in the domain of *f*.

EXERCISES

1 If $f(x) = x^3 + 4x - 3$, find: **a** $f(1)$, **b** $f(-1)$, **c** $f(0)$, **d** $f(\sqrt{2})$.

2 Same as Exercise 1 if $f(x) = 2x^4 - 3x^2 + 5$.

3 If $f(x) = 3x^2 - x + 2$, and if $a, h \in \mathbf{R}$, find: **a** $f(a)$, **b** $f(-a)$, **c** $-f(a)$, **d** $f(a + h)$, **e** $f(a) + f(h)$, **f** $\dfrac{f(a + h) - f(a)}{h}$, $h \neq 0$.

4 Same as Exercise 3 if $f(x) = \dfrac{1}{x^2 + 1}$

5 If $g(x) = \dfrac{1}{x^2 + 4}$, find: **a** $g(1/a)$, **b** $1/g(a)$, **c** $g(a^2)$, **d** $(g(a))^2$, **e** $g(\sqrt{a})$, **f** $\sqrt{g(a)}$.

6 Same as Exercise 5 if $g(x) = 1/x$

7 If $f(x) = \sqrt{x + 5}$, what number maps into 4? If $a > 0$, what number maps into a? Find the range of f.

8 Same as Exercise 7 if $f(x) = x^3 + 1$

In each of Exercises 9–12 find the largest subset of **R** that can serve as domain of the function *f*.

9 $f(x) = \sqrt{2x - 3}$

10 $f(x) = \sqrt{x^2 - 1}$

11 $f(x) = \dfrac{1}{x^2 - 9}$

12 $f(x) = \dfrac{x^2 - 2x + 1}{3x^2 - 2x - 1}$

In each of Exercises 13–16 determine whether the function *f* is one-to-one.

13 $f(x) = 3x - 8$

14 $f(x) = 3 - 7x^2$

15 $f(x) = 2x^2 + 3x - 4$

16 $f(x) = |x|$

In Exercises 17–24 determine which of the given relations on **R** are functions in the sense of (3.13).

17 $W = \{(x, y) : y = 3x - 5\}$

18 $W = \{(x, y) : y = x^2 + 1\}$

19 $W = \{(x, y) : x^2 + y^2 = 1\}$

20 $W = \{(x, y) : y^2 = x^2\}$

21 $W = \{(x, y) : x = 4\}$

22 $W = \{(x, y) : y = 1\}$

23 $W = \{(x, y) : x + y = 0\}$

24 $W = \{(x, y) : xy = 0\}$

A function *f* is termed **even** if $f(-a) = f(a)$ for all *a* in the domain *X* of *f*, whereas f is **odd** if $f(-a) = -f(a)$ for all $a \in X$. In each of Exercises 25–30 determine whether f is even, odd, or neither even nor odd.

25 $f(x) = 2x^3 - 9x$ **26** $f(x) = 1 - 4x^2$

27 $f(x) = 2x^4 - 3x^2 - 1$ **28** $f(x) = 3$

29 $f(x) = 4x^2 - 2x + 6$ **30** $f(x) = (x + 1)^2$

31 Find a formula which expresses the radius r of a circle as a function of its circumference C. If the circumference of *any* circle is increased by 12 inches, determine how much the radius increases.

32 Find a formula which expresses the volume of a cube as a function of its surface area. Find the volume if the surface area is 12 square inches.

33 An open box is to be made from a rectangular piece of cardboard having dimensions 24 inches by 36 inches by cutting out identical squares of area x^2 from each corner and turning up the sides. Express the volume V of the box as a function of x.

34 Find a formula which expresses the area A of an equilateral triangle as a function of the length a of a side.

35 Express the perimeter P of a square as a function of its area A.

36 Express the surface area S of a sphere as a function of its volume V.

4 GRAPHS OF FUNCTIONS

In many applications it is necessary to determine the behavior of $f(x)$ as x varies through the domain of a function f. A useful tool for describing this behavior is provided by the following definition.

(3.14)   **Definition**

The **graph of a function** f is the graph of the relation

$$\{(x, f(x)) : x \text{ is in the domain of } f\}.$$

The graph of f can also be defined as the set of all points $P(x, y)$ in a coordinate plane such that $y = f(x)$. Hence the graph of f is the same as the graph of the equation $y = f(x)$. If $P(x, y)$ is on the graph of f, then the ordinate y of P is the functional value $f(x)$. It is important to note that since there is a unique $f(x)$ for each x in the domain, there is only *one* point on the graph with abscissa x.

EXAMPLE 1   Sketch the graph of the function f if $f(x) = 2x - 1$.

    Solution

    By (3.14) the graph of f is the graph of $\{(x, 2x - 1) : x \in \mathbf{R}\}$

and hence is identical with the graph of the relation considered in Example 3 of Section 2. The graph is sketched in Fig. 3.7.

It is often useful to determine the points at which the graph of a function intersects the x-axis. The abscissas of these points are called the **x-intercepts** of the graph and are found by locating all points with zero ordinates, that is, all points $(x, f(x))$ such that $f(x) = 0$. In Example 1 the x-intercept is $\frac{1}{2}$. A number a such that $f(a) = 0$ is also called a **zero of the function** f.

If the number 0 is in the domain of f, then $f(0)$ is called the **y-intercept** of the graph of f. It is the ordinate of the point at which the graph intersects the y-axis. The graph of a function can have at most one y-intercept. The y-intercept in Example 1 is -1.

If f is the function in Example 1 and if $x_1 < x_2$, then $2x_1 - 1 < 2x_2 - 1$, that is, $f(x_1) < f(x_2)$. This means that as abscissas of points increase, then ordinates also increase. When this happens, the function f is said to be *increasing*, according to (3.15) below. If f is increasing, then the graph rises as x increases. For certain functions, we have $f(x_1) > f(x_2)$ whenever $x_1 < x_2$. In this case the graph of f falls as x increases and so the function is called a *decreasing* function. In general we shall speak of functions which increase or decrease on intervals, as in the following definition.

(3.15) **Definition**

(i) A function f is **increasing** on an interval I if $f(x_1) < f(x_2)$ whenever $x_1 < x_2$ in I.

(ii) A function f is **decreasing** on an interval I if $f(x_1) > f(x_2)$ whenever $x_1 < x_2$ in I.

EXAMPLE 2 Sketch the graph of f if $f(x) = x^2 - 3$.

Solution

We list coordinates $(x, f(x))$ of some points on the graph of f in tabular form, as shown below.

x	-3	-2	-1	0	1	2	3
$f(x)$	6	1	-2	-3	-2	1	6

The x-intercepts are the solutions of the equation $f(x) = 0$, that is, of $x^2 - 3 = 0$. These are $\pm\sqrt{3}$. The y-intercept is $f(0) = -3$. Plotting the points given by the table and using the x-intercepts gives us the configuration shown in (i) of Fig. 3.14. Beginning students are sometimes tempted to connect adjacent points with line segments. However, it can be shown using more advanced methods that if $f(x)$ is a

113

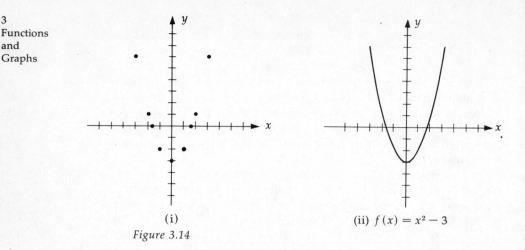

(i)

(ii) $f(x) = x^2 - 3$

Figure 3.14

polynomial, then no part of the graph of f is a line segment. Since this is true, we draw a smooth curve through the points, obtaining the sketch shown in (ii) of Fig. 3.14. In arriving at this sketch we assumed that f is decreasing on the interval $(-\infty, 0]$ and increasing on $[0, \infty)$. To prove these facts we note that if $0 \le x_1 < x_2$, then $x_1^2 < x_2^2$ and hence $x_1^2 - 3 < x_2^2 - 3$ that is, $f(x_1) < f(x_2)$. Hence, by (i) of (3.15), f is increasing on $[0, \infty)$. It can be shown in similar fashion that f is decreasing on $(-\infty, 0]$. It follows that $f(x)$ takes on its least value at $x = 0$. This smallest value, -3, is called the **minimum value** of f. The corresponding point $(0, -3)$ is the lowest point on the graph. Clearly $f(x)$ does not attain a **maximum value**, that is, a *largest* value. The graph in this example is called a **parabola**.

In future examples on graphs we shall not go into as much detail as in the preceding solution. For most of the expressions we shall encounter, the technique used will consist of plotting a sufficient number of points until some pattern emerges, and then sketching the graph accordingly. This is obviously a crude (and often inaccurate) way to arrive at the graph; however, it is the method usually employed in elementary courses. In order to give accurate descriptions of graphs, especially when complicated expressions are involved, it is necessary to use more advanced mathematical tools of the types introduced in the study of calculus.

EXAMPLE 3 Sketch the graph of f if $f(x) = |x|$.

Solution

If $x > 0$, then $f(x) = x$ and we obtain the set of points (x, x) on the graph of f. Negative values of x give rise to the following table:

x	-1	-2	-3	-4
$f(x)$	1	2	3	4

More generally, we obtain all points of the form $(-a, a)$ where $a > 0$. Plotting points leads to the sketch shown in Fig. 3.15. As in Example 2, this function decreases on $(-\infty, 0]$ and increases on $[0, \infty)$, with a minimum value 0 at $x = 0$.

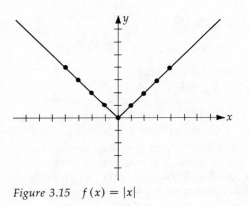

Figure 3.15 $f(x) = |x|$

Notice that no portion of the graph in Example 3 appears below the x-axis. A region of the coordinate plane in which there is no graph is called an **excluded region.**

EXAMPLE 4 Sketch the graph of f if $f(x) = \sqrt{x-1}$.

Solution

The domain of f does not include values of x such that $x - 1 < 0$, since $f(x)$ is not real in this case. Consequently the set of points (x, y) with $x < 1$ is an excluded region for the graph. The following table lists some points $(x, f(x))$ on the graph.

x	1	2	3	4	5	6
$f(x)$	0	1	$\sqrt{2}$	$\sqrt{3}$	2	$\sqrt{5}$

Plotting points leads to the sketch shown in Fig. 3.16. The function is increasing throughout its domain. The x-intercept is 1 and there is no y-intercept.

115

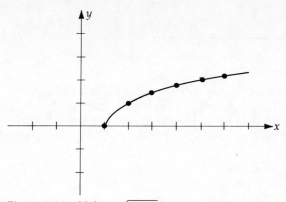

Figure 3.16 $f(x) = \sqrt{x-1}$

EXAMPLE 5 Sketch the graph of f if $f(x) = 1/x$.

Solution

The domain of f is the set of all nonzero real numbers. Before constructing a table, let us make some general observations. First, if x is positive, so is $f(x)$, and hence quadrant IV is an excluded region. Quadrant II is also excluded since if $x < 0$, then $f(x) < 0$. If x is close to zero, the ordinate $1/x$ is very large numerically. As x increases through positive values, $1/x$ decreases and is close to zero when x is large. Similarly, if we let x take on numerically large negative values, the ordinate $1/x$ is close to zero. Using these facts and the table at the top of the next page, we obtain the sketch given in Fig. 3.17. Note that this graph has neither an x nor a y-intercept.

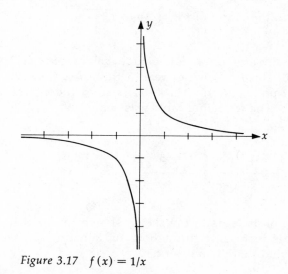

Figure 3.17 $f(x) = 1/x$

x	$\frac{1}{10}$	$\frac{1}{2}$	1	2	5	10	$-\frac{1}{2}$	-1	-2	-4
$f(x)$	10	2	1	$\frac{1}{2}$	$\frac{1}{5}$	$\frac{1}{10}$	-2	-1	$-\frac{1}{2}$	$-\frac{1}{4}$

Moreover, f does not take on a maximum or minimum value. The graph in this example is called a **hyperbola**. We shall discuss hyperbolas further in Section 8.

EXAMPLE 6   Sketch the graph of a constant function with domain **R**.

Solution

From (3.12), f is a constant function if there is some (fixed) real number c such that $f(x) = c$ for all x. The graph of f consists of all points with coordinates (x, c), where $x \in \mathbf{R}$. In particular this includes $(-1, c)$, $(0, c)$, $(2, c)$, and so on. Since all the ordinates equal c, the graph is a line parallel to the x-axis with y-intercept c. A sketch for the case $c > 0$ is shown in Fig. 3.18. A constant function is neither increasing nor decreasing.

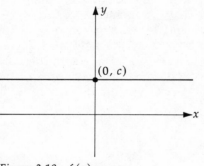

Figure 3.18 $f(x) = c$

Graphs of functions do not always consist of one unbroken line or curve. This is illustrated in the next example.

EXAMPLE 7   If x is any real number, then there exist consecutive integers n and $n + 1$ such that $n \leq x < n + 1$. Let f be the function from **R** to **R** defined as follows: if $x \in \mathbf{R}$ and $n \leq x < n + 1$, then $f(x) = n$. Sketch the graph of f.

Solution

Abscissas and ordinates of points on the graph may be listed as follows:

Values of x	$f(x)$
$\cdots$	$\cdot$
$-2 \le x < -1$	-2
$-1 \le x < 0$	-1
$0 \le x < 1$	0
$1 \le x < 2$	1
$2 \le x < 3$	2
$\cdots$	$\cdot$
$\cdots$	$\cdot$

Since f behaves in the same manner as a constant function when x is between integral values, the corresponding part of the graph is a segment of a horizontal line. Part of the graph of f is sketched in Fig. 3.19. A function of this type is sometimes referred to as a **step function.** The steps continue indefinitely in both directions. The graph has one y-intercept and an infinite number of x-intercepts.

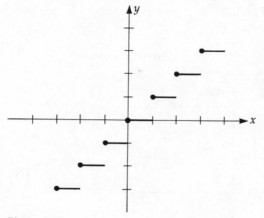

Figure 3.19

The symbol $[x]$ is often used to denote the largest integer z such that $z \le x$. For example,

$$[1.6] = 1, \ [\sqrt{5}] = 2, \ [\pi] = 3, \ [-3.5] = -4.$$

Using this notation, the function f of Example 8 may be defined by $f(x) = [x]$. It is customary to call f the **greatest integer function.**

EXERCISES

What subset of **R** is assumed to be the domain of f if $f(x)$ is defined as in Exercises 1–6?

1 $f(x) = \dfrac{2x - 3}{x^2 - x}$
                                              2 $f(x) = \dfrac{x - 1}{x^2 - 1}$

3 $f(x) = \sqrt{x^2 - 4}$
                                              4 $f(x) = \dfrac{x}{\sqrt{16 - x^2}}$

5 $f(x) = \dfrac{1}{\sqrt{x - 5}\ \sqrt{7 - x}}$
                      6 $f(x) = \dfrac{1}{\sqrt{x}\ (x - 2)}$

In each of the following exercises sketch the graph of the function f. Determine the intervals on which f is increasing and the intervals on which f is decreasing.

7 $f(x) = 4x + 3$
                                               8 $f(x) = 4x - 3$

9 $f(x) = -4x + 3$
                                        10 $f(x) = -4x - 3$

11 $f(x) = -3$
                                              12 $f(x) = 3$

13 $f(x) = 4 - x^2$
                                    14 $f(x) = -(4 + x^2)$

15 $f(x) = \sqrt{4 - x^2}$
                              16 $f(x) = \sqrt{x^2 - 4}$

17 $f(x) = \dfrac{1}{x - 4}$
                                  18 $f(x) = \dfrac{1}{4 - x}$

19 $f(x) = \dfrac{1}{(x - 4)^2}$
                          20 $f(x) = \dfrac{-1}{(x - 4)^2}$

21 $f(x) = |x - 4|$
                                    22 $f(x) = |x| - 4$

23 $f(x) = \dfrac{x}{|x|}$
                                    24 $f(x) = x + |x|$

25 $f(x) = \sqrt{x - 4}$
                                  26 $f(x) = 2 - \sqrt{x}$

27 $f(x) = \begin{cases} -1 & \text{if } x < 0 \\ 1 & \text{if } x \geq 0 \end{cases}$
        28 $f(x) = \begin{cases} 1 \text{ if } x \text{ is an integer} \\ 0 \text{ if } x \text{ is not an integer} \end{cases}$

29 $f(x) = \begin{cases} -5 & \text{if } x < -5 \\ x & \text{if } -5 \leq x \leq 5 \\ 5 & \text{if } x > 5 \end{cases}$
    30 $f(x) = \begin{cases} -x & \text{if } x < 0 \\ 2 & \text{if } 0 \leq x < 1 \\ x^2 & \text{if } x \geq 1 \end{cases}$

5 LINEAR FUNCTIONS

A function f is called a **polynomial function** if $f(x)$ can be written in the form

$$(3.16) \quad f(x) = a_n x^n + a_{n-1} x^{n-1} + \cdots + a_1 x + a_0,$$

where the coefficients $a_0, a_1, \cdots, a_n$ are real numbers and the exponents are nonnegative integers. Of course, if some of the coefficients are negative we may use minus signs instead of plus signs in the appropriate places. As in Section 7 of Chapter 1, the expression on the right in (3.16) is called a **polynomial in** x (with real coefficients) and each $a_k x^k$ is called a **term** of the polynomial. If $a_n \neq 0$, then a_n is called the **leading coefficient** of $f(x)$ and we say that $f(x)$ (or f) has **degree** n. For example, the polynomial $5x^3 - 2x^2 + 3x - 7$ has degree 3. Expressions such as $3x + 1/x$, $(x - 5)/(x^2 + 2)$ and $5x^2 + \sqrt{x} - 2$ are not polynomials since they cannot be written in the form (3.16).

If a polynomial function f has degree 0, then $f(x) = c$ for some $c \neq 0$ and hence f is a constant function. If any coefficient a_i equals 0 we may abbreviate (3.16) by deleting the term $a_i x^i$. If *all* the coefficients are 0 we denote the polynomial by 0 and call it the **zero polynomial.**

If $f(x)$ is a polynomial of degree 1, then

(3.17) $f(x) = ax + b,$

where a and b are real numbers and $a \neq 0$. In this section we shall show that the graph of a function of this type is always a line. For this reason f is called a **linear function.**

Let us begin by introducing several fundamental concepts pertaining to lines. All lines referred to are considered to be in some fixed coordinate plane.

(3.18)   Definition

If l is a line which is not parallel to the y-axis, and if $P_1(x_1, y_1)$ and $P_2(x_2, y_2)$ are distinct points on l, then the **slope m** of l is given by

$$m = \frac{y_2 - y_1}{x_2 - x_1}.$$

If l is parallel to the y-axis, then it has no slope.

The numerator $y_2 - y_1$ in the formula for m is sometimes called the **rise** from P_1 to P_2. It measures the vertical change in direction in proceeding from P_1 to P_2 and may be positive, negative, or zero. The denominator $x_2 - x_1$ is called the **run** from P_1 to P_2. It measures the amount of horizontal change in going from P_1 to P_2. The run may be positive or negative but is never zero because l is not parallel to the y-axis. Using this terminology we could write (3.18) as

$$\text{slope of } l = \frac{\text{rise from } P_1 \text{ to } P_2}{\text{run from } P_1 \text{ to } P_2}.$$

In finding the slope of a line it is immaterial which point is labeled P_1 and which is labeled P_2, since

$$\frac{y_2 - y_1}{x_2 - x_1} = \frac{y_1 - y_2}{x_1 - x_2}.$$

Consequently we may as well assume that the points are labeled so that $x_1 < x_2$, as in Fig. 3.20. In this event $x_2 - x_1 > 0$, and hence the slope is positive, negative, or zero, depending on whether $y_2 > y_1$, $y_2 < y_1$, or $y_2 = y_1$. The slope of the line shown in (i) of Fig. 3.20 is positive, whereas the slope of the line shown in (ii) of the figure is negative. The slope is zero if and only if the line is horizontal.

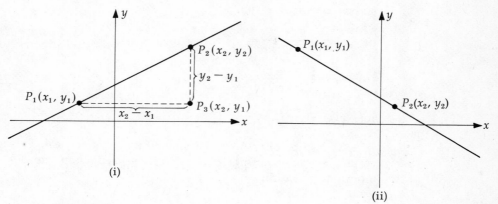

Figure 3.20

It is important to note that the definition of slope is independent of the two points that are chosen on l, for if other points $P_1'(x_1', y_1')$ and $P_2'(x_2', y_2')$ are used, then as in Fig. 3.21, the triangle with vertices P_1', P_2', and $P_3'(x_2', y_1')$ is similar to the triangle with vertices P_1, P_2, $P_3(x_2, y_1)$. Since the ratios of corresponding sides are equal, it follows that

$$\frac{y_2 - y_1}{x_2 - x_1} = \frac{y_2' - y_1'}{x_2' - x_1'}.$$

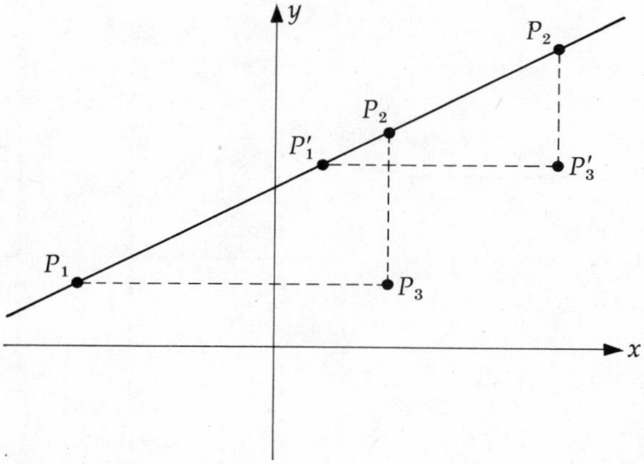

Figure 3.21

EXAMPLE 1 Sketch the lines through the following pairs of points and find their slopes.

(a) $A(-1, 4)$ and $B(3, 2)$ (b) $A(2, 5)$ and $B(-2, -1)$
(c) $A(4, 3)$ and $B(-2, 3)$ (d) $A(4, -1)$ and $B(4, 4)$.

Solution

The lines are sketched in Fig. 3.22. Using (3.18) gives the slopes for parts (a)–(c).

(a) $m = \dfrac{2 - 4}{3 - (-1)} = \dfrac{-2}{4} = -\dfrac{1}{2}$,

(b) $m = \dfrac{5 - (-1)}{2 - (-2)} = \dfrac{6}{4} = \dfrac{3}{2}$

(c) $m = \dfrac{3 - 3}{-2 - 4} = \dfrac{0}{-6} = 0$

(d) The slope does not exist since the line is vertical. This is also seen by noting that if (3.18) is used, then the denominator is zero.

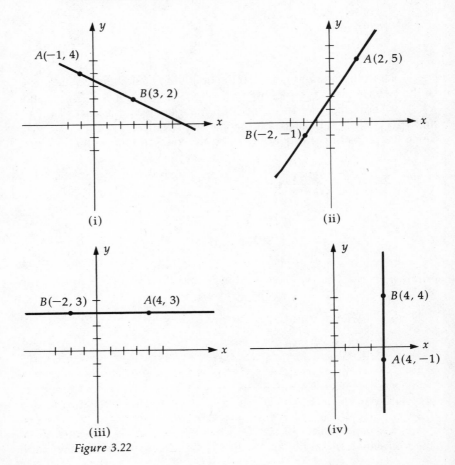

(i)

(ii)

(iii)

(iv)

Figure 3.22

(a)   5/3   (b)   −5/3.

Solutions

If the slope of a line is a/b where b is positive, then for every b units change in horizontal direction the line rises or falls a units, depending on whether the quotient a/b is positive or negative, respectively. If $P(2, 1)$ is on the line and $m = 5/3$, we can obtain another point on the line by starting at P and moving 3 units to the right and 5 units upward. This gives us the point $Q(5, 6)$ and the line is determined (see (i) of Fig. 3.23). Similarly, if $m = -5/3$ we move 3 units to the right and 5 units downward, obtaining $Q(5, -4)$ as in (ii) of Fig. 3.23.

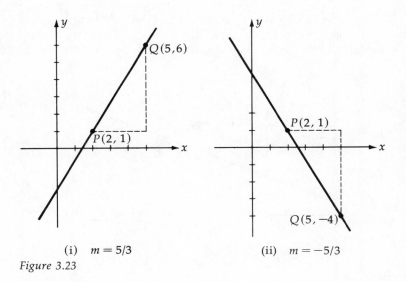

(i) $m = 5/3$ (ii) $m = -5/3$

Figure 3.23

The equation $y = b$, where b is a real number, may be considered as an equation in two variables x and y by writing it in the form

$$(0 \cdot x) + y = b.$$

Some typical solutions of this equation are $(-2, b)$, $(1, b)$, and $(3, b)$. The solution set consists of *all* pairs of the form (x, b) where x may have any value and b is fixed. It follows that the graph of $y = b$ is a line parallel to the x-axis with y-intercept b. This was to be expected since the graph is the same as the graph of the constant function f, where $f(x) = b$ (see Fig. 3.18). Conversely, every horizontal line is the graph of an equation of the form $y = b$. A similar argument can be used to show that the graph of the equation $x = a$ is a line parallel to the y-axis with x-intercept a. The graphs of these lines are illustrated in Fig. 3.24.

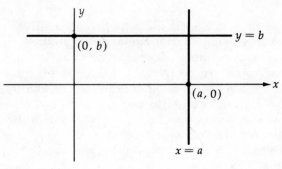

Figure 3.24

Let us now find an equation of a line l through a point $P_1(x_1, y_1)$ with slope m (only one such line exists). If $P(x, y)$ is any point with $x \neq x_1$ (see Fig. 3.25), then P is on l if and only if the slope of the line

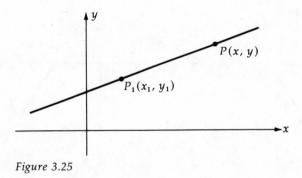

Figure 3.25

through P_1 and P is m, that is, if and only if

$$\frac{y - y_1}{x - x_1} = m.$$

This equation may be written in the form

$$y - y_1 = m(x - x_1).$$

Note that (x_1, y_1) is also a solution of the latter equation and hence the points on l are precisely the points which correspond to the solution set. This equation for l is referred to as the **point-slope form.** Our discussion may be summarized as follows:

(3.19)   Point-Slope Form for the Equation of a Line.

An equation for the line through the point $P(x_1, y_1)$ with slope m is

$$y - y_1 = m(x - x_1).$$

EXAMPLE 3   Find an equation of the line through the points $A(1, 7)$ and $B(-3, 2)$.

Solution

By (3.18) the slope m of the line is $m = \dfrac{7 - 2}{1 - (-3)} = \dfrac{5}{4}$. Using the coordinates of A in the point-slope form (3.19) gives us

$$y - 7 = \tfrac{5}{4}(x - 1),$$

which is equivalent to

$$4y - 28 = 5x - 5$$

or

$$5x - 4y + 23 = 0.$$

The same equation would have been obtained if the coordinates of point B had been substituted in (3.19).

Equation (3.19) may be rewritten as $y = mx - mx_1 + y_1$, which is of the form

(3.20) $y = mx + b,$

where $b = -mx_1 + y_1$. The real number b is the y-intercept of the graph, as may be seen by setting $x = 0$. Since (3.20) displays the slope and y-intercept of l, it is called the **slope-intercept form** for the equation of a line. Conversely, given an equation of the form (3.20) with $m \neq 0$, we may write

$$y - 0 = m\left(x + \frac{b}{m}\right).$$

Comparing with (3.19) we see that the graph is a line with slope m. This gives us the next result.

(3.21)   Slope-Intercept Form for the Equation of a Line.

The graph of the equation $y = mx + b$ is a line having slope m and y-intercept b.

The work we have done shows that every line is the graph of an equation of the form

(3.22) $ax + by + c = 0,$

where $a, b, c \in \mathbf{R}$ and a and b are not both zero. (In this equation, b is not necessarily the y-intercept.) We call (3.22) a **linear equation** in x and y. Let us show conversely that the graph of (3.22) is always a line. On the one hand, if $b \neq 0$ we may write (3.22) as

$$y = (-a/b)x + (-c/b),$$

which, by the slope-intercept form (3.21), is an equation of a line with slope $-a/b$ and y-intercept $-c/b$. On the other hand, if $b = 0$ but $a \neq 0$, then we may write (3.22) as $x = -c/a$, which is the equation of a line parallel to the y-axis with x-intercept $-c/a$. This establishes the following theorem:

(3.23)   Theorem

The graph of a linear equation $ax + by + c = 0$ is a line, and conversely every line is the graph of a linear equation.

As a corollary to (3.23) it follows that the graph of a linear function f is always a line, for if $f(x) = mx + b$, then the graph of f is the same as the graph of the equation $y = mx + b$. This proves the remark made near the beginning of this section.

For simplicity we shall often use the terminology *the line ax + by + c = 0* instead of the more accurate phrase *the line with equation* $ax + by + c = 0$.

EXAMPLE 4   Sketch the graph of $2x - 5y = 8$.

Solution

We know from (3.23) that the graph is a line and hence it is sufficient to find two points on the graph. Let us find the x- and y-intercepts. Substituting $y = 0$ in the given equation we obtain the x-intercept 4. Substituting $x = 0$ we see that the y-intercept is $-8/5$. This leads to the graph in Fig. 3.26. Another method of solution is to express the given equation in the slope-intercept form

$$y = (2/5)x + (-8/5).$$

By comparison with (3.21) we obtain the slope $m = 2/5$ and the y-intercept $b = -8/5$. We may then sketch a line through the point $(0, -8/5)$ with slope 2/5.

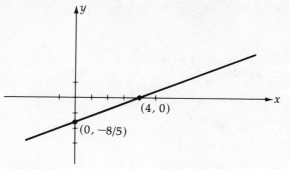

Figure 3.26 $2x - 5y = 8$

It can be shown geometrically that *two nonvertical lines are parallel if and only if they have the same slope.* This fact is used in the next example.

EXAMPLE 5   Find an equation of a line through the point $(5, -7)$ which is parallel to the line $6x + 3y - 4 = 0$.

Solution

The equation of the given line may be rewritten as

$$y = -2x + 4/3.$$

This is in slope-intercept form (3.21) with $m = -2$ and hence the slope is -2. From the preceding remark the required line also has slope -2. Applying the point-slope form gives us

$$y + 7 = -2(x - 5),$$

or equivalently,

$$2x + y - 3 = 0.$$

EXERCISES

In each of Exercises 1–6 plot the points A and B and find the slope of the line through A and B.

1   $A(-4, 6)$, $B(-1, 18)$ 2 $A(6, -2)$, $B(-3, 5)$

3   $A(\frac{2}{3}, \frac{1}{2})$, $B(2, 1)$ 4 $A(\sqrt{8}, \sqrt{12})$, $B(-\sqrt{2}, -\sqrt{27})$

5   $A(-1, -3)$, $B(-1, 2)$ 6 $A(-3, 4)$, $B(2, 4)$

7   Use slopes to show that $A(-3, 1)$, $B(5, 3)$, $C(3, 0)$, and $D(-5, -2)$ are vertices of a parallelogram.

8 Use slopes to show that $A(1, -3)$, $B(-3, -11)$, and $C(3, 1)$ lie on a straight line.

In each of Exercises 9–20 find an equation for the line satisfying the given conditions.

9 Through $A(2, -6)$; slope $\frac{1}{2}$

10 Slope -3; y-intercept 5

11 Through $A(-5, -7)$ and $B(3, -4)$

12 x-intercept -4; y-intercept 8

13 Through $A(8, -2)$; y-intercept -3

14 Slope 6; x-intercept -2

15 Through $A(10, -6)$ parallel to the **a** x-axis. **b** y-axis

16 Through $A(-5, 1)$ perpendicular to the **a** y-axis, **b** x-axis

17 Bisecting the second and fourth quadrants

18 Coinciding with the y-axis

19 Through $A(-7, 2)$, parallel to the line through $B(0, 4)$ and $C(-6, -6)$

20 Through $P(-3/4, -1/2)$, parallel to the line with equation $x + 3y = 1$

In each of Exercises 21–28 use the slope-intercept form to find the slope and y-intercept of the line with the given equation and sketch the graph.

21 $3x - 4y + 8 = 0$ **22** $2y - 5x = 1$

23 $x + 2y = 0$ **24** $8x = 1 - 4y$

25 $y = 4$ **26** $x + 2 = (\frac{1}{2})y$

27 $5x + 4y = 20$ **28** $y = 0$

Sketch the graph of each of the equations in Exercises 29–32.

29 $2x^2 - 5xy - 3y^2 = 0$ (*Hint:* Factor the expression.)

30 $xy(x + y - 1) = 0$

31 $(x + y)^2 = 1$ **32** $y^2 = 6 - y$

33 Find a real number k such that the point $P(4, -1)$ is on the line $kx + 3y - 5 = 0$.

34 Find a real number k such that the line $4x - ky + 1 = 0$ has y-intercept -5.

35 If $a > 0$, prove that the linear function f defined by $f(x) = ax + b$ is an increasing function throughout its domain. If $a < 0$, prove that f is decreasing throughout its domain.

36 Prove that the graph of the equation $ax + by = 0$, where a and b are not both zero, is a straight line passing through the origin.

37 If a line l has nonzero x- and y-intercepts a and b, respectively, prove that an equation for l is

$$\frac{x}{a} + \frac{y}{b} = 1.$$

(This is called the **intercept form** for the equation of a line.) Express the equation $4x - 3y = 8$ in intercept form.

38 Prove that an equation of the line through $P_1(x_1, y_1)$ and $P_2(x_2, y_2)$ is

$$(y - y_1)(x_2 - x_1) = (y_2 - y_1)(x - x_1).$$

(This is called the **two-point form** for the equation of a line). Use the two-point form to find an equation of the line through $A(4, -5)$ and $B(-1, 1)$.

39 Find all values of r such that the slope of the line through the points $(r, 4)$ and $(1, 3 - 2r)$ is less than 5.

40 Find all values of t such that the slope of the line through $(t, 3t + 1)$ and $(1 - 2t, t)$ is greater than 4.

6 POLYNOMIAL FUNCTIONS OF DEGREE GREATER THAN 1

If f is a polynomial function of degree 2, then

(3.24) $f(x) = ax^2 + bx + c$

where a, b, and c are real numbers and $a \neq 0$. Functions of this type are called **quadratic functions.** In Example 2 of Section 4 we considered $f(x) = x^2 - 3$. This is the special case of (3.24) where $a = 1$, $b = 0$, and $c = -3$. The graph of f is the parabola illustrated in (ii) of Fig. 3.14. The low point on the graph is called the **vertex** of the parabola. It can be shown that if $a > 0$, then the graph of (3.24) is always a parabola that *opens upward* as in Fig. 3.14. The position of the vertex depends on the coefficients a, b, and c. If $a < 0$, the parabola *opens downward,* that is, the curve in Fig. 3.14 would be inverted and the vertex would be the highest point on the graph. An illustration of this is given in Example 2 of this section.

Let us analyze the graph of the general quadratic function given by (3.24) when $a > 0$. We begin by writing

$$f(x) = a \left[x^2 + \frac{b}{a} x + \frac{c}{a} \right].$$

Next we complete the square in the first two terms within the brackets by adding and subtracting $(b/2a)^2$ as follows:

$$f(x) = a \left[\left(x^2 + \frac{b}{a} x + \frac{b^2}{4a^2} \right) + \left(\frac{c}{a} - \frac{b^2}{4a^2} \right) \right],$$

or

(3.25) $f(x) = a \left[\left(x + \frac{b}{2a} \right)^2 + \left(\frac{4ac - b^2}{4a^2} \right) \right].$

129

Observe that substituting numbers for x in (3.25) has no effect on the fixed term $(4ac - b^2)/4a^2$ within the brackets. Consequently, in order to determine where f increases or decreases it is sufficient to study the expression $(x + b/2a)^2$. Since $(x + b/2a)^2 = 0$ if $x = -b/2a$, and since $(x + b/2a)^2 > 0$ if $x \neq -b/2a$, it follows that $f(x)$ has a minimum value at $x = -b/2a$. This gives us the abscissa of the vertex illustrated in Fig. 3.27.

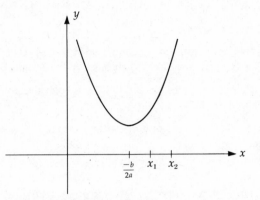

Figure 3.27 $f(x) = ax^2 + bx + c, a > 0$

Let us next consider any two numbers x_1, x_2 such that

$$-\frac{b}{2a} < x_1 < x_2.$$

(See Fig. 3.27.) Adding $b/2a$ to each side gives us

$$0 < x_1 + \frac{b}{2a} < x_2 + \frac{b}{2a}$$

and hence

$$\left(x_1 + \frac{b}{2a}\right)^2 < \left(x_2 + \frac{b}{2a}\right)^2.$$

It follows from (3.25) that $f(x_1) < f(x_2)$, that is, f is increasing on the interval $(-b/2a, \infty)$. It can be shown in like manner that if $x_1 < x_2 < -b/2a$, then $f(x_1) > f(x_2)$, that is, f is decreasing on $(-\infty, -b/2a)$. This behavior of f is illustrated in Fig. 3.27.

A similar analysis can be given if $a < 0$ in (3.24). In this case, f increases on $(-\infty, -b/2a)$ and decreases on $(-b/2a, \infty)$. Hence the abscissa of the vertex is again $x = -b/2a$, but the parabola opens downward.

To find the x-intercepts of the graph of (3.24) we solve the equation $ax^2 + bx + c = 0$. By the quadratic formula (2.3), if $b^2 - 4ac > 0$ there are two real and unequal roots and the graph has two x-intercepts. If $b^2 - 4ac = 0$ there is one (double) root and the graph is tangent to the x-axis. Finally, if $b^2 - 4ac < 0$ there are no x-intercepts. We have sketched these cases in Fig. 3.28 under the assumption that $a > 0$. A similar situation occurs if $a < 0$, but in this case the parabolas open downward.

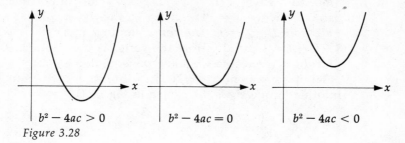

$b^2 - 4ac > 0$ $b^2 - 4ac = 0$ $b^2 - 4ac < 0$

Figure 3.28

EXAMPLE 1   Sketch the graph of f if $f(x) = 2x^2 - 6x + 4$.

Solution

The polynomial $f(x)$ has the form (3.24) with $a = 2$, $b = -6$, and $c = 4$. From the preceding discussion we know that the graph is a parabola having vertex with abscissa

$$-\frac{b}{2a} = \frac{6}{4} = \frac{3}{2}.$$

The corresponding ordinate is

$$f(\tfrac{3}{2}) = 2(\tfrac{9}{4}) - 6(\tfrac{3}{2}) + 4 = -\tfrac{1}{2}.$$

Hence the vertex is $(\tfrac{3}{2}, -\tfrac{1}{2})$. Since $a = 2 > 0$, the parabola opens upward. The y-intercept is $f(0) = 4$. To find the x-intercepts we solve $2x^2 - 6x + 4 = 0$ or equivalently $(2x - 2)(x - 2) = 0$, obtaining $x = 2$ and $x = 1$. The four points we have found are sufficient for a reasonably accurate sketch (see Fig. 3.29).

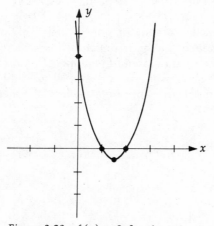

Figure 3.29 $f(x) = 2x^2 - 6x + 4$

EXAMPLE 2   Sketch the graph of the equation $y = 8 + 2x - x^2$.

Solution

The graph of the equation is the same as the graph of the function f where $f(x) = 8 + 2x - x^2$. Hence the graph is a parabola. Since $a = -1$ and $b = 2$, the vertex has abscissa $-b/2a = -2/(-2) = 1$. The corresponding ordinate is $f(1) = 8 + 2 - 1 = 9$ and therefore the vertex is the point $(1, 9)$. To find the x-intercepts we solve the equation $8 + 2x - x^2 = 0$ obtaining $x = 4$ and $x = -2$. The y-intercept is 8. Using this information gives us the sketch in Fig. 3.30.

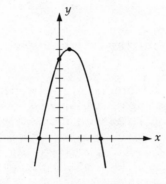

Figure 3.30 $y = 8 + 2x - x^2$

Parabolas (and hence quadratic functions) are very useful in applications of mathematics to the physical world. It can be shown that if a projectile is fired and is acted upon only by the force of gravity, that is, air resistance and other outside factors are inconsequential, then the path of the projectile is a parabola. Properties of parabolas are used in the design of mirrors for telescopes and searchlights. They are also employed in the design of field microphones used in television broadcasts of football games. These are only a few of many physical applications.

As the degree increases, graphs of polynomial functions become more complicated. Indeed, methods developed in calculus are needed to obtain accurate information about the graph of a polynomial function of degree greater than 2. In the next example we shall give a partial description of the graph of a specific polynomial function of degree 3.

EXAMPLE 3   Sketch the graph of f if $f(x) = -x^3 - x^2 + 2x$.

Solution

We may express $f(x)$ as a product thus:

$$f(x) = -x(x^2 + x - 2) = -x(x + 2)(x - 1).$$

By considering $f(x) = 0$ we determine the x-intercepts -2, 0, and 1. These in turn may be used to obtain the following intervals on the x-axis:

$$(-\infty, -2), (-2, 0), (0, 1), (1, \infty).$$

We shall now investigate the sign of $f(x)$ in each interval by using the technique introduced in Section 6 of Chapter 2. Specifically, let us consider the sign of each of the factors $-x$, $x + 2$, and $x - 1$, as illustrated in Fig. 3.31.

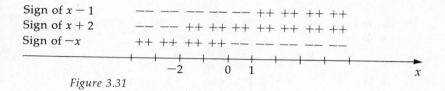

Sign of $x - 1$

Sign of $x + 2$

Sign of $-x$

Figure 3.31

Since $f(x)$ is the product of the three factors, we see that $f(x)$ is positive (and hence the graph is above the x-axis) if x is in either of the intervals $(-\infty, -2)$ or $(0, 1)$. The value of $f(x)$ is negative (and hence the graph is below the x-axis) if x is in $(-2, 0)$ or $(1, \infty)$. Coordinates of several points on the graph are $(-3, 12)$, $(-1, -2)$, $(1/2, 5/8)$, and $(3/2, -21/8)$. The graph is sketched in Fig. 3.32.

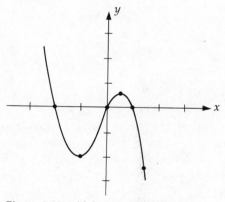

Figure 3.32 $f(x) = -x^3 - x^2 + 2x$

It can be shown that the graph of a polynomial function of degree 3 always has an S-shaped pattern resembling either that shown in Fig. 3.32 or inversions of this type. However, the S-curve may be very sharp or very shallow.

The algebraic theory of polynomials will be discussed further in Chapter 9, after we have introduced the system of complex numbers.

EXERCISES

Sketch the graph of f in each of Exercises 1–10.

1 $f(x) = \frac{1}{9}x^2$ **2** $f(x) = -4x^2$

3 $f(x) = 4x^2 - 9$ **4** $f(x) = 9 - 4x^2$

5 $f(x) = x^2 + 5x + 4$ **6** $f(x) = x^2 - 6x$

7 $f(x) = 8x - 12 - x^2$ **8** $f(x) = 10 + 3x - x^2$

9 $f(x) = x^2 + x + 3$ **10** $f(x) = x^2 + 2x + 5$

In each of Exercises 11–14, sketch the graphs of (a), (b), and (c) on the same coordinate plane.

11 **a** $y = x^2$ **b** $y = x^2 + 4$ **c** $y = x^2 - 4$

12 **a** $y = x^3$ **b** $y = x^3 + 2$ **c** $y = x^3 - 2$

13 **a** $y = -x^3$ **b** $y = 2 - x^3$ **c** $y = -2 - x^3$

14 **a** $y = x^2$ **b** $y = x^4$ **c** $y = x^6$

Sketch the graph of the equation in each of Exercises 15–18.

15 $12x^2 - 2y - 8x + 1 = 0$ **16** $y = 2x^2 + x - 10$

17 $y = x^3 - 3x^2 - x + 3$ **18** $y = x^3 + 4x^2 - 4x - 16$

19 Find a real number k such that the graph of $y = x^2 + kx + 3$ passes through the point $P(-1, -2)$.

20 Prove that a parabola which has an equation of the form $y = ax^2 + c$ has its vertex on the y-axis.

7 COMPOSITE AND INVERSE FUNCTIONS

We shall now describe an important method for using two functions f and g to obtain a third function. Suppose X, Y, and Z are subsets of **R** and let f be a function from X to Y and g a function from Y to Z. In terms of the arrow notation introduced in Section 3 we have

$$X \xrightarrow{f} Y \xrightarrow{g} Z;$$

that is, f maps X into Y and g maps Y into Z. This is illustrated schematically in Fig. 3.33 by the solid curved arrows. A function from X to Z may then be defined in a natural way. For every x in X, the number $f(x)$ is in Y. Since the domain of g is Y we may then find the image of $f(x)$ under g. Of course, this element of Z is written as $g(f(x))$. By assigning $g(f(x))$ to x, we obtain a function from X to Z called the *composite func-*

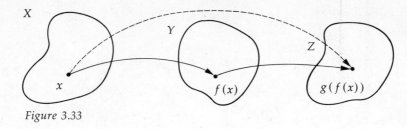

X

Y

Z

x

$f(x)$

$g(f(x))$

Figure 3.33

tion of g by f. The dashes in Fig. 3.33 indicate the correspondence we have defined from X to Z. Since we have used two functions f and g to obtain a third function, we shall employ the operational symbol ∘ and denote the latter function by g ∘ f. The following definition summarizes our remarks.

(3.26)   Definition

Let X, Y, and Z be subsets of **R**. If f is a function from X to Y and g is a function from Y to Z, then the **composite function** g ∘ f is the function from X to Z defined by

$$(g \circ f)(x) = g(f(x)),$$

for every x in X.

EXAMPLE 1   If $f(x) = x^3$ and $g(x) = 5x^2 + 2x + 1$, find $(g \circ f)(x)$.

Solution

Using (3.26) and the definition of f gives us

$$(g \circ f)(x) = g(f(x)) = g(x^3).$$

Since $g(x^3)$ means that x^3 should be substituted for x in the expression for g(x) we have

$$g(x^3) = 5(x^3)^2 + 2(x^3) + 1.$$

Consequently

$$(g \circ f)(x) = 5x^6 + 2x^3 + 1.$$

In applying (3.26) it is not essential that the domain of g be all of Y but merely that the domain of g *contain* the range of f, since it is only necessary to find g(f(a)) for each a in X. In certain cases we may wish to restrict x to some subset of X so that f(x) is in the domain of g. This is illustrated in the next example.

EXAMPLE 2 If $f(x) = x - 2$ and $g(x) = 5x + \sqrt{x}$ find $(g \circ f)(x)$.

Solution

Formal substitutions give us the following:

$$
\begin{aligned}
(g \circ f)(x) &= g(f(x)) & \text{(definition of } g \circ f) \\
&= g(x - 2) & \text{(definition of } f) \\
&= 5(x - 2) + \sqrt{x - 2} & \text{(definition of } g) \\
&= 5x - 10 + \sqrt{x - 2} & \text{(simplifying)}.
\end{aligned}
$$

The domain X of f is the set of all real numbers; however, the last equality implies that $(g \circ f)(x)$ is a real number only if $x \geq 2$. Thus, when working with the composite function $g \circ f$ it is necessary to restrict x to the interval $[2, \infty)$.

If $X = Y = Z$ in (3.26) then it is possible to find $f(g(x))$. In this case we first obtain the image of x under g and then apply f to $g(x)$. This gives us a function from Z to X called the *composite function* of f by g and denoted by $f \circ g$. Thus, by definition,

$$(f \circ g)(x) = f(g(x)),$$

for all x in Z.

EXAMPLE 3 If $f(x) = x^2 - 1$ and $g(x) = 3x + 5$, find $(f \circ g)(x)$ and $(g \circ f)(x)$.

Solution

We have the following:

$$
\begin{aligned}
(f \circ g)(x) &= f(g(x)) & \text{(definition of } f \circ g) \\
&= f(3x + 5) & \text{(definition of } g) \\
&= (3x + 5)^2 - 1 & \text{(definition of } f) \\
&= 9x^2 + 30x + 24 & \text{(simplifying)}.
\end{aligned}
$$

Also,

$$
\begin{aligned}
(g \circ f)(x) &= g(f(x)) & \text{(definition of } g \circ f) \\
&= g(x^2 - 1) & \text{(definition of } f) \\
&= 3(x^2 - 1) + 5 & \text{(definition of } g) \\
&= 3x^2 + 2 & \text{(simplifying)}.
\end{aligned}
$$

We see from Example 3 that $f(g(x))$ and $g(f(x))$ are not always the same, that is, $f \circ g \neq g \circ f$. In certain cases it may happen that equality *does* occur. Of major importance is the case in which $f(g(x))$ and $g(f(x))$ are not only identical, but both are equal to x. Needless to say, f and g must be very special functions in order for this to happen. In the following discussion we indicate the manner in which they must be restricted.

Suppose f is a *one-to-one function* with domain X and range Y. As mentioned in Section 3 this implies that each element of Y is the image of precisely one element of X. Another way of phrasing this is to say that *each element of Y can be written in one and only one way in the form f(x), where x $\in$ X.* We may then define a function g from Y to X by demanding that

$$g(f(x)) = x, \quad \text{for every } x \text{ in } X.$$

This amounts to *reversing* the correspondence given by f. If f is represented geometrically by drawing arrows as in (i) of Fig. 3.34, then g can be represented by simply *reversing* these arrows as illustrated in (ii) of the figure. It follows that g is a one-to-one function with domain Y and range X.

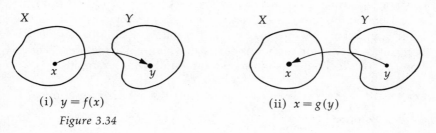

(i) $y = f(x)$ (ii) $x = g(y)$

Figure 3.34

As illustrated in Fig. 3.34, if $f(x) = y$, then $x = g(y)$. This means that

$$f(g(y)) = y, \quad \text{for every } y \text{ in } Y.$$

Since the notation used for the variable is immaterial, we may write

$$f(g(x)) = x, \quad \text{for every } x \text{ in } Y.$$

The functions f and g are called *inverse functions* of one another, according to the following definition.

(3.27) Definition

If f is a one-to-one function with domain X and range Y, then a function g with domain Y and range X is called the **inverse function of f** if

$$f(g(x)) = x, \quad \text{for every } x \text{ in } Y,$$

and

$$g(f(x)) = x, \quad \text{for every } x \text{ in } X.$$

There can be only one inverse function of f. Moreover, if g is the inverse function of f, then by (3.27), $(g \circ f)(x) = x$, that is, $g \circ f$ is the

137

identity function on X (see (3.11)). Similarly, since $(f \circ g)(x) = x$, for all x in Y, $f \circ g$ is the identity function on Y. For this reason the symbol f^{-1} is often used to denote the inverse function of f. Employing this notation,

(3.28)
$$f^{-1}(f(x)) = x, \quad \text{for every } x \text{ in } X, \text{ and}$$
$$f(f^{-1}(x)) = x, \quad \text{for every } x \text{ in } Y.$$

The symbol -1 in (3.28) should not be mistaken for an exponent. It is used here to represent inverse functions.

Inverse functions are very important in the study of trigonometry. In Chapter Four we shall discuss two other important classes of inverse functions.

An algebraic method can sometimes be used to find the inverse of a one-to-one function f with domain X and range Y. Given any x in X, its image y in Y may be found by means of the equation $y = f(x)$. In order to determine the inverse function f^{-1} we wish to *reverse* this procedure, in the sense that given y, the element x may be found. Since x and y are related by means of $y = f(x)$, it follows that if the latter equation can be solved for x in terms of y, we may arrive at the inverse function f^{-1}. This technique is illustrated in the following examples.

EXAMPLE 4   If $f(x) = 3x - 5$ for all x, find the inverse function of f.

Solution

It is not difficult to show that f is a one-to-one function with domain and range **R**, and hence the inverse function g exists. If we let $y = 3x - 5$ and then solve for x in terms of y, we get $x = (y + 5)/3$. That equation enables us to find x when given y. Letting $g(y) = (y + 5)/3$ gives us a function g from Y to X that reverses the correspondence determined by f. Since the symbol used for the independent variable is immaterial, we may replace y by x in the expression for g, obtaining

(3.29) $g(x) = (x + 5)/3$.

To verify that g is actually the inverse function of f, we must verify that the two conditions stated in (3.27) are fulfilled. Thus

$$f(g(x)) = f\left(\frac{x + 5}{3}\right) \qquad \text{(3.29)}$$

$$= 3\left(\frac{x + 5}{3}\right) - 5 \qquad \text{(definition of } f)$$

$$= x \qquad \text{(simplifying)}.$$

Also,

$$g(f(x)) = g(3x - 5) \qquad \text{(definition of } f)$$

$$= \frac{(3x - 5) + 5}{3} \qquad \text{(3.29)}$$

$$= x \qquad \text{(simplifying)}.$$

This proves that (3.29) defines the inverse function g of f. Using the notation of (3.28),

$$f^{-1}(x) = \frac{x+5}{3}.$$

EXAMPLE 5 Find the inverse function of f if the domain X is the set of nonnegative real numbers and $f(x) = x^2 - 3$ for all x in X.

Solution

The domain has been restricted so that f is one-to-one. The range of f is $\{y : y \geq -3\}$. Considering the equation $y = x^2 - 3$ and solving for x gives us $x = \pm\sqrt{y+3}$. Since x is nonnegative we reject $x = -\sqrt{y+3}$. As in the preceding example we let

$$g(y) = \sqrt{y+3}$$

or equivalently,

$$g(x) = \sqrt{x+3}.$$

We now check the two conditions in (3.27), obtaining

$$f(g(x)) = f(\sqrt{x+3}) = (\sqrt{x+3})^2 - 3$$
$$= (x+3) - 3 = x$$

and

$$g(f(x)) = g(x^2 - 3) = \sqrt{(x^2-3)+3} = x.$$

This proves that

$$f^{-1}(x) = \sqrt{x+3}, \quad \text{where } x \geq -3.$$

There is an interesting relationship between the graphs of a function f and its inverse function f^{-1}. We first note that f maps a into b if and only if f^{-1} maps b into a; that is, $b = f(a)$ means the same thing as $a = f^{-1}(b)$. These equations imply that the point (a, b) is on the graph of f if and only if the point (b, a) is on the graph of f^{-1}. As an illustration, in Example 5 we found that the functions f and f^{-1} given by $f(x) = x^2 - 3$ and $f^{-1}(x) = \sqrt{x+3}$ are inverse functions of one another, provided x is suitably restricted. Some points on the graph of f are $(0, -3)$, $(1, -2)$, $(2, 1)$, and $(3, 6)$. Corresponding points on the graph of f^{-1} are $(-3, 0)$, $(-2, 1)$, $(1, 2)$, and $(6, 3)$. The graphs of f and f^{-1} are sketched on the same coordinate axes in Fig. 3.35. If the page is folded along the line l which bisects quadrants I and III (as indicated by the dashes in the figure), then the graphs of f and f^{-1} coincide. The two graphs are said to be *reflections* of one another through the line l. This is typical of the graph of every function f that has an inverse function f^{-1}.

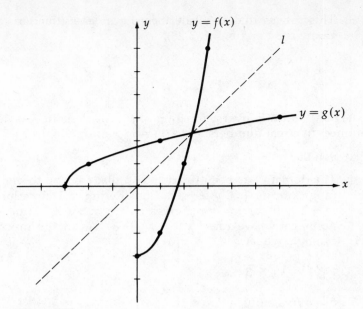

Figure 3.35

EXERCISES

In Exercises 1–14 find $(f \circ g)(x)$ and $(g \circ f)(x)$.

1 $f(x) = 2x + 1, \quad g(x) = 3x - 5$

2 $f(x) = 4x - 3, \quad g(x) = x + 5$

3 $f(x) = 2x^2 + 5, \quad g(x) = 3x$

4 $f(x) = x^2 - 5, \quad g(x) = x + 1$

5 $f(x) = x^2 - 2x, \quad g(x) = 3x + 1$

6 $f(x) = 1/(3x + 1), \quad g(x) = 2/x^2$

7 $f(x) = 1/(3x + 2), \quad g(x) = x - 5$

8 $f(x) = x^3, \quad g(x) = x + 1$

9 $f(x) = x^2 + 4, \quad g(x) = \sqrt{x - 1}$

10 $f(x) = \sqrt{x^2 + 4}, \quad g(x) = 7x^2 + 1$

11 $f(x) = 3x^2 + 2, \quad g(x) = 1/(3x^2 + 2)$

12 $f(x) = 7, \quad g(x) = 4$

13 $f(x) = \sqrt{2x + 1}, \quad g(x) = x^2 + 3$

14 $f(x) = 6x - 12, \quad g(x) = \frac{1}{6}x + 2$

In each of Exercises 15–18 prove that f and g are inverse functions of one another.

15 $f(x) = 9x + 2$; $g(x) = \frac{1}{9}x - \frac{2}{9}$

16 $f(x) = x^3 + 1$; $g(x) = \sqrt[3]{x - 1}$

17 $f(x) = \sqrt{2x + 1}$; $g(x) = \frac{1}{2}x^2 - \frac{1}{2}$, $x \geq 0$

18 $f(x) = 1/(x - 1)$, $x > 1$; $g(x) = (1 + x)/x$, $x > 0$

In each of Exercises 19–28 find the inverse function of f.

19 $f(x) = 8 + 11x$ **20** $f(x) = \dfrac{1}{8 + 11x}$, $x > -11/8$

21 $f(x) = 6 - x^2$, $0 \leq x$ **22** $f(x) = 2x^3 - 5$

23 $f(x) = \sqrt{7x - 2}$, $x \geq \frac{2}{7}$ **24** $f(x) = \sqrt{1 - 4x^2}$, $0 \leq x \leq \frac{1}{2}$

25 $f(x) = 7 - 3x^3$ **26** $f(x) = x$

27 $f(x) = (x^3 + 8)^5$ **28** $f(x) = x^{1/3} + 2$

29 Sketch the graphs of the functions f and g in Exercise 15 on the same coordinate plane. Do the same for the functions defined in Exercise 17.

30 Sketch the graphs of the functions f and g in Exercise 16 on the same coordinate plane.

31   **a**   Prove that the linear function f defined by $f(x) = ax + b$, where $a \neq 0$, has an inverse function.

     b   Does a quadratic function have an inverse function?

     c   Does a constant function have an inverse?

     d   Does an identity function have an inverse?

32 If f is a one-to-one function with domain X and range Y, prove that f^{-1} is a one-to-one function with domain Y and range X.

33 Prove that a one-to-one function can have at most one inverse function.

34 If $f(x) = ax + b$ and $g(x) = cx + d$, find conditions on c and d in terms of a and b which will guarantee that $f \circ g = g \circ f$. Discuss the case where $a = b = 1$.

8 CONIC SECTIONS

We know from our work in Section 6 that if $a \neq 0$, then the graph of the equation $y = ax^2 + bx + c$ is a parabola having vertex with abscissa $-b/2a$, and opening upward if $a > 0$ or downward if $a < 0$. By interchanging the variables x and y we obtain the equation

$$(3.30) \quad x = ay^2 + by + c.$$

It can be shown that the graph of (3.30) is a parabola having vertex with *ordinate* $-b/2a$ and opening to the *right* if $a > 0$ or to the *left* if $a < 0$. (See Fig. 3.36.) Note that these are not graphs of functions since there may be two values of y for each x.

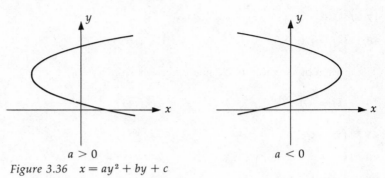

$$a > 0 \qquad\qquad a < 0$$

Figure 3.36 $x = ay^2 + by + c$

EXAMPLE 1   Sketch the graph of $x = 3y^2 + 8y - 3$.

Solution

The equation has the form (3.30) with $a = 3$, $b = 8$, and $c = -3$. Since $a > 0$ the graph is a parabola which opens to the right. The vertex has ordinate $-b/2a = -8/6 = -4/3$. The abscissa of the vertex may be found by substituting $-4/3$ for y in the given equation. Doing this gives us

$$x = 3(-\tfrac{4}{3})^2 + 8(-\tfrac{4}{3}) - 3 = -\tfrac{25}{3}.$$

Hence the vertex is the point $(-25/3, -4/3)$. To find the y-intercepts of the graph we let $x = 0$ in the given equation obtaining

$$0 = 3y^2 + 8y - 3 = (3y - 1)(y + 3).$$

Consequently, the y-intercepts are $\tfrac{1}{3}$ and -3. The x-intercept -3 is found by setting $y = 0$ in the equation of the parabola. Plotting the vertex and the points corresponding to the intercepts leads to the sketch in Fig. 3.37.

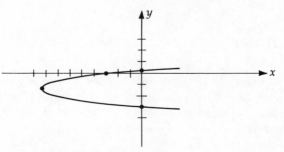

Figure 3.37 $x = 3y^2 + 8y - 3$

The graph of the equation $x^2 + y^2 = a^2$, where $a > 0$, is a circle of radius a with center at the origin (see (3.8)). A generalization of this equation is

(3.31) $$\frac{x^2}{a^2} + \frac{y^2}{b^2} = 1,$$

where a and b are positive real numbers. Note that (3.31) reduces to the circle equation when $a = b$. To find the x-intercepts of the graph we let $y = 0$ in (3.31), obtaining $\pm a$. Similarly, letting $x = 0$ gives us the y-intercepts $\pm b$. We may solve (3.31) for y in terms of x as follows:

$$\frac{y^2}{b^2} = 1 - \frac{x^2}{a^2} = \frac{a^2 - x^2}{a^2}$$

$$y^2 = \frac{b^2}{a^2}(a^2 - x^2)$$

(3.32) $$y = \pm \frac{b}{a}\sqrt{a^2 - x^2}.$$

In order to obtain points on the graph, the radicand $a^2 - x^2$ must be nonnegative. This will be true if $-a \leq x \leq a$. Consequently the entire graph of (3.31) lies between the vertical lines $x = -a$ and $x = a$.

For each permissible value of x in (3.32) there correspond two values for y. Let us consider the nonnegative values given by

$$y = \frac{b}{a}\sqrt{a^2 - x^2}.$$

If we let x vary from $-a$ to 0, we see that y increases from 0 to b. As x varies from 0 to a, y decreases from b to 0. This gives us the upper half of the graph in Fig. 3.38.

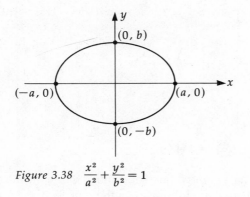

Figure 3.38 $\frac{x^2}{a^2} + \frac{y^2}{b^2} = 1$

The lower half is the graph of $y = (-b/a)\sqrt{a^2 - x^2}$. The graph of (3.31) is called an **ellipse** with center at the origin.

Multiplying both sides of (3.31) by a^2b^2 we obtain

$$b^2x^2 + a^2y^2 = a^2b^2,$$

which may be written in the form

(3.33) $Ax^2 + By^2 = C$

where A, B, and C are positive real numbers. Knowing that (3.33) is the equation of an ellipse, the graph may be readily sketched by using the x and y-intercepts and plotting several points.

EXAMPLE 2   Sketch the graph of $9x^2 + 16y^2 = 144$.

Solution

The equation has form (3.33) with $A = 9$, $B = 16$, and $C = 144$. Hence the graph is an ellipse with center at the origin. The x-intercepts (found by letting $y = 0$) are ± 4, and the y-intercepts (found by letting $x = 0$) are ± 3. Let us next locate the points on the graph with abscissa 2. Substituting $x = 2$ in the given equation we obtain

$$9(2)^2 + 16y^2 = 144, \quad \text{or} \quad y^2 = \tfrac{108}{16} = \tfrac{27}{4},$$

and hence

$$y = \pm \frac{\sqrt{27}}{2} \approx 2.6.$$

Consequently the points $(2, \pm \sqrt{27}/2)$ are on the graph. Similarly $(-2, \pm \sqrt{27}/2)$ are in the solution set of the given equation. The graph is sketched in Fig. 3.39.

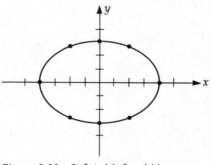

Figure 3.39 $9x^2 + 16y^2 = 144$

The horizontal and vertical line segments which join the x and y-intercepts in Fig. 3.39 are called the **axes** of the ellipse. The longer of the two axes may be on the y-axis, as in the graph of $16x^2 + 9y^2 = 144$ sketched in Fig. 3.40.

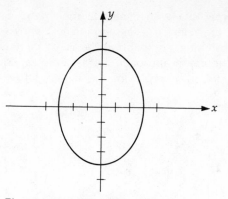

Figure 3.40 $16x^2 + 9y^2 = 144$

An equation similar to (3.31) is

(3.34) $\dfrac{x^2}{a^2} - \dfrac{y^2}{b^2} = 1.$

The x-intercepts of the graph are $\pm a$; however, there are no y-intercepts since the equation $-y^2/b^2 = 1$ has no solutions (Why?). Solving (3.34) for y in terms of x we obtain

(3.35) $y = \pm \dfrac{b}{a}\sqrt{x^2 - a^2}.$

To obtain points on the graph we must have $x^2 - a^2 \geq 0$ (Why?). This will be true if $x \geq a$ or $x \leq -a$. We shall not discuss the details of determining the shape of the graph of (3.34). For any specific equation we could plot points; however, this is a cumbersome and time-consuming task. It can be shown that the graph has the general appearance illustrated in Fig. 3.41.

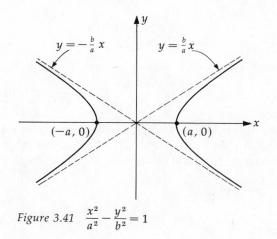

Figure 3.41 $\dfrac{x^2}{a^2} - \dfrac{y^2}{b^2} = 1$

145

The graph is called a **hyperbola** with center at the origin. The two dashed lines in the figure are referred to as **asymptotes.** They serve as excellent guidelines for sketching the graph since as $|x|$ increases, the corresponding points on the graph approach the asymptotes. Equations of the asymptotes may be found by replacing the 1 in (3.34) by 0. This gives us

$$\frac{x^2}{a^2} - \frac{y^2}{b^2} = 0, \quad \text{or} \quad \frac{y^2}{b^2} = \frac{x^2}{a^2},$$

and hence

$$y = \pm \frac{b}{a} x.$$

If we multiply both sides of (3.34) by a^2b^2, we obtain an equation of the form

(3.36) $Ax^2 - By^2 = C,$

where A, B, and C are positive real numbers. The x-intercepts and the equations of the asymptotes (obtained from $Ax^2 - By^2 = 0$) can be used to obtain a rough sketch of the graph.

EXAMPLE 3 Sketch the graph of $4x^2 - 9y^2 = 36$.

Solution

Since the equation has form (3.36), the graph is a hyperbola with center at the origin. The x-intercepts (obtained by letting $y = 0$) are ± 3 and there are no y-intercepts. The equations of the asymptotes may be found by replacing the number 36 by 0 in the given equation. This gives us

$$4x^2 - 9y^2 = 0, \quad \text{or} \quad y^2 = \tfrac{4}{9}x^2,$$

and hence

$$y = \pm \tfrac{2}{3}x.$$

Plotting the x-intercepts and using the asymptotes as guidelines leads to the sketch in Fig. 3.42. The student may find it instructive to plot several other points on the graph.

The graph of the equation

(3.37) $Ay^2 - Bx^2 = C,$

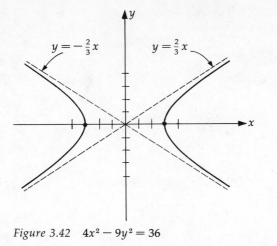

Figure 3.42 $4x^2 - 9y^2 = 36$

where A, B, and C are positive real numbers, is also a hyperbola. To sketch the graph we begin by finding the y-intercepts. There are no x-intercepts (Why?). Equations for the asymptotes are obtained by replacing C by 0 in (3.37). It is left to the reader to show that the graph of $4y^2 - 9x^2 = 36$ has the shape illustrated in Fig. 3.43.

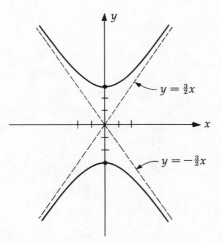

Figure 3.43 $4y^2 - 9x^2 = 36$

Each of the geometric figures considered in this section can be obtained by intersecting a double-napped right circular cone with a plane, as illustrated in Fig. 3.44. For this reason they are called **conic sections,** or simply **conics.** The conic sections were studied extensively by the early Greek mathematicians who used the methods of Euclidean geometry. A remarkable fact about conic sections is that although they were studied thousands of years ago, they are far from being obsolete. Indeed, they are important tools for current investigations in outer space

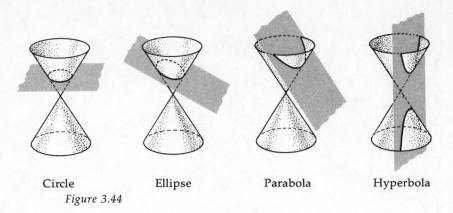

| Circle | Ellipse | Parabola | Hyperbola |

Figure 3.44

and for the study of the behavior of atomic particles. Several applications of parabolas were mentioned in Section 6. There are also numerous applications involving ellipses and hyperbolas. For example, orbits of planets are ellipses. If the ellipse is very flat, the curve resembles the path of a comet. Elliptic gears or cams are sometimes used in machines. The hyperbola is useful for describing the path of an alpha particle in the electric field of the nucleus of an atom. The interested person can find many other applications of conic sections.

EXERCISES

Sketch the graph of the equation in each of Exercises 1–24.

1 $x = 4y^2$

2 $2x + y^2 = 0$

3 $x = y^2 - 4y + 3$

4 $x = y^2 + y - 2$

5 $x = -2y^2 + 7y - 5$

6 $x = 2y - 3y^2$

7 $y = 2x^2 + 5x + 2$

8 $y = x^2 - 7x + 6$

9 $4x^2 + 9y^2 = 36$

10 $16x^2 + 25y^2 = 400$

11 $4x^2 + y^2 = 16$

12 $y^2 + 9x^2 = 9$

13 $5x^2 + 2y^2 = 10$

14 $x^2 + 4y^2 = 16$

15 $4x^2 + 25y^2 = 1$

16 $10y^2 + x^2 = 5$

17 $16y^2 - 9x^2 = 144$

18 $16y^2 - 49x^2 = 784$

19 $9x^2 - 16y^2 = 144$

20 $25x^2 - 16y^2 = 400$

21 $y^2 - 4x^2 = 16$

22 $x^2 - 2y^2 = 8$

23 $x^2 - y^2 = 1$

24 $y^2 - 16x^2 = 1$

25 Find an equation of an ellipse with center at the origin, x-intercepts ±4, and y-intercepts ±2.

26 Determine B so that the point $(-2, 1)$ is on the conic $3x^2 + By^2 = 4$. Is the conic an ellipse or a hyperbola?

27 If a square with sides parallel to the coordinate axes is inscribed in the ellipse having equation (3.31), express the area A of the square in terms of a and b.

28 The **eccentricity** e of the ellipse having equation (3.31) is defined as the ratio $\sqrt{a^2 - b^2}/a$. Prove that $0 < e < 1$. If a is fixed and b varies, describe the general shape of the ellipse when the eccentricity is close to 1 and when it is close to 0.

29 The graphs of the equations

$$\frac{x^2}{a^2} - \frac{y^2}{b^2} = 1 \quad \text{and} \quad \frac{x^2}{a^2} - \frac{y^2}{b^2} = -1$$

are called **conjugate hyperbolas.** Sketch the graphs of both equations on the same coordinate plane if $a = 2$ and $b = 5$. Describe the relationship between the two graphs.

30 Find an equation of a hyperbola having x-intercepts ± 3 and asymptotes $y = \pm 2x$.

9 VARIATION

In this section we shall introduce terminology which is used in science for describing relationships among variables quantities. In the following definitions we shall use the letters u, v, w, and s for variables. The domains and ranges of the variables will not be specified. In any particular problem these should be evident.

(3.38)   Definition

The phrase u **varies directly as** v, or u **is directly proportional to** v, means that $u = kv$, for some real number k.

The number k is sometimes called the **constant of variation** or the **constant of proportionality.** For example, if an automobile is moving at a rate of 50 miles per hour, then the distance d it travels in t hours is given by $d = 50t$. Hence the distance d is directly proportional to the time t and the constant of proportionality is 50.

(3.39)   Definition

If n is a positive real number, then the phrase u **varies directly as the** **nth power of** v or u **is directly proportional to the nth power of** v means that $u = kv^n$, for some real number k.

149

To illustrate, the formula $A = \pi r^2$ for the area of a circle states that the area A varies directly as the square of the radius r. The constant of proportionality is π. As another illustration, the formula $V = \frac{4}{3}\pi r^3$ for the volume of a sphere of radius r states that the volume V is directly proportional to the cube of the radius. The constant of proportionality in this case is $\frac{4}{3}\pi$.

(3.40) Definition

The phrase u **varies inversely as** v means that $u = k/v$, where k is a real number. If $u = k/v^n$ for some positive real number n, then u **varies inversely as the nth power of v.**

Note that in direct variation the dependent variable increases numerically as the independent variable increases, whereas in inverse variation the absolute value of the dependent variable decreases as the independent variable increases.

In many applied problems the constant of proportionality can be determined by examining experimental facts. This is illustrated in the following example.

EXAMPLE 1 If the temperature remains constant, then the pressure of an enclosed gas is inversely proportional to the volume. The pressure of a certain gas within a spherical balloon of radius 9 inches is 20 pounds per square inch. If the radius of the balloon increases to 12 inches, find the new pressure of the gas.

Solution

The original volume is $\frac{4}{3}\pi(9)^3 = 972\pi$ cubic inches. If we denote the pressure by P and the volume by V, then from (3.40) we have

$$P = \frac{k}{V},$$

for some real number k. Since $P = 20$ when $V = 972\pi$,

$$20 = \frac{k}{972\pi},$$

and hence $k = 20(972\pi) = 19440\pi$. Consequently a formula for P is

$$P = \frac{19440\pi}{V}.$$

If the radius is 12 inches, then $V = \frac{4}{3}(12)^3\pi = 2304\pi$ cubic inches. Substituting this number for V in the previous equation gives us

$$P = \frac{19440\pi}{2304\pi} = \frac{135}{16} = 8.4375.$$

Thus the pressure is 8.4375 pounds per square inch when the radius is 12 inches.

Several independent variables often occur in the same problem. Some of these situations are considered below.

(3.41) Definition

The phrase w **varies jointly as** u **and** v means $w = kuv$, for some real number k. If $w = ku^n v^m$ for positive real numbers n and m, then w **varies jointly as the** n**th power of** u **and the** m**th power of** v.

It should be obvious how (3.41) may be extended to the case of more than two variables.

Finally, combinations of the above types of variation may occur. It is inconvenient to discuss all possible situations, but perhaps the following illustration will suffice. If a variable s varies jointly as u and the cube of v and inversely as the square of w, then

$$s = k \frac{uv^3}{w^2},$$

where k is some real number.

EXAMPLE 2 The weight that can be safely supported by a beam with a rectangular cross section varies jointly as the width and square of the depth of the cross section and inversely as the length of the beam. If a 2-inch by 4-inch beam which is 8 feet long safely supports a load of 500 pounds, what weight can be safely supported by a 2-inch by 8-inch beam which is 10 feet long? (Assume that the width is the *shorter* dimension of the cross section.)

Solution

If the width, depth, length, and weight are denoted by $w, d, l,$ and W respectively, then

$$W = k \frac{wd^2}{l}.$$

According to the given data,

$$500 = k \frac{2(4^2)}{8}.$$

151

Solving for k we obtain $k = 125$, and hence the formula for W is

$$W = 125 \left(\frac{wd^2}{l}\right).$$

To answer the question we substitute $w = 2$, $d = 8$, and $l = 10$, obtaining

$$W = 125 \left(\frac{2 \cdot 8^2}{10}\right) = 1600 \text{ pounds.}$$

EXERCISES

In each of Exercises 1–6 express each statement as a formula and determine the constant of variation by means of the given conditions.

1   s is directly proportional to t. If $t = 10$, then $s = 4$.

2   y varies directly as x and inversely as z. If $x = 3$ and $z = -2$, then $y = \frac{1}{4}$.

3   s varies directly as t and inversely as d. If $t = 8$ and $d = 2$, then $s = 12$.

4   y varies directly as the square of x and inversely as the cube of z. If $x = 3$ and $z = 2$, then $y = 5$.

5   w varies jointly as the square of u and the cube of v. If $u = 2$ and $v = -3$, then $w = 9$.

6   r varies jointly as p and v and inversely as the square of s. If $p = 4$, $v = 3$ and $s = 2$, then $r = 20$.

7   Hooke's Law states that the force required to stretch a spring x units beyond its natural length is directly proportional to x. If a weight of 4 pounds stretches a spring from its natural length of 10 inches to a length of 10.3 inches, what weight will stretch it to a length of 11.5 inches?

8   If an object is dropped from a position near the surface of the earth, then the distance s that it falls in t seconds varies as the square of t. If $s = 36$ when $t = 1.5$, find a formula for s as a function of t. If an object is dropped from a height of 1000 feet, how long does it take to reach the ground?

9   The volume V of a gas varies directly as the temperature T and inversely as the pressure P. Find a formula for V in terms of T and P, if $V = 50$ when $P = 40$ and $T = 300$. What is V when $T = 320$ and $P = 30$?

10   The kinetic energy K of an object in motion varies jointly as its mass m and the square of its velocity v. If an object weighing 25 pounds and moving with a velocity of 90 feet per second has a kinetic energy of 400 foot-pounds, find its kinetic energy when the velocity is 130 feet per second.

11   The electrical resistance of a wire varies directly as its length and inversely

as the square of its diameter. If a wire 100 feet long of diameter 0.01 inches has a resistance of 25 ohms, find the resistance in a wire made of the same material which has a diameter of 0.015 inches and is 50 feet long.

12 The intensity of illumination *I* from a source of light varies inversely as the square of the distance *d* from the source. If a searchlight has an intensity of 1,000,000 candlepower at 50 feet, what is the intensity at a distance of 1 mile?

13 The period of a simple pendulum, that is, the time required for one complete oscillation, varies directly as the square root of its length. If a pendulum 2 feet long has a period of 1.5 seconds, find the period of a pendulum 6 feet long.

14 The distance a ball rolls down an inclined plane varies as the square of the time. If a ball rolls 5 feet in the first second, how far will it roll in 4 seconds?

10 REVIEW EXERCISES

Oral

Define or discuss each of the following.

1   Ordered pair

2   Rectangular coordinate system in a plane

3   The abscissa and ordinate of a point

4   The distance formula   **5**   The midpoint formula

6   The product set of two sets   **7**   A relation on a set

8   The graph of a relation   **9**   Unit circle

10 The graph of an equation in x and y

11 The standard equation of a circle

12 Function

13 The domain and range of a function

14 The graph of a function   **15** Increasing function

16 Decreasing function   **17** One-to-one function

18 Identity function   **19** Constant function

20 Slope of a line

21 The point-slope form for the equation of a line

22 The slope-intercept form   **23** Linear equation in x and y

24 Polynomial function   **25** Linear function

Written

1 Given the points $A(2, 1)$, $B(-1, 4)$, and $C(-2, -3)$:

    a Prove that A, B, and C are the vertices of a right triangle and find its area.

    b Find the coordinates of the midpoint of AB.

    c Find the slope of the line through B and C.

    d Find an equation of the line through C that is parallel to the line through A and B.

2 Sketch the graph of each of the following equations.

    a $3x - 5y = 10$ **b** $x + 3 = 0$

    c $x^2 + y = 4$ **d** $y = 2x^2 + 3x - 5$

    e $x = 24 - 2y - y^2$ **f** $y = \sqrt{4 - x^2}$

    g $(x - 2)^2 + y^2 = 25$ **h** $y = x^3 + 5x^2 - x - 5$

3 Sketch the graph of each of the following relations:

    a $W = \{(x, y) : x > 0\}$ **b** $W = \{(x, y) : y > x\}$

    c $W = \{(x, y) : x^2 + y^2 < 1\}$ **d** $W = \{(x, y) : x^2 + y^2 = 0\}$

4 Find an equation of the circle:

    a with center $C(4, -7)$ and containing the origin

    b with center $C(-4, -3)$ and tangent to the line $x = 5$

    c with center $C(6, -2)$ and passing through the vertex of the parabola having equation $y = 3x^2 + 12x - 7$.

5 Find the center and radius of each of the following circles:

    a $x^2 + y^2 - 10x + 4y + 20 = 0$

    b $x^2 + y^2 - 8x - 9 = 0$

    c $x^2 + y^2 + 6x - 6y = 0$

6 In each of the following, sketch the graph of f and determine the intervals on which f is increasing or decreasing.

    a $f(x) = 1 - 4x^2$ **b** $f(x) = 100$

    c $f(x) = -1/(x + 1)$ **d** $f(x) = |x + 5|$

    e $f(x) = 12x^2 + 5x - 3$ **f** $f(x) = 36x - 4x^2$

7 If $f(x) = 1/\sqrt{x+1}$ find:

a $f(1)$　　　　**b** $f(3)$　　　　**c** $f(0)$　　　　**d** $f(\sqrt{2}-1)$

e $f(-x)$　　　**f** $-f(x)$　　　**g** $f(x^2)$　　　**h** $(f(x))^2$

8 **a** Find an equation of the line through $A(-7, 2)$ that is parallel to the line $5x - 2y + 1 = 0$.

b Find an equation of a line with slope $\frac{2}{3}$ and x-intercept -6.

c Find an equation of the line having slope 0 and y-intercept 4.

9 **a** Find an equation for the line through the points $A(4, -5)$ and $B(1, 6)$.

b Find an equation for the line through $P(12, -6)$ that is parallel to the y-axis.

c Express the equation $3x + 8y - 24 = 0$ in slope-intercept form.

10 Find $(f \circ g)(x)$ and $(g \circ f)(x)$ if:

a $f(x) = x^2 + 3x + 1,\quad g(x) = 2x - 1$

b $f(x) = x^2 + 4,\quad g(x) = \sqrt{2x + 5}$

c $f(x) = 5x + 2,\quad g(x) = 1/x^2$

11 Find the inverse function of f if

a $f(x) = 5 - 7x$

b $f(x) = 4x^2 + 3,\quad x \geq 0$

12 Sketch the graph of each of the following equations.

a $25x^2 + 9y^2 = 225$　　　　　**b** $4x^2 + 7y^2 = 28$

c $4y^2 - 25x^2 = 100$　　　　　**d** $y^2 - x^2 = 4$

13 If the altitude and radius of a right circular cylinder are equal, express the volume V as a function of the circumference C of the base.

14 Find a formula which expresses the fact that w varies jointly as x and the square of y, and inversely as z, if $w = 20$ when $x = 2$, $y = 3$, and $z = 4$.

4

Exponential and Logarithmic Functions

Most of the functions considered in the preceding chapter were **algebraic functions** *in the sense that f(x) was an algebraic expression in x. Functions which are not algebraic are termed* **transcendental.** *In this chapter we shall define two important types of transcendental functions and investigate some of their properties.*

1 EXPONENTIAL FUNCTIONS

Throughout this section the letter a will denote a positive real number. In Chapter One we defined a^r, where r is any rational number. It is also possible to define a unique real number a^x for every real number x (rational or irrational) in such a way that the laws of exponents (1.22) remain valid. To illustrate, given a number such as a^π, we may use the nonterminating decimal representation $3.14159 \cdots$ for π and consider the numbers a^3, $a^{3.1}$, $a^{3.14}$, $a^{3.141}$, $a^{3.1415}$, $\cdots$. We might expect that each successive power gets closer to a^π. This is precisely what happens if a^x is properly defined. However, the definition requires deeper concepts than are available to us; consequently it is better to leave it for a more advanced mathematics course such as calculus.

Although definitions and proofs are omitted we shall assume, henceforth, that formulas such as $a^x a^y = a^{x+y}$, $(a^x)^y = a^{xy}$, and so on, are valid for all real numbers x and y.

Before continuing the discussion of real exponents, we shall establish several results about rational exponents. Let us first show that if $a > 1$, then $a^r > 1$ for every positive rational number r. If $a > 1$, then

156

multiplying both sides by a we obtain $a^2 > a$, and hence $a^2 > a > 1$. Multiplying by a again we obtain $a^3 > a^2 > a > 1$. Continuing, we see that $a^n > 1$ for all positive integers n. (A complete proof requires the method of mathematical induction to be discussed in Chapter 10.) Similarly, if $0 < a < 1$, it follows that $a^n < 1$ for all positive integers n. Next suppose that $a > 1$ and $r = p/q$, where p and q are positive integers. If it were true that $a^{p/q} \leq 1$, then from the previous discussion $(a^{p/q})^q \leq 1$, or $a^p \leq 1$, which contradicts the fact that $a^n > 1$ for all positive integers n. Consequently, $a^r > 1$ for every positive rational number r. We may use that fact to prove the following theorem.

(4.1)   Theorem

If $a > 1$ and r, s are rational numbers such that $r < s$, then $a^r < a^s$.

   Proof

   If $r < s$, then $s - r$ is a positive rational number and from above, $a^{s-r} > 1$. Multiplying both sides by a^r we obtain $a^s > a^r$, which is what we wished to prove.

   Since to each real number x there corresponds a unique real number a^x, we can define a function as follows.

(4.2)   Definition

If $a > 0$, then the **exponential function f with base** a is defined by

$$f(x) = a^x,$$

where x is any real number.

   It is possible to extend Theorem (4.1) to the case of real exponents r and s. Specifically, if $a > 1$ and x_1, x_2 are real numbers such that $x_1 < x_2$, then it can be shown that $a^{x_1} < a^{x_2}$, that is, $f(x_1) < f(x_2)$. This means that if $a > 1$, then the exponential function f with base a is *increasing* for all real numbers. It can also be shown that if $0 < a < 1$, then f is decreasing for all real numbers.

EXAMPLE 1   Sketch the graph of f if $f(x) = 2^x$.

   Solution

   Coordinates of some points on the graph are listed in the following table.

x	-3	-2	-1	0	1	2	3	4
2^x	$\frac{1}{8}$	$\frac{1}{4}$	$\frac{1}{2}$	1	2	4	8	16

Plotting, and using the fact that f is increasing, gives us the sketch in Fig. 4.1.

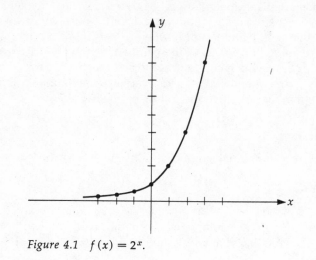

Figure 4.1 $f(x) = 2^x$.

The graph in Fig. 4.1 is typical of the exponential function (4.2) if $a > 1$. Since $a^0 = 1$, the y-intercept is always 1. Observe that as x decreases through negative values, the graph approaches the x-axis but never intersects it, since $a^x > 0$ for all x. The x-axis is called a **horizontal asymptote** for the graph. As x increases through positive values the graph rises very rapidly. Indeed, if we begin with $x = 0$ and consider successive unit changes in x, then the corresponding changes in y are 1, 2, 4, 8, 16, 32, 64, and so on. This type of variation is very common in nature and is characteristic of the **exponential law of growth.** At the end of this section we shall give several practical illustrations of this behavior.

EXAMPLE 2   If $f(x) = (3/2)^x$ and $g(x) = 3^x$, sketch the graphs of f and g on the same coordinate plane.

Solution

The table below displays coordinates of several points on the graphs.

x	-2	-1	0	1	2	3	4
$(3/2)^x$	$\frac{4}{9}$	$\frac{2}{3}$	1	$\frac{3}{2}$	$\frac{9}{4}$	$\frac{27}{8}$	$\frac{81}{16}$
3^x	$\frac{1}{9}$	$\frac{1}{3}$	1	3	9	27	81

With the aid of these points we obtain Figure 4.2, where dashes have been used for the graph of *f* in order to distinguish it from the graph of *g*.

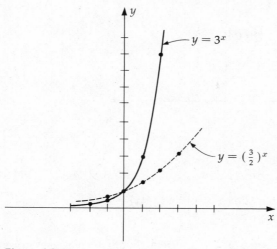

Figure 4.2

Example 2 brings out the fact that if $1 < a < b$, then $a^x < b^x$ for positive values of *x* and $b^x < a^x$ for negative values of *x*. In particular, since $\frac{3}{2} < 2 < 3$, this tells us that the graph in Example 1 lies between the graphs of the functions in Example 2.

EXAMPLE 3 Sketch the graph of the equation $y = (\frac{1}{2})^x$.

Solution

Some points on the graph may be obtained from the following table.

x	-3	-2	-1	0	1	2	3
$(\frac{1}{2})^x$	8	4	2	1	$\frac{1}{2}$	$\frac{1}{4}$	$\frac{1}{8}$

The graph is sketched in Fig. 4.3. Since $(\frac{1}{2})^x = 2^{-x}$, the graph is the same as the graph of the equation $y = 2^{-x}$.

In advanced mathematics and applications it is often necessary to consider a function *f* such that $f(x) = a^p$, where *p* is some expression in *x*. We do not intend to study such functions in detail; however, let us consider one example.

159

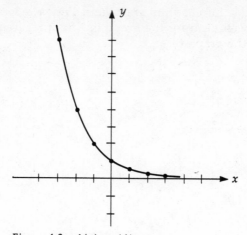

Figure 4.3 $f(x) = (\frac{1}{2})^x = 2^{-x}$

EXAMPLE 4 Sketch the graph of f if $f(x) = 2^{-x^2}$.

Solution

Since $f(x) = 1/2^{x^2}$, it follows that if x increases numerically the corresponding point $(x, f(x))$ on the graph approaches the x-axis. Thus the x-axis is a horizontal asymptote for the graph. The maximum value of $f(x)$ occurs at $x = 0$. Tabulating some coordinates of points on the graph, we have the table shown below.

x	-2	-1	0	1	2
$f(x)$	$\frac{1}{16}$	$\frac{1}{2}$	1	$\frac{1}{2}$	$\frac{1}{16}$

The graph is sketched in Fig. 4.4. Functions of this type arise in the study of the branch of mathematics called *probability*.

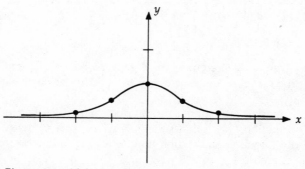

Figure 4.4 $f(x) = 2^{-x^2}$

The variation of many physical quantities can be described by means of an exponential function. One of the most common examples occurs in the growth of certain populations. As an illustration it might be observed experimentally that the number of bacteria in a culture doubles every hour. If there are 1000 bacteria present at the start of the experiment, then the experimenter would obtain the readings listed below, where t is the time in hours and $f(t)$ is the bacteria count at time t.

t	0	1	2	3	4
$f(t)$	1000	2000	4000	8000	16000

It appears that $f(t) = (1000)2^t$. This formula makes it possible to predict the number of bacteria present at any time t. For example, at $t = 1.5$ we have

$$f(t) = (1000)2^{3/2} = 1000\ \sqrt{2^3} = 1000\sqrt{8} \approx 2828.$$

Exponential growth patterns of this type may also be observed in some human and animal populations.

Compound interest provides another illustration of exponential growth. If a sum of money P (called the **principal**) is invested at a simple interest rate of i percent, then the interest at the end of one interest period is Pi. For example, if $P = \$100$ and i is 8% per year, then the interest at the end of one year is $\$100(.08)$, or $\$8$. If the interest is reinvested at the end of this period then the new principal is

$$P + Pi, \quad \text{or} \quad P(1 + i).$$

Note that to find the new principal we multiply the original principal by $(1 + i)$. In the illustration above, the new principal is $\$100(1.08)$, or $\$108$.

If another time period elapses then the new principal may be found by multiplying $P(1 + i)$ by $(1 + i)$. Thus the principal after two time periods is $P(1 + i)^2$. If we reinvest and continue this process, the principal after three periods is $P(1 + i)^3$; after four it is $P(1 + i)^4$; and in general, after n time periods the principal P_n is given by

$$P_n = P(1 + i)^n.$$

Interest accumulated in this way is called **compound interest.** We see that the principal is given in terms of an exponential function whose base is $1 + i$ and exponent is n. The time period may vary, being measured in years, months, weeks, days, or any other suitable unit of time. When the formula for P_n above is employed, it must be remembered that i is the interest rate per time period. For example, if the rate is stated as 9% *per year compounded monthly*, then the rate per month is $\frac{9}{12}\%$ or $\frac{3}{4}\%$. In this case $i = .75\% = .0075$ and n is the number of months. The sketch in Fig. 4.5 illustrates the growth of $100 invested at this rate over a period of 15 years. We have connected the principal amounts by a smooth curve in order to indicate the growth during this time.

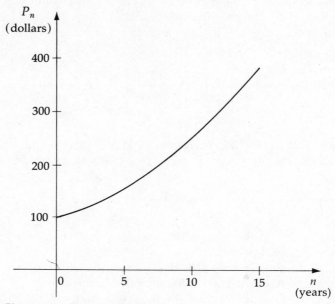

Figure 4.5   Compound Interest

Certain physical quantities may *decrease* exponentially. In this case the base a of the exponential function is between 0 and 1. One of the most common examples is the decay of a radioactive substance. As an illustration, the polonium isotope $^{210}P_0$ has a half-life of approximately 140 days; that is, given any amount, one-half of it will disintegrate in 140 days. If there is initially 20 mg present, then the table on p. 163 indicates the amount remaining after various intervals of time. The sketch in Fig. 4.6 illustrates the exponential nature of the disintegration.

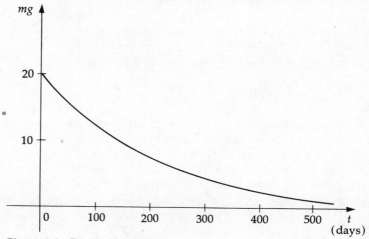

Figure 4.6   Decay of Polonium

t (days)	0	140	280	420	560
Amount remaining	20	10	5	2.5	1.25

The behavior of an electrical condenser can be used to illustrate exponential decay. If the condenser is allowed to discharge, the initial rate of discharge is relatively high, but it then tapers off as in the preceding example of radioactive decay.

EXERCISES

In Exercises 1–16 sketch the graph of the function f defined by the given expression.

1 $f(x) = 4^x$

2 $f(x) = (\frac{2}{3})^x$

3 $f(x) = 3^{-x}$

4 $f(x) = 5^x$

5 $f(x) = 10^x$

6 $f(x) = 10^{-x}$

7 $f(x) = 4^{x/2}$

8 $f(x) = 4^{-(x/2)}$

9 $f(x) = 2^{x+1}$

10 $f(x) = 2^{x-1}$

11 $f(x) = 2^{1-x}$

12 $f(x) = 2^{-x-1}$

13 $f(x) = 2^{3-x^2}$

14 $f(x) = 2^{-(3+x)^2}$

15 $f(x) = 3^x + 3^{-x}$

16 $f(x) = 3^x - 3^{-x}$

17 The half-life of radium is 1600 years; that is, given any quantity, one-half of it will disintegrate in 1600 years. If the initial amount is q_0 milligrams, then it can be shown that the quantity $q(t)$ remaining after t years is given by $q(t) = q_0 2^{kt}$. Find k.

18 The number of bacteria in a certain culture at time t is given by $Q(t) = 2(3^t)$, where t is measured in hours and $Q(t)$ in thousands. What is the initial number of bacteria? What is the number after 10 minutes? After 30 minutes? After 1 hour?

19 If $1,000 is invested at a rate of 12% per year compounded monthly, what is the principal after 1 month? 2 months? 6 months? 1 year?

20 If 10 pounds of salt is added to a certain quantity of water, then the amount $q(t)$ which is undissolved after t minutes is given by $q(t) = 10(\frac{4}{5})^t$. Sketch a graph which shows the value $q(t)$ at any time from $t = 0$ to $t = 10$.

21 Why was $a < 0$ ruled out in the discussion of a^x?

22 Use (4.1) to prove that if $0 < a < 1$ and r and s are rational numbers such that $r < s$, then $a^r > a^s$.

23 How does the graph of $y = a^x$ compare with the graph of $y = -a^x$?

24 If $a > 1$, how does the graph of $y = a^x$ compare with the graph of $y = a^{-x}$?

2 LOGARITHMS

Throughout this section and the next, it is assumed that a is a positive real number different from 1. Let us begin by examining the graph of f, where $f(x) = a^x$ (see Fig. 4.7 for the case $a > 1$). It appears that every

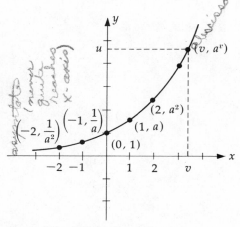

Figure 4.7 $f(x) = a^x, a > 1$

positive real number u is the ordinate of some point on the graph; that is, there is a number v such that $u = a^v$. Indeed, v is the abscissa of the point where the line $y = u$ intersects the graph. Moreover, since f is increasing on **R**, the number u can occur as an ordinate only once. A similar situation exists if $0 < a < 1$. This makes the following theorem plausible.

(4.3)   Theorem

For each positive real number u there is a unique real number v such that $a^v = u$.

A rigorous proof of (4.3) is beyond the scope of this text.

(4.4)   Definition

If u is any positive real number, then the (unique) exponent v such that $a^v = u$ is called the **logarithm of u with base a** and is denoted by $\log_a u$.

A convenient way to memorize (4.4) is by means of the following equivalent equations:

(4.5) $v = \log_a u$ *if and only if* $a^v = u.$

As illustrations of (4.5) we may write

$$3 = \log_2 8, \text{ since } 2^3 = 8,$$
$$-2 = \log_5 \tfrac{1}{25}, \text{ since } 5^{-2} = \tfrac{1}{25},$$
$$4 = \log_{10} 10,000, \text{ since } 10^4 = 10,000.$$

The fact that $\log_a u$ is the exponent v such that $a^v = u$ gives us the following important identity:

(4.6) $a^{\log_a u} = u.$

Since $a^r = a^r$ for every real number r, we may use (4.5) with $v = r$ and $u = a^r$ to obtain

(4.7) $r = \log_a a^r.$

In particular, letting $r = 1$ in (4.7) we see that

$$\log_a a = 1.$$

Since $a^0 = 1$, it follows from (4.5) that

(4.8) $\log_a 1 = 0.$

EXAMPLE 1 In each of the following find s.

(a) $s = \log_4 2.$ (b) $\log_5 s = 2.$ (c) $\log_s 8 = 3.$

Solutions

We may use (4.5) to solve each part as follows:

(a)   If $s = \log_4 2$, then $4^s = 2$ and hence $s = \tfrac{1}{2}$.
(b)   If $\log_5 s = 2$, then $5^2 = s$ and hence $s = 25$.
(c)   If $\log_s 8 = 3$, then $s^3 = 8$ and hence $s = \sqrt[3]{8} = 2$.

EXAMPLE 2 Solve the equation $\log_4 (5 + x) = 3$.

Solution

If $\log_4 (5 + x) = 3$, then by (4.5)

$$5 + x = 4^3, \quad \text{or} \quad 5 + x = 64.$$

Hence the solution set is $\{59\}$.

165

EXAMPLE 3 Solve the inequality $2 < \log_{10} x$.

Solution

Since the exponential function with base 10 is increasing, it follows that if $2 < \log_{10} x$, then

$$10^2 < 10^{\log_{10} x}.$$

By (4.6) this implies that $100 < x$. Hence the solution set is the infinite interval $(100, \infty)$.

The following laws are fundamental for all work with logarithms, where it is assumed that u and w are positive real numbers.

(4.9) **Laws of Logarithms**

    (i) $\log_a (uw) = \log_a u + \log_a w,$

    (ii) $\log_a (u/w) = \log_a u - \log_a w,$

    (iii) $\log_a (u^c) = c \log_a u,$ for every real number c.

Proof

Let

(4.10) $r = \log_a u$ and $s = \log_a w.$

By (4.5) this implies that $a^r = u$ and $a^s = w$. Consequently

$$a^r a^s = uw,$$

and then by a law of exponents,

$$a^{r+s} = uw.$$

Again using (4.5) we see that the last equation is equivalent to

$$r + s = \log_a (uw).$$

Substituting for r and s from (4.10) gives us

$$\log_a u + \log_a w = \log_a uw.$$

This proves (i).

To prove (ii) we begin as in the proof of (i), but this time we *divide* a^r by a^s, obtaining

$$\frac{a^r}{a^s} = \frac{u}{w}, \quad \text{or} \quad a^{r-s} = \frac{u}{w}.$$

According to (4.5) the latter equation is equivalent to

$r - s = \log_a (u/w).$

Substituting for r and s from (4.10) gives us (ii).

Finally, if c is any real number, then using the same notation as above,

$(a^r)^c = u^c, \quad \text{or} \quad a^{cr} = u^c.$

According to the definition of logarithm, the last equality implies that

$cr = \log_a u^c.$

Substituting for r from (4.10) we obtain

$c \log_a u = \log_a u^c.$

This proves Law (iii).

EXAMPLE 4 If $\log_a 3 = .4771$ and $\log_a 2 = .3010$, find each of the following:

(a) $\log_a 6.$ (b) $\log_a (\tfrac{3}{2}).$
(c) $\log_a \sqrt{2}.$ (d) $(\log_a 3)/(\log_a 2).$

Solutions

(a) Since $6 = 2 \cdot 3$, we may use (i) of (4.9) to obtain

$\log_a 6 = \log_a (2 \cdot 3) = \log_a 2 + \log_a 3$
$= .4771 + .3010 = .7781.$

(b) By (ii) of (4.9),

$\log_a (\tfrac{3}{2}) = \log_a 3 - \log_a 2$
$= .4771 - .3010 = .1761$

(c) Using (iii) of (4.9), we get

$\log_a \sqrt{2} = \log_a 2^{1/2} = (\tfrac{1}{2}) \log_a 2$
$= (\tfrac{1}{2})(.3010) = .1505.$

(d) There is no law of logarithms which allows us to simplify $(\log_a 3)/(\log_a 2)$. Consequently we *divide* .4771 by .3010, obtaining the approximation 1.585. It is important to notice the difference between this problem and the one stated in part (b).

EXAMPLE 5 Solve each of the following equations.

(a) $\log_5 (2x + 3) = \log_5 11 + \log_5 3$.

(b) $2 \log_7 x = \log_7 36$.

(c) $\log_4 (x+6) - \log_4 10 = \log_4 (x-1) - \log_4 2$.

Solutions

(a) Using (i) of (4.9) the given equation may be written as

$$\log_5 (2x + 3) = \log_5 (11 \cdot 3) = \log_5 33.$$

Consequently $2x + 3 = 33$, or $2x = 30$, and therefore the solution set is $\{15\}$.

(b) In order for $\log_7 x$ to exist, x must be positive. If $x > 0$, then by (iii) of (4.9) $2 \log_7 x = \log_7 x^2$ and hence the given equation is equivalent to

$$\log_7 x^2 = \log_7 36.$$

It follows that $x^2 = 36$, and hence the solution set is $\{6\}$.

(c) The given equation is equivalent to

$$\log_4 (x + 6) - \log_4 (x - 1) = \log_4 10 - \log_4 2.$$

Applying (ii) of (4.9) gives us

$$\log_4 \left(\frac{x + 6}{x - 1}\right) = \log_4 \frac{10}{2} = \log_4 5,$$

and hence

$$\frac{x + 6}{x - 1} = 5.$$

The last equation implies that

$$x + 6 = 5x - 5, \quad \text{or} \quad 4x = 11.$$

Thus the solution set is $\{11/4\}$.

Sometimes it is necessary to *change the base* of a logarithm by expressing $\log_b u$ in terms of $\log_a u$, for some positive real number b different from 1. This can be accomplished as follows. We begin with the equivalent equations

$$v = \log_b u \quad \text{and} \quad b^v = u.$$

Taking the logarithm, base a, of both sides of the second equation gives us

$$\log_a b^v = \log_a u.$$

Applying (iii) of (4.9), we get

$$v \log_a b = \log_a u.$$

Solving for v (that is, $\log_b u$) we obtain

(4.11) $\quad \log_b u = \dfrac{\log_a u}{\log_a b}.$

The most important special case of (4.11) is obtained by letting $u = a$. Since $\log_a a = 1$, this gives us

(4.12) $\quad \log_b a = \dfrac{1}{\log_a b}.$

EXAMPLE 6 Verify (4.12) if $a = 2$ and $b = 4$.

Solution

From part (a) of Example 1, $\log_4 2 = \frac{1}{2}$. We also have the following equivalent equations

$$s = \log_2 4; \quad 2^s = 4; \quad s = 2.$$

Hence

$$\log_4 2 = \frac{1}{2} = \frac{1}{\log_2 4}.$$

The laws of logarithms are often used as in the next example.

EXAMPLE 7 Express $\log_a \dfrac{x^3 \sqrt{y}}{z^2}$ in terms of the logarithms of x, y, and z.

Solution

Using the three laws stated in (4.9), we have

$$\log_a \frac{x^3 \sqrt{y}}{z^2} = \log_a (x^3 \sqrt{y}) - \log_a z^2$$
$$= \log_a x^3 + \log_a \sqrt{y} - \log_a z^2$$
$$= 3 \log_a x + (\tfrac{1}{2}) \log_a y - 2 \log_a z.$$

169

The procedure in Example 7 can also be reversed; that is, beginning with the final equation we can retrace our steps to obtain the original expression (see Exercises 75–78).

As a final remark, note that there is no general law for expressing $\log_a (u + w)$ in terms of simpler logarithms. It is evident that this does not always equal $\log_a u + \log_a w$, since the latter equals $\log_a (uw)$.

EXERCISES

Use (4.5) to change each of the equations in Exercises 1–8 to logarithmic form.

1 $4^2 = 16$

2 $3^4 = 81$

3 $2^4 = 16$

4 $3^2 = 9$

5 $10^{-4} = 0.0001$

6 $10^{-3} = 0.001$

7 $r^s = t$

8 $u^v = w$

Use (4.5) to change each of the equations in Exercises 9–16 to exponential form.

9 $\log_5 125 = 3$

10 $\log_4 64 = 3$

11 $\log_4 \left(\frac{1}{64}\right) = -3$

12 $\log_3 (1/27) = -3$

13 $\log_{10} 1 = 0$

14 $\log_a 1 = 0$

15 $\log_p s = t$

16 $\log_s p = t$

Find each of the numbers in Exercises 17–30.

17 $\log_3 \left(\frac{1}{9}\right)$

18 $\log_2 (64)$

19 $\log_{10} (1000)$

20 $\log_9 (81)$

21 $10^{\log_{10} 7}$

22 $\log_{10} (0.001)$

23 $\log_8 \sqrt[5]{8}$

24 $10^{3 \log_{10} 2}$

25 $\log_{10} \left(\frac{1}{10}\right)$

26 $\log_{10} (10{,}000)$

27 $\log_{1/2} 4$

28 $\log_2 (1/8)$

29 $10^{5 \log_{10} 2}$

30 $\log_4 \sqrt[3]{4}$

Given $\log_a 3 = .5$ and $\log_a 2 = .3$, change the expressions in Exercises 31–40 to decimal form.

31 $\log_a (2/3)$

32 $\log_a (3/2)$

33 $\log_a (3^2)$

34 $\log_a (2^3)$

35 $\log_a (1/2)$ .3

36 $\log_a 18$

37 $\log_a (1/\sqrt[3]{2})$

38 $\log_a (3 - 2)$

39 $\log_2 a$

40 $(\log_2 a)(\log_a 2)$

Find the solution sets in Exercises 41–66.

41 $\log_4 (x - 3) = 2$

42 $\log_6 x = \frac{1}{3}$

43 $\log_8 x = 2/3$

44 $\log_3 (x + 5) = -1$

45 $\log_{10} x^2 = -2$

46 $\log_{10} (x^2) = -5$

47 $\log_2 (x^2 + 3x + 4) = 1$

48 $\log_2 (x^2 - x - 2) = 2$

49 $\log_x 5 = 2$

50 $10^{\log_{10} x} = 2/13$

51 $\log_5 (x + 1) < 2$

52 $\log_3 (2x - 1) > 1$

53 $1 < \log_{10} x < 2$

54 $2 < \log_{10} x < 3$

55 $\log_3 2x = \log_3 3 + \log_3 5$

56 $\log_5 2x = \log_5 8 - \log_5 3$

57 $\log_6 (3x + 1) = \log_6 10 - \log_6 2$

58 $3 \log_2 x = 2 \log_2 8$

59 $\log_4 x - \log_4 (x - 1) = 2 \log_4 3$

60 $2 \log_5 \sqrt{x} = 3$

61 $\log_{10} x^2 - \log_{10} 5 = \log_{10} 7 + \log_{10} 2x$

62 $\log_{10} x^2 = \log_{10} x$

63 $\log_5 (x - 1) + \log_5 (x - 2) = 2 \log_5 \sqrt{6}$

64 $\frac{1}{3} \log_8 (x + 1) = 2 \log_8 3 - \frac{2}{3} \log_8 (x + 1)$

65 $\log_{10} |3x - 1| < 2$

66 $\log_4 \sqrt{x} > 2$

In each of Exercises 67–74 express the logarithm in terms of logarithms of x, y, and z.

67 $\log_a \dfrac{xy^3}{z^2}$

68 $\log_a \dfrac{x^2 y^3}{z^4}$

69 $\log_a \dfrac{\sqrt[3]{x}\, z^2}{y}$

70 $\log_a \dfrac{z^2 \sqrt{x}}{y^3}$

71 $\log_a \sqrt[4]{\dfrac{x}{y^2 z^5}}$

72 $\log_a \dfrac{\sqrt[3]{x}\, \sqrt{y}}{z}$

73 $\log_a \sqrt{x \sqrt{y}}$

74 $\log_a \sqrt{x \sqrt{yz}}$

In each of Exercises 75–78 write the expression as one logarithm.

75 $3 \log_a x + \frac{1}{2} \log_a (x + 3) - 4 \log_a (x - 2)$

76 $2 \log_a x + (1/5) \log_a (x + 1) - 3 \log_a (x - 1)$

171

77 $\log_a (yx^2) + 2 \log_a \left(\dfrac{y}{x}\right) - 3 \log_a y \sqrt[3]{x}$

78 $\log_a (y/x) + 2 \log_a x - \log_a (xy^2)$

3 LOGARITHMIC FUNCTIONS

The concept of logarithm can be used to introduce a new function whose domain is the set of positive real numbers.

(4.13)   Definition

The function f defined by

$$f(x) = \log_a x$$

for all positive real numbers x is called the **logarithmic function with base** a.

The graph of f is the same as the graph of the equation $y = \log_a x$ which, by (4.5), is equivalent to

$$x = a^y.$$

In order to find some pairs in the solution set of the preceding equation, we may substitute for y and find the corresponding values of x, as illustrated in the following table.

y	-3	-2	-1	0	1	2	3
x	$\dfrac{1}{a^3}$	$\dfrac{1}{a^2}$	$\dfrac{1}{a}$	1	a	a^2	a^3

If $a > 1$, we obtain the sketch in Fig. 4.8. In this case f is an increasing function throughout its domain. If $0 < a < 1$, then the graph has the general shape shown in Fig. 4.9, and hence f is a decreasing function. Note that for every a under consideration the region to the left of the y-axis is excluded. There is no y-intercept and the x-intercept is 1.

Functions defined by expressions of the form $\log_a p$, where p is some expression in x, often occur in mathematics and its applications. Functions of this type are classified as members of the logarithmic family; however, the graphs may differ from those sketched in Figs. 4.8 and 4.9, as is illustrated in the following examples.

EXAMPLE 1   Sketch the graph of f if $f(x) = \log_3 (-x)$, $x < 0$.

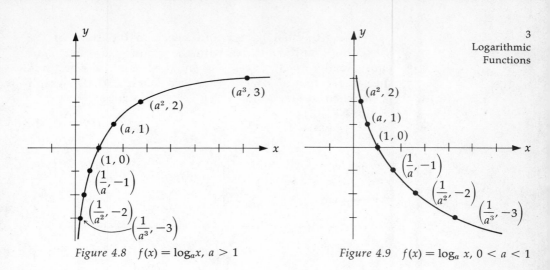

Figure 4.8 $f(x) = \log_a x,\ a > 1$

Figure 4.9 $f(x) = \log_a x,\ 0 < a < 1$

Solution

If $x < 0$, then $-x > 0$ and hence $\log_3 (-x)$ is defined. We wish to sketch the graph of the equation $y = \log_3 (-x)$, or equivalently, $3^y = -x$. The following table displays coordinates of some points on the graph, which is sketched in Fig. 4.10.

y	-2	-1	0	1	2
x	$-\frac{1}{9}$	$-\frac{1}{3}$	-1	-3	-9

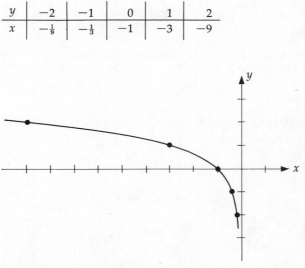

Figure 4.10 $f(x) = \log_3 (-x)$

EXAMPLE 2   Sketch the graph of the equation $y = \log_3 |x|,\ x \neq 0$.

Solution

Since $|x| > 0$ for all $x \neq 0$, there are points on the graph corresponding to negative values of x as well as to positive val-

173

ues. If $x > 0$, then $|x| = x$ and hence to the right of the y-axis the graph coincides with the graph of $y = \log_3 x$, or equivalently, $x = 3^y$. If $x < 0$, then $|x| = -x$ and the graph is the same as that of $y = \log_3 (-x)$ (see Example 1). The graph is sketched in Fig. 4.11.

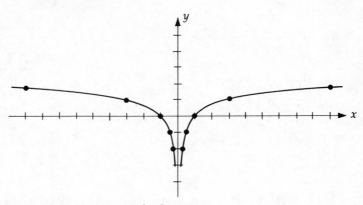

Figure 4.11 $f(x) = \log_3 |x|$

There is a close relationship between exponential and logarithmic functions. This is to be expected since the logarithmic function was defined in terms of the exponential function. A precise description of the relationship is stated in the following theorem.

(4.14)   Theorem

The exponential and logarithmic functions with base a are inverse functions of one another.

Proof

If $f(x) = a^x$ and $g(x) = \log_a x$, then according to (3.27) we must show that

$$f(g(x)) = x \quad \text{and} \quad g(f(x)) = x.$$

For all positive real numbers x we have

$$
\begin{aligned}
f(g(x)) &= a^{g(x)} & \text{(definition of } f) \\
&= a^{\log_a x} & \text{(definition of } g) \\
&= x & (4.6).
\end{aligned}
$$

For all real numbers x, we have

$$
\begin{aligned}
g(f(x)) &= \log_a f(x) & \text{(definition of } g) \\
&= \log_a a^x & \text{(definition of } f) \\
&= x & (4.7).
\end{aligned}
$$

This proves the theorem.

Logarithmic functions occur very frequently in applications. Indeed, if two variables u and v are related so that u is an exponential function of v, then it follows from (4.14) that v is a logarithmic function of u. As a specific example, if \$1.00 is invested at a rate of 9% per year compounded monthly, then from the discussion in Section 1 the principal P after n interest periods is

$$P = (1.075)^n.$$

By (4.5), the logarithmic form of this equation is

$$n = \log_{1.075} P.$$

Thus, the number of periods n required for \$1.00 to grow to an amount P is related logarithmically to P.

Another example we considered in Section 1 was that of population growth. In particular, the equation for the number N of bacteria in a certain culture after t years was

$$N = (1000)2^t, \quad \text{or} \quad 2^t = N/1000.$$

Changing to logarithmic form we obtain

$$t = \log_2 (N/1000).$$

Hence the time t is a logarithmic function of N.

In calculus and its applications a certain irrational number, denoted by e, is used for the logarithmic base. To five decimal places, $e \approx 2.71828$. This base arises naturally in the theory of certain mathematical concepts and in the description of many types of physical phenomena. For this reason, logarithms with base e are called **natural logarithms.** The notation **ln** x is used as an abbreviation for $\log_e x$. Since $e \approx 3$, the graph of $y = \ln x$ is similar in appearance to the graph of $y = \log_3 x$.

EXAMPLE 3 According to Newton's Law of Cooling, the rate at which an object cools is directly proportional to the difference in temperature between the object and the surrounding medium. Newton's Law can be used to show that under certain conditions the temperature T of an object is given by

$$T = 75e^{-2t},$$

where t is time. Express t as a function of T.

Solution

The given equation may be rewritten

$$e^{-2t} = T/75.$$

Using logarithms with base e yields

$$-2t = \log_e (T/75) = \ln (T/75).$$

175

Consequently

$$t = -\tfrac{1}{2}\ln\,(T/75),$$

or

$$t = -\tfrac{1}{2}[\ln\,T - \ln\,75].$$

We shall not have many occasions to work with natural logarithms in this text. For computational purposes the base 10 is used. This base gives us the system of common logarithms discussed in subsequent sections.

EXERCISES

In Exercises 1–12 sketch the graph of the function f.

1 $f(x) = \log_3 x$ **2** $f(x) = \log_4 x$

3 $f(x) = \log_3\,(3x)$ **4** $f(x) = \log_{1/2} x$

5 $f(x) = \log_2 \sqrt{x}$ **6** $f(x) = \log_2 x^2$

7 $f(x) = \log_2\,(2 + x)$ **8** $f(x) = \log_2\,(2 - x)$

9 $f(x) = \log_2 |2 + x|$ **10** $f(x) = \log_2\left(\dfrac{1}{x}\right)$

11 $f(x) = |\log_3 x|$ **12** $f(x) = 1/(\log_2 x)$

13 Sketch the graphs of $y = \log_2 x$ and $y = 2^x$ on the same coordinate plane. What is the geometric relationship between the graphs?

14 Sketch the graphs of $y = \log_2 x$, $y = \log_3 x$, and $y = \log_4 x$ on the same coordinate plane. Formulate a conjecture based on your graphs. How does the graph of $y = \ln x$ compare with these graphs?

15 An electrical condenser with initial charge Q_0 is allowed to discharge. After t seconds the charge Q is given by

$$Q = Q_0 e^{kt},$$

where k is a constant of proportionality. Use natural logarithms to solve for t in terms of Q_0, Q, and k.

16 Under certain conditions the atmospheric pressure p at altitude h is given by

$$p = 29e^{0.000034h}.$$

Use natural logarithms to solve for h as a function of p.

Logarithms with base 10 are useful for certain numerical computations. It is customary to refer to such logarithms as **common logarithms** and to use the symbol **log** x as an abbreviation for $\log_{10} x$.

Base 10 is used in computational work because every positive real number can be written in the scientific form $c \cdot 10^k$, where $1 \le c < 10$ and k is an integer (see Section 6 of Chapter 1). For example,

$$513 = (5.13)10^2, \qquad 2375 = (2.375)10^3, \quad 720000 = (7.2)10^5,$$
$$0.641 = (6.41)10^{-1}, \quad 0.00000438 = (4.38)10^{-6}, \quad 4.601 = (4.601)10^0.$$

If x is any positive real number and we write

$$x = c \cdot 10^k,$$

where $1 \le c < 10$ and k is an integer, then applying (i) of (4.9) gives us

$$\log x = \log c + \log 10^k.$$

Next, using (4.7) with $a = 10$ and $r = k$ we obtain

(4.15) $\log x = \log c + k$

Equation (4.15) means that to find $\log x$ for any positive real number x it is sufficient to know the logarithms of numbers between 1 and 10. In (4.15) the number $\log c$, where $1 \le c < 10$, is called the **mantissa,** and the integer k is called the **characteristic** of $\log x$.

If $1 \le c < 10$, then since $\log x$ increases as x increases,

$$\log 1 \le \log c < \log 10,$$

or, by (4.8) and (4.7),

$$0 \le \log c < 1.$$

Hence the mantissa of a logarithm is a number between 0 and 1. When working numerical problems it is usually necessary to approximate logarithms. For example, it can be shown that

$$\log 2 = .3010299957 \cdots ,$$

where the decimal is nonrepeating and nonterminating. We shall round off such logarithms to four decimal places and write

$$\log 2 \approx .3010.$$

If a number between 0 and 1 is written as a finite decimal, it is referred to as a **decimal fraction.** Thus (4.15) tells us that if x is any positive real number, then *log x may be approximated by the sum of a positive decimal fraction (the mantissa) and an integer k (the characteristic).* We shall refer to this representation as the **standard form** for log x.

Logarithms of many of the numbers between 1 and 10 have been calculated. Table 2 contains four-decimal-place approximations for logarithms of numbers between 1.00 and 9.99 at intervals of .01. This table can be used to find the logarithm of any three-digit number to four-decimal-place accuracy. There exist far more extensive tables which provide logarithms of many additional numbers to much greater accuracy than four decimal places. The use of Table 2 is illustrated in the following examples.

EXAMPLE 1 Find

(a)   log 43.6,   (b)   log 43,600,   (c)   log 0.0436.

Solutions

(a)   Since $43.6 = (4.36)10^1$, the characteristic of log 43.6 is 1. Referring to Table 2 we find that the mantissa of log 4.36 may be approximated by .6395. Hence, as in (4.15),

log 43.6 $\approx$ .6395 + 1 = 1.6395.

(b)   Since $43,600 = (4.36)10^4$, the mantissa is the same as in part (a), however the characteristic is 4. Consequently

log 43,600 $\approx$ .6395 + 4 = 4.6395.

(c)   If we write $0.0436 = (4.36)10^{-2}$, then

log 0.0436 = log 4.36 + (−2).

Hence

(4.16)   log 0.0436 $\approx$ .6395 + (−2).

We could subtract 2 from .6395 obtaining

log 0.0436 $\approx$ −1.3605,

but this is not standard form, since −1.3605 = −.3605 + (−1), a number in which the decimal fraction is *negative.* A common error is to write (4.16) as −2.6395. This is incorrect since −2.6395 = −.6395 + (−2), which is not the same as .6395 + (−2).

If a logarithm has negative characteristic it is customary either to leave it in standard form as in (4.16) or to rewrite the logarithm, keeping the decimal part positive. To illustrate the latter technique, let us add and subtract 8 on the right side of (4.16). This gives us

$$\log 0.0436 \approx .6395 + (8 - 8) + (-2),$$

or

$$\log 0.0436 \approx 8.6395 - 10.$$

We could also write

$$\log 0.0436 \approx 18.6395 - 20 = 43.6395 - 45,$$

and so on, as long as the *integral part* of the logarithm is -2.

EXAMPLE 2 Find

(a) $\log (0.00652)^2$,
(b) $\log (0.00652)^{-2}$,
(c) $\log (0.00652)^{1/2}$.

Solutions

(a) By (iii) of (4.9),

$$\log (0.00652)^2 = 2 \log 0.00652.$$

Since $0.00652 = (6.52)10^{-3}$,

$$\log 0.00652 = \log 6.52 + (-3).$$

Referring to Table 2 we see that $\log 6.52$ is approximately .8142 and therefore

$$\log 0.00652 \approx .8142 + (-3).$$

Hence

$$\begin{aligned}
\log (0.00652)^2 &= 2 \log 0.00652 \\
&\approx 2[.8142 + (-3)] \\
&= 1.6284 + (-6).
\end{aligned}$$

The standard form is $.6284 + (-5)$.

(b) Using (iii) of (4.9) and the value for $\log 0.00652$ found in part (a),

$$\begin{aligned}
\log (0.00652)^{-2} &= -2 \log 0.00652 \\
&\approx -2[.8142 + (-3)] \\
&= -1.6284 + 6.
\end{aligned}$$

It is important to note that -1.6284 means $-.6284 + (-1)$ and consequently the decimal part is negative. To obtain the standard form, we may write

$$-1.6284 + 6 = 6.0000 - 1.6284$$
$$= 4.3716.$$

This shows that the mantissa is .3716 and the characteristic is 4.

(c) By (iii) of (4.9)

$$\log (0.00652)^{1/2} = (\tfrac{1}{2}) \log 0.00652$$
$$\approx \tfrac{1}{2}[.8142 + (-3)].$$

If we multiply by $\tfrac{1}{2}$, the standard form is not obtained since neither number in the resulting sum is the characteristic. In order to avoid this, we may adjust the expression within brackets by adding and subtracting a suitable number. If we use 1 in this way we obtain

$$\log (0.00652)^{1/2} \approx \tfrac{1}{2}[1.8142 + (-4)]$$
$$= .9021 + (-2),$$

which is in standard form. We could also have added and subtracted a number other than 1. For example,

$$\tfrac{1}{2}[.8142 + (-3)] = \tfrac{1}{2}[17.8142 + (-20)]$$
$$= 8.9021 + (-10).$$

Table 2 can be used to find an approximation to x if $\log x$ is given. This is illustrated in the following example.

EXAMPLE 3 Find a decimal approximation to x if

(a) $\log x = 1.7959$,   (b) $\log x = -3.5918$.

Solutions

(a) The mantissa .7959 determines the sequence of digits in x and the characteristic determines the position of the decimal point. Referring to the *body* of Table 2, we see that the mantissa .7959 is the logarithm of 6.25. Since the characteristic is 1, x lies between 10 and 100; therefore $x \approx 62.5$.

(b) In order to find x from Table 2, $\log x$ must be written in standard form. In order to change $\log x = -3.5918$ to standard form, we may add and subtract 4, obtaining

$$\log x = (4 - 3.5918) - 4$$
$$= .4082 - 4.$$

Referring to Table 2 we see that the mantissa .4082 is the logarithm of 2.56. Since the characteristic of log x is -4, it follows that $x \approx .000256$.

In each of the previous examples the mantissas appeared in Table 2. If this is not the case, then x can be approximated by the method of interpolation discussed in the next section.

EXERCISES

In Exercises 1–14 use Table 2 and the laws of logarithms to approximate the common logarithms of the given numbers.

1 931; 0.00931; 9.31

2 72.4; 7,240; 0.724

3 0.37; 370; 370,000

4 402; 4.02; 40,200

5 50.4; 0.0000504; 504

6 6; 0.6; 0.0006

7 $(88.5)^2$; $(88.5)^{1/2}$; $(88.5)^{-2}$

8 $(5470)^4$; $(5470)^{40}$; $(5470)^{1/4}$

9 $(0.243)^3$; $(0.243)^{-3}$; $(0.243)^{1/3}$

10 $(0.014)^{10}$; $10^{0.014}$; $10^{1.632}$

11 $\dfrac{437}{26.1}$

12 $\dfrac{(6.93)(30.1)}{537}$

13 $\sqrt{10.2}\,(412)^3$

14 $\dfrac{(0.0061)^{10}}{\sqrt{0.29}}$

In Exercises 15–28 use Table 2 to find a decimal approximation for x. If log x does not appear in the table, find the entry in the table which most nearly approximates x.

15 log $x = 3.8102$

16 log $x = 1.4183$

17 log $x = 0.0212$

18 log $x = 0.9533$

19 log $x = 5.6972$

20 log $x = 6.3444 - 10$

21 log $x = 9.2923 - 10$

22 log $x = 7.7774 - 10$

23 log $x = 8.8789 - 10$

24 log $x = 5.5386$

25 log $x = -1.1399$

26 log $x = -3.6421$

27 log $x = -0.0191$

28 log $x = -5.2235$

5 LINEAR INTERPOLATION

The only logarithms that can be found *directly* from Table 2 are logarithms of numbers that contain at most three nonzero digits. If *four* nonzero digits are involved, then it is possible to obtain an approximation by using the method of linear interpolation described in this section. In order to illustrate this process and at the same time give some justification for it, let us consider the specific example log 12.64. Since the logarithmic function with base 10 is increasing, this number lies between log 12.60 ≈ 1.1004 and log 12.70 ≈ 1.1038. Examining the graph of $y = \log x$, we have the situation shown in Fig. 4.12, where we

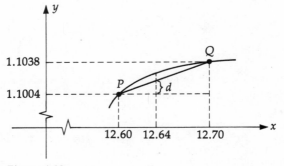

Figure 4.12

have distorted the units on the x- and y-axes and also the portion of the graph shown. A more accurate drawing would indicate that the graph of $y = \log x$ is much closer to the line segment joining $P(12.60, 1.1004)$ to $Q(12.70, 1.1038)$ than is shown in the figure. Since log 12.64 is the ordinate of the point on the graph having abscissa 12.64, it can be approximated by the ordinate of the point with abscissa 12.64 on the *line segment PQ*. Referring to Fig. 4.12 we see that the latter ordinate is $1.1004 + d$. The number d can be approximated by using similar triangles. Referring to Fig. 4.13, where the graph of $y = \log x$ has been deleted, we may form the following proportion.

$$\frac{d}{.0034} = \frac{.04}{.1}.$$

Hence

$$d = \frac{(.04)(.0034)}{.1} = .00136.$$

When using this technique, we always round off decimals to the same number of places as appear in the body of the table. Consequently $d \approx .0014$ and

$$\log 12.64 \approx 1.1004 + .0014 = 1.1018.$$

The process we have demonstrated is referred to as **linear interpolation** since a straight line is used to approximate the graph of $y = \log x$. The process is not restricted to tables of logarithms: It can be used

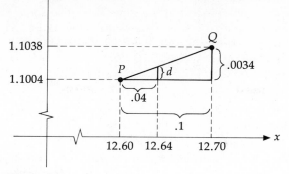

Figure 4.13

for any table in which entries vary in a manner similar to the behavior of a linear function.

Hereafter we shall not sketch a graph when interpolating. Instead we shall use the scheme illustrated in the next example.

EXAMPLE 1   Approximate log 572.6.

Solution

It is convenient to arrange our work as follows:

$$1.0 \left\{ .6 \left\{ \begin{matrix} \log 572.0 \approx 2.7574 \\ \log 572.6 = ? \\ \log 573.0 \approx 2.7582 \end{matrix} \right\} d \right\} .0008$$

where we have indicated differences by appropriate symbols alongside of the braces. This leads to the proportion

$$\frac{d}{.0008} = \frac{.6}{1.0} = \frac{6}{10}$$

or

$$d = (\tfrac{6}{10})(.0008) = .00048 \approx .0005.$$

Hence

$$\log 572.6 \approx 2.7574 + .0005 = 2.7579.$$

Another way of working this type of problem is to reason that since 572.6 is $\tfrac{6}{10}$ of the way from 572.0 to 573.0, then log 572.6 is (approximately) $\tfrac{6}{10}$ of the way from 2.7574 to 2.7582. Hence

$$\log 572.6 \approx 2.7574 + (\tfrac{6}{10})(.0008) \approx 2.7574 + .0005 = 2.7579.$$

183

EXAMPLE 2 Approximate log 0.003678.

Solution

We begin by arranging our work as in the solution of Example 1. Thus

$$10 \left\{ 8 \begin{cases} \log 0.003670 \approx .5647 + (-3) \\ \log 0.003678 = ? \\ \log 0.003680 \approx .5658 + (-3) \end{cases} d \right\} .0011.$$

Since we are only interested in ratios, we have used the numbers 8 and 10 on the left side because their ratio is the same as the ratio of .000008 to .000010. This leads to the proportion

$$\frac{d}{.0011} = \frac{8}{10} = .8,$$

or

$$d = (.0011)(.8) = .00088 \approx .0009.$$

Hence

$$\log 0.003678 \approx [.5647 + (-3)] + .0009$$
$$= .5656 + (-3).$$

If a number x is written in the form $x = c \cdot 10^k$, where $1 \le c < 10$, then before using Table 2 to find log x by interpolation, c should be rounded off to three decimal places. Another way of saying this is that x should be rounded off to four **significant figures.** Some examples will help to clarify the procedure. If $x = 36.4635$, we round off to 36.46 before approximating log x. The number 684,279 should be rounded off to 684,300. For a decimal such as 0.096202 we write 0.09620, and so on. The reason for doing this is that Table 2 does not guarantee more than four-digit accuracy, since the mantissas which appear in it are approximations. This means that if *more* than four-digit accuracy is required in a problem, then Table 2 cannot be used. If, in more extensive tables, the logarithm of a number containing n digits can be found directly, then interpolation is allowed for numbers involving $n + 1$ digits and numbers should be rounded off accordingly.

The method of interpolation can also be used to find x when we are given log x. If we use Table 2, then x may be found to four significant figures. In this case we are given the *ordinate* of a point on the graph of $y = \log x$ and are asked to find the *abscissa*. A geometric argument similar to the one given earlier can be used to justify the procedure illustrated in the next example.

EXAMPLE 3 Find x to four significant figures if log $x = 1.7949$.

Solution

The mantissa .7949 does not appear in Table 2, but it can be isolated between adjacent entries, namely the mantissas corresponding to 6.230 and 6.240. We shall arrange our work as follows:

$$.1 \left\{ r \begin{cases} \log 62.30 \approx 1.7945 \\ \log x \quad = 1.7949 \\ \log 62.40 \approx 1.7952 \end{cases} \begin{matrix} .0004 \\ \end{matrix} \right\} .0007.$$

This leads to the proportion

$$\frac{r}{.1} = \frac{.0004}{.0007} = \frac{4}{7} \quad \text{or} \quad r = (.1)(\tfrac{4}{7}) \approx .06.$$

Hence

$$x \approx 62.30 + .06 = 62.36.$$

If we are given log x, then the number x is called the **antilogarithm** of log x. In Example 3 the antilogarithm of log $x = 1.7949$ is $x \approx 62.36$. Sometimes the notation antilog $(1.7949) \approx 62.36$ is used.

EXERCISES

Use the method of linear interpolation to approximate the common logarithms of the numbers in Exercises 1–18.

1	27.34	**2**	365.8	**3**	6483
4	0.4217	**5**	0.001769	**6**	78,350
7	148,900	**8**	0.02002	**9**	0.8888
10	1.101	**11**	293.8	**12**	35.74
13	0.8464	**14**	7463	**15**	55,550
16	0.001693	**17**	0.01234	**18**	200,700

In Exercises 19–36 use linear interpolation to approximate x.

19	log $x = 1.4397$	**20**	log $x = 2.8165$
21	log $x = 4.6312$	**22**	log $x = 0.9392$
23	log $x = 9.0186 - 10$	**24**	log $x = 8.4461 - 10$
25	log $x = 3.7777 - 6$	**26**	log $x = 5.9292 - 9$

27 $\log x = 2.4397$ **28** $\log x = 4.8165$

29 $\log x = 0.4461$ **30** $\log x = 0.0186$

31 $\log x = 8.6312 - 10$ **32** $\log x = 2.9392 - 5$

33 $\log x = -1.6312$ **34** $\log x = -2.9382$

35 $\log x = -8.6312$ **36** $\log x = -1.9382$

6 COMPUTATIONS WITH LOGARITHMS

The importance of logarithms for numerical computations has diminished in recent years because of the development of computing machines and portable calculators. However, since mechanical devices are not always available, it is worthwhile to have some familiarity with the use of logarithms for solving arithmetic problems. At the same time practice in working numerical problems leads to a deeper understanding of the theoretical aspects of logarithms. The following examples illustrate some computational techniques.

EXAMPLE 1   Approximate $N = \dfrac{(59700)\,(0.0163)}{41.7}$.

Solution

Using (4.9) and Table 2 leads to

$$\begin{aligned}
\log N &= \log 59700 + \log 0.0163 - \log 41.7 \\
&\approx 4.7760 + (.2122 - 2) - (1.6201) \\
&= 4.9882 - 3.6201 \\
&= 1.3681.
\end{aligned}$$

Referring to Table 2 for the antilogarithm we have, to three significant figures,

$$N \approx 23.3.$$

EXAMPLE 2   Approximate $N = \sqrt[3]{56.11}$ to four significant figures.

Solution

Writing $N = (56.11)^{1/3}$ and using (iii) of (4.9) gives us

$$\log N = \tfrac{1}{3} \log 56.11.$$

To find $\log 56.11$ we interpolate from Table 2 as follows:

$$10 \left\{ 1 \left\{ \begin{array}{l} \log 56.10 \approx 1.7490 \\ \log 56.11 = ? \\ \log 56.20 \approx 1.7497 \end{array} \right\} d \right\} .0007,$$

$$\frac{d}{.0007} = \frac{1}{10}$$
$$d = .00007 \approx .0001.$$

Hence

$$\log 56.11 \approx 1.7490 + .0001 = 1.7491,$$

and therefore

$$\log N \approx \tfrac{1}{3}(1.7491) \approx .5830.$$

The antilogarithm may be found by interpolation from Table 2 as follows

$$.01 \left\{ r \begin{cases} \log 3.820 \approx .5821 \\ \log N \quad\; \approx .5830 \end{cases} 9 \right\} 11$$
$$\log 3.830 \approx .5832$$

$$\frac{r}{.01} = \frac{9}{11}$$
$$r = \tfrac{9}{11}(.01) \approx .008.$$

Consequently

$$N \approx 3.820 + .008 = 3.828.$$

If we were interested in only *three* significant figures, then the interpolations in Example 2 could have been avoided. In the remaining examples we shall, for simplicity, work only with three-digit numbers.

EXAMPLE 3 Approximate $N = \dfrac{(1.32)^{10}}{\sqrt[5]{0.0268}}$.

Solution

Using (4.9) and Table 2,

$$\begin{aligned} \log N &= 10 \log 1.32 - \tfrac{1}{5} \log 0.0268 \\ &\approx 10(.1206) - \tfrac{1}{5}(3.4281 - 5) \\ &= 1.206 - .6856 + 1 \\ &= 1.5204. \end{aligned}$$

Finding the antilogarithm (to three significant figures) we obtain

$$N \approx 33.1.$$

EXAMPLE 4   Find x if $x^{2.1} = 6.5$.

Solution

Taking the common logarithm of both sides and using (iii) of (4.9) gives us

$$(2.1) \log x = \log 6.5.$$

Hence

$$\log x = (\log 6.5)/(2.1),$$

and by (4.5),

$$x = 10^{(\log\ 6.5)/(2.1)}$$

If an approximation to x is desired, then from the second equation above we have

$$\log x \approx \frac{.8129}{2.1} \approx .3871.$$

Using Table 2, the antilogarithm (to three significant figures) is

$$x \approx 2.44.$$

EXAMPLE 5   Approximate $N = \dfrac{69.3 + \sqrt[3]{56.1}}{\log 807}$.

Solution

Since we have no formula for the logarithm of a sum, the two terms in the numerator must be *added* before the logarithm can be found. From Example 2, $\sqrt[3]{56.1} \approx 3.8$ and hence the numerator is approximately 73.1. Using Table 2 we see that

$$\log 807 \approx 2.9069 \approx 2.91.$$

Hence the given expression may be approximated by

$$N \approx \frac{73.1}{2.91}.$$

It is now an easy matter to find N, either by using logarithms or by long division. It is left to the reader to verify that $N \approx 25.1$.

EXERCISES

7

Exponential
and
Logarithmic
Equations and
Inequalities

Use logarithms to approximate each of Exercises 1–20 to three significant figures.

1 $(35.7)(0.484)$ **2** $(0.00639)(0.0127)$

3 $\dfrac{74,600}{9,230}$ **4** $\dfrac{2.07}{58.1}$

5 $(5.17)^4$ **6** $(0.162)^5$

7 $\sqrt[4]{0.267}$ **8** $\sqrt[6]{46.3}$

9 $\dfrac{(6.23)(17.4)^2}{81.5}$ **10** $\dfrac{(7.06)^3}{(31.3)\,\sqrt{1.05}}$

11 $\sqrt[3]{(4.61)(1.29)^2}$ **12** $\left[\dfrac{(13.1)^4}{\sqrt{2.41}}\right]^{-1/2}$

13 $\dfrac{(-6.43)^2(80.6)^{-1/3}}{-74.3}$ **14** $\sqrt[5]{\dfrac{(127)^2(69.2)}{42,700}}$

15 $10^{-4.623}$ **16** $(10^{0.764})(0.764)^{10}$

17 $(2.07)^{0.36}$ **18** $\sqrt[3]{1.62}+\sqrt[3]{1.62}$

19 $\dfrac{\log 46.3}{\log 3.14}$ **20** $\dfrac{8.31+\log 9.42}{\sqrt[10]{0.666}}$

21 The area A of a triangle with sides a, b, and c may be calculated from the formula $A = \sqrt{s(s-a)(s-b)(s-c)}$, where s is one-half the perimeter. Use logarithms to approximate the area of a triangle with sides 12.6, 18.2, and 14.1.

22 The volume V of a right circular cone of altitude h and radius of base r is $V = \frac{1}{3}\pi r^2 h$. Use logarithms to approximate the volume of a cone of radius 2.43 and altitude 7.28.

23 The formula used in physics to approximate the period T (seconds) of a simple pendulum of length L (feet) is $T = 2\pi\sqrt{L/(32.2)}$. Approximate the period of a pendulum 33 inches long.

24 The pressure p (pounds per cubic foot) and volume v (cubic feet) of a certain gas are related by the formula $pv^{1.4} = 600$. Approximate the pressure if $v = 8.22$ cubic feet.

7 EXPONENTIAL AND LOGARITHMIC EQUATIONS AND INEQUALITIES

The variables in certain equations and inequalities appear as exponents or logarithms. Some of these may be solved as illustrated in the examples on the following pages.

EXAMPLE 1 Solve the equation $3^x = 21$.

Solution

Taking the common logarithm of both sides and using (iii) of (4.9), we obtain

$$x \log 3 = \log 21,$$

and therefore

$$x = \frac{\log 21}{\log 3}.$$

If an approximation is desired we may use Table 2 to obtain

$$x \approx \frac{1.3222}{0.4771} \approx 2.77,$$

where the last number is obtained by dividing 1.3222 by 0.4771. A partial check on the solution is to note that since $3^2 = 9$ and $3^3 = 27$, the number x such that $3^x = 21$ should lie between 2 and 3, somewhat closer to 3 than to 2.

EXAMPLE 2 Solve $5^{2x+1} = 6^{x-2}$.

Solution

Taking the common logarithm of both sides and using (iii) of (4.9) we obtain

$$(2x + 1) \log 5 = (x - 2) \log 6.$$

We may now solve for x as follows:

$$2x \log 5 + \log 5 = x \log 6 - 2 \log 6$$
$$2x \log 5 - x \log 6 = -\log 5 - 2 \log 6$$
$$x(2 \log 5 - \log 6) = -(\log 5 + \log 6^2)$$
$$x = \frac{-(\log 5 + \log 36)}{2 \log 5 - \log 6}$$
$$x = \frac{-\log (5 \cdot 36)}{\log 5^2 - \log 6}$$
$$x = \frac{-\log 180}{\log (25/6)}.$$

If an approximation to the solution is desired, we may proceed as in Example 1.

EXAMPLE 3   Solve the inequality $(0.4)^x > 3$.

Solution

Writing $3 < (0.4)^x$ and using the fact that the logarithmic function with base 10 is increasing, we obtain

$$\log 3 < \log (0.4)^x,$$

and so by (iii) of (4.9),

$$\log 3 < x \log 0.4.$$

If we divide both sides by the negative number log 0.4, the sense of the inequality is reversed. Hence

$$\frac{\log 3}{\log 0.4} > x.$$

Thus the solution set is

$$\{x : x < \log 3 / \log 0.4\}$$

Since

$$\frac{\log 3}{\log 0.4} \approx \frac{.4771}{.6021 - 1} = \frac{.4771}{-.3979} \approx -1.2,$$

an *approximate* solution is the set of all real numbers x such that $x < -1.2$.

EXAMPLE 4   Solve the equation $\log (5x - 1) - \log (x - 3) = 2$.

Solution

The given equation may be written as

$$\log \frac{5x - 1}{x - 3} = 2.$$

Using the definition of logarithm (4.5) with $a = 10$ gives us

$$\frac{5x - 1}{x - 3} = 10^2.$$

Consequently

$$5x - 1 = 100x - 300, \quad \text{or} \quad 299 = 95x.$$

Hence the solution set is $\{299/95\}$.

191

EXAMPLE 5 Solve the equation $\dfrac{5^x - 5^{-x}}{2} = 3$.

Solution

Multiplying both sides of the given equation by $2(5^x)$ gives us

$$5^{2x} - 1 = 6(5^x),$$

which may be written

$$(5^x)^2 - 6(5^x) - 1 = 0.$$

Letting $u = 5^x$ gives us a quadratic equation in the variable u. Applying the quadratic formula, we obtain

$$5^x = \frac{6 \pm \sqrt{36 + 4}}{2} = 3 \pm \sqrt{10}.$$

Since 5^x is never negative, the number $3 - \sqrt{10}$ must be discarded; therefore

$$5^x = 3 + \sqrt{10}.$$

Taking the common logarithm of both sides we have

$$x \log 5 = \log (3 + \sqrt{10}),$$

or

$$x = \frac{\log (3 + \sqrt{10})}{\log 5}.$$

To obtain an approximate solution we may write $3 + \sqrt{10} \approx 6.16$ and use Table 2. This gives us

$$x \approx \frac{\log 6.16}{\log 5} \approx \frac{.7896}{.6990} \approx 1.13.$$

EXERCISES

Find the solution sets in Exercises 1–22.

1 $10^x = 5$

2 $4^x = 7$

3 $5^x = 4$

4 $10^x = 2$

5 $4^{3-x} = 2$

6 $(\tfrac{1}{2})^x = 100$

7 $2^{1-2x} = 3^{x+5}$

8 $4^{3x+2} = 5^{x-1}$

9 $6^x = 2$ **10** $3^{-x^2} = 4$

11 $\log (x - 15) = 2 - \log x$

12 $\log (x + 3) = 1 + \log (3x - 10)$

13 $\log (x^2 + 1) - \log (x + 1) = 1 + \log (x - 1)$

14 $\log (x + 2) - \log (4x + 3) = \log \left(\dfrac{1}{x}\right)$

15 $(\tfrac{1}{3})^x > 3$ **16** $(2.7)^x < 4.1$

17 $\log (x^2) = (\log x)^2$ **18** $\log \sqrt{x} = \sqrt{\log x}$

19 $\log (\log x) = 3$ **20** $\log \sqrt{x^2 - 1} = 2$

21 $x^{\sqrt{\log x}} = 10^8$ **22** $\log (x^3) = (\log x)^3$

In Exercises 23 and 24 solve for x in terms of y.

23 $y = \dfrac{10^x + 10^{-x}}{2}$ **24** $y = \dfrac{10^x - 10^{-x}}{2}$

In Exercises 25–28 use natural logarithms to solve for x in terms of y.

25 $y = \dfrac{e^x - e^{-x}}{2}$ **26** $y = \dfrac{e^x + e^{-x}}{2}$

27 $y = \dfrac{e^x - e^{-x}}{e^x + e^{-x}}$ **28** $y = \dfrac{e^x + e^{-x}}{e^x - e^{-x}}$

29 The current i in a certain electrical circuit is given by

$$i = \frac{E}{R} (1 - e^{-Rt/L}).$$

Use natural logarithms to solve for t in terms of the remaining symbols.

30 If a sum of money P_0 is invested at an interest rate of $100r$ percent per year, compounded m times per year, then the principal P at the end of t years is given by

$$P = P_0 \left(1 + \frac{r}{m}\right)^{mt}.$$

Solve for t in terms of the other symbols.

8 REVIEW EXERCISES

Oral

Define or discuss the following.

1 The exponential function with base a

2 The logarithm of u with base a

3 The laws of logarithms

4 The logarithmic function with base a

5 Common logarithms

6 Mantissa

7 Characteristic

8 Linear interpolation

Written

Find the numbers in Exercises 1–6.

1 $\log_2 \left(\frac{1}{8}\right)$ **2** $\log_5 \sqrt{5}$

3 $15^{\log_{15} 7}$ **4** $10^{2 \log 3}$

5 $\log 100,000$ **6** $\ln e$

In Exercises 7–12 sketch the graph of f.

7 $f(x) = 3^{2-x}$ **8** $f(x) = 3^{x/2}$

9 $f(x) = 3^{-x^2}$ **10** $f(x) = -4^{3-x}$

11 $f(x) = \left(\frac{3}{2}\right)^x$ **12** $f(x) = \left(\frac{2}{3}\right)^x$

Find the solution sets in Exercises 13–18.

13 $\log_6 (x + 5) = \frac{1}{2}$ **14** $\log_2 (x^2 - x - 2) = 2$

15 $\log_4 (x + 2) < 3$ **16** $|\log x| < 1$

17 $2 \log_3 (x + 1) - \log_3 (x + 4) = 2 \log_3 2$

18 $3 < \log x < 5$

In Exercises 19–22 sketch the graph of f.

19 $f(x) = \log_2 x$ **20** $f(x) = \log_3 x^2$

21 $f(x) = 2 \log_3 x$ **22** $f(x) = \log_3 2x$

23 Express $\log \sqrt[5]{\dfrac{x^2 y}{z^4}}$ in terms of logarithms of x, y, and z.

24 Express $\log (x/y^2) + 3 \log y - 2 \log xy^3$ as one logarithm.

Use linear interpolation to approximate the logarithms in Exercises 25–28.

25 $\log 35.74$ **26** $\log 0.001693$

27 $\log 200,700$ **28** $\log 0.1234$

Use linear interpolation to approximate x in Exercises 29–32.

29 $\log x = 2.6291$

30 $\log x = 8.8973 - 10$

31 $\log x = .3442$

32 $\log x = -2.6348$

Use logarithms to approximate the numbers in Exercises 33–36.

33 $\dfrac{(27.4)^2}{\sqrt[3]{948}}$

34 $\sqrt[5]{\dfrac{64.7}{86.1}}$

35 $(1.89)^{3.4}$

36 $(1.01)^{1000}$

Find the solution sets in Exercises 37–42.

37 $2^{3-x} = 4$

38 $2^{x^2} = 6$

39 $3^{1-2x} = 2^{x+5}$

40 $\log x = 1 - \log (x - 9)$

41 $\left(\tfrac{1}{2}\right)^x < 2$

42 $\log (\log (\log x)) = 2$

Solve the following equations for x in terms of y.

43 $y = \dfrac{10^x + 10^{-x}}{10^x - 10^{-x}}$

44 $y = (10^x + 10^{-x})^{-1}$

5

The Trigonometric Functions

*In this chapter we shall lay the foundation for work in the area of mathematics called trigonometry. The approach we use disguises the fact that the origins of trigonometry had to do with measurement of angles and triangles. The reason for using this approach is motivated by modern applications in which the notion of angle either is secondary or does not enter into the picture. Our main objective is to study the so-called **trigonometric** or **circular functions**. Later in the chapter we shall turn to angular aspects of trigonometry. These ideas have many practical uses and also provide additional insight into the nature of the trigonometric functions.*

1 THE WRAPPING FUNCTION

Let U be a unit circle, that is, a circle of radius 1 with center at the origin of a rectangular coordinate system. If A is the point with coordinates (1, 0) and if P is any other point on U, then measuring along U in a *counterclockwise direction* from A (see Fig. 5.1), there is a unique positive real number t called the **length of the arc** $\widehat{AP}$.

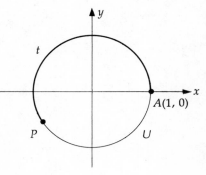

Figure 5.1

Since the circumference of U is 2π, we have $0 < t < 2\pi$. If we let $t = 0$ when $A = P$, then for each point P on U there is associated precisely one real number in the half-open interval $[0, 2\pi)$.

Conversely, it can be shown that for each real number t in $[0, 2\pi)$ there is one and only one point P on U such that the length of $\overset{\frown}{AP}$ (measured in a counterclockwise direction) is t. This establishes a one-to-one correspondence between the real numbers in the interval $[0, 2\pi)$ and the points on U. Since the position of P depends on t, we sometimes use the functional notation and denote it by $P(t)$.

The preceding discussion can be extended so that with *every* real number there corresponds exactly one point on U. A convenient way to demonstrate this fact is to consider, as in Fig. 5.2, a real axis w

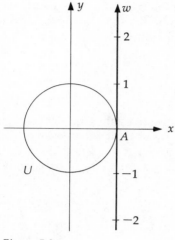

Figure 5.2

with origin at the point $A(1, 0)$ and tangent to U at A, with positive direction upward. Let us regard the w-axis as perfectly flexible and think of wrapping the positive part of the w-axis in a counterclockwise manner about U, as we would wrap thread about a spool. This is illustrated in Fig. 5.3, where the dashed line indicates the initial position of w.

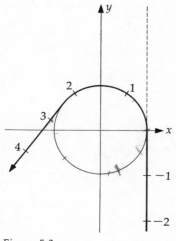

Figure 5.3

One revolution about U gives us a one-to-one correspondence between the real numbers in the interval $[0, 2\pi)$ and the points on U. If we continue the wrapping process, a second revolution of the w-axis about U produces a one-to-one correspondence between the numbers in the interval $[2\pi, 4\pi)$ and the points on U. At this stage, for each point on U there correspond two real numbers in $[0, 4\pi)$ and these two numbers differ numerically by 2π units. A third revolution associates a point on U with each number in $[4\pi, 6\pi)$. Continuing this wrapping procedure, we see that for every nonnegative real number t, there corresponds a unique point $P(t)$ on U. Moreover, since the circumference of U has length 2π, two such real numbers t_1 and t_2 are associated with the same point if and only if $t_2 - t_1$ is a multiple of 2π; that is,

(5.1) $t_2 = t_1 + 2\pi n$

for some integer n.

A similar correspondence can be established between the negative real numbers and points on U by wrapping the negative part of the w-axis in a *clockwise* direction about U, as illustrated in Fig. 5.4. In this case the real numbers in the interval $[-2\pi, 0)$ are associated with specific points of U, and so on.

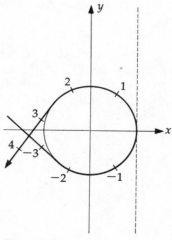

Figure 5.4

We have shown that to each real number t there corresponds a unique point $P(t)$ on the unit circle U. Moreover, two real numbers t_1 and t_2 are associated with the same point if and only if (5.1) is true for some integer n. This gives us the important formula

(5.2) $P(t) = P(t + 2\pi n)$

for every real number t and every integer n.

The correspondence we have described determines a function with domain **R** and range U which we shall refer to as the **wrapping function.** To reiterate, the point $P(t)$ that the wrapping function associates with the real number t can be located as follows. If $t > 0$ we measure a distance t around U in the counterclockwise direction. If $t < 0$, we measure $|t|$ units around U in the clockwise direction. If $t = 0$, we take $P(t) = A$, the point with coordinates $(1, 0)$. Unless t is suitably specialized it is difficult to find the rectangular coordinates of the point $P(t)$ on U. The following examples contain several important cases.

EXAMPLE 1 (a) Locate $P(\pi/2)$ and find its rectangular coordinates.
(b) Find two other values of t such that $P(t) = P(\pi/2)$.

Solution

(a) We first observe that since the radius of U is 1, the circumference is 2π. Next we note that

$$\frac{\pi}{2} = \frac{1}{4}(2\pi).$$

Consequently the point $P(\pi/2)$ may be found by starting at $A(1, 0)$ and moving one-fourth of the way around the circle in the counterclockwise direction. This leads to the point $B(0, 1)$ shown in Fig. 5.5.

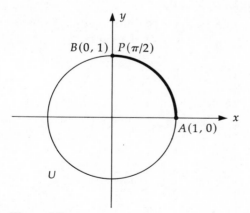

Figure 5.5

(b) By (5.2),

$$P\left(\frac{\pi}{2}\right) = P\left(\frac{\pi}{2} + 2\pi n\right)$$

for every integer n. In particular, if we let $n = 1$ and $n = -1$, we obtain

$$P\left(\frac{\pi}{2}\right) = P\left(\frac{\pi}{2} + 2\pi\right) = P\left(\frac{\pi}{2} - 2\pi\right),$$

or

$$P\left(\frac{\pi}{2}\right) = P\left(\frac{5\pi}{2}\right) = P\left(-\frac{3\pi}{2}\right).$$

The student should check these answers by measuring $5\pi/2$ units along U in the counterclockwise direction and $3\pi/2$ units in the clockwise direction.

EXAMPLE 2 (a) Locate $P(\pi/4)$ and find its rectangular coordinates. (b) Find the coordinates of $P(3\pi/4)$, $P(5\pi/4)$, and $P(7\pi/4)$.

Solution

(a) Since

$$\frac{\pi}{4} = \frac{1}{8}(2\pi),$$

the point $P(\pi/4)$ may be located by starting at $A(1, 0)$ and moving one-eighth of the way around U in the counterclockwise direction. This is equivalent to going along U one-half the way from A to the point $P(\pi/2) = B(0, 1)$ found in Example 1. Since the arc $\overset{\frown}{AB}$ is bisected by $P(\pi/4)$, it follows from plane geometry that P has coordinates of the form (c, c) for some positive real number c, as illustrated in Fig. 5.6. By the Pythagorean Theorem, $c^2 + c^2 = 1^2$, or

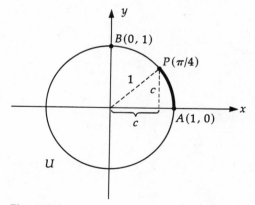

Figure 5.6

$2c^2 = 1$. Hence $c = \sqrt{1/2} = \sqrt{2}/2$. Consequently

$$P\left(\frac{\pi}{4}\right) = \left(\frac{\sqrt{2}}{2}, \frac{\sqrt{2}}{2}\right).$$

(b) Since

$$\frac{3\pi}{4} = \frac{\pi}{4} + \frac{\pi}{2},$$

we may locate $P(3\pi/4)$ by starting at $P(\pi/4)$ and traveling $\pi/2$ units along U in the counterclockwise direction to the point shown in Fig. 5.7. The points $P(5\pi/4)$ and $P(7\pi/4)$

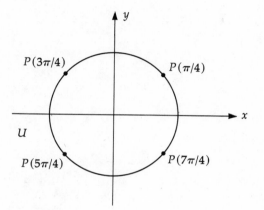

Figure 5.7

plotted in the figure are determined in like manner. Employing arguments similar to that in part (a), or by using the symmetry of the circle, we see that

$$P\left(\frac{3\pi}{4}\right) = \left(-\frac{\sqrt{2}}{2}, \frac{\sqrt{2}}{2}\right)$$

$$P\left(\frac{5\pi}{4}\right) = \left(-\frac{\sqrt{2}}{2}, -\frac{\sqrt{2}}{2}\right)$$

$$P\left(\frac{7\pi}{4}\right) = \left(\frac{\sqrt{2}}{2}, -\frac{\sqrt{2}}{2}\right).$$

EXAMPLE 3 Find the rectangular coordinates of $P(\pi/6)$.

Solution

Since $\pi/6 = (1/3)(\pi/2)$ the point $P(\pi/6)$ on U is one-third the way from $A(1, 0)$ to $B(0, 1)$. If we denote the rectangular coordinates of P by (c, d), then as in Fig. 5.8, the point P' which corresponds to $-\pi/6$ has coordinates $(c, -d)$. The length of $\overline{P'P}$ is the same as the length of $\overline{PB}$, and hence $d(P', P) = d(P, B)$. Employing the distance formula gives us

$$2d = \sqrt{(c - 0)^2 + (d - 1)^2}.$$

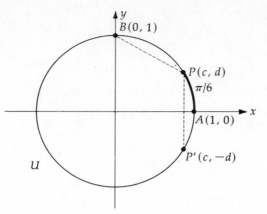

Figure 5.8

Squaring both sides, we have

(5.3) $4d^2 = c^2 + d^2 - 2d + 1.$

Since (c, d) are coordinates of a point on U, and since an equation of U is $x^2 + y^2 = 1$, it follows that $c^2 + d^2 = 1$. Substituting 1 for $c^2 + d^2$ in (5.3) leads to

$4d^2 = 1 - 2d + 1,$

or

$4d^2 + 2d - 2 = 0.$

This may be written as

$2(2d - 1)(d + 1) = 0.$

Since d is positive, we get $d = \frac{1}{2}$. Using $c = \sqrt{1 - d^2}$, we obtain $c = \sqrt{1 - (1/4)} = \sqrt{3/4} = \sqrt{3}/2$, and consequently

$$P\left(\frac{\pi}{6}\right) = \left(\frac{\sqrt{3}}{2}, \frac{1}{2}\right).$$

EXAMPLE 4 Find the rectangular coordinates of $P(\pi/3)$.

Solution

Since $\pi/3 = (2/3)(\pi/2)$, the point $P(\pi/3)$ lies two-thirds the way from $A(1, 0)$ to $B(0, 1)$. An argument similar to that in Example 3 may be used to find its coordinates. Referring to Fig. 5.9, we see that if the rectangular coordinates of $P(\pi/3)$

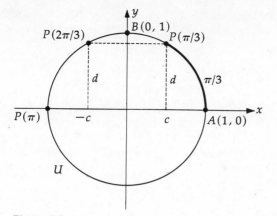

Figure 5.9

are (c, d), then the rectangular coordinates of $P(2\pi/3)$ are $(-c, d)$. Moreover, since

$$d(P(\pi/3), P(2\pi/3)) = d(A, P(\pi/3)),$$

we have by the distance formula,

$$2c = \sqrt{(c-1)^2 + d^2}.$$

Squaring leads to

$$4c^2 = c^2 - 2c + 1 + d^2.$$

If we now proceed as in our work with (5.3), we obtain $c = 1/2$ and $d = \sqrt{3}/2$. The verification of this is left to the reader. Consequently,

$$P\left(\frac{\pi}{3}\right) = \left(\frac{1}{2}, \frac{\sqrt{3}}{2}\right).$$

We may use Examples 2-4 together with geometric symmetries of the unit circle to find the rectangular coordinates of $P(t)$, where t is any integral multiple of $\pi/6$. For example, if the rectangular coordinates of $P(\pi/6)$ are labeled (c, d), then as in Fig. 5.8, the coordinates of $P(-\pi/6)$ are $(c, -d)$, that is,

$$P\left(-\frac{\pi}{6}\right) = \left(\frac{\sqrt{3}}{2}, -\frac{1}{2}\right).$$

Moreover, since $P(11\pi/6) = P(-\pi/6)$,

$$P\left(\frac{11\pi}{6}\right) = \left(\frac{\sqrt{3}}{2}, -\frac{1}{2}\right).$$

203

In similar fashion, by plotting the point $P(5\pi/6)$ on the circle in Fig. 5.8, the reader may verify that its coordinates are $(-c, d)$. Consequently

$$P\left(\frac{5\pi}{6}\right) = \left(-\frac{\sqrt{3}}{2}, \frac{1}{2}\right).$$

In all of the previous illustrations we have chosen rational multiples of π for values of t. We may also approximate the positions of points such as $P(1)$, $P(2)$, $P(3)$, $P(0.5)$, $P(-1)$, $P(-2.5)$, and so on, as illustrated in Fig. 5.10. However, advanced methods are needed to find their rectangular coordinates.

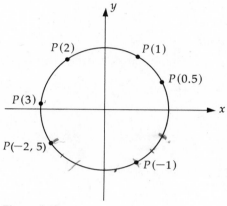

Figure 5.10

EXERCISES

In Exercises 1–4 determine the quadrants in which the given points lie and plot their approximate positions on the unit circle U.

1 $P(2)$; $P(1.5)$; $P(5)$; $P(-3)$; $P(7)$; $P(4\pi/3)$; $P(-\pi/6)$

2 $P(3.2)$; $P(-2)$; $P(6)$; $P(-5)$; $P(-\frac{1}{2})$; $P(7\pi/6)$; $P(-\pi/4)$

3 $P(3)$; $P(-1)$; $P(4.5)$; $P(11\pi/6)$; $P(-5\pi/4)$

4 $P(6)$; $P(-4)$; $P(5)$; $P(5\pi/3)$; $P(13\pi/4)$

In each of Exercises 5-8 find the quadrant in which the point P lies.

5 $P(43)$ 6 $P(65)$

7 $P(-16)$ 8 $P(-21)$

In Exercises 9 and 10 find the rectangular coordinates of the given points.

9 $P(3\pi)$; $P(3\pi/2)$; $P(-6\pi)$; $P(-5\pi/2)$; $P(7\pi/2)$; $P(43\pi)$

10 $P(5\pi)$; $P(5\pi/2)$; $P(-8\pi)$; $P(-7\pi/2)$; $P(27\pi)$; $P(-100\pi)$

In Exercises 11–16 use the coordinates found in Examples 2–4 together with symmetries of the unit circle to find the rectangular coordinates of the given point.

11 $P(7\pi/6)$ **12** $P(-5\pi/6)$ **13** $P(2\pi/3)$

14 $P(4\pi/3)$ **15** $P(7\pi/3)$ **16** $P(-7\pi/6)$

17 Find two positive and two negative values of t such that $P(t)$ has rectangular coordinates $(\sqrt{3}/2, \frac{1}{2})$. (*Hint:* Use Example 3 and (5.2).)

18 Same as Exercise 17 for $(-\sqrt{2}/2, -\sqrt{2}/2)$

19 If $P(t)$ has rectangular coordinates $(\frac{4}{5}, \frac{3}{5})$, find the rectangular coordinates of

  a $P(t+\pi)$ **b** $P(-t)$

  c $P(t-\pi)$ **d** $P(-\pi-t)$

20 Same as Exercise 19 for $(-\frac{3}{5}, \frac{4}{5})$

21 Same as Exercise 19 for $(\frac{8}{17}, -\frac{15}{17})$

22 Generalize Exercise 19 to the case where $P(t)$ is any point on U with coordinates (a, b).

2 THE TRIGONOMETRIC FUNCTIONS

The wrapping function can be used to define the six **trigonometric** (or **circular**) **functions.** These functions are referred to as the **sine, cosine, tangent, cotangent, secant,** and **cosecant functions,** and are designated by the symbols **sin, cos, tan, cot, sec,** and **csc,** respectively. If t is a real number, then the real number which the sine function associates with t will be denoted by either sin (t) or sin t, and similarly for the other five functions.

(5.4)   **Definition**

If t is any real number, let $P(t)$ be the point on the unit circle U that the wrapping function associates with t. If the rectangular coordinates of $P(t)$ are (x, y), then

$$\sin t = y \qquad\qquad \csc t = \frac{1}{y} \quad \text{(if } y \neq 0)$$

$$\cos t = x \qquad\qquad \sec t = \frac{1}{x} \quad \text{(if } x \neq 0)$$

$$\tan t = \frac{y}{x} \quad \text{(if } x \neq 0) \qquad \cot t = \frac{x}{y} \quad \text{(if } y \neq 0).$$

To find the values of the trigonometric functions that correspond to a real number t, it is necessary to determine the rectangular coordinates (x, y) of the point $P(t)$ and then to substitute in the appropriate formula. This is illustrated in the next example.

EXAMPLE 1 Find the values of the trigonometric functions for $t = 0$, $\pi/6$, $\pi/4$, $\pi/3$, and $\pi/2$.

Solution

The rectangular coordinates of $P(0)$ are $(1, 0)$. Since the ordinate y of $P(0)$ is 0, we see from (5.4) that the cosecant and cotangent functions are undefined. The remaining functional values may be found by substituting 1 for x and 0 for y in (5.4). Thus

$$\sin 0 = 0, \quad \cos 0 = 1, \quad \tan 0 = \tfrac{0}{1} = 0, \quad \sec 0 = \tfrac{1}{1} = 1.$$

From Example 3 of the preceding section, $P(\pi/6)$ has rectangular coordinates $(\sqrt{3}/2, 1/2)$. Hence the functional values for $t = \pi/6$ may be obtained by substituting $\sqrt{3}/2$ for x and $1/2$ for y in (5.4). This gives us

$$\sin \frac{\pi}{6} = \frac{1}{2} \qquad\qquad \csc \frac{\pi}{6} = \frac{1}{1/2} = 2$$

$$\cos \frac{\pi}{6} = \frac{\sqrt{3}}{2} \qquad\qquad \sec \frac{\pi}{6} = \frac{1}{\sqrt{3}/2} = \frac{2}{\sqrt{3}} = \frac{2\sqrt{3}}{3}$$

$$\tan \frac{\pi}{6} = \frac{1/2}{\sqrt{3}/2} = \frac{1}{\sqrt{3}} = \frac{\sqrt{3}}{3} \quad \cot \frac{\pi}{6} = \frac{\sqrt{3}/2}{1/2} = \sqrt{3}.$$

To find the values of the trigonometric functions for $t = \pi/4$, we may use the coordinates $(\sqrt{2}/2, \sqrt{2}/2)$ of $P(\pi/4)$ calculated in Example 2 of the preceding section. In like manner, the values for $t = \pi/3$ may be determined from the coordinates $(\tfrac{1}{2}, \sqrt{3}/2)$ obtained in Example 4 of Section 1. Finally, the values for $t = \pi/2$ are found by using $(0, 1)$. These values are arranged in tabular form below, where all denominators have been rationalized. The reader should check each entry in the table. A dash indicates that the function is undefined for the indicated number t.

Table A

t	$\sin t$	$\cos t$	$\tan t$	$\csc t$	$\sec t$	$\cot t$
0	0	1	0	$-$	1	$-$
$\dfrac{\pi}{6}$	$\dfrac{1}{2}$	$\dfrac{\sqrt{3}}{2}$	$\dfrac{\sqrt{3}}{3}$	2	$\dfrac{2\sqrt{3}}{3}$	$\sqrt{3}$
$\dfrac{\pi}{4}$	$\dfrac{\sqrt{2}}{2}$	$\dfrac{\sqrt{2}}{2}$	1	$\sqrt{2}$	$\sqrt{2}$	1
$\dfrac{\pi}{3}$	$\dfrac{\sqrt{3}}{2}$	$\dfrac{1}{2}$	$\sqrt{3}$	$\dfrac{2\sqrt{3}}{3}$	2	$\dfrac{\sqrt{3}}{3}$
$\dfrac{\pi}{2}$	1	0	$-$	1	$-$	0

EXAMPLE 2 Find the values of the trigonometric functions for $t = 3\pi/4$.

Solution

From Example 2 of Section 1, the point $P(3\pi/4)$ has coordinates $(-\sqrt{2}/2, \sqrt{2}/2)$. Using $x = -\sqrt{2}/2$ and $y = \sqrt{2}/2$ in (5.4) we obtain

$$\sin \frac{3\pi}{4} = \frac{\sqrt{2}}{2} \qquad\qquad \csc \frac{3\pi}{4} = \frac{2}{\sqrt{2}} = \sqrt{2}$$

$$\cos \frac{3\pi}{4} = -\frac{\sqrt{2}}{2} \qquad\qquad \sec \frac{3\pi}{4} = -\frac{2}{\sqrt{2}} = -\sqrt{2}$$

$$\tan \frac{3\pi}{4} = \frac{\sqrt{2}/2}{-\sqrt{2}/2} = -1 \qquad \cot \frac{3\pi}{4} = \frac{-\sqrt{2}/2}{\sqrt{2}/2} = -1.$$

It is important to remember the signs of the functional values for the trigonometric functions when $P(t)$ is in various quadrants. We see from (5.4) that if $P(t)$ is in quadrant I, all functional values are positive. If $P(t)$ is in quadrant II, then y is positive and x is negative, and consequently the sine and cosecant values are positive and all others are negative. If $P(t)$ is in quadrant III, then both x and y are negative, and therefore the tangent and cotangent have positive values and all other functions have negative values. Finally, if $P(t)$ is in quadrant IV, the cosine and secant values are positive and all others are negative. These facts are brought out schematically in Fig. 5.11, where the dots indicate various positions of $P(t)$. The functions which are unlisted in each of quadrants II, III and IV of the figure are negative if $P(t)$ is in that quadrant.

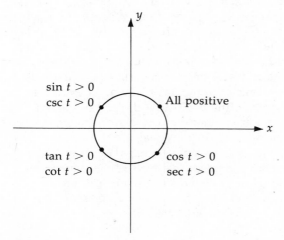

Figure 5.11

EXAMPLE 3 Find the quadrant in which $P(t)$ lies if both $\sin t < 0$ and $\cos t > 0$.

Solution

Referring to Fig. 5.11, we see that $\sin t < 0$ if $P(t)$ is in quadrants III or IV, and $\cos t > 0$ if $P(t)$ is in quadrants I or IV. Hence, for both conditions to be satisfied, $P(t)$ must lie in quadrant IV.

The domain of each trigonometric function can be obtained by referring to (5.4). If $P(t)$ has coordinates (x, y), then the sine and cosine functions always exist and hence each has domain **R**. For the tangent and secant functions we must exclude values of t for which x is 0 (Why?). These correspond to the points $(0, 1)$ and $(0, -1)$. It follows that the domains of the tangent and secant functions consist of all numbers *except* those of the form $\pi/2 + n\pi$, where n is an integer. In particular, we exclude $\pm\pi/2$, $\pm3\pi/2$, $\pm5\pi/2$, and so on.

Similarly, the domain of the cotangent and cosecant functions is the set of all real numbers except those numbers t for which the ordinate y of $P(t)$ is 0. The latter include the numbers 0, $\pm\pi$, $\pm2\pi$, $\pm3\pi$, and, in general, all numbers of the form $n\pi$, where n is an integer. When we work with $\tan t$, $\cot t$, $\sec t$, and $\csc t$ in the future, it will always be assumed that t is in the appropriate domain, even though this fact will not always be pointed out explicitly.

Let us now investigate the range of each trigonometric function. Since the pair (x, y) in (5.4) gives coordinates of a point on the unit circle U, we have $|x| \le 1$ and $|y| \le 1$. Since $\sin t = y$ and $\cos t = x$, it follows that the range of both the sine and cosine functions is the set of all numbers in the closed interval $[-1, 1]$. On the other hand, the range of the cosecant and secant functions consists of all real numbers of absolute value greater than or equal to 1 (Why?).

To determine the range of the tangent function, let us first note that if (x, y) are coordinates of the point $P(t)$ on U, then $x^2 + y^2 = 1$, and hence either $y = \sqrt{1 - x^2}$ or $y = -\sqrt{1 - x^2}$. If we restrict $P(t)$ to the first or second quadrants, then

$$\tan t = \frac{y}{x} = \frac{\sqrt{1 - x^2}}{x}.$$

If a is *any* real number, there exists a real number x such that $\sqrt{1 - x^2}/x = a$, for we may let x equal $1/\sqrt{1 + a^2}$ or $-1/\sqrt{1 + a^2}$, depending on whether a is positive or negative. This shows that $\tan t$ takes on all real values and hence the range of the tangent function is all of **R**. Similarly, the range of the cotangent function is **R**.

We see from (5.4) that the following equations are true for all values of t for which denominators are not zero:

(5.5)　　$\csc t = \dfrac{1}{\sin t}$

(5.6)　　$\sec t = \dfrac{1}{\cos t}$

(5.7) $\cot t = \dfrac{1}{\tan t}$

(5.8) $\tan t = \dfrac{\sin t}{\cos t}$

(5.9) $\cot t = \dfrac{\cos t}{\sin t}$

These formulas may be proved by using (5.4). For example,

$$\tan t = \frac{y}{x} = \frac{\sin t}{\cos t} \qquad \csc t = \frac{1}{x} = \frac{1}{\sin t}$$

and so on. Equations (5.5)–(5.9) are identities since they are true for every allowable value of t.

Another useful relationship is a consequence of the fact that if (x, y) are coordinates of a point on U, then $y^2 + x^2 = 1$. Since $y = \sin t$ and $x = \cos t$, this gives us the identity

$$(\sin t)^2 + (\cos t)^2 = 1.$$

Powers such as $(\cos t)^n$, where $n \neq -1$, are written in the form $\cos^n t$. The symbol $\cos^{-1} t$ is reserved for a special situation to be discussed in the next chapter; the same is true for the other trigonometric functions. With this agreement on notation, the preceding identity may be written

(5.10) $\sin^2 t + \cos^2 t = 1.$

We may use (5.10) to express $\sin t$ in terms of $\cos t$, or vice versa. For example,

$$\sin t = \pm \sqrt{1 - \cos^2 t},$$

where the $+$ sign is used if $P(t)$ is in quadrants I or II, and the $-$ sign is used if $P(t)$ is in quadrants III or IV. Similarly,

$$\cos t = \pm \sqrt{1 - \sin^2 t},$$

where "+" is used if $P(t)$ is in quadrants I or IV and "−" is used if $P(t)$ is in quadrants II or III.

EXAMPLE 4 If $\sin t = 3/5$ and $P(t)$ is in quadrant II, find the values of the other trigonometric functions.

Solution

Since $\cos t$ is negative in quadrant II,

$$\cos t = -\sqrt{1 - \sin^2 t} = -\sqrt{1 - (3/5)^2} = -\sqrt{16/25} = -4/5.$$

Next, using (5.8),

$$\tan t = \frac{\sin t}{\cos t} = \frac{3/5}{-4/5} = -\frac{3}{4}.$$

Applying (5.5)–(5.7), the remaining values are

$$\csc t = \frac{5}{3}, \qquad \sec t = -\frac{5}{4}, \qquad \cot t = -\frac{4}{3}.$$

Two other very important identities, may be established without difficulty. If $\cos t \neq 0$, then dividing both sides of (5.10) by $\cos^2 t$ gives us

$$\frac{\sin^2 t}{\cos^2 t} + 1 = \frac{1}{\cos^2 t}.$$

By (5.8) and (5.6), this equation is equivalent to

(5.11) $\tan^2 t + 1 = \sec^2 t.$

We shall leave it as an exercise to prove the identity

(5.12) $1 + \cot^2 t = \csc^2 t.$

Identities (5.5)–(5.12) are often referred to as **fundamental identities.** They are the basis for much of our future work in trigonometry and should be memorized.

EXAMPLE 5   Use the fundamental identities to simplify the expression

$$\sec t - \sin t \tan t.$$

Solution

The reader should give reasons for each of the following steps:

$$\sec t - \sin t \tan t = \frac{1}{\cos t} - \sin t \left(\frac{\sin t}{\cos t}\right)$$

$$= \frac{1 - \sin^2 t}{\cos t}$$

$$= \frac{\cos^2 t}{\cos t}$$

$$= \cos t$$

Example 5 illustrates how the fundamental identities may be used to simplify trigonometric expressions. In the next chapter much more will be done along these lines.

Let us return briefly to the wrapping function discussed in Section 1. If we let t range from 0 to 2π, then the point $P(t)$ traces the unit circle U once in the counterclockwise direction, whereas $P(-t)$ traces U once in the clockwise direction. Moreover, as illustrated in Fig. 5.12, if $P(t)$ has coordinates (x, y), then $P(-t)$ has coordinates $(x, -y)$.

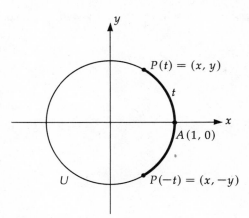

Figure 5.12

Hence by the definition of the trigonometric functions,

$$\sin(-t) = -y, \qquad \cos(-t) = x, \qquad \text{and} \qquad \tan(-t) = -\frac{y}{x}.$$

The previous three equalities may be written

$$(5.13) \quad \begin{array}{l} \sin(-t) = -\sin t, \\ \cos(-t) = \cos t, \\ \tan(-t) = -\tan t, \end{array}$$

where t is any real number in the domain of the indicated function. Similar formulas are true for the other trigonometric functions (see Exercise 31).

Incidentally, according to the instructions for Exercise 25 of Section 3 in Chapter Three, (5.13) implies that the sine and tangent are odd functions and the cosine is an even function.

EXAMPLE 6 Find $\sin(-\pi/6)$ and $\cos(-\pi/4)$.

Solution

Using (5.13) and Table A on p. 206,

$$\sin\frac{-\pi}{6} = -\sin\frac{\pi}{6} = -\frac{1}{2} \quad \text{and} \quad \cos\frac{-\pi}{4} = \cos\frac{\pi}{4} = \frac{\sqrt{2}}{2}.$$

EXERCISES

1 Verify all the entries in Table A on p. 206.

2 Write out the proofs of (5.5)–(5.9).

In Exercises 3–10 find the quadrant in which $P(t)$ lies if the given conditions are true.

3 $\cos t < 0$ and $\sin t > 0$

4 $\tan t < 0$ and $\cos t < 0$

5 $\sin t < 0$ and $\tan t < 0$

6 $\sec t > 0$ and $\cot t < 0$

7 $\csc t < 0$ and $\sec t > 0$

8 $\csc t > 0$ and $\tan t < 0$

9 $\sec t < 0$ and $\cot t > 0$

10 $\sin t > 0$ and $\sec t < 0$

In Exercises 11–18 find the values of all six trigonometric functions if the given conditions are true.

11 $\tan t = \frac{3}{4}$ and $\sin t < 0$

12 $\cot t = -\frac{3}{4}$ and $\cos t > 0$

13 $\sec t = -\frac{13}{5}$ and $\tan t < 0$

14 $\cos t = -\frac{1}{2}$ and $\sin t > 0$

15 $\sin t = -\frac{2}{3}$ and $\sec t > 0$

16 $\sec t = -5$ and $\csc t < 0$

17 $\csc t = 8$ and $\cos t < 0$

18 $\cos t = 0$ and $\sin t = -1$

In Exercises 19–24 use the rectangular coordinates of $P(t)$ to find the values of the six trigonometric functions for each of the given values of t.

19 **a** $t = \pi$ **b** $t = 3\pi/2$

20 **a** $t = 3\pi$ **b** $t = -5\pi/2$

21 **a** $t = -3\pi$ **b** $t = 5\pi/3$

22 **a** $t = 21\pi$ **b** $t = 7\pi/4$

23 **a** $t = 5\pi/6$ **b** $t = 3\pi/4$

24 **a** $t = -\pi/6$ **b** $t = 2\pi/3$

In Exercises 25–30 use the fundamental identities to transform the first
expression into the second.

3
Values of the
Trigonometric
Functions

25 $\cos t\ \csc t,\ \cot t$

26 $\sec t/\tan t,\ \csc t$

27 $\sin t(\csc t - \sin t),\quad \cos^2 t$

28 $\tan t\ \sin t + \cos t,\quad \sec t$

29 $\sec x/\csc x,\ \tan x$

30 $\tan^2 x/(\sec x + 1),\quad \sec x - 1$

31 Use (5.13) to establish the following identities: **a** $\csc (-t) = - \csc t,$
b $\sec (-t) = \sec t,$ **c** $\cot (-t) = -\cot t.$

32 Show that the range of the cotangent function is **R**.

33 Is there a real number t such that $4 \sin t = 5$? Explain.

34 Is there a real number t such that $\csc t = -1/2$? Explain.

35 Find the solution set of the inequality $\cos t > \sec t$.

36 Find the solution set of the inequality $\sec t < 1$.

37 If $f(t) = \cos t$ and $g(t) = t/6$, find the following:

   a $f(g(\pi))$,   **b** $g(f(\pi))$.

38 If $f(t) = \tan t$ and $g(t) = t/2$, find the following:

   a $f(g(\pi/3))$,   **b** $g(f(\pi/3))$.

39 Do the functions defined in (5.4) have inverse functions? Explain.

40 Prove that the secant function is an even function and the cosecant function is odd.

3 VALUES OF THE TRIGONOMETRIC FUNCTIONS

In the previous section several values of the trigonometric functions
were calculated. We now wish to discuss the problem of determining *all*
values. For the moment let us consider only the sine function. We
know from (5.4) that for every real number t there corresponds a
unique real number $\sin t$ which lies between -1 and 1. In general, the
methods used to determine this number are beyond the scope of this
book, since we do not have the processes necessary for measuring arc

length. We cannot, by elementary methods, find points $P(t)$ on the unit circle U that the wrapping function associates with values of t such as 0.74619, $-\frac{4}{7}$, $\sqrt{5}$, $\sqrt[3]{-11}$, or for that matter, integers such as 1, 2, 3, 4, and so on. We may, however, make some general observations. If t varies from 0 to $\pi/2$, then the coordinates (x, y) of $P(t)$ vary from $(1, 0)$ to $(0, 1)$. In particular, the ordinate y, that is, the value of the sine function, increases from 0 to 1. Moreover, this function takes on *all* values between 0 and 1. If we let t range from $\pi/2$ to π, then the coordinates (x, y) of $P(t)$ vary from $(0, 1)$ to $(-1, 0)$, and hence the sine function, that is, the ordinate y of $P(t)$, decreases from 1 to 0. In similar fashion, we see that as t varies from π to $3\pi/2$, $\sin t$ decreases from 0 to -1, and as t varies from $3\pi/2$ to 2π, $\sin t$ increases from -1 to 0. The following table partially indicates this behavior of $\sin t$ in the interval $[0, 2\pi]$. For a more complete description we would have to insert many more values of t and $\sin t$.

t	0	$\dfrac{\pi}{2}$	π	$\dfrac{3\pi}{2}$	2π
$\sin t$	0	1	0	-1	0

If we let t range through the interval $[2\pi, 4\pi]$, the identical pattern for $\sin t$ is repeated. The same is true for other intervals of length 2π. Indeed, by (5.2),

(5.14) $\sin (t + 2\pi n) = \sin t$,

for every real number t and every integer n. According to the next definition, this implies that the sine function is periodic.

(5.15)   Definition

A function f is **periodic** if there exists a positive real number k such that

$$f(x + k) = f(x)$$

for every x in the domain of f. If a least such positive real number k exists, it is called the **period** of f.

If a function f has period k, then the ordinate of the point with abscissa x is the same as the ordinate of the point with abscissa $x + k$ for every x. This means that the graph repeats itself in intervals of k units along the x-axis.

All the trigonometric functions are periodic. In particular, the next theorem shows that the sine function is periodic.

(5.16) Theorem

The sine function is periodic with period 2π.

Proof

Since $\sin(t + 2\pi n) = \sin t$ for all integers n, we have, if $n = 1$, $\sin(t + 2\pi) = \sin t$. Hence by (5.15) with $k = 2\pi$, the sine function is periodic. According to (5.15), it is sufficient to prove that there is no smaller positive number k such that $\sin(t + k) = \sin t$, for all t. We shall give an indirect proof. Suppose there is a positive number k less than 2π such that $\sin(t + k) = \sin t$ for all t. Letting $t = 0$, we obtain $\sin k = \sin 0 = 0$. Since $0 < k < 2\pi$ it follows that $P(k)$ has coordinates $(-1, 0)$. Consequently, $k = \pi$, and we may write $\sin(t + \pi) = \sin t$ for all t. In particular, if $t = \pi/2$, then $\sin(3\pi/2) = \sin \pi/2$, or $-1 = 1$, an absurdity. This completes the proof.

Similar theorems may be established for the other trigonometric functions. The variation of the cosine function in $[0, 2\pi]$ can be determined by observing the behavior of the abscissa x of $P(t)$ as t varies from 0 to 2π. The miniature table of values given below indicates that the cosine function decreases from 1 to 0 in the interval $[0, \pi/2]$, decreases from 0 to -1 in $[\pi/2, \pi]$, increases from -1 to 0 in $[\pi, 3\pi/2]$, and increases from 0 to 1 in $[3\pi/2, 2\pi]$. This pattern is then repeated in successive intervals of length 2π. It is left as an exercise to prove that the cosine function has period 2π.

t	0	$\dfrac{\pi}{2}$	π	$\dfrac{3\pi}{2}$	2π
$\cos t$	1	0	-1	0	1

By employing (5.5) and (5.6) it is easy to show that the secant and cosecant functions are periodic with period 2π. We shall discuss the variation of these functions further in Section 4, where their behavior is shown rather strikingly by means of graphs.

If $C(x, y)$ is any point on the unit circle U, then the point $C'(-x, -y)$ is diametrically opposite C, as is illustrated in Fig. 5.13.

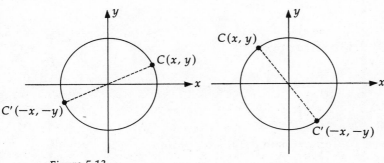

Figure 5.13

Since

$$\tan t = \frac{y}{x} = \frac{-y}{-x},$$

it follows that the tangent function has the same value at C' as at C. Let t be a real number such that $P(t) = C$, where as usual, $P(t)$ is the point which the wrapping function associates with t. Since the arc length $\overset{\frown}{CC'}$ (measured in the counterclockwise direction) is π, we have $P(t + \pi) = C'$ and hence

$$\tan (t + \pi) = \tan t.$$

It can be shown that there is no positive real number k smaller than π such that $\tan (t + k) = \tan t$, and hence *the tangent function is periodic with period π*. The variation of $\tan t$ will be discussed further in Section 4.

Let us return to the problem of finding all values of the trigonometric functions. Since the sine function has period 2π, it is sufficient to know the values of $\sin t$ for $0 \le t \le 2\pi$, because these same values are repeated in intervals of length 2π. The same is true for the other trigonometric functions. As a matter of fact, the values of any trigonometric function can be determined if its values in the t-interval $[0, \pi/2]$ are known. In order to prove this, suppose t is any real number and let $P(t)$ be the point that the wrapping function associates with t. The shortest distance t' between $P(t)$ and the x-axis, measured along U, will be called the **reference number** associated with t. Figure 5.14 illustrates arcs of length t' for positions of $P(t)$ in various quadrants.

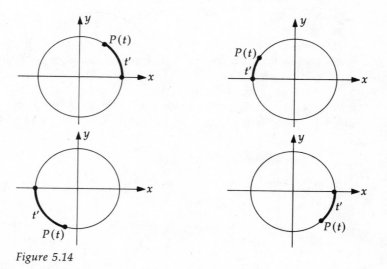

Figure 5.14

EXAMPLE 1 Approximate the reference number t' if t equals

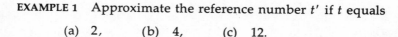

(a) 2,   (b) 4,   (c) 12.

(a)   Using $\pi \approx 3.1416$ and $\pi/2 \approx 1.5708$, we see that $\pi/2 <$ $2 < \pi$. Hence $P(2)$ lies in quadrant II (see Fig. 5.15) and the reference number t' for 2 is

$$t' = \pi - 2 \approx 3.1416 - 2 = 1.1416.$$

(b)   Since $P(4)$ is in quadrant III (see Fig. 5.15)

$$t' = 4 - \pi \approx 0.8584.$$

(c)   Since $2\pi \approx 6.2832$, the point $P(12)$ is found by making one complete counterclockwise revolution around U plus a partial revolution of length

$$12 - 2\pi \approx 12 - 6.2832 = 5.7168.$$

Since $3\pi/2 \approx 4.7$ we see that

$$3\pi/2 < 5.7168 < 2\pi$$

and therefore $P(12)$ is in quadrant IV (see Fig. 5.15) Hence the reference number is

$$t' \approx 2\pi - 5.7168 = 0.5664.$$

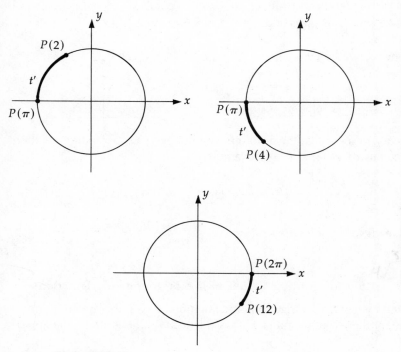

Figure 5.15

From the preceding discussion we see that if $P(t)$ is not on a coordinate axis and t' is the reference number for t, then $0 < t' < \pi/2$. Suppose (x, y) are the rectangular coordinates of $P(t)$, so that $P(t) = P(x, y)$. Consider the point $A(1, 0)$ and let $P'(x', y')$ be the point on U in quadrant I such that $\widehat{AP'} = t'$. Illustrations in which $P(x, y)$ lies in quadrants II, III, or IV are given in Fig. 5.16.

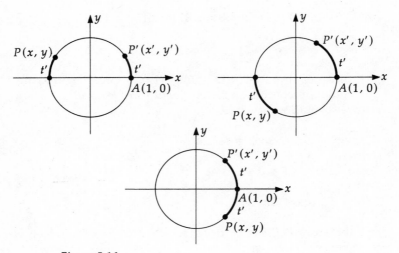

Figure 5.16

We see that in all cases

$$x' = |x| \quad \text{and} \quad y' = |y|,$$

and hence we have

$$|\cos t| = |x| = x' = \cos t',$$
$$|\sin t| = |y| = y' = \sin t'.$$

It is easy to show that the absolute value of *every* trigonometric function at t equals its value at t'. For example,

$$|\tan t| = \left|\frac{y}{x}\right| = \frac{|y|}{|x|} = \frac{y'}{x'} = \tan t'.$$

This leads to the following rule:

(5.17) Rule for Finding Values of the Trigonometric Functions

To find the value of a trigonometric function at a number t, determine its value for the reference number t' associated with t and prefix the appropriate sign.

EXAMPLE 2   Find sin (7π/4) and sec (−7π/6).

Solution

Since $7\pi/4 = 2\pi - \pi/4$ and $-7\pi/6 = -\pi - \pi/6$, the reference numbers are $\pi/4$ and $\pi/6$, respectively (see Fig. 5.17). Hence by (5.17) and Table A of Section 2,

$$\sin \frac{7\pi}{4} = -\sin \frac{\pi}{4} = -\frac{\sqrt{2}}{2},$$

$$\sec \frac{-7\pi}{6} = -\sec \frac{\pi}{6} = -\frac{2\sqrt{3}}{3}.$$

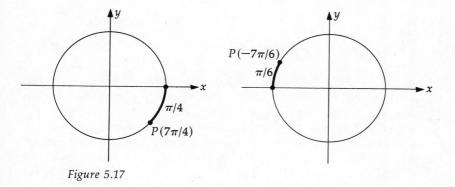

Figure 5.17

By employing more advanced techniques it is possible to compute, to any degree of accuracy, all the values of the trigonometric functions in the *t*-interval [0, π/2]. Table 3 gives approximations to such values for the sine, cosine, tangent, and cotangent functions. Note that the range of *t* in Table 3 is from 0 to 1.5708. The latter number is a four-decimal-place approximation to π/2. Although tables for the secant and cosecant are available, we shall not include them in this book. If values of these functions are required, they can be computed by using sec *t* = 1/cos *t* and csc *t* = 1/sin *t*. Table 3 also includes columns labeled *degrees*. The significance and use of these columns will be discussed in a later section. It will be seen that the inclusion of the degree columns is the reason that *t* varies at intervals of approximately 0.0029. For the time being, the reader should ignore the degree columns when using Table 3.

To find the values of trigonometric functions at a real number *t*, where 0 ≤ *t* ≤ .7854, the labels at the *top* of the columns in Table 3 should be used. For example,

cos (.1338) ≈ .9911,   sin (.4654) ≈ .4488,
tan (.5789) ≈ .6536,

and so on. On the other hand, if .7854 ≤ *t* ≤ 1.5708, then the labels at the *bottom* of the columns should be employed. For example,

219

$$\sin (1.2363) \approx .9446, \quad \tan (1.5213) \approx 20.206,$$
$$\cos (.8639) \approx .6494.$$

The reason that the table can be arranged in this way follows from the fact, to be proved later (see (6.4)), that

$$\sin t = \cos (\pi/2 - t) \quad \text{and} \quad \cot t = \tan (\pi/2 - t)$$

for all real numbers t. In particular, since $\pi/2 \approx 1.5708$,

$$\sin (.1047) \approx \cos (1.5708 - .1047),$$

or

$$\sin (.1047) \approx \cos (1.4661).$$

Likewise, the value of the sine function at 1.4661 is the same as the value of the cosine function at .1047. Similar remarks are true for the tangent and cotangent functions.

If it is necessary to find functional values when t lies *between* numbers given in the table, the method of linear interpolation used for logarithms may be employed. Similarly, given a value such as $\sin t = .6371$, we may refer to the body of Table 3 and use linear interpolation, if necessary, to obtain an approximation to t.

EXAMPLE 3 Find approximations for

    (a) $\tan (2.3824)$,   (b) $\cos (.4)$.

Solutions

(a)   Since $P(2.3824)$ is in quadrant II, the reference number t' is $\pi - 2.3824$ or $t' \approx 3.1416 - 2.3824 = .7592$. Using (5.17) and Table 3, we have

$$\tan (2.3824) \approx - \tan (.7592)$$
$$\approx -.9490.$$

(b)   To find $\cos (.4)$ we locate the number .4000 between successive values of t in Table 3 and interpolate as follows:

$$.0029 \left\{ .0015 \left\{ \begin{array}{l} \cos (.3985) \approx .9216 \\ \cos (.4000) \approx \quad ? \end{array} \right\} d \atop \cos (.4014) \approx .9205 \right\} .0011$$

$$\frac{.0015}{.0029} = \frac{d}{.0011}$$

or

$$d = \tfrac{15}{29}(.0011) \approx .0006.$$

Hence

$$\cos (.4000) \approx .9216 - .0006 = .9210.$$

Note that since the cosine function is decreasing in the given interval we must *subtract* d from .9216.

EXAMPLE 4 Approximate the smallest positive real number t such that sin $t = .6635$.

Solution

We locate .6635 between successive entries in the sine column of Table 3 and interpolate as follows:

$$.0029 \left\{ d \left\{ \begin{matrix} \sin (.7243) \approx .6626 \\ \sin t \quad = .6635 \end{matrix} \right\} .0009 \right\} .0022 \\ \sin (.7272) \approx .6648 $$

$$\frac{d}{.0029} = \frac{.0009}{.0022}$$

or

$$d = \tfrac{9}{22}(.0029) \approx .0012.$$

Hence

$$t \approx .7243 + .0012 = .7255.$$

EXERCISES

In Exercises 1–8 find the reference number t' if t has the given value.

1 **a** $7\pi/6$ **b** $5\pi/6$ **c** $-\pi/6$

2 **a** $-4\pi/3$ **b** $4\pi/3$ **c** $5\pi/3$

3 **a** $3\pi/4$ **b** $-5\pi/4$ **c** $9\pi/4$

4 **a** $11\pi/4$ **b** $-\pi/3$ **c** $-5\pi/6$

5 1.9   **6** -2.8

7 5   **8** 6

9–12 Use (5.17) and Table A of Section 2 to find the values of the sine, cosine, and tangent functions at the numbers t given in Exercises 1–4.

In Exercises 13–26 use Table 3 to approximate the indicated numbers.

13 sin (.5760)   **14** sin (1.4399)

15 cos (1.1781)   **16** cos (.6952)

17 tan (1.8733)   **18** cot (2.9002)

19 cot (7.1355)

20 tan (−0.2763)

21 sin (1.4952)

22 sin (.3403)

23 cos (.6545)

24 cos (.9687)

25 tan (.7476)

26 cot (.1600)

In Exercises 27–32 use interpolation to approximate the given numbers.

27 tan (.5824)

28 cos (.2167)

29 sin (.53)

30 cot (1.4)

31 cos (4)

32 tan (10)

In each of Exercises 33–38 use interpolation to approximate the smallest positive real number t for which the given equality is true.

33 cos $t = .8392$

34 sin $t = .1174$

35 tan $t = .3947$

36 cot $t = .7150$

37 sin $t = .7592$

38 tan $t = 2.7623$

In Exercises 39–42 approximate all values of t between 0 and 2π for which the given equation is true.

39 sin $t = .6648$

40 cos $t = .9793$

41 tan $t = 2.1123$

42 cos $t = .9713$

43 Prove that the cosine function has period 2π.

44 Prove that the tangent function has period π.

4 GRAPHS OF THE TRIGONOMETRIC FUNCTIONS

In our previous work with graphs we used the symbols x and y as labels for the coordinate axes. In the present chapter x has been used primarily for the abscissa of a point on the unit circle U; hence in this section we shall use the symbol t for the horizontal axis. In the ty-coordinate system the graph of the sine function is the same as the graph of the equation $y = \sin t$.

It is not difficult to sketch the graphs of the trigonometric functions. For example since $|\sin t| \leq 1$ for all t, the graph of the sine function lies between the horizontal lines $y = 1$ and $y = -1$. Moreover, since the sine function is periodic with period 2π, it is sufficient to determine the graph in the interval $[0, 2\pi]$ on the t-axis, for the same pattern is repeated in intervals of 2π over the entire t-axis. In the preceding section we discussed the variation of $\sin t$ in the interval $[0, 2\pi]$. Several values of $\sin t$ are given in the following table.

t	0	$\dfrac{\pi}{4}$	$\dfrac{\pi}{2}$	$\dfrac{3\pi}{4}$	π	$\dfrac{5\pi}{4}$	$\dfrac{3\pi}{2}$	$\dfrac{7\pi}{4}$	2π
$\sin t$	0	$\dfrac{\sqrt{2}}{2}$	1	$\dfrac{\sqrt{2}}{2}$	0	$\dfrac{-\sqrt{2}}{2}$	-1	$\dfrac{-\sqrt{2}}{2}$	0

To obtain a rough sketch we may plot several points, draw a smooth curve through them, and extend the configuration to the right and left in periodic fashion. This gives us the portion of the graph shown in Fig. 5.18. Of course, the graph does not terminate, but continues indefinitely in both directions. If greater accuracy is desired, additional points could be plotted using, for example, $\sin \pi/6 = \tfrac{1}{2}$, $\sin \pi/3 = \sqrt{3}/2 \approx 0.86$, and so on. Table 3 could also be used to obtain many additional points on the graph. We refer to the part of the graph in the interval $[0, 2\pi]$ as a **sine wave.**

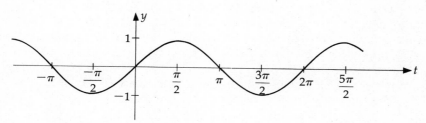

Figure 5.18 $y = \sin t$

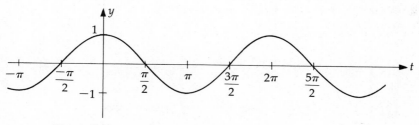

Figure 5.19 $y = \cos t$

It can be shown that the graph of the cosine function has the appearance shown in Fig. 5.19. The verification of this is left as an exercise.

A partial table of values for the tangent function is given below.

t	$-\dfrac{\pi}{3}$	$-\dfrac{\pi}{4}$	$-\dfrac{\pi}{6}$	0	$\dfrac{\pi}{6}$	$\dfrac{\pi}{4}$	$\dfrac{\pi}{3}$
$\tan t$	$-\sqrt{3}$	-1	$\dfrac{-\sqrt{3}}{3}$	0	$\dfrac{\sqrt{3}}{3}$	1	$\sqrt{3}$

The corresponding points are plotted in Fig. 5.20. The values of $\tan t$ near $t = \pi/2$ demand special consideration. As t increases through posi-

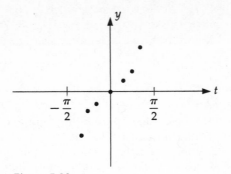

Figure 5.20

tive values toward $\pi/2$, the point $P(t)$ assigned by the wrapping function approaches the point $(0, 1)$. Hence the abscissa x of $P(t)$ gets close to 0. Since $\tan t = y/x$ it follows that if $t \approx \pi/2$ and $t < \pi/2$, then $\tan t$ is a large positive number. Indeed, $\tan t$ can be made arbitrarily large by choosing t sufficiently close to $\pi/2$. The terminology "$\tan t$ *increases without bound*" or "$\tan t$ *becomes positively infinite*" as t approaches $\pi/2$ is used to describe this situation. If t approaches $-\pi/2$ through values larger than $-\pi/2$, then $\tan t$ *decreases without bound,* that is, $\tan t$ *becomes negatively infinite.* This behavior is illustrated in Fig. 5.21. The

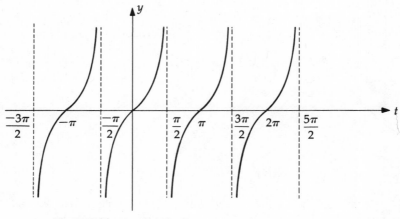

Figure 5.21 $y = \tan t$

vertical lines that are indicated by dashes are not part of the graph but merely serve as guide lines for sketching. These lines are called **vertical asymptotes** for the graph. Since the tangent function has period π, the pattern given in the interval $(-\pi/2, \pi/2)$ is repeated in other similar intervals of length π.

The graphs of the remaining three trigonometric functions can be obtained from those we have found. For example, since $\csc t = 1/\sin t$ we may find an ordinate of a point on the graph of the cosecant function by taking the reciprocal of the corresponding ordinate of the sine graph. This is possible except for $t = n\pi$, where n is an integer, for in this case sin $t = 0$. As an aid to sketching the graph of the cosecant function it is convenient to sketch the graph of the sine function with dashes (see Fig. 5.22) and then to take reciprocals of ordinates

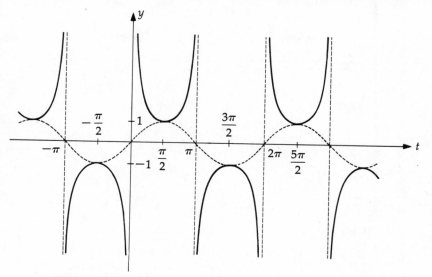

Figure 5.22 $y = \csc t$

to obtain points on the cosecant graph. Notice the manner in which the cosecant function increases or decreases without bound as t approaches $n\pi$, where n is an integer. The graph has vertical asymptotes as indicated in the figure.

The graphs of the secant and cotangent functions may be obtained in similar fashion. These are shown in Figs. 5.23 and 5.24. We leave their verifications as exercises.

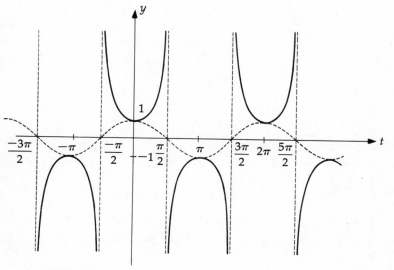

Figure 5.23 $y = \sec t$

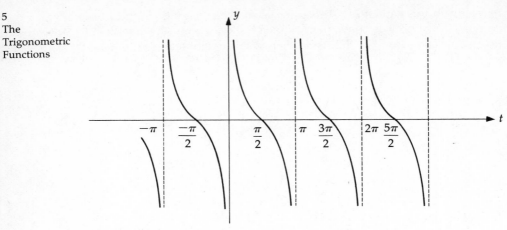

Figure 5.24 $y = \cot t$

EXAMPLE 1 Sketch the graph of the function f if $f(t) = 2 \sin t$.

Solution

Although the graph could be obtained by plotting points, note that for each t_1 the ordinate $f(t_1)$ is always twice that of the corresponding ordinate on the sine graph. A simple graphical technique is to sketch the graph of $y = \sin t$ with dashes and then double each ordinate to find points on the graph of $y = 2 \sin t$ (see Fig. 5.25).

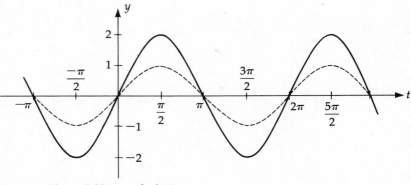

Figure 5.25 $y = 2 \sin t$

EXERCISES

1 **a** Verify the graphs of the cosine and secant functions.

   b Describe the intervals between -2π and 2π in which the secant function increases.

   c Describe the intervals between -2π and 2π in which the secant function decreases.

2 **a** Verify the graph of the cotangent function.

b In what intervals does the cotangent function increase?

c In what intervals does the cotangent function decrease?

In Exercises 3–10 use the method illustrated in Example 1 to sketch the graphs of the functions f which are defined as indicated.

3 $f(t) = 3 \sin t$

4 $f(t) = 4 \sin t$

5 $f(t) = 2 \cos t$

6 $f(t) = \frac{1}{2} \cos t$

7 $f(t) = \frac{1}{2} \sin t$

8 $f(t) = 2 \tan t$

9 $f(t) = -\sin t$

10 $f(t) = -\cos t$

11 Sketch the graph of the equation $y = \sin(-t)$ and describe how it is related to the graph of $y = \sin t$.

12 In what way are the graphs of the equations $y = \cos(-t)$ and $y = \cos t$ related?

In Exercises 13–16 sketch the graphs of the given equations after plotting a sufficient number of points.

13 $y = \sin(2t)$

14 $y = \sin(\frac{1}{2}t)$

15 $y = \cos(\frac{1}{2}t)$

16 $y = \cos(2t)$

5 TRIGONOMETRIC GRAPHS

In this section we shall consider the graphs of

(5.18) $\quad f(x) = a \sin(bx + c),$

where a, b, and c are real numbers, and the graphs of similar equations that involve other trigonometric functions. Instead of using a ty-coordinate system as in Section 4, we shall now use the conventional xy-coordinate system. It is important to remember that in this situation the letter x is used in place of t and hence is not to be regarded as the x used in Definition (5.4). In order to sketch the graph of f in (5.18) we could begin by plotting many points; however, it is generally easier to use information about the graphs of the trigonometric functions discussed in the preceding section. Let us consider the special case of (5.18) in which $c = 0$, $b = 1$, and $a > 0$. Thus we wish to find the graph of the equation $y = a \sin x$. We may find the ordinate of a point on the graph by multiplying the corresponding ordinate on the graph of $y = \sin x$ by a. Thus if $y = 2 \sin x$, we multiply by 2; if $y = \frac{1}{2} \sin x$, we multiply by $\frac{1}{2}$; and so on. The graph of $y = 2 \sin x$ is sketched in Fig. 5.25. The graph of $y = \frac{1}{2} \sin x$ is sketched in Fig. 5.26, where for comparison we have indicated the graph of $y = \sin x$ with dashes.

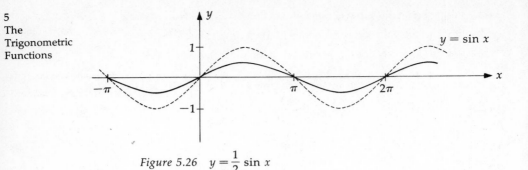

Figure 5.26 $y = \dfrac{1}{2} \sin x$

EXAMPLE 1 Sketch the graph of the equation $y = 3 \cos x$.

Solution

The graph of the given equation may be obtained from the graph of $y = \cos x$ by multiplying ordinates of points by 3. We first sketch $y = \cos x$ with dashes and then we triple the ordinates. This gives us the sketch in Fig. 5.27.

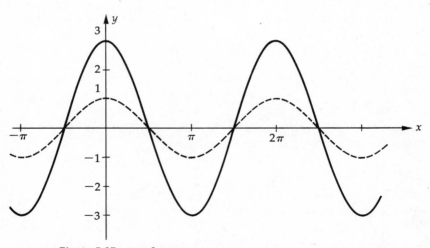

Figure 5.27 $y = 3 \cos x$

If $a < 0$, then the ordinates of points on the graph of $y = a \sin x$ are negatives of the corresponding ordinates of points on the graph of $y = |a| \sin x$. This is illustrated in the next example.

EXAMPLE 2 Sketch the graph of $y = -2 \sin x$.

Solution

As a guide we first sketch the graph of $y = \sin x$ with dashes and then we multiply each ordinate by -2. This gives us the sketch shown in Fig. 5.28. The sketch may also

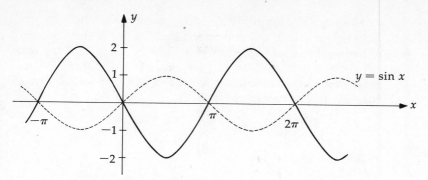

Figure 5.28 $y = -2 \sin x$

be obtained by multiplying ordinates of the graph of $y = 2 \sin x$ by -1. We sometimes refer to the graph of $y = -2 \sin x$ as a *reflection through the x-axis* of the graph of $y = 2 \sin x$.

If $f(x) = a \sin (bx + c)$, then the **amplitude** of f (or of its graph) is defined as the maximum ordinate of points on the graph. If $a > 0$, the maximum ordinate occurs if $\sin (bx + c) = 1$, whereas if $a < 0$, we must take $\sin (bx + c) = -1$. In either case the amplitude is $|a|$. For example, if $f(x) = \frac{1}{2} \sin x$, the amplitude is $\frac{1}{2}$; if $f(x) = -2 \sin x$, the amplitude is 2, and so on.

The same is true if f is given by $f(x) = a \cos (bx + c)$. In general, given any periodic function, the amplitude is defined as $|M - m|/2$, where M is the maximum value and m is the minimum value of the function, provided they exist. Thus the amplitude of $y = a \sin x$ is $|a - (-a)|/2 = |2a|/2 = |a|$.

The next theorem provides some information about periods.

(5.19) Theorem

If $f(x) = \sin bx$, where $b \neq 0$, then the period of f is $2\pi/b$.

Proof

If $b > 0$, then as bx ranges from 0 to 2π we obtain exactly one sine wave for the graph of f. The conclusion of the theorem follows from the fact that bx ranges from 0 to 2π if and only if x ranges from 0 to $2\pi/b$. A similar proof may be given if $b < 0$.

EXAMPLE 3 Find the period and sketch the graph of f if

(a) $f(x) = \sin 2x$.   (b) $f(x) = \sin \frac{1}{2}x$.

Solution

Both functions have the form given in (5.19). For (a) the period is $2\pi/2 = \pi$, which means that there is exactly one

Trigonometric
Graphs

sine wave of amplitude 1 corresponding to the interval $[0, \pi]$. The graph is sketched in Fig. 5.29, where for convenience we have used different scales on the x and y-axes. For (b) the period is $2\pi/(1/2) = 4\pi$ and hence there is one sine wave of amplitude 1 corresponding to the interval $[0, 4\pi]$, as illustrated in Fig. 5.30.

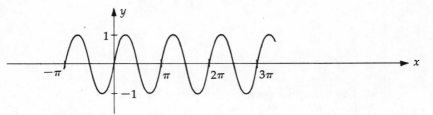

Figure 5.29 $y = \sin 2x$

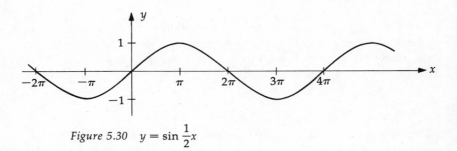

Figure 5.30 $y = \sin \dfrac{1}{2}x$

Note in general that if b is large, then $2\pi/b$ is small and the waves are close together. As a matter of fact, there are b sine waves in an interval of 2π units. However if b is small, then $2\pi/b$ is large and the waves are shallow. For example, if $y = \sin\left(\tfrac{1}{10}\right)x$, then an interval 20π units long is required for one complete wave.

If $b < 0$, we can use the fact that $\sin(-u) = -\sin u$ to obtain the graph. To illustrate, the graph of $y = \sin(-2x)$ is the same as the graph of $y = -\sin 2x$.

By combining the above discussions we can arrive at a technique for sketching the graph of a function f defined by $f(x) = a \sin bx$. The graph has the basic sine wave pattern. However, the amplitude is $|a|$ and the period is $2\pi/|b|$. If $a < 0$ or $b < 0$, we make adjustments on the signs of ordinates, as discussed earlier.

EXAMPLE 4 Sketch the graph of $f(x) = 3 \sin 2x$.

Solution

From the preceding discussion we see that the amplitude is 3 and the period is $2\pi/2 = \pi$. The graph is readily obtained by first sketching the graph of $y = \sin 2x$ with dashes and then multiplying the ordinates of each point by 3. This leads to the sketch in Fig. 5.31.

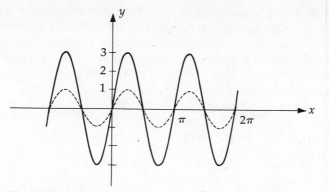

Figure 5.31 $f(x) = 3 \sin 2x$

EXAMPLE 5   Sketch the graph of $f(x) = 2 \sin(-3x)$.

Solution

Since $f(x) = -2 \sin 3x$, we see that the amplitude is 2 and the period is $2\pi/3$. Thus there is one sine wave every $2\pi/3$ units. The minus sign indicates a reflection through the x-axis. If we mark off an interval of length $2\pi/3$ and sketch a sine wave of amplitude 2 (reflected through the x-axis), the shape of the graph is apparent. The configuration given in the interval $[0, 2\pi/3]$ is carried along periodically as illustrated in Fig. 5.32.

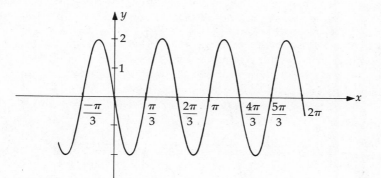

Figure 5.32 $y = -2 \sin 3x$

Similar discussions can be given if f is defined by $f(x) = a \cos bx$ or by $f(x) = a \tan bx$. In the latter case the period is π/b since the tangent function has period π. There are no maximum or minimum ordinates for points on the graph of the tangent function and hence we do not refer to its amplitude; however, we may still use the process of multiplying tangent ordinates by a in order to obtain points on the graph of $y = a \tan bx$.

Let us conclude this section by considering the general situation $f(x) = a \sin(bx + c)$. We have already observed that the amplitude is $|a|$.

231

We also see that one complete sine wave is obtained if $bx + c$ ranges from 0 to 2π, that is, if bx ranges from $-c$ to $2\pi - c$. In turn, the latter variation is obtained by letting x range from $-c/b$ to $(2\pi - c)/b$. If $-c/b > 0$, this amounts graphically to *shifting* the graph of $y = a \sin bx$ to the right $-c/b$ units. If $-c/b < 0$, the shift is to the left. The number $-c/b$ is sometimes called the **phase shift** associated with the function. Similar remarks can be made for the other functions.

EXAMPLE 6   Sketch the graph of $f(x) = 3 \sin (2x - \pi/2)$.

Solution

This equation is of the form (5.18) with $a = 3$, $b = 2$, and $c = -\pi/2$. It follows from the preceding discussion that the graph has the sine wave pattern with amplitude 3 and period $2\pi/2 = \pi$. Since

$$-c/b = -(-\pi/2)/2 = \pi/4,$$

the phase shift is $\pi/4$. Hence we may obtain the graph by shifting the graph of $f(x) = 3 \sin 2x$ (see Fig. 5.31) to the right $\pi/4$ units. This gives us Fig. 5.33.

The graph could also be sketched without memorizing the result about phase shifts, by reasoning as follows. In order to obtain an interval containing one sine wave, we let $2x - \pi/2$ range from 0 to 2π. Since $2x - \pi/2 = 0$ when $x = \pi/4$, and $2x - \pi/2 = 2\pi$ when $x = 5\pi/4$, we see that one sine wave of amplitude 3 will occur in the interval $[\pi/4, 5\pi/4]$. Note that the period is $5\pi/4 - \pi/4 = \pi$.

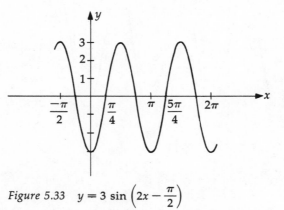

Figure 5.33 $y = 3 \sin \left(2x - \dfrac{\pi}{2}\right)$

EXERCISES

1   Without plotting many points, sketch the graph and determine the amplitude and period of each function f defined as follows:

a $f(x) = 4 \sin x$ **b** $f(x) = \sin 4x$

c $f(x) = \frac{1}{4} \sin x$ **d** $f(x) = \sin (x/4)$

e $f(x) = 2 \sin (x/4)$ **f** $f(x) = \frac{1}{2} \sin 4x$

g $f(x) = -4 \sin x$ **h** $f(x) = \sin (-4x)$

2 Sketch the graphs of the functions which involve the cosine and which are analogous to those defined in Exercise 1.

3 Without plotting many points, sketch the graph of each function f defined as follows, determining the amplitude and period in each case:

a $f(x) = 3 \cos x$ **b** $f(x) = \cos 3x$ **c** $f(x) = \frac{1}{3} \cos x$

d $f(x) = \cos (x/3)$ **e** $f(x) = 2 \cos (x/3)$ **f** $f(x) = \frac{1}{3} \cos (2x)$

g $f(x) = -3 \cos x$ **h** $f(x) = \cos (-3x)$

4 Sketch the graphs of the functions which involve the sine and which are analogous to those defined in Exercise 3.

In Exercises 5–22 sketch the graphs of the given equations.

5 $y = 4 \sin (x - \pi/2)$ **6** $y = 2 \cos (x + \pi/4)$

7 $y = 3 \sin (2x + \pi/2)$ **8** $y = 2 \sin (3x - \pi/2)$

9 $y = 4 \cos (3x + \pi)$ **10** $y = \frac{1}{2} \cos (2x - \pi)$

11 $y = 6 \cos \pi x$ **12** $y = 3 \sin \frac{1}{2} \pi x$

13 $y = 2 \sec x$ **14** $y = \csc (x/2)$

15 $y = \frac{1}{2} \csc x$ **16** $y = \sec 2x$

17 $y = \tan (x/2)$ **18** $y = \cot 2x$

19 $y = - \cot x$ **20** $y = - \tan (x/3)$

6 ADDITIONAL GRAPHICAL TECHNIQUES

In mathematical applications it is common to encounter functions that are defined in terms of sums and products of expressions such as

$$f(x) = \sin 2x + \cos x,$$

$$f(x) = 2^{-x} \sin x,$$

and so on. When working with a sum of two expressions a graphical technique called **addition of ordinates** is useful. This method applies not only to trigonometric expressions but to arbitrary expressions as well. Thus, suppose f is defined by

$$f(x) = g(x) + h(x),$$

where the functions g and h have domain D. The graph of f may be obtained from the graphs of g and h as follows. We begin by sketching the graphs of the equations $y = g(x)$ and $y = h(x)$ on the same coordinate axes, as illustrated by the dashes in Fig. 5.34. Since $f(x_1) = g(x_1) + h(x_1)$ for every x_1 in D, the ordinate of the point on the graph of $y = g(x) + h(x)$ with abscissa x_1 is the *sum* of the corresponding ordinates of points on the graphs of g and h. If we draw a vertical line at the point with coordinates $(x_1, 0)$, then the ordinates $g(x_1)$ and $h(x_1)$ may be added geometrically by means of a compass or ruler, as is illustrated in Fig. 5.34. If either $g(x_1)$ or $h(x_1)$ is negative, then a *subtraction* of ordinates may be employed. By using this technique for all x we obtain the graph of f, as illustrated in Fig. 5.34.

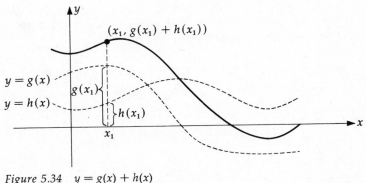

Figure 5.34 $y = g(x) + h(x)$

EXAMPLE 1   If $f(x) = \cos x + \sin x$, use the method of addition of ordinates to sketch the graph of f.

Solution

We begin by sketching, with dashes, the graphs of the equations $y = \cos x$ and $y = \sin x$. Next, for various numbers x_1, we add ordinates as indicated by the sketch in Fig. 5.35.

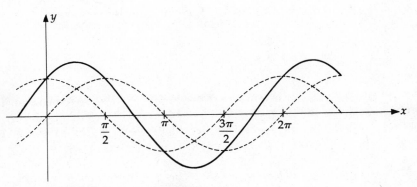

Figure 5.35 $y = \cos x + \sin x$

After a sufficient number of ordinates are added and a pattern emerges, we draw a smooth curve through the points. As a check, it would be worthwhile to plot some points on the graph by substituting numbers for x. We shall leave such verifications to the reader. It can be seen from the graph that the function f is periodic with period 2π.

EXAMPLE 2 Sketch the graph of the equation $y = \cos x + \sin 2x$.

Solution

We sketch, with dashes, the graphs of the equations $y = \cos x$ and $y = \sin 2x$ on the same coordinate axes and use the method of addition of ordinates. The graph is illustrated in Fig. 5.36. Evidently f is periodic with period 2π.

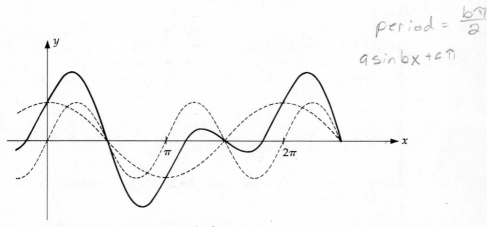

period $= \dfrac{b\pi}{a}$

$a \sin bx + c \pi$

Figure 5.36 $y = \cos x + \sin 2x$

EXAMPLE 3 Sketch the graph of $f(x) = 2^{-x} \sin x$.

Solution

Since $|f(x)| = |2^{-x}|\,|\sin x|$ and since $|\sin x| \le 1$ and $2^{-x} > 0$, it follows that $|f(x)| \le 2^{-x}$ for all x. Consequently,

$$-2^{-x} \le f(x) \le 2^{-x}$$

for all x, which implies that the graph of f lies between the graphs of the equations $y = -2^{-x}$ and $y = 2^{-x}$. The graph of f will coincide with one of the latter graphs when $|\sin x| = 1$, that is, when $x = \pi/2 + n\pi$, where n is an integer. Since $2^{-x} > 0$, the x-intercepts on the graph of f occur at $\sin x = 0$, that is, at $x = n\pi$. With this information we obtain the sketch shown in Fig. 5.37.

235

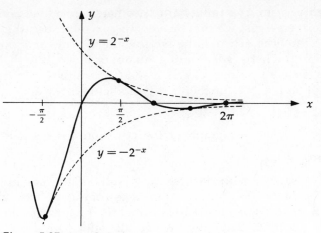

$y = 2^{-x}$

$y = -2^{-x}$

Figure 5.37 $y = 2^{-x} \sin x$

The graph of $y = 2^{-x} \sin x$ is called a **damped sine wave** and 2^{-x} is called the **damping factor.** By using different damping factors, we may obtain other variations of such compressed or expanded sine waves. The analysis of such graphs is important in electrical theory.

EXERCISES

Sketch the graphs of the equations in Exercises 1–16 by the method of addition of ordinates.

1 $y = \cos x + 3 \sin x$ **2** $y = \sin x + 3 \cos x$

3 $y = 2 \cos x + 3 \sin x$ **4** $y = 2 \sin x + 3 \cos x$

5 $y = \sin x + \cos 2x$ **6** $y = 2 \cos x + \sin 2x$

7 $y = \cos x - \sin x$ **8** $y = 2 \sin x - \cos x$

9 $y = 2 \cos x - \frac{1}{2} \sin 2x$ **10** $y = 2 \cos x + \cos \frac{1}{2}x$

11 $y = \frac{1}{2}x + \sin x$ **12** $y = |x + 1| + \cos 2x$

13 $y = e^x + \sin x$ **14** $y = 1 + \cos x$

15 $y = 2 + \sec x$ **16** $y = \csc x - 1$

In each of Exercises 17–22 sketch the graph of f.

17 $f(x) = 2^{-x} \cos x$ **18** $f(x) = 2^x \sin x$

19 $f(x) = |x| \sin x$ **20** $f(x) = (1 + x^2)\cos x$

21 $f(x) = \frac{1}{2}x^2 \sin x$ **22** $f(x) = \cot x \tan x$

7 ANGLES AND THEIR MEASUREMENT

The definitions of the trigonometric functions can also be based upon the notion of angles. This more traditional approach is quite common in applications of mathematics and hence should not be obscured by our version in terms of the wrapping function. Indeed, for a thorough appreciation of the trigonometric functions, it is probably best to blend the two ideas.

In geometry an angle is often thought of as the geometric configuration formed by two rays or half-lines, l_1 and l_2, having the same initial point O. If A and B are points on l_1 and l_2 respectively (see Fig. 5.38), then by definition, **angle AOB** is the union of the points on the

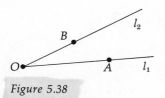

Figure 5.38

rays. The same is true for finite line segments with a common endpoint. For trigonometric purposes it is convenient to regard an angle as generated by starting with a fixed ray l_1 with endpoint O and rotating it about O, in a plane, to a position specified by a ray l_2. As above, if A and B are points on l_1 and l_2 respectively, the resulting geometric figure is referred to as angle AOB. We call l_1 the **initial side,** l_2 the **terminal side,** and O the **vertex** of the angle. The amount or direction of rotation is not restricted in any way. Thus we might let l_1 make several revolutions in either direction about O before coming to the position l_2.

If a rectangular coordinate system is introduced, then the **standard position** of an angle is obtained by taking the vertex at the origin and letting l_1 coincide with the positive x-axis. If l_1 is rotated in a counterclockwise direction to position l_2, then the angle is considered **positive,** whereas if l_1 is rotated in a clockwise direction, the angle is **negative.** We often denote angles by lower-case Greek letters and specify the direction of rotation by means of a circular arc or spiral with an arrow attached. Fig. 5.39 contains sketches of two positive angles α and β and

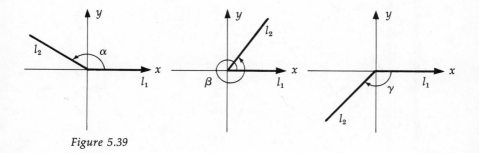

Figure 5.39

a negative angle γ. If the terminal side of an angle is in a certain quadrant, we speak of the *angle* as being in that quadrant. In Fig. 5.39, α is in quadrant II, β is in quadrant I, and γ is in quadrant III. If the terminal side coincides with a coordinate axis, then the angle is referred to as a **quadrantal angle.** It is important to observe that there are many different angles in standard position which have the same terminal side. Any two such angles are called **coterminal.**

We shall now consider the problem of assigning a measure to a given angle. Let U be a unit circle with the center at the origin O of a rectangular coordinate system, and let θ be an angle in standard position. As in the preceding discussion we regard θ as generated by rotating the positive x-axis about O. As the axis rotates to its terminal position, its point of intersection with U travels a certain distance t before arriving at its final position P, as illustrated in Fig. 5.40.

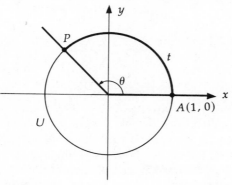

Figure 5.40

If t is considered positive for a counterclockwise rotation and negative for a clockwise rotation, then P is precisely the point which the wrapping function associates with the real number t. A natural way of assigning a measure to θ is to use the number t. When this is done, we say that **θ is an angle of t radians.** Notationally, it is customary to write $\theta = t$ instead of $\theta = t$ *radians*. In particular, if $\theta = 1$, then θ is an angle that subtends an arc of unit length on the unit circle U. The notation $\theta = -7.5$ means that θ is the angle generated by a clockwise rotation in which the point of intersection of the terminal side of θ with the unit circle U travels 7.5 units. Several angles, measured in radians, are sketched in Fig. 5.41.

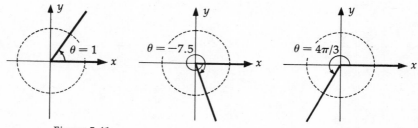

Figure 5.41

The radian measure of an angle can be found by using a circle of *any* radius. Thus suppose that θ is a central angle of a circle of radius r and that θ subtends an arc of length s, where $0 \le s < 2\pi r$. Let us show that the radian measure of θ is given by the formula

$$(5.20) \quad \theta = \frac{s}{r}.$$

To prove (5.20), let us place θ in standard position on a rectangular coordinate system and superimpose a unit circle U, as shown in Fig. 5.42. If t is the length of arc subtended by θ on U, then by definition we

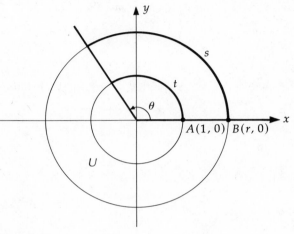

Figure 5.42

may write $\theta = t$. From geometry, the ratio of the arcs is the same as the ratio of the radii; that is,

$$\frac{t}{s} = \frac{1}{r}, \quad \text{or} \quad t = \frac{s}{r}.$$

Substituting θ for t gives us (5.20).

Formula (5.20) indicates that the radian measure of an angle is independent of the size of the circle. For example, if the radius of the circle is $r = 4$ inches and the arc subtended by a central angle θ is 8 inches, then by (5.20) the radian measure is

$$\theta = \frac{8 \text{ inches}}{4 \text{ inches}} = 2.$$

If the radius of the circle is 10 feet and the subtended arc is 5 feet, then

$$\theta = \frac{10 \text{ feet}}{5 \text{ feet}} = 2.$$

239

These calculations indicate that the radian measure of an angle is dimensionless, and hence may be regarded as a real number. Indeed, it is for this reason that we use the notation $\theta = t$ instead of $\theta = t$ radians.

Formula (5.20) may also be written

(5.21) $\quad s = r\theta.$

This gives us a formula for finding the length of arc subtended by a central angle of radian measure θ on a circle of radius r.

Another useful formula can be obtained by using the result from plane geometry which states that if θ and θ_1 are the radian measures for central angles of a circle of radius r, and if A and A_1 are the areas of the sectors determined by θ and θ_1 respectively, then

$$\frac{A}{A_1} = \frac{\theta}{\theta_1}.$$

In particular, if we let $\theta_1 = 2\pi$ then $A_1 = \pi r^2$, and substituting in the previous equation we obtain

$$\frac{A}{\pi r^2} = \frac{\theta}{2\pi}.$$

Multiplying both sides by πr^2 gives us

(5.22) $\quad A = \frac{1}{2} r^2 \theta.$

This formula may be used to find the area A of a sector of a circle of radius r determined by a central angle of radian measure θ.

EXAMPLE 1   A central angle θ subtends an arc 10 inches long on a circle of radius 4 inches. Find the radian measure of θ and the area A of the circular sector determined by θ.

Solution

Substituting in (5.20) gives us the radian measure $\theta = \frac{10}{4} = 2.5$. Applying (5.22) we obtain, for the area,

$A = \frac{1}{2}(4)^2(2.5) = 20$ square inches.

Another unit of measurement for angles is the **degree.** If the angle is placed in standard position on a rectangular coordinate system, then an angle of 1 degree is by definition the measure of the angle formed by $\frac{1}{360}$ of a complete revolution in the counterclockwise direction. The symbol "∘" is used to denote the number of degrees in the measure of an angle. In Fig. 5.43 several angles measured in degrees are shown in standard position on a rectangular coordinate system. It is customary to refer to an angle of measure 90° as a **right angle.** An angle is

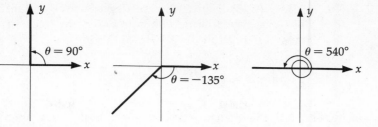

Figure 5.43

acute if its degree measure is between 0° and 90°. If its measure is between 90° and 180° an angle is **obtuse.**

If smaller measurements than those afforded by the degree and radian are required, we can use tenths, hundredths, or thousandths of radians or degrees. If degrees are used, then another method is to divide each degree into 60 equal parts called **minutes** (denoted by "′") and each minute into 60 equal parts called **seconds** (denoted by "″"). Thus 1′ is $\frac{1}{60}$ of 1°, and 1″ is $\frac{1}{60}$ of 1′. A notation such as $\theta = 73°56'18''$ refers to an angle θ of measure 73 degrees, 56 minutes, and 18 seconds.

It is not difficult to transform angular measure from one system to another. If we consider the angle θ, in standard position, generated by one-half of a complete counterclockwise rotation, then $\theta = 180°$. Also, by (5.20), the radian measure of θ is π. This gives us the basic relation

$$180° = \pi \text{ radians.}$$

Equivalent formulas are

(5.23) $1° = \pi/180$ radians   and   1 radian $= (180/\pi)°$.

By long division we obtain the fact that

$$1° \approx 0.01745 \text{ radians,} \quad 1 \text{ radian} \approx 57.296°.$$

We may also calculate

$$1' \approx 0.00029 \text{ radians,} \quad 1'' \approx 0.00000485 \text{ radians.}$$

EXAMPLE 2   (a)   Find the radian measure of θ if $\theta = 150°$. (b)   Find the degree measure of θ if $\theta = 7\pi/4$.

Solutions

(a)   By (5.23) there are $\pi/180$ radians in each degree and hence the number of radians in 150° can be found by multiplying 150 by $\pi/180$. Thus

$$150° = 150\left(\frac{\pi}{180}\right) = \frac{5\pi}{6} \text{ radians.}$$

241

(b) By (5.23) the number of degrees in 1 radian is $180/\pi$. Consequently, to find the number of degrees in $7\pi/4$ radians, we multiply by $180/\pi$, obtaining

$$\frac{7\pi}{4} \text{ radians} = \frac{7\pi}{4}\left(\frac{180}{\pi}\right) = 315°.$$

EXAMPLE 3 If the measure of an angle θ is 3 radians, find the approximate measure of θ in terms of degrees, minutes, and seconds.

Solution

Since 1 radian $\approx 57.296°$, we have

3 radians $\approx 171.888° = 171° + .888°.$

Since there are 60' in each degree, the number of minutes in .888° is 60(.888), or 53.28'. Hence

3 radians $\approx 171°53.28'.$

Finally, $.28' = (.28)60'' \approx 17''.$ Therefore

3 radians $\approx 171°53'17''.$

EXERCISES

In each of Exercises 1–12 sketch the angle with the indicated measure in standard position on a rectangular coordinate system, and find the measure of two positive angles and two negative angles which are coterminal with the given angle.

1	135°	**2**	225°
3	210°	**4**	330°
5	−60°	**6**	−120°
7	580°	**8**	610°
9	$3\pi/4$ radians	**10**	$2\pi/3$ radians
11	$-5\pi/6$ radians	**12**	7π radians

In Exercises 13 and 14 find the radian measure that corresponds to the given degree measure.

13	**a** 60°	**b** −150°		**c** 270°
	d 225°	**e** 85°		**f** 200°

14 a 150°   **b** 315°   **c** −450°

    d 135°   **e** 40°   **f** −20°

In Exercises 15 and 16 find the degree measure that corresponds to the given radian measure.

15 a $3\pi/4$ **b** $2\pi/3$ **c** $-3\pi/2$

    d $7\pi/6$ **e** 5π **f** $\pi/5$

16 a $4\pi/3$ **b** $-5\pi/4$ **c** $5\pi/6$

    d $7\pi/2$ **e** -7π **f** $\pi/8$

17 If $\theta = 4$, find the approximate measure of θ in terms of degrees, minutes, and seconds. $229° 11' 2''$

18 Work Exercise 17 if $\theta = 5$.

19 A central angle θ subtends an arc 5 inches long on a circle of radius 3 inches.

    a Find an approximation to the measure of θ in radians and in degrees.

    b Find the area of the circular sector determined by θ.

20 Work Exercise 19 if θ subtends an arc 2 feet long on a circle of radius 16 inches.

21 a Find an approximation to the length of arc subtended by a central angle of 40° on a circle of radius 6 inches.

    b Find the area of the circular sector determined by the angle in part (a).

22 Work Exercise 21 if the central angle is 1.75 radians and the radius of the circle is 3 yards.

8 TRIGONOMETRIC FUNCTIONS OF ANGLES

In certain applications it is convenient to change the domain of the trigonometric functions from a subset of **R** to the set of angles. This is very easy to do. If θ is an angle, we merely have to agree on the values sin θ, cos θ, and so on. The usual way of assigning these values is to use the radian measure of θ as in the following definition.

(5.24) Definition

If θ is an angle and if the radian measure of θ is t, then the value of each trigonometric function at θ is its value at the real number t.

We see from (5.24) that if t is the radian measure of θ, then

$$\sin \theta = \sin t, \quad \cos \theta = \cos t, \quad \tan \theta = \tan t,$$

and likewise for the other functions. For convenience we shall use the terminology *trigonometric functions* regardless of whether angles or real numbers are employed for the domain. To make the unit of angular measure clear we shall use the degree symbol and write sin 65°, tan 150°, and so on, if the angle is measured in degrees. Numerals without any symbol attached, such as cos 3 and csc $\pi/6$, will indicate that radian measure is being used. This is not in conflict with our previous work where, for example, cos 3 meant the value of the cosine function at the real number 3, since by (5.24) the cosine of 3 radians is identical with the cosine of the real number 3.

In Section 3 it was pointed out that all the values of the trigonometric functions can be found if the values are known in the *t*-interval $[0, \pi/2]$. Since $\pi/2$ radians is the same as 90°, it follows that a table of functional values corresponding to the degree interval $[0°, 90°]$ is sufficient for finding all values of the trigonometric functions. It is convenient to introduce the following analogue of the reference number defined in Section 3. If θ is an angle in standard position and θ is not a quadrantal angle, then the **reference angle associated with θ** is the acute angle θ' that the terminal side of θ makes with the x-axis. If θ lies in quadrant I, then $\theta = \theta'$. The situations in which θ lies in quadrants II, III, or IV are illustrated in Fig. 5.44. We have not included the angle θ in the sketches since there are an infinite number of angles with a given terminal side. It is important to note that $0 < \theta' < 90°$. If a unit circle U is introduced as in Fig. 5.45, and if t' is the reference number associated with $P(t)$, then evidently sin $\theta' = $ sin t', cos $\theta' = $ cos t', and so on. The following rule is a consequence of our discussion in Section 3 (see (5.17)).

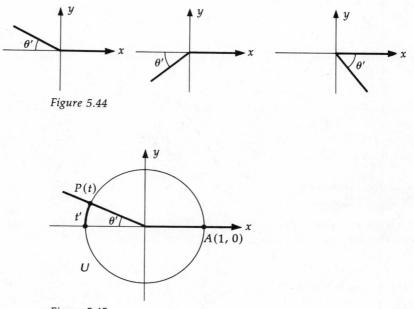

Figure 5.44

Figure 5.45

(5.25) **Rule for Finding Values of the
Trigonometric Functions**

To find the value of a trigonometric function at an angle θ, find its value at the reference angle θ' associated with θ and prefix the appropriate sign.

EXAMPLE 1 Find each of the following:

(a) $\sin 150°$,   (b) $\tan 315°$,   (c) $\sec (-240°)$.

Solutions

The angles and their reference angles are shown in Fig. 5.46. By (5.25), (5.24), and Table A of Section 2, we have the following results:

(a) $\sin 150° \quad = \quad \sin 30° \quad = \quad \sin \pi/6 = \tfrac{1}{2}$,

(b) $\tan 315° \quad = - \tan 45° \quad = - \tan \pi/4 = -1$,

(c) $\sec (-240°) \quad = - \sec 60° \quad = - \sec \pi/3 = -2$.

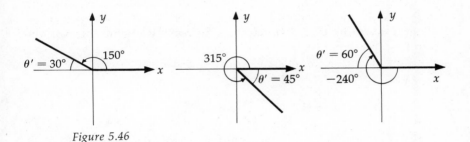

Figure 5.46

Table 3 is arranged so that functional values which correspond to angles in degree measure may be found directly. Angular measures are given in 10′ intervals from 0° to 90° and interpolation may be used to approximate values for angles between those listed. If Table 3 is used, we shall round off answers to the nearest minute. As in our work in Section 3, if the angular measure is between 0° and 45°, labels at the tops of the columns are used, whereas if the measure is between 45° and 90°, labels at the bottoms of the columns are employed.

EXAMPLE 2 Approximate $\tan 155°44'$.

Solution

Since the angle is in quadrant II, the reference angle is $180° - 155°44' = 24°16'$ and   by   (5.25), $\tan 155°44' = - \tan 24°16'$. We now consult Table 3 and interpolate as follows:

245

$$10' \left\{ 6' \left\{ \begin{matrix} \tan 24°10' \approx .4487 \\ \tan 24°16' = \quad ? \\ \tan 24°20' \approx .4522 \end{matrix} \right\} d \right\} .0035$$

$$\frac{d}{.0035} = \frac{6}{10} \quad \text{or} \quad d = \frac{6}{10}(.0035) \approx .0021$$

$$\tan 24°16' \approx .4487 + .0021 = .4508$$

$$\tan 155°44' \approx -.4508.$$

EXAMPLE 3   Approximate cos $(-117°47')$.

Solution

The angle is in quadrant III. The reader should check that the reference angle is 62°13'. Applying (5.25), cos $(-117°47') = -\cos 62°13'$. Interpolating, we have

$$10' \left\{ 3' \left\{ \begin{matrix} \cos 62°10' \approx .4669 \\ \cos 62°13' = \quad ? \\ \cos 62°20' \approx .4643 \end{matrix} \right\} d \right\} .0026$$

$$\frac{d}{.0026} = \frac{3}{10} \quad \text{or} \quad d \approx .0008.$$

Since the cosine function is decreasing in this interval, we have

$$\cos 62°13' \approx .4669 - .0008 = .4661.$$

Hence cos$(-117°47') \approx -.4661$.

EXAMPLE 4   If sin $\theta = -.7963$, approximate the degree measure of all angles θ which lie in the interval $[0°, 360°]$.

Solution

Let θ' be the reference angle, so that sin $\theta' = .7963$. Interpolating in Table 3,

$$10' \left\{ d \left\{ \begin{matrix} \sin 52°40' \approx .7951 \\ \sin \theta' \quad = .7963 \end{matrix} \right\} .0012 \right\} .0018$$
$$\sin 52°50' \approx .7969$$

$$\frac{d}{10} = \frac{.0012}{.0018} \quad \text{or} \quad d \approx 7'$$

$$\theta' \approx 52°47'.$$

Since sin θ is negative, θ lies in quadrant III or IV. Using the reference angle 52°47' we have

$$\theta \approx 180° + 52°47' = 232°47'$$

$$\theta \approx 360° - 52°47' = 307°13'.$$

The values of the trigonometric functions at an angle θ may be
determined by means of an arbitrary point on the terminal side of θ. To
prove this, let θ be an angle in standard position and let $P(x, y)$ be any
point on the terminal side of θ, where $d(O, P) = r > 0$. Figure 5.47
illustrates the case in which the terminal side lies in quadrant III; however, our discussion applies to any angle.

8
Trigonometric
Functions
of Angles

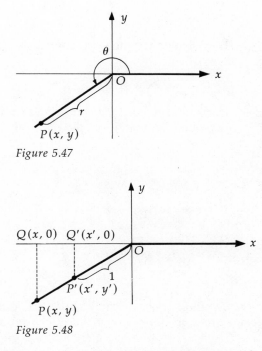

Figure 5.47

Figure 5.48

The point $P(x, y)$ is not necessarily a point assigned by the wrapping
function, since r may be different from 1. Let $P'(x', y')$ be the point on
the terminal side of θ such that $d(O, P') = 1$. Hence P' is on the unit
circle U. If t is the radian measure of θ, by (5.24) and (5.4) we have

$$\sin \theta = \sin t = y',$$
$$\cos \theta = \cos t = x',$$

and so on. As in Fig. 5.48, let us consider vertical lines through P' and P
intersecting the x-axis at $Q'(x', 0)$ and $Q(x, 0)$ respectively. Since triangles $OP'Q'$ and OPQ are similar, we have

$$\frac{d(Q', P')}{d(O, P')} = \frac{d(Q, P)}{d(O, P)}$$

or

$$\frac{|y'|}{1} = \frac{|y|}{r}.$$

Since y and y' always have the same sign, this gives us

$$y' = \frac{y}{r} \quad \text{and hence} \quad \sin \theta = \frac{y}{r}.$$

In similar fashion we obtain

$$\cos \theta = \frac{x}{r}.$$

We may now use (5.5)–(5.9) to obtain the following theorem.

(5.26) Theorem

Let θ be an angle in standard position on a rectangular coordinate system and let $P(x, y)$ be any point other than O on the terminal side of θ. If $d(O, P) = r$, then

$$\sin \theta = \frac{y}{r} \qquad\qquad \csc \theta = \frac{r}{y} \quad \text{(if } y \neq 0)$$

$$\cos \theta = \frac{x}{r} \qquad\qquad \sec \theta = \frac{r}{x} \quad \text{(if } x \neq 0)$$

$$\tan \theta = \frac{y}{x} \quad \text{(if } x \neq 0) \qquad \cot \theta = \frac{x}{y} \quad \text{(if } y \neq 0).$$

It can be shown by using similar triangles that the formulas given in (5.26) are independent of the point $P(x, y)$ that is chosen on the terminal side of θ. Note that if $r = 1$, then (5.26) reduces to (5.4). In some books on trigonometry (5.26) is used to *define* the trigonometric functions. Since our approach has been nonangular, (5.26) appears here as a theorem.

Theorem (5.26) is extremely important for certain applications of trigonometry. In the next section we shall use it to solve problems concerned with right triangles. Another reason for its importance is that it can be used to obtain values of trigonometric functions without resorting to the rather cumbersome wrapping process involving arc length on a unit circle. Indeed, by (5.26) it is sufficient to find *one* point (other than O) on the terminal side of an angle which is in standard position on a rectangular coordinate system.

EXAMPLE 5 If θ is an angle in standard position on a rectangular coordinate system, and if the point $P(-15, 8)$ is on the terminal side of θ, find the values of the trigonometric functions of θ.

Solution

By (3.3) the distance r from the origin O to any point $P(x, y)$ is $r = \sqrt{(x - 0)^2 + (y - 0)^2} = \sqrt{x^2 + y^2}$. Hence for $P(-15, 8)$, we have

$$r = \sqrt{(-15)^2 + 8^2} = \sqrt{225 + 64} = \sqrt{289} = 17.$$

Applying (5.26) with $x = -15$, $y = 8$, and $r = 17$, we obtain

$$\sin \theta = \frac{8}{17} \qquad \csc \theta = \frac{17}{8}$$

$$\cos \theta = -\frac{15}{17} \qquad \sec \theta = -\frac{17}{15}$$

$$\tan \theta = -\frac{8}{15} \qquad \cot \theta = -\frac{15}{8}.$$

EXAMPLE 6 Find the values of the trigonometric functions of θ if the terminal side of θ lies on the line $4y = 3x$ and θ is in quadrant III.

Solution

We begin by choosing any point, say $P(-4, -3)$, on the terminal side of θ. Since $d(O, P) = 5$, we have, by (5.26)

$$\sin \theta = -\frac{3}{5} \qquad \csc \theta = -\frac{5}{3}$$

$$\cos \theta = -\frac{4}{5} \qquad \sec \theta = -\frac{5}{4}$$

$$\tan \theta = \frac{3}{4} \qquad \cot \theta = \frac{4}{3}.$$

In the proof of (5.26) we considered θ as a nonquadrantal angle; however, the formulas we derived are also true if the terminal side of θ lies on either the x or y-axis. This is illustrated by the next example.

EXAMPLE 7 Find the values of the trigonometric functions of θ if $\theta = 270°$.

Solution

The terminal side of θ coincides with the negative y-axis. To use (5.26) we may choose any point on the terminal side of θ. For simplicity we consider $P(0, -1)$. In this case $r = 1$ and by (5.26),

$$\sin \theta = \frac{-1}{1} = -1 \qquad \csc \theta = \frac{1}{-1} = -1$$

$$\cos \theta = \frac{0}{1} = 0 \qquad \cot \theta = \frac{0}{-1} = 0.$$

The tangent and secant functions are undefined, since the meaningless expressions $\tan \theta = (-1)/0$ and $\sec \theta = 1/0$ arise when we substitute in the appropriate formulas.

EXERCISES

In Exercises 1 and 2 find the reference angle θ' if θ has the given measure.

1 **a** 220°   **b** −155°   **c** 413°

    d 127°12′   **e** −358°   **f** 1200°

    g 212°37′42″

2 **a** 295°   **b** 158°   **c** −87°

    d 523°   **e** 149°27′   **f** 2150°

    g 169°22′15″

In Exercises 3 and 4 use Table 3 to find the given numbers.

3 **a** cos 33°40′   **b** sin 73°10′

    c tan 11°50′ .   **d** cot (−215°)

    e sin 335°30′   **f** cos 840°20′

4 **a** sin 41°20′   **b** tan 80°10′

    c cos 6°40′   **d** cot 110°

    e sin (−117°10′)   **f** tan 1000°

In Exercises 5–12 approximate the numbers by using interpolation.

5 sin 23°37′   **6** cos 37°23′   **7** cos 51°53′

8 tan 69°7′   **9** cot 75°42′   **10** sin 11°58′

11 tan 349°28′   **12** cot 253°34′

In Exercises 13–20 use interpolation in Table 3 to approximate, to the nearest minute, the degree measure of all angles θ which lie in the interval [0°, 360°].

13 sin θ = 0.4759   **14** cos θ = 0.1088

15 tan θ = 1.1822   **16** sin θ = 0.4444

17 cot θ = −2.2390   **18** tan θ = −0.1180

19 cos θ = −0.7302   **20** cot θ = −0.8441

In Exercises 21–34 use (5.26) to find the values of the six trigonometric functions of θ if θ is in standard position and satisfies the given condition.

21 The point $P(-3, -4)$ is on the terminal side of θ.

22 The point $P(8, -15)$ is on the terminal side of θ.

23 The point $P(-2, 1)$ is on the terminal side of θ.

24 The point $P(-4, -5)$ is on the terminal side of θ.

25 The terminal side of θ lies in quadrant IV on the line with equation $y = -2x$.

26 The terminal side of θ lies in quadrant II on the line with equation $3y + 5x = 0$.

27 The terminal side is in quadrant III and is parallel to the line $5x - 2y - 6 = 0$.

28 The terminal side is in quadrant IV and is parallel to the line through the points $A(-6, 3)$ and $B(2, -4)$.

29 $\theta = 90°$ **30** $\theta = 180°$ **31** $\theta = 135°$

32 $\theta = 315°$ **33** $\theta = -90°$ **34** $\theta = -270°$

35 Prove geometrically that the formulas given in (5.26) are independent of the point $P(x, y)$ that is chosen on the terminal side of θ.

36 If (5.26) were used for the definition of the trigonometric functions, how could one prove that $\sin^2 \theta + \cos^2 \theta = 1$ for all angles θ?

9 RIGHT TRIANGLE TRIGONOMETRY

A triangle is called a **right triangle** if one of its angles is a right angle. If θ is an acute angle, then it can be regarded as an angle of a right triangle and we may refer to the lengths of the **hypotenuse,** the **opposite side,** and the **adjacent side** in the usual way. For convenience we shall use **hyp, opp,** and **adj,** respectively, to denote these numbers. Introducing a rectangular coordinate system as in Fig. 5.49, we see that

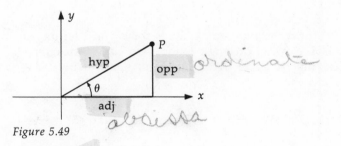

Figure 5.49

the lengths of the adjacent side and the opposite side for θ are the abscissa and ordinate, respectively, of a point P on the terminal side of θ and, by (5.26), we have

$$\sin \theta = \frac{\text{opp}}{\text{hyp}} \qquad \csc \theta = \frac{\text{hyp}}{\text{opp}}$$

$$(5.27) \quad \cos \theta = \frac{\text{adj}}{\text{hyp}} \qquad \sec \theta = \frac{\text{hyp}}{\text{adj}}$$

$$\tan \theta = \frac{\text{opp}}{\text{adj}} \qquad \cot \theta = \frac{\text{adj}}{\text{opp}}$$

The formulas given in (5.27) are very useful in work with triangles and should be memorized.

EXAMPLE 1 Use (5.27) to find the values of sin θ, cos θ, and tan θ, if

(a) $\theta = 60°$,   (b) $\theta = 30°$,   (c) $\theta = 45°$.

Solutions

Let us consider an equilateral triangle having sides of length 2. The median from one vertex to the opposite side bisects the angle at that vertex, as illustrated in (i) of Fig. 5.50. Using (5.27) to calculate functional values we obtain

(a) $\sin 60° = \dfrac{\sqrt{3}}{2}$, $\cos 60° = \dfrac{1}{2}$, $\tan 60° = \dfrac{\sqrt{3}}{1} = \sqrt{3}$;

(b) $\sin 30° = \dfrac{1}{2}$, $\cos 30° = \dfrac{\sqrt{3}}{2}$, $\tan 30° = \dfrac{1}{\sqrt{3}} = \dfrac{\sqrt{3}}{3}$.

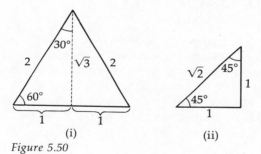

(i)   (ii)

Figure 5.50

To find the functional values for $\theta = 45°$, let us consider an isosceles right triangle whose two equal sides have length 1, as illustrated in (ii) of Fig. 5.50. Again using (5.27), we obtain

(c) $\sin 45° = \dfrac{1}{\sqrt{2}} = \dfrac{\sqrt{2}}{2} = \cos 45°$, $\tan 45° = \dfrac{1}{1} = 1$.

In our work with triangles in this section and in Chapter Six we shall often use the following notation. The vertices of the triangle will be denoted by A, B, and C. The angles of the triangle at A, B, and C will be denoted by α, β, and γ, respectively, and the lengths of the sides opposite these angles by a, b, and c, respectively. The triangle itself will often be referred to as *triangle ABC*. To *solve* a triangle means to find all of its parts, that is, the lengths of the three sides and the measures of the three angles. If the triangle is a right triangle and if one of the acute angles and a side are known, or if two sides are given, then (5.27) may be used to find the remaining parts.

EXAMPLE 2 If, in triangle *ABC*, $\gamma = 90°$, $\alpha = 34°$, and $b = 10.5$, 9
approximate the remaining parts of the triangle.

Right
Triangle
Trigonometry

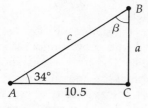

Figure 5.51

Solution

Since the sum of the angles is 180° it follows that $\beta = 56°$.
Referring to Figure 5.51 and using (5.27),

$$\tan 34° = \frac{a}{10.5} \quad \text{or} \quad a = (10.5) \tan 34°.$$

Substituting from Table 3 we obtain

$$a \approx (10.5)\,(.6745) \approx 7.1.$$

Similarly,

$$\cos 34° = \frac{10.5}{c} \quad \text{or} \quad c = \frac{10.5}{\cos 34°}.$$

Again using Table 3,

$$c \approx (10.5)/(.8290) \approx 12.7.$$

If a table of values for the secant function were available, we
could have avoided the division in the calculation of *c* by
writing $c = (10.5)\sec 34°$.

As illustrated in Example 2, when working with triangles we
shall round off answers. There are several reasons for this. In applications, lengths of sides of triangles and measures of angles are usually
found by mechanical devices and hence are only approximations to the
exact values. Because of this, the number 10.5 in Example 2 is assumed
to have been rounded off to the nearest tenth. One cannot expect more
accuracy in the calculated values for the remaining sides, and consequently they should also be rounded off to the nearest tenth.

In some problems a large number of digits, such as 13,647.29,
may be given for a number. If Table 3 is used, a number of this type
should be rounded off to four significant figures and written as 13,650
before beginning any calculations. The situation here is similar to that

discussed in Chapter 4 with regard to the use of the table of logarithms. Since the values of the trigonometric functions given in Table 3 have been rounded off to four significant figures, we cannot expect more than four-figure accuracy in our computations.

As a final remark on approximations, answers should also be rounded off when Table 3 is used to find angles. In general, we shall use the following rules: if the sides of a triangle are known to four significant figures, then measures of angles calculated from Table 3 should be rounded off to the nearest minute; if the sides are known to three significant figures, then calculated measures of angles should be rounded off to the nearest multiple of ten minutes; and if the sides are known to only two significant figures, then calculations should be rounded off to the nearest degree. In order to justify these rules, we would have to make a much deeper analysis of problems involving approximate data.

EXAMPLE 3   If, in triangle ABC, $\gamma = 90°$, $a = 12.3$, and $b = 31.6$, find the remaining parts.

Solution

Applying (5.27) to the triangle shown in Fig. 5.52 we have

$$\tan \alpha = \frac{12.3}{31.6} \approx .3892.$$

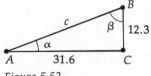

Figure 5.52

By the rule just stated, α should be rounded off to the nearest multiple of $10'$. Referring to Table 3, we see that $\alpha \approx 21°20'$. Consequently

$$\beta \approx 90° - 21°20' = 68°40'.$$

Again referring to Fig. 5.52, $\cos \alpha = (31.6)/c$, and hence $c = (31.6)/\cos 21°20' \approx (31.6)/(.9315)$. Dividing and rounding off, we obtain $c \approx 33.9$.

In certain triangle problems it is convenient to use logarithms. Such problems could be worked by first using Table 3 to find values of the trigonometric functions and then employing Table 2 to find the logarithms. The work can be shortened, however, by using Table 4, which contains the logarithms of $\sin \theta$, $\cos \theta$, $\tan \theta$, and $\cot \theta$ for values of θ from $0°$ to $90°$ at $10'$ intervals. If θ is between $0°$ and $90°$ then $0 < \sin \theta < 1$ and $0 < \cos \theta < 1$, and consequently the characteristic of log sin θ and log cos θ is negative. The same is true for log tan θ if $0 < \theta < 45°$. In order to conserve space, Table 4 is constructed so that -10 must be added to each entry. For example,

$$\log \sin 38°40' \approx 9.7957 - 10,$$

$$\log \cot 86°20' \approx 8.8067 - 10,$$

and so on.

If we wish to solve the problem given in Example 2 by means of logarithms we have, from $a = (10.5)\tan 34°$

$$\log a = \log 10.5 + \log \tan 34°.$$

Using Tables 2 and 4, we obtain

$$\log a \approx 1.0212 + [9.8290 - 10]$$
$$= .8502,$$

and from Table 2, $a \approx 7.1$.

Similarly, the measure of angle α of Example 3 may be found from $\tan \alpha = 12.3/31.6$ by writing

$$\log \tan \alpha = \log 12.3 - \log 31.6$$
$$\approx 1.0899 - 1.4997$$
$$= (11.0899 - 10) - 1.4997$$
$$= 9.5902 - 10.$$

Consulting Table 4 we find $\alpha \approx 21°20'$.

Needless to say, if an angle between two of those in Table 4 is given, then an approximation to the logarithm of a functional value may be determined by interpolation.

Right triangles are useful in solving various types of applied problems. The following examples give two illustrations; others will be found in the exercises.

EXAMPLE 4 From a point on level ground 135 feet from the foot of a tower, the angle of elevation of the top of the tower is 57°20'. Find the height of the tower.

Solution

The **angle of elevation** is the angle which the line of sight makes with the horizontal. Referring to Fig. 5.53 we may write

$$\tan 57°20' = d/135, \quad \text{or} \quad d = (135) \tan 57°20'.$$

Substituting from Table 3 gives us

$$d = (135)(1.5597) \approx 211.$$

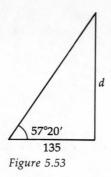

57°20′

135

Figure 5.53

EXAMPLE 5 From the top of a building which overlooks an ocean, a man sees a boat sailing directly toward him. If the man is 100 feet above sea level and if the angle of depression of the boat changes from 25° to 40° during the period of observation, find the approximate distance the boat travels during this time.

Solution

The **angle of depression** is the angle between the line of sight and the horizontal. Let A and B be the positions of the boat which correspond to the 25° and 40° angles, respectively. Suppose the man is at point D and let C be the point 100 feet directly below him. Let d denote the distance the boat travels and let g denote the distance from B to C. This gives us the drawing in Fig. 5.54, where α and β denote angles DAC and DBC, respectively. It follows that $\alpha = 25°$ and $\beta = 40°$ (Why?).

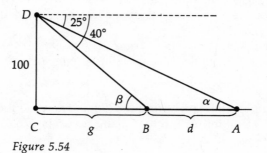

Figure 5.54

From triangle BCD we have

$$\cot \beta = \frac{g}{100} \quad \text{or} \quad g = 100 \cot \beta.$$

From triangle DAC we have

$$\cot \alpha = \frac{d + g}{100} \quad \text{or} \quad d + g = 100 \cot \alpha.$$

Consequently

$$d = 100 \cot \alpha - g$$
$$= 100 \cot \alpha - 100 \cot \beta$$
$$= 100(\cot \alpha - \cot \beta)$$
$$= 100(\cot 25° - \cot 40°)$$
$$\approx 100(2.1445 - 1.1918)$$
$$= 100(.9527).$$

Hence $d \approx 95$ feet.

In certain navigation and surveying problems the **direction,** or **bearing,** from a point P to a point Q is often specified by stating the acute angle which the half-line from P through Q varies to the east or west from the north-south line. Figure 5.55 illustrates four such lines. The north-south and east-west lines are labeled NS and WE, respectively. The bearing from P to Q_1 is 25° east of north and is denoted by N25°E. We also refer to the *direction* N25°E, meaning the direction from P to Q_1. The bearings from P to Q_2, Q_3, and Q_4 are represented in a similar manner in the figure.

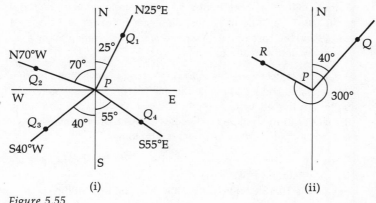

(i)   (ii)

Figure 5.55

In air navigation, directions and bearings are specified by measuring from the north in a clockwise direction. In this particular situation a positive measure is assigned to the angle instead of the negative measure to which we are accustomed for clockwise rotations. Thus, referring to (ii) of Fig. 5.55, we see that the direction of PQ is 40°, whereas the direction of PR is 300°.

We shall use the foregoing notations in several of the following exercises.

EXERCISES

Given the indicated parts of triangle ABC with $\gamma = 90°$, approximate the remaining parts of each triangle in Exercises 1–12.

1 $\alpha = 60°$, $b = 10$ **2** $\beta = 45°$, $b = 15$

3 $\beta = 42°$, $a = 19$ **4** $\alpha = 27°$, $b = 34$

5 $\alpha = 19°20'$, $a = 7.3$ **6** $\beta = 73°40'$, $a = 29.7$

7 $\beta = 68°24'$, $b = 132$ **8** $\alpha = 33°33'$, $a = 333$

9 $a = 14$, $b = 39$ **10** $a = 27$, $b = 8$

11 $a = 569$, $b = 341$ **12** $a = 38$, $b = 63$

13 Approximate the angle of elevation of the sun if a man 6 feet tall casts a shadow 10 feet long on level ground.

14 From a point 234 feet above level ground an observer measures the angle of depression of an object on the ground as $57°20'$. Approximately how far is the object from the point on the ground directly beneath the observer?

15 The string on a kite is taut and makes an angle of $27°40'$ with the horizontal. Find the approximate height of the kite above level ground if 170 feet of string are out and the end of the string is held 4 feet above the ground.

16 The side of a regular pentagon is 16 inches. Approximate the radius of the circumscribed circle.

17 From a point P on level ground the angle of elevation of the top of a tower is $36°10'$. From a point 78 feet closer to the tower and on the same line with P and the base of the tower, the angle of elevation of the top is $48°20'$. Approximate the height of the tower.

18 A ladder 24 feet long leans against the side of a building. If the angle between the ladder and the building is $18°10'$, approximately how far is the bottom of the ladder from the building? If the distance from the bottom of the ladder to the building is increased 2 feet, approximately how far does the top of the ladder move down the building?

19 From a point A, 20 feet above level ground, the angle of elevation of the top of a building is $41°50'$ and the angle of depression of the base of the building is $19°10'$. Approximate the height of the building.

20 As a balloon rises vertically its angle of elevation from a point P on the level ground 350 feet from the point Q directly underneath the balloon changes from $49°10'$ to $67°40'$. Approximately how far does the balloon rise during this period?

21 A ship leaves port at 1:00 P.M. and sails in the direction N34°W at a rate of 24 miles per hour. Another ship leaves port at 1:30 P.M. and sails in the direction N56°E at a rate of 18 miles per hour. Approximately how far apart are the ships at 3:00 P.M.?

22 From an observation point A a forest ranger sights a fire in the direction S48°20'W. From a point B, 5 miles due west of A, another ranger sights the same fire in the direction S54°10'E. Approximate the distance of the fire from A.

23 An airplane flying at a speed of 360 miles per hour flies from a point A in the direction 137° for 30 minutes and then flies in the direction 227° for 45 minutes. Approximately how far is the airplane from A?

24 Generalize Exercise 15 to the case where the angle is α, the number of feet of string out is d, and the end of the string is held c feet above the ground. Express the height h of the kite in terms of α, d, and c.

25 Generalize Exercise 19 to the case where point A is d feet above ground and the angles of elevation and depression are α and β, respectively. Express the height h of the building in terms of d, α, and β.

26 Generalize Exercise 20 to the case where the distance from P to Q is d feet and the angle of elevation changes from α to β.

10 REVIEW EXERCISES

Oral

Define or discuss the following.

1 The wrapping function

2 The trigonometric functions

3 Graphs of the trigonometric functions

4 The fundamental identities

5 Periodic function; period

6 Angles

7 Standard position of an angle

8 Terminal side of an angle

9 Positive and negative angles

10 Radian measure of an angle

11 The relationship between radian measure and degree measure of an angle

12 Trigonometric functions of angles

Written

In Exercises 1–3, $P(t)$ denotes the point on the unit circle U which the wrapping function associates with the real number t.

1 Find the rectangular coordinates of $P(-3\pi)$, $P(5\pi/2)$, $P(7\pi/2)$, $P(5\pi/6)$, $P(-2\pi/3)$, and $P(5\pi/4)$.

2 If $P(t)$ has coordinates $(-\frac{8}{17}, -\frac{15}{17})$, find the rectangular coordinates of $P(t + \pi)$, $P(t - \pi)$, $P(-t)$, and $P(2\pi - t)$.

3 Find the quadrant which contains $P(t)$ if:

a $\cos t > 0$ and $\sin t < 0$

b $\tan t > 0$ and $\cos t < 0$

c $\sec t < 0$ and $\cot t < 0$

4 Find the values of all six trigonometric functions at t if:

a $\cot t = \frac{3}{4}$ and $\cos t < 0$

b $\csc t = -\frac{13}{5}$ and $\cot t < 0$

c $\sin t = \frac{1}{3}$ and $\cos t < 0$

5 Use fundamental identities to transform the first expression into the second:

a $\cos t \tan t$, $\sin t$

b $(\cot^2 t)/(\csc t - 1)$, $\csc t + 1$

c $(1 - \sin^2 x)/\cos^2 x$, 1

6 Find the amplitude and period, and sketch the graph of f if:

a $f(t) = 3 \cos t$ **b** $f(t) = \frac{1}{3} \sin t$ **c** $f(t) = -\tan t$

d $f(x) = 4 \sin 2x$ **e** $f(x) = 2 \sin 4x$ **f** $f(x) = -3 \sin \frac{1}{2}x$

g $f(x) = -\frac{1}{2} \cos 3x$

7 Sketch the graph of $y = 2^{-x} \sin 2x$.

8 Sketch the graph of each of the following equations:

a $y = 4 \cos (3x - \pi/2)$ **b** $y = -2 \sin (2x + \pi/4)$

c $y = 2 \sin x + \sin 2x$ **d** $y = \tan \left(x - \dfrac{\pi}{4}\right)$

e $y = \csc 2x$ **f** $y = \frac{1}{4} \sec x$

9 **a** Find the reference number t' if t equals: $13\pi/4$, $-7\pi/6$, $-5\pi/3$.

b Find the reference angle θ' if θ has measure: $250°$, $157°12'$, $1111°$.

10 **a** Find the radian measures that correspond to the following degree measures: $30°$, $-120°$, $450°$, $135°$, $65°$.

b Find the degree measures that correspond to the following radian measures: $5\pi/4$, $-2\pi/3$, $3\pi/2$, $11\pi/6$, $\pi/10$.

11 Find, without the aid of tables, the value of:

a $\sin 225°$ **b** $\tan \pi/3$ **c** $\cos (-\pi/6)$

d $\sec \pi$ **e** $\cot 330°$ **f** $\csc (-210°)$

12 Use interpolation in Table 3 to approximate:

a $\cos 21°33'$ **b** $\tan 74°52'$ **c** $\sin 248°19'$

13 Use interpolation to approximate, to the nearest minute, the degree measure of all angles θ which are in the interval $[0°, 360°)$ if:

a $\cos \theta = .6318$ **b** $\sin \theta = -.8412$

c $\tan \theta = .9635$

14 Find the values of the six trigonometric functions of θ if θ is in standard position and satisfies the stated condition.

a The point $P(-8, -15)$ is on the terminal side of θ.

b The terminal side of θ is in quadrant II on the line having equation $y = -5x$.

c $\theta = 270°$

15 Given the following parts of triangle *ABC* with $\gamma = 90°$, approximate the remaining parts.

a $\alpha = 60°$, $a = 10$ **b** $\beta = 52°10'$, $a = 32.0$

c $a = 64$, $b = 41$

16 If the side of a rectangular octagon is 12 inches, approximate the radius of the circumscribed circle.

6

Analytic Trigonometry

In this chapter we shall examine various algebraic and geometric aspects of trigonometry. In Sections 1 through 6
the emphasis is on identities
and equations. In Section 7 we
consider inverse functions for
the trigonometric functions.
Methods for solving oblique
triangles are discussed in
Sections 8 and 9.

1 TRIGONOMETRIC IDENTITIES

Any mathematical expression that
contains symbols such as sin x, cos β, tan v,
and so on, where the letters x, β, v are variables, is referred to as a **trigonometric expression.** Trigonometric expressions may be
very simple or very complicated. Some examples are

$$x + \sin x, \quad \frac{\cos (3y + 1)}{x^2 + \tan^2 (z - y^2)},$$

$$\frac{\sec (\cot \theta)}{|\theta| \log \sin \theta}.$$

As usual, we assume that the domain of the
variables is the set of real numbers (or
angles) for which the expressions are meaningful. Many trigonometric expressions can
be simplified or changed in form by employing the fundamental identities (5.5)–
(5.12). This is illustrated in the next example.

EXAMPLE 1 Simplify the expression
$(\sec \theta + \tan \theta)(1 - \sin \theta)$.

Solution

The reader should supply reasons
for the following steps.

262

$$(\sec \theta + \tan \theta)(1 - \sin \theta) = \left(\frac{1}{\cos \theta} + \frac{\sin \theta}{\cos \theta}\right)(1 - \sin \theta)$$

$$= \left(\frac{1 + \sin \theta}{\cos \theta}\right)(1 - \sin \theta)$$

$$= \frac{1 - \sin^2 \theta}{\cos \theta}$$

$$= \frac{\cos^2 \theta}{\cos \theta}$$

$$= \cos \theta.$$

There are other ways to treat the expression in Example 1. We could begin by multiplying the two factors and then proceed to simplify and combine terms. The method we employed—of changing all expressions to expressions which involve only sines and cosines—is often worthwhile. However, this technique does not always lead to the shortest possible simplification.

A **trigonometric equation** is an equation which contains trigonometric expressions. We define **solution** and **solution set** of a trigonometric equation exactly as in our discussion of algebraic equations. **Conditional equations** and **identities** have the same meaning as before. When working with trigonometric equations we may express solution sets either in terms of real numbers or in terms of angles.

We encountered several trigonometric identities in Chapter Five. Some of these, such as the fundamental identities, are basic and should be memorized. Other identities are introduced merely to supply practice in manipulating trigonometric expressions and therefore should not be memorized. We shall take the latter point of view in this section. Thus the identities to be investigated are unimportant in their own right. The important thing is the manipulative practice that is gained. The ability to carry out trigonometric manipulations is essential for problems which are encountered in advanced courses in mathematics and science.

We shall usually use the phrase "verify an identity" instead of "prove that an equation is an identity." When verifying an identity we shall not use (5.4) to return to the coordinates (x, y) of a point on the unit circle U. Although many of the identities could be proved in this way, it is not the type of practice we desire. Rather, we shall use the fundamental identities and algebraic manipulations to change the form of trigonometric expressions in a manner similar to the solution of Example 1. The preferred method of showing that an equation is an identity is to transform one side into the other. This method is illustrated in the next three examples. The reader should supply reasons for all steps in the solutions.

EXAMPLE 2 Verify the identity $\dfrac{\tan t + \cos t}{\sin t} = \sec t + \cot t$.

Solution

We shall transform the left side into the right side. Thus,

$$\frac{\tan t + \cos t}{\sin t} = \frac{\tan t}{\sin t} + \frac{\cos t}{\sin t}$$

$$= \frac{(\sin t/\cos t)}{\sin t} + \cot t$$

$$= \frac{1}{\cos t} + \cot t$$

$$= \sec t + \cot t.$$

EXAMPLE 3 Verify the identity $\sec \alpha - \cos \alpha = \sin \alpha \tan \alpha$.

Solution

$$\sec \alpha - \cos \alpha = \frac{1}{\cos \alpha} - \cos \alpha$$

$$= \frac{1 - \cos^2 \alpha}{\cos \alpha}$$

$$= \frac{\sin^2 \alpha}{\cos \alpha}$$

$$= \sin \alpha \left(\frac{\sin \alpha}{\cos \alpha} \right)$$

$$= \sin \alpha \tan \alpha.$$

EXAMPLE 4 Verify the identity $\dfrac{\cos x}{1 - \sin x} = \dfrac{1 + \sin x}{\cos x}$.

Solution

We begin by multiplying the numerator and denominator of the fraction on the left by $1 + \sin x$. Thus,

$$\frac{\cos x}{1 - \sin x} = \frac{\cos x}{1 - \sin x} \cdot \frac{1 + \sin x}{1 + \sin x}$$

$$= \frac{\cos x \, (1 + \sin x)}{1 - \sin^2 x}$$

$$= \frac{\cos x \, (1 + \sin x)}{\cos^2 x}$$

$$= \frac{1 + \sin x}{\cos x}.$$

Another technique for showing that an equation $p = q$ is an identity is to begin by transforming the left side p into another expression s, making sure that each step is *reversible* in the sense that it is possible to transform s back into p by reversing the procedure which has been used. In this case the equation $p = s$ is an identity. Next, as a *separate* exercise, it is shown that the right side q can also be transformed to the expression s by means of reversible steps and hence that $q = s$ is an identity. It then follows that $p = q$ is an identity. This method is illustrated in the next example.

EXAMPLE 5 Verify the identity $(\tan \theta - \sec \theta)^2 = \dfrac{1 - \sin \theta}{1 + \sin \theta}.$

Solution

We shall verify the identity by showing that each side of the equation can be transformed into the same expression. Starting with the left side we may write

$$(\tan \theta - \sec \theta)^2 = \tan^2 \theta - 2 \tan \theta \sec \theta + \sec^2 \theta$$

$$= \frac{\sin^2 \theta}{\cos^2 \theta} - \frac{2 \sin \theta}{\cos^2 \theta} + \frac{1}{\cos^2 \theta}$$

$$= \frac{\sin^2 \theta - 2 \sin \theta + 1}{\cos^2 \theta}.$$

The right-hand side of the given equation may be changed by multiplying numerator and denominator by $1 - \sin \theta$. Thus

$$\frac{1 - \sin \theta}{1 + \sin \theta} = \frac{1 - \sin \theta}{1 + \sin \theta} \cdot \frac{1 - \sin \theta}{1 - \sin \theta}$$

$$= \frac{1 - 2 \sin \theta + \sin^2 \theta}{1 - \sin^2 \theta}$$

$$= \frac{1 - 2 \sin \theta + \sin^2 \theta}{\cos^2 \theta},$$

which is the same expression as that obtained for $(\tan \theta - \sec \theta)^2$. Since all steps are reversible, it follows that the given equation is an identity.

EXERCISES

Verify the following identities.

1 $\cos \theta \sec \theta = 1$

2 $\tan \alpha \cot \alpha = 1$

3 $\sin \theta \sec \theta = \tan \theta$

4 $\sin \alpha \cot \alpha = \cos \alpha$

5 $\csc x/\sec x = \cot x$

6 $\tan \beta \cos \beta = \sin \beta$

7 $\dfrac{\sin t}{\csc t} + \dfrac{\cos t}{\sec t} = 1$

8 $1 - 2 \sin^2 x = 2 \cos^2 x - 1$

9 $(1 + \sin \alpha)(1 - \sin \alpha) = 1/\sec^2 \alpha$

10 $(1 - \sin^2 t)(1 + \tan^2 t) = 1$

11 $\csc^2 \theta/(1 + \tan^2 \theta) = \cot^2 \theta$

12 $\sin x + \cos x \cot x = \csc x$

13 $\sin t(\csc t - \sin t) = \cos^2 t$

14 $\cot t + \tan t = \csc t \sec t$

15 $\csc \theta - \sin \theta = \cot \theta \cos \theta$

16 $\cos \theta(\tan \theta + \cot \theta) = \csc \theta$

17 $\dfrac{\sec^2 u - 1}{\sec^2 u} = \sin^2 u$

18 $(\tan u + \cot u)(\cos u + \sin u) = \sec u + \csc u$

19 $(\cos^2 x - 1)(\tan^2 x + 1) = 1 - \sec^2 x$

20 $\dfrac{1 + \cos^2 y}{\sin^2 y} = 2 \csc^2 y - 1$

21 $\sec^2 \theta \csc^2 \theta = \sec^2 \theta + \csc^2 \theta$

22 $\dfrac{\sec x - \cos x}{\tan x} = \dfrac{\tan x}{\sec x}$

23 $\dfrac{1 + \cos t}{\sin t} + \dfrac{\sin t}{1 + \cos t} = 2 \csc t$

24 $\tan^2 \alpha - \sin^2 \alpha = \tan^2 \alpha \sin^2 \alpha$

25 $\dfrac{1 + \tan^2 v}{\tan^2 v} = \csc^2 v$

26 $\dfrac{\cos x \cot x}{\cot x - \cos x} = \dfrac{\cot x + \cos x}{\cos x \cot x}$

27 $(\sec u - \tan u)(\csc u + 1) = \cot u$

28 $\dfrac{\cot \theta - \tan \theta}{\sin \theta + \cos \theta} = \csc \theta - \sec \theta$

29 $\dfrac{\cot \alpha - 1}{1 - \tan \alpha} = \cot \alpha$

30 $\dfrac{1 + \sec \beta}{\tan \beta + \sin \beta} = \csc \beta$

31 $\cot^4 t - \csc^4 t = -\cot^2 t - \csc^2 t$

32 $\cos^4 \theta + \sin^2 \theta = \sin^4 \theta + \cos^2 \theta$

33 $\dfrac{\cos \beta}{1 - \sin \beta} = \sec \beta + \tan \beta$ **34** $\dfrac{1}{\csc y - \cot y} = \csc y + \cot y$

35 $\dfrac{\tan^2 x}{\sec x + 1} = \dfrac{1 - \cos x}{\cos x}$ **36** $\dfrac{\cot x}{\csc x + 1} = \dfrac{\csc x - 1}{\cot x}$

37 $\dfrac{\cot u - 1}{\cot u + 1} = \dfrac{1 - \tan u}{1 + \tan u}$ **38** $\dfrac{1 + \sec x}{\sin x + \tan x} = \csc x$

39 $\sec^2 \gamma + \tan^2 \gamma = (1 - \sin^4 \gamma)(\sec^4 \gamma)$

40 $\dfrac{\sin t}{1 - \cos t} = \csc t + \cot t$

41 $(\sin^2 \theta + \cos^2 \theta)^3 = 1$

42 $\left(\dfrac{\sin^2 x}{\tan^4 x}\right)^3 \left(\dfrac{\csc^3 x}{\cot^6 x}\right)^2 = 1$

43 $\dfrac{\cos^3 x - \sin^3 x}{\cos x - \sin x} = 1 + \sin x \cos x$

44 $\dfrac{\sin \theta + \cos \theta}{\tan^2 \theta - 1} = \dfrac{\cos^2 \theta}{\sin \theta - \cos \theta}$

45 $(\csc t - \cot t)^4 (\csc t + \cot t)^4 = 1$

46 $(a \cos t - b \sin t)^2 + (a \sin t + b \cos t)^2 = a^2 + b^2$

47 $\sin^6 v + \cos^6 v = 1 - 3 \sin^2 v \cos^2 v$

48 $\dfrac{\sin \alpha \cos \beta + \cos \alpha \sin \beta}{\cos \alpha \cos \beta - \sin \alpha \sin \beta} = \dfrac{\tan \alpha + \tan \beta}{1 - \tan \alpha \tan \beta}$

49 $\sqrt{\dfrac{1 - \cos t}{1 + \cos t}} = \dfrac{1 - \cos t}{|\sin t|}$

50 $-\log |\sec \theta - \tan \theta| = \log |\sec \theta + \tan \theta|$

2 CONDITIONAL EQUATIONS

If a trigonometric equation is not an identity, then methods similar to those used for algebraic equations may be employed for finding solution sets. The main difference here is that we usually solve for sin x, cos θ, and so on, and *then* find x and θ. Some methods of solution are illustrated in the following examples.

EXAMPLE 1   Solve the equation $\sin \theta \tan \theta = \sin \theta$.

Solution

Each of the following is equivalent to the given equation:

$\sin \theta \tan \theta - \sin \theta = 0$
$\sin \theta (\tan \theta - 1) = 0.$

As in our work with algebraic equations in Chapter 2, the solution set of the second equation is the union of the solution sets of

$\sin \theta = 0 \quad \text{and} \quad \tan \theta = 1.$

The solutions of $\sin \theta = 0$ consist of all multiples of π, that is, $\theta = n\pi$, where n is any integer.

Since the tangent function has period π, it is sufficient to find the solutions of the equation $\tan \theta = 1$ which are in the interval $[0, \pi)$, for once these are known, we may obtain all the others by adding multiples of π. The only solution of $\tan \theta = 1$ in $[0, \pi)$ is $\pi/4$ and hence every solution has the form

$$\theta = \frac{\pi}{4} + n\pi,$$

where n is an integer.

It follows that the solution set of the given equation consists of all numbers of the form

$$n\pi \quad \text{and} \quad \frac{\pi}{4} + n\pi,$$

where $n \in \mathbf{Z}$. Some particular solutions are $0, \pm\pi, \pm 2\pi$, $\pm 3\pi, \pi/4, 5\pi/4, -3\pi/4$, and $-7\pi/4$.

The solution in Example 1 was obtained by factoring. It would have been incorrect to begin by dividing both sides by $\sin\theta$, for this manipulation would lose the solutions of $\sin\theta = 0$.

EXAMPLE 2 Solve the equation $2\sin^2 t - \cos t - 1 = 0$.

Solution

First we change the equation to an equation which involves only $\cos t$ and then we factor as follows.

$$2(1 - \cos^2 t) - \cos t - 1 = 0,$$
$$-2\cos^2 t - \cos t + 1 = 0,$$
$$2\cos^2 t + \cos t - 1 = 0,$$
$$(2\cos t - 1)(\cos t + 1) = 0.$$

As in Example 1, the solution set of the last equation is the union of the solution sets of

$$\cos t = \tfrac{1}{2} \quad \text{and} \quad \cos t = -1.$$

It is sufficient to find the solutions which are in the interval $[0, 2\pi)$, for once these are known, all solutions may be found by adding multiples of 2π.

If $\cos t = \tfrac{1}{2}$, then the reference number (or reference angle) is $\pi/3$ (or $60°$). Since $\cos t$ is positive, $P(t)$ lies in quadrant I or quadrant IV. Hence in the interval $[0, 2\pi)$ we have

$$t = \frac{\pi}{3} \quad \text{or} \quad t = 2\pi - \frac{\pi}{3} = \frac{5\pi}{3}.$$

If $\cos t = -1$, then $t = \pi$. It follows that the totality of solutions for the given equation consists of all the numbers

$$\frac{\pi}{3} + 2\pi n, \quad \frac{5\pi}{3} + 2\pi n, \quad \pi + 2\pi n,$$

where $n \in \mathbf{Z}$. If we wish to express the solutions in terms of degrees, we may write

268

$$60° + 360°n, \quad 300° + 360°n, \quad 180° + 360°n.$$

EXAMPLE 3   Find the solutions of $4 \sin^2 x \tan x - \tan x = 0$ in the interval $[0, 2\pi)$.

Solution

Factoring the left side gives us

$$\tan x \, (4 \sin^2 x - 1) = 0$$

and hence the solution set is the union of the solution sets of

$$\tan x = 0 \quad \text{and} \quad \sin^2 x = \tfrac{1}{4},$$

or of

$$\tan x = 0, \quad \sin x = \tfrac{1}{2}, \quad \text{and} \quad \sin x = -\tfrac{1}{2}.$$

The equation $\tan x = 0$ has solutions 0 and π in the interval $[0, 2\pi)$. The equation $\sin x = \tfrac{1}{2}$ has solutions $\pi/6$ and $5\pi/6$. The equation $\sin x = -\tfrac{1}{2}$ leads to the numbers in $[\pi, 2\pi)$ which have reference number $\pi/6$. These are $\pi + \pi/6 = 7\pi/6$ and $2\pi - \pi/6 = 11\pi/6$. Hence the solutions of the given equation in the interval $[0, 2\pi)$ are

$$\left\{ 0, \pi, \frac{\pi}{6}, \frac{5\pi}{6}, \frac{7\pi}{6}, \frac{11\pi}{6} \right\}.$$

EXAMPLE 4   Find the solution set of the equation $\csc^4 2u - 4 = 0$.

Solution

Factoring we have

$$(\csc^2 2u - 2)(\csc^2 2u + 2) = 0.$$

Setting each factor equal to 0 leads to

$$\csc^2 2u = 2 \quad \text{and} \quad \csc^2 2u = -2.$$

The solution set of $\csc^2 2u = -2$ is $\varnothing$ (Why?). The solution set of $\csc^2 2u = 2$ is the union of the solution sets of

$$\csc 2u = \sqrt{2} \quad \text{and} \quad \csc 2u = -\sqrt{2}.$$

If $\csc 2u = \sqrt{2}$, then

$$2u = \frac{\pi}{4} + 2\pi n \quad \text{and} \quad 2u = \frac{3\pi}{4} + 2\pi n,$$

where $n \in \mathbf{Z}$. Dividing both sides of these equations by 2 gives us the solutions

$$u = \frac{\pi}{8} + \pi n \quad \text{and} \quad u = \frac{3\pi}{8} + \pi n.$$

Similarly, from $\csc 2u = -\sqrt{2}$ we obtain

$$2u = \frac{5\pi}{4} + 2\pi n \quad \text{and} \quad 2u = \frac{7\pi}{4} + 2\pi n$$

where $n \in \mathbf{Z}$. This implies that

$$u = \frac{5\pi}{8} + \pi n \quad \text{and} \quad u = \frac{7\pi}{8} + \pi n.$$

Collecting all the above solutions gives us the solution set:

$$\left\{ \frac{\pi}{8} + \frac{\pi}{4} n : n \in \mathbf{Z} \right\}.$$

EXAMPLE 5 Approximate the solutions of $5 \sin \theta \tan \theta - 10 \tan \theta + 3 \sin \theta - 6 = 0$ in the degree interval $[0°, 360°)$.

Solution

The equation may be factored by grouping terms as follows:

$$5 \tan \theta \, (\sin \theta - 2) + 3(\sin \theta - 2) = 0.$$
$$(5 \tan \theta + 3)(\sin \theta - 2) = 0.$$

Since the solution set of $\sin \theta = 2$ is $\varnothing$ (Why?), the solution set of the given equation is the same as the solution set of

$$\tan \theta = -\tfrac{3}{5} = -0.6000.$$

Approximations to θ may be found by interpolating in Table 3. We first approximate the reference angle θ' as follows:

$$10' \left\{ d \begin{cases} \tan 30°50' \approx .5969 \\ \tan \theta' \quad = .6000 \end{cases} .0031 \right\} .0040$$
$$\tan 31°00' \approx .6009$$

$$\frac{d}{10} = \frac{31}{40} \quad \text{or} \quad d = \frac{31}{4} \approx 8'$$

$$\theta' \approx 30°58'.$$

Since θ lies in quadrant II or quadrant IV, this implies that

$$\theta \approx 180° - 30°58' = 149°2'$$

or

$$\theta \approx 360° - 30°58' = 329°2'.$$

EXERCISES

Find the solution sets of the equations in Exercises 1–12.

1 $2 \sin \theta + 1 = 0$ **2** $\tan \theta = -1$

3 $(\sin t - 1)(\cos t + 1) = 0$ **4** $2 \cos x - \sqrt{3} = 0$

5 $\csc^2 \alpha - 4 = 0$ **6** $\cot^2 \beta = 3$

7 $\tan^2 x - 3 = 0$ **8** $4 \cos^2 t - 3 = 0$

9 $(2 \cos \theta + 1)(2 \sin \theta + \sqrt{2}) = 0$ **10** $(\sin u - 1)(\cos u - 1) = 0$

11 $\tan 2x (\sec 2x - 2) = 0$ **12** $\cot^2 \theta + \cot \theta = 0$

In each of Exercises 13–32 find the solutions of the equation in the interval $[0, 2\pi)$ and also find the degree measure of each solution.

13 $4 \cos^2 \theta - 1 = 0$ **14** $\cot^2 x + \cot x = 0$

15 $2 \sin^2 t + \sin t - 1 = 0$ **16** $2 \cos^2 \theta + 3 \cos \theta + 1 = 0$

17 $\tan^2 x \cos x = \cos x$ **18** $\sec \alpha \csc \alpha = 2 \sec \alpha$

19 $\sin \beta + 2 \sin^2 \beta = 0$ **20** $\cos \theta - \sin \theta = 0$

21 $\cos^2 u + \cos u - 6 = 0$ **22** $2 \sin^2 x + \sin x - 6 = 0$

23 $1 - \cos t = \sqrt{3} \sin t$ **24** $\sin \theta - \cos \theta = 1$

25 $\sin \theta + \cos \theta = 1$ **26** $\sec^2 v - 2 \tan v = 0$

27 $\sec \alpha + \tan \alpha = 1$ **28** $\tan x + \cot x = \sec x \csc x$

29 $2 \cos^3 u + \cos^2 u - 2 \cos u - 1 = 0$

30 $\csc^5 x - 4 \csc x = 0$

31 $2 \sec \theta \tan \theta + 2 \sec \theta + \tan \theta + 1 = 0$

32 $2 \csc \alpha \cos \alpha - 4 \cos \alpha - \csc \alpha + 2 = 0$

In each of Exercises 33–36 use Table 3 to approximate, to the nearest multiple of ten minutes, the solutions of the given equation in the interval $[0°, 360°)$.

33 $\sin^2 t - 4 \sin t + 1 = 0$ **34** $\tan^2 \theta + 3 \tan \theta - 2 = 0$

35 $12 \sin^2 u - 5 \sin u - 2 = 0$ **36** $5 \cos^2 \alpha + 3 \cos \alpha - 2 = 0$

3 THE ADDITION FORMULAS

In this section we shall establish several identities for the trigonometric functions that are extremely important for advanced work in mathematics.

Let t_1 and t_2 be any real numbers and let $P(t_1)$ and $P(t_2)$ be the corresponding points on the unit circle U which are assigned by the wrapping function. We shall denote these points in terms of rectangular coordinates $P_1(x_1, y_1)$ and $P_2(x_2, y_2)$, respectively. Let us next consider the real number $t_1 - t_2$ and denote the rectangular coordinates of $P(t_1 - t_2)$ by $P_3(x_3, y_3)$. Applying (5.4) we have

(6.1)
$$\cos t_1 = x_1, \quad \cos t_2 = x_2, \quad \cos (t_1 - t_2) = x_3,$$
$$\sin t_1 = y_1, \quad \sin t_2 = y_2, \quad \sin (t_1 - t_2) = y_3.$$

Our goal is to obtain a formula for $\cos (t_1 - t_2)$ in terms of functional values of t_1 and t_2. For convenience let us assume that t_1 and t_2 are in the interval $[0, 2\pi]$ and $0 \le t_1 - t_2 < t_2$. In this case, $t_2 \le t_1$ and hence the length t_1 of $\widehat{AP_1}$ is greater than or equal to the length t_2 of $\widehat{AP_2}$. Also, the length $t_1 - t_2$ of $\widehat{AP_3}$ is less than the length t_2 of $\widehat{AP_2}$. Figure 6.1 illustrates one arrangement of points P_1 and P_2 under these conditions.

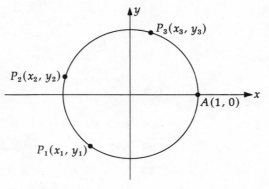

Figure 6.1

It is possible to extend our discussion to cover all values of t. Since the arc lengths of $\widehat{P_2P_1}$ and $\widehat{AP_3}$ both equal $t_1 - t_2$, the line segments P_2P_1 and AP_3 also have the same length, that is,

$$d(A, P_3) = d(P_1, P_2).$$

By the distance formula this implies that

$$\sqrt{(x_3 - 1)^2 + (y_3 - 0)^2} = \sqrt{(x_2 - x_1)^2 + (y_2 - y_1)^2}.$$

Squaring both sides and expanding the terms underneath the radical gives us

$$x_3^2 - 2x_3 + 1 + y_3^2 = x_2^2 - 2x_1x_2 + x_1^2 + y_2^2 - 2y_1y_2 + y_1^2.$$

Since P_1, P_2, and P_3 lie on U, we may substitute 1 for $x_1^2 + y_1^2$, $x_2^2 + y_2^2$, and $x_3^2 + y_3^2$ in the last equation. Doing this and simplifying, we obtain

$$2 - 2x_3 = 2 - 2x_1x_2 - 2y_1y_2,$$

which reduces to

$$x_3 = x_1 x_2 + y_1 y_2.$$

Substituting from (6.1) gives us the desired formula, namely

$$\cos (t_1 - t_2) = \cos t_1 \cos t_2 + \sin t_1 \sin t_2.$$

In order to make the formula less cumbersome, we shall eliminate subscripts by changing the symbols for the variables from t_1 and t_2 to u and v, respectively. Our identity then takes on the form

(6.2) $\cos (u - v) = \cos u \cos v + \sin u \sin v.$

Note that, in most cases, $\cos (u - v) \neq \cos u - \cos v$.

EXAMPLE 1   Find the exact value of cos 15°.

Solution

Writing $15° = 60° - 45°$ and using (6.2),

$$\cos 15° = \cos (60° - 45°)$$
$$= \cos 60° \cos 45° + \sin 60° \sin 45°$$
$$= \left(\frac{1}{2}\right) \left(\frac{\sqrt{2}}{2}\right) + \left(\frac{\sqrt{3}}{2}\right) \left(\frac{\sqrt{2}}{2}\right)$$
$$= \frac{\sqrt{2} + \sqrt{6}}{4}.$$

It is now easy to obtain a formula for $\cos (u + v)$. We simply write $u + v = u - (-v)$ and employ (6.2). Thus

$$\cos (u + v) = \cos [u - (-v)]$$
$$= \cos u \cos (-v) + \sin u \sin (-v).$$

By (5.13), $\cos (-v) = \cos v$ and $\sin (-v) = -\sin v$ for all v, and hence

(6.3) $\cos (u + v) = \cos u \cos v - \sin u \sin v.$

EXAMPLE 2   Find $\cos 7\pi/12$ by using $\pi/3$ and $\pi/4$.

Solution

Since $7\pi/12 = \pi/3 + \pi/4$ we have from (6.3),

$$\cos \frac{7\pi}{12} = \cos \left(\frac{\pi}{3} + \frac{\pi}{4}\right)$$
$$= \cos \frac{\pi}{3} \cos \frac{\pi}{4} - \sin \frac{\pi}{3} \sin \frac{\pi}{4}$$
$$= \frac{1}{2} \frac{\sqrt{2}}{2} - \frac{\sqrt{3}}{2} \frac{\sqrt{2}}{2}$$
$$= \frac{\sqrt{2} - \sqrt{6}}{4}.$$

Identities similar to (6.2) and (6.3) may be proved for the sine function. Let us first establish the following identities, which are of some interest in themselves:

$$\cos\left(\frac{\pi}{2} - u\right) = \sin u$$

(6.4) $$\sin\left(\frac{\pi}{2} - u\right) = \cos u$$

$$\tan\left(\frac{\pi}{2} - u\right) = \cot u$$

Similar identities are true for the other trigonometric functions (see Exercise 34). To prove the first identity we use (6.2) as follows.

$$\cos\left(\frac{\pi}{2} - u\right) = \cos\frac{\pi}{2}\cos u + \sin\frac{\pi}{2}\sin u$$

$$= 0 \cdot \cos u + 1 \cdot \sin u$$

$$= \sin u$$

Substituting $\pi/2 - v$ for u in this identity, we obtain

$$\cos\left[\frac{\pi}{2} - \left(\frac{\pi}{2} - v\right)\right] = \sin\left(\frac{\pi}{2} - v\right),$$

or

$$\cos v = \sin\left(\frac{\pi}{2} - v\right),$$

for all v. This leads to the second identity of (6.4).

The third formula in (6.4) may be established as follows:

$$\tan(\pi/2 - u) = \frac{\sin(\pi/2 - u)}{\cos(\pi/2 - u)} = \frac{\cos u}{\sin u} = \cot u.$$

If θ denotes an angle expressed in degree measure, then formulas (6.4) may be written in the form

$$\sin(90° - \theta) = \cos\theta$$
(6.5) $$\cos(90° - \theta) = \sin\theta$$
$$\tan(90° - \theta) = \cot\theta$$

The angles θ and $90° - \theta$ are **complementary,** since their sum is 90°. It is customary to refer to the sine and cosine functions as **cofunctions** of one another. Similarly, the tangent and cotangent functions are cofunctions, as are the secant and cosecant. Consequently (6.5) is a partial description of the fact that *any functional value of θ equals the cofunction of the complementary angle $90° - \theta$.*

The following relationships for the sine and tangent may now be proved.

(6.6) $\sin (u + v) = \sin u \cos v + \cos u \sin v,$

(6.7) $\sin (u - v) = \sin u \cos v - \cos u \sin v,$

(6.8) $\tan (u + v) = \dfrac{\tan u + \tan v}{1 - \tan u \tan v},$

(6.9) $\tan (u - v) = \dfrac{\tan u - \tan v}{1 + \tan u \tan v}.$

We shall prove (6.6) and (6.8) and leave the others as exercises. Thus

$$\sin (u + v) = \cos \left[\frac{\pi}{2} - (u + v) \right] \qquad (6.4)$$

$$= \cos \left[\left(\frac{\pi}{2} - u \right) - v \right] \qquad \text{(Why?)}$$

$$= \cos \left(\frac{\pi}{2} - u \right) \cos v$$

$$+ \sin \left(\frac{\pi}{2} - u \right) \sin v \qquad (6.2)$$

$$= \sin u \cos v + \cos u \sin v \qquad (6.4).$$

To prove (6.8) we begin as follows:

$$\tan (u + v) = \frac{\sin (u + v)}{\cos (u + v)}$$

$$= \frac{\sin u \cos v + \cos u \sin v}{\cos u \cos v - \sin u \sin v}.$$

Next, dividing numerator and denominator by $\cos u \cos v$ (assuming, of course, that $\cos u \cos v \neq 0$), we obtain

$$\tan (u + v) = \frac{\left(\dfrac{\sin u}{\cos u} \right) \left(\dfrac{\cos v}{\cos v} \right) + \left(\dfrac{\cos u}{\cos u} \right) \left(\dfrac{\sin v}{\cos v} \right)}{\left(\dfrac{\cos u}{\cos u} \right) \left(\dfrac{\cos v}{\cos v} \right) - \left(\dfrac{\sin u}{\cos u} \right) \left(\dfrac{\sin v}{\cos v} \right)},$$

which is equivalent to (6.8). The formulas derived in this section are known as the **addition formulas**.

EXAMPLE 3   Given $\sin \alpha = \frac{4}{5}$, where α is an angle in quadrant I, and $\cos \beta = -\frac{12}{13}$, where β is in quadrant II, find $\sin (\alpha + \beta)$, $\tan (\alpha + \beta)$, and the quadrant in which $\alpha + \beta$ lies.

Solution

It is convenient to represent α and β geometrically, as illustrated in Fig. 6.2.

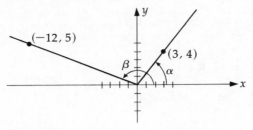

Figure 6.2

There is no loss of generality in picturing α and β as positive angles between 0 and 2π as we have done in the figure. Since $\sin \alpha = \frac{4}{5}$, the point $(3, 4)$ is on the terminal side of α (Why?). Similarly, since $\cos \beta = -\frac{12}{13}$ we may choose the point $(-12, 5)$ on the terminal side of β. Referring to Fig. 6.2 and using (5.26) we see that

$$\cos \alpha = \tfrac{3}{5}, \quad \tan \alpha = \tfrac{4}{3}, \quad \sin \beta = \tfrac{5}{13}, \quad \text{and} \quad \tan \beta = -\tfrac{5}{12}.$$

Using (6.6) and (6.8) gives us

$$\sin(\alpha + \beta) = \sin \alpha \cos \beta + \cos \alpha \sin \beta$$
$$= (\tfrac{4}{5})(-\tfrac{12}{13}) + (\tfrac{3}{5})(\tfrac{5}{13})$$
$$= -\tfrac{33}{65}.$$

$$\tan(\alpha + \beta) = \frac{\tan \alpha + \tan \beta}{1 - \tan \alpha \tan \beta}$$

$$= \frac{\tfrac{4}{3} + (-\tfrac{5}{12})}{1 - (\tfrac{4}{3})(-\tfrac{5}{12})}$$

$$= \tfrac{33}{56}.$$

Since $\sin(\alpha + \beta)$ is negative and $\tan(\alpha + \beta)$ is positive, it follows that $\alpha + \beta$ lies in quadrant III.

The type of simplification used in the next example is required in the study of calculus.

EXAMPLE 4   If $f(x) = \sin x$ and $h \neq 0$, prove that

$$\frac{f(x+h) - f(x)}{h} = \sin x \left(\frac{\cos h - 1}{h} \right) + \cos x \left(\frac{\sin h}{h} \right).$$

Solution

Using the definition of f and (6.6) we obtain

$$\frac{f(x+h)-f(x)}{h} = \frac{\sin(x+h)-\sin x}{h}$$

$$= \frac{\sin x \cos h + \cos x \sin h - \sin x}{h}$$

$$= \frac{\sin x(\cos h - 1) + \cos x \sin h}{h}$$

$$= \sin x \left(\frac{\cos h - 1}{h}\right) + \cos x \left(\frac{\sin h}{h}\right).$$

EXERCISES

Use (6.4) and (6.5) to write the expressions in Exercises 1 and 2 in terms of cofunctions of complementary angles.

1   **a** $\sin 33°24'$ **b** $\cos \pi/3$ **c** $\tan 67°10'$

2   **a** $\tan 51°37'23''$ **b** $\sin \pi/6$ **c** $\cos 18°51'$

Find the exact functional values in Exercises 3–8.

3   **a** $\sin \pi/4 + \sin \pi/3$ **b** $\sin 7\pi/12$

4   **a** $\cos 2\pi/3 + \cos \pi/4$ **b** $\cos 11\pi/12$

5   **a** $\tan 45° + \tan 300°$ **b** $\tan 345°$

6   **a** $\sin 135° - \sin 60°$ **b** $\sin 75°$

7   **a** $\cos 3\pi/4 - \cos \pi/6$ **b** $\cos 7\pi/12$

8   **a** $\tan 5\pi/4 - \tan 5\pi/6$ **b** $\tan 5\pi/12$

9   If α and β are acute angles such that $\cos \alpha = \frac{3}{5}$ and $\tan \beta = \frac{15}{8}$, calculate $\cos(\alpha + \beta)$, $\sin(\alpha + \beta)$, and find the quadrant in which $\alpha + \beta$ lies.

10   If $\sin \alpha = -\frac{3}{5}$ and $\sec \beta = \frac{5}{4}$, where α is a third-quadrant angle and β is a first-quadrant angle, find $\sin(\alpha + \beta)$, $\tan(\alpha + \beta)$ and the quadrant in which $\alpha + \beta$ lies.

11   If $\sec \alpha = -\frac{25}{24}$ and $\cot \beta = \frac{4}{3}$, where α is in the second quadrant and β is in the third quadrant, find $\sin(\alpha + \beta)$, $\cos(\alpha + \beta)$, $\tan(\alpha + \beta)$, $\sin(\alpha - \beta)$, $\cos(\alpha - \beta)$, and $\tan(\alpha - \beta)$.

12   If $P(t_1)$ and $P(t_2)$ are in quadrant III such that $\cos t_1 = -\frac{1}{3}$ and $\cos t_2 = -\frac{2}{3}$, find $\sin(t_1 - t_2)$, $\cos(t_1 - t_2)$, and the quadrant in which $P(t_1 - t_2)$ lies.

Verify the identities in Exercises 13–28.

13 $\sin(x + \pi/2) = \cos x$ **14** $\cos(x + \pi/2) = -\sin x$

15 $\cos(\theta + 3\pi/2) = \sin \theta$ **16** $\sin(\alpha - 3\pi/2) = \cos \alpha$

17 $\sin(\theta + \pi/4) = (\sqrt{2}/2)(\sin \theta + \cos \theta)$

18 $\cos(\theta + \pi/4) = (\sqrt{2}/2)(\cos \theta - \sin \theta)$

277

19 $\tan (u + \pi/4) = \dfrac{1 + \tan u}{1 - \tan u}$ **20** $\tan (x - \pi/4) = \dfrac{\tan x - 1}{\tan x + 1}$

21 $\tan (u + \pi/2) = -\cot u$ **22** $\cot (t - \pi/3) = \dfrac{\sqrt{3} \tan t + 1}{\tan t - \sqrt{3}}$

23 $\sin (u + v) \cdot \sin (u - v) = \sin^2 u - \sin^2 v$

24 $\cos (u + v) \cdot \cos (u - v) = \cos^2 u - \sin^2 v$

25 $\cos (u + v) + \cos (u - v) = 2 \cos u \cos v$

26 $\sin (u + v) + \sin (u - v) = 2 \sin u \cos v$

27 $\sin 2u = 2 \sin u \cos u$ (*Hint:* $2u = u + u$.)

28 $\cos 2u = \cos^2 u - \sin^2 u$

29 Express $\sin (u + v + w)$ in terms of functions of u, v, and w. (*Hint:* Write $\sin (u + v + w) = \sin [(u + v) + w]$ and use (6.6) two times.)

30 Same as Exercise 25 for $\tan (u + v + w)$

31 Use (6.8) and (5.7) to derive the formula

$$\cot (u + v) = \frac{\cot u \cot v - 1}{\cot u + \cot v}.$$

32 Verify the identity

$$\frac{\sin (u + v)}{\sin (u - v)} = \frac{\tan u + \tan v}{\tan u - \tan v}.$$

33 Prove (6.7) and (6.9).

34 Prove the following analogues of (6.4):   **a** $\sec (\pi/2 - u) = \csc u$,
   b $\csc (\pi/2 - u) = \sec u$,   **c** $\cot (\pi/2 - u) = \tan u$.

35 If $f(x) = \cos x$, prove that

$$\frac{f(x + h) - f(x)}{h} = \cos x \left(\frac{\cos h - 1}{h} \right) - \sin x \left(\frac{\sin h}{h} \right).$$

36 If $f(x) = \tan x$, prove that

$$\frac{f(x + h) - f(x)}{h} = \sec^2 x \left(\frac{\sin h}{h} \right) \frac{1}{\cos h - \sin h \tan x}.$$

4 MULTIPLE ANGLE FORMULAS

In the preceding section we were interested primarily in trigonometric identities which involve values of $u \pm v$. In this section we shall obtain formulas for values of nu, where n represents certain integers or rational numbers. The formulas are referred to as **multiple angle formulas.** Let us list some of these formulas for reference before proceeding to the proofs. As usual, the following equations are true for all values of u for which the expressions on both sides are meaningful.

(6.10) $\sin 2u = 2 \sin u \cos u,$

(6.11) $\cos 2u = \cos^2 u - \sin^2 u$
$= 1 - 2 \sin^2 u$
$= 2 \cos^2 u - 1,$

(6.12) $\tan 2u = \dfrac{2 \tan u}{1 - \tan^2 u}.$

These identities may be proved by letting $u = v$ in the appropriate addition formulas of Section 3. In particular, using (6.6), we have

$\sin 2u = \sin (u + u)$
$= \sin u \cos u + \cos u \sin u$
$= 2 \sin u \cos u.$

Similarly, using (6.3) gives us

$\cos 2u = \cos (u + u)$
$= \cos u \cos u - \sin u \sin u$
$= \cos^2 u - \sin^2 u.$

The other forms in (6.11) may be obtained from the first by using the fundamental identity (5.10). Thus

$\cos 2u = \cos^2 u - \sin^2 u$
$= (1 - \sin^2 u) - \sin^2 u$
$= 1 - 2 \sin^2 u.$

Similarly, if we substitute for $\sin^2 u$ instead of $\cos^2 u$, we obtain

$\cos 2u = \cos^2 u - (1 - \cos^2 u)$
$= 2 \cos^2 u - 1.$

The final formula (6.12) is obtained by taking $u = v$ in (6.8).

Because of the factor 2, the formulas in (6.10)–(6.12) are often called the **double angle formulas.**

EXAMPLE 1   Find $\sin 2\alpha$ and $\cos 2\alpha$ if $\sin \alpha = \frac{4}{5}$ and α is in quadrant I.

Solution

As in Example 3 of the preceding section, $\cos \alpha = \frac{3}{5}$. Substitution in (6.10) and (6.11) gives us

$$\sin 2\alpha = 2 \sin \alpha \cos \alpha = 2(\tfrac{4}{5})(\tfrac{3}{5}) = \tfrac{24}{25}$$
$$\cos 2\alpha = \cos^2 \alpha - \sin^2 \alpha = \tfrac{9}{25} - \tfrac{16}{25} = -\tfrac{7}{25}.$$

EXAMPLE 2 Express $\cos 3\theta$ in terms of $\cos \theta$.

Solution

$$\begin{aligned}
\cos 3\theta &= \cos (2\theta + \theta) \\
&= \cos 2\theta \cos \theta - \sin 2\theta \sin \theta \\
&= (2 \cos^2 \theta - 1) \cos \theta - (2 \sin \theta \cos \theta) \sin \theta \\
&= 2 \cos^3 \theta - \cos \theta - 2 \cos \theta \sin^2 \theta \\
&= 2 \cos^3 \theta - \cos \theta - 2 \cos \theta (1 - \cos^2 \theta) \\
&= 2 \cos^3 \theta - \cos \theta - 2 \cos \theta + 2 \cos^3 \theta \\
&= 4 \cos^3 \theta - 3 \cos \theta.
\end{aligned}$$

If we solve the second and third equations of (6.11) for $\sin^2 u$ and $\cos^2 u$, respectively, the following identities are obtained:

(6.13)
$$\sin^2 u = \frac{1 - \cos 2u}{2},$$
$$\cos^2 u = \frac{1 + \cos 2u}{2}.$$

Since $\tan^2 u = \sin^2 u / \cos^2 u$, (6.13) leads to

(6.14) $$\tan^2 u = \frac{1 - \cos 2u}{1 + \cos 2u}.$$

We may use (6.13) to reduce exponents, as illustrated in the next two examples.

EXAMPLE 3 Verify the identity

$$\sin^2 x \cos^2 x = \tfrac{1}{8}(1 - \cos 4x).$$

Solution

Using (6.13) with $u = x$ we obtain

$$\begin{aligned}
\sin^2 x \cos^2 x &= \left(\frac{1 - \cos 2x}{2}\right)\left(\frac{1 + \cos 2x}{2}\right) \\
&= \tfrac{1}{4}(1 - \cos^2 2x) \\
&= \tfrac{1}{4} \sin^2 2x.
\end{aligned}$$

Applying (6.13) again, with $u = 2x$, gives us

$$\begin{aligned}
\sin^2 x \cos^2 x &= \tfrac{1}{4}\left(\frac{1 - \cos 4x}{2}\right) \\
&= \tfrac{1}{8}(1 - \cos 4x).
\end{aligned}$$

Another method of proof is to use the fact that $\sin 2x = 2 \sin x \cos x$ and hence that

$$\sin x \cos x = \tfrac{1}{2} \sin 2x.$$

Squaring both sides we have

$$\sin^2 x \cos^2 x = \tfrac{1}{4} \sin^2 2x.$$

The remainder of the solution is the same as that given above.

EXAMPLE 4 Express $\cos^4 t$ in terms of values of cos with exponent 1.

Solution

The reader should supply reasons for the following steps.

$$\begin{aligned}
\cos^4 t &= (\cos^2 t)^2 \\
&= \left(\frac{1 + \cos 2t}{2}\right)^2 \\
&= \tfrac{1}{4}(1 + 2 \cos 2t + \cos^2 2t) \\
&= \tfrac{1}{4}\left(1 + 2 \cos 2t + \frac{1 + \cos 4t}{2}\right) \\
&= \tfrac{3}{8} + \tfrac{1}{2} \cos 2t + \tfrac{1}{8} \cos 4t.
\end{aligned}$$

The following alternate forms for (6.13) and (6.14) are obtained by substituting $v/2$ for u.

$$\sin^2 \frac{v}{2} = \frac{1 - \cos v}{2}$$

$$\cos^2 \frac{v}{2} = \frac{1 + \cos v}{2}$$

$$\tan^2 \frac{v}{2} = \frac{1 - \cos v}{1 + \cos v}$$

If we take the square root of both sides of these equations and use the fact that $\sqrt{a^2} = |a|$ for all real numbers a, the following identities result:

$$\left|\sin \frac{v}{2}\right| = \sqrt{\frac{1 - \cos v}{2}}$$

(6.15) $$\left|\cos \frac{v}{2}\right| = \sqrt{\frac{1 + \cos v}{2}}$$

$$\left|\tan \frac{v}{2}\right| = \sqrt{\frac{1 - \cos v}{1 + \cos v}}$$

The absolute value signs may be eliminated if more information is known about $v/2$. For example, if the point $P(v/2)$ assigned by the wrapping function is in either quadrant I or II (or, equivalently, if the angle determined by $v/2$ lies in one of these quadrants), then $\sin v/2$ is positive and we may write

$$\sin \frac{v}{2} = \sqrt{\frac{1 - \cos v}{2}} \, .$$

However, if $v/2$ leads to a point (or angle) in either quadrant III or IV, then

$$\sin \frac{v}{2} = -\sqrt{\frac{1 - \cos v}{2}} \, .$$

Similar remarks are true for the other formulas.

An alternate form for $\tan v/2$ can be obtained. Multiplying numerator and denominator of the radicand in the last formula of (6.15) by $1 - \cos v$ gives us

$$\left| \tan \frac{v}{2} \right| = \sqrt{\frac{1 - \cos v}{1 + \cos v} \cdot \frac{1 - \cos v}{1 - \cos v}}$$

$$= \sqrt{\frac{(1 - \cos v)^2}{\sin^2 v}}$$

$$= \frac{1 - \cos v}{|\sin v|} \, .$$

The absolute value sign is unnecessary in the numerator since $1 - \cos v$ is never negative. It can be shown that $\tan v/2$ and $\sin v$ always have the same sign. For example, if $0 < v < \pi$, then $0 < v/2 < \pi/2$, and hence $\sin v$ and $\tan v/2$ are both positive. If $\pi < v < 2\pi$, then $\pi/2 < v/2 < \pi$, and hence $\sin v$ and $\tan v/2$ are both negative. It is possible to generalize these remarks to all values of v for which the expressions $\tan v/2$ and $(1 - \cos v)/|\sin v|$ have meaning. Because of this we may write

$$(6.16) \quad \tan \frac{v}{2} = \frac{1 - \cos v}{\sin v} \, .$$

If we multiply numerator and denominator on the right-hand side of (6.16) by $1 + \cos v$ and simplify, we obtain

$$(6.17) \quad \tan \frac{v}{2} = \frac{\sin v}{1 + \cos v} \, .$$

Formulas (6.15)–(6.17) are called the **half-angle formulas.**

EXAMPLE 5   Find the exact value of $\sin 22.5°$ and $\cos 22.5°$.

Solution

From (6.15) and the fact that $22.5°$ is in quadrant I, we have

$$\sin 22.5° = \sin \frac{45°}{2} = \sqrt{\frac{1 - \cos 45°}{2}}$$

$$= \sqrt{\frac{1 - \sqrt{2}/2}{2}}$$

$$= \frac{\sqrt{2 - \sqrt{2}}}{2};$$

$$\cos 22.5° = \sqrt{\frac{1 + \cos 45°}{2}}$$

$$= \frac{\sqrt{2 + \sqrt{2}}}{2}.$$

EXAMPLE 6 If $\tan \alpha = -\frac{4}{3}$, where α is in quadrant IV, find $\tan \alpha/2$.

Solution

If we choose the point $(3, -4)$ on the terminal side of α as illustrated in Fig. 6.3, then by (5.26), $\sin \alpha = -\frac{4}{5}$ and $\cos \alpha = \frac{3}{5}$. Apply (6.16) we have

$$\tan \frac{\alpha}{2} = \frac{1 - \cos \alpha}{\sin \alpha} = \frac{1 - \frac{3}{5}}{-\frac{4}{5}} = -\frac{1}{2}.$$

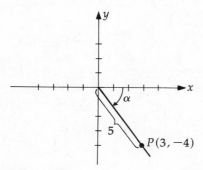

Figure 6.3

EXERCISES

In Exercises 1–4 find the exact values of $\sin 2\theta$, $\cos 2\theta$, and $\tan 2\theta$, subject to the given conditions.

1 $\sin \theta = \frac{3}{5}$ and θ acute

2 $\tan \theta = \frac{4}{3}$ and $180° < \theta < 270°$

3 $\sec \theta = -\frac{17}{8}$ and $90° < \theta < 180°$

4 $\cos \theta = \frac{4}{5}$ and $270° < \theta < 360°$

In Exercises 5–8 find the exact values of sin $\theta/2$, cos $\theta/2$, and tan $\theta/2$ subject to the given conditions.

5 csc $\theta = \frac{5}{4}$ and θ acute

6 sin $\theta = -\frac{3}{5}$ and $-90° < \theta < 0°$

7 cot $\theta = 1$ and $-180° < \theta < -90°$

8 sec $\theta = -3$ and $180° < \theta < 270°$

9 Find the exact value of   **a** sin $67°30'$,   **b** cos $15°$,   **c** tan $3\pi/8$.

10 Find the exact value of   **a** sin $165°$ **b** cos $157°30'$ **c** tan $\pi/8$.

Verify the identities given in Exercises 11–20.

11 $(\sin t + \cos t)^2 = 1 + \sin 2t$

12 $\csc 2u = \frac{1}{2} \sec u \csc u$

13 $\sin 3u = \sin u(3 - 4 \sin^2 u)$

14 $\sin 4t = 4 \cos t \sin t(1 - 2 \sin^2 t)$

15 $\cos 4\theta = 8 \cos^4 \theta - 8 \cos^2 \theta + 1$

16 $\cos 6t = 32 \cos^6 t - 48 \cos^4 t + 18 \cos^2 t - 1$

17 $\sin^4 t = \frac{3}{8} - \frac{1}{2} \cos 2t + \frac{1}{8} \cos 4t$ **18** $\cos^4 x - \sin^4 x = \cos 2x$

19 $\sec 2\theta = \dfrac{\sec^2 \theta}{2 - \sec^2 \theta}$ **20** $\cot 2u = \dfrac{\cot^2 u - 1}{2 \cot u}$

In Exercises 21–28 find all solutions of the given equations in the interval $[0, 2\pi]$. Express the solutions both in radian measure and degree measure.

21 $\sin 2t + \sin t = 0$ **22** $\cos t - \sin 2t = 0$

23 $\cos u + \cos 2u = 0$ **24** $\cos 2\theta - \tan \theta = 1$

25 $\tan 2x = \tan x$ **26** $\tan 2t - 2 \cos t = 0$

27 $\sin \frac{1}{2}u + \cos u = 1$ **28** $2 - \cos^2 x = 4 \sin^2 \frac{1}{2}x$

29 If $a > 0$, $b > 0$, and $0 < u < \pi/2$, prove that

$$a \sin u + b \cos u = \sqrt{a^2 + b^2} \sin (u + v),$$

where $0 < v < \pi/2$, sin $v = b/\sqrt{a^2 + b^2}$, and cos $v = a/\sqrt{a^2 + b^2}$.

30 Use Exercise 29 to express $8 \sin u + 15 \cos u$ in the form $c \sin (u + v)$.

5 SUM AND PRODUCT FORMULAS

The addition formulas of Section 3 can be used to obtain several other useful identities. If the expressions on the left and right-hand sides of

(6.6) and (6.7) are added, the term involving cos *u* sin *v* drops out and we obtain

(6.18) $\sin (u + v) + \sin (u - v) = 2 \sin u \cos v.$

In like manner,

(6.19) $\sin (u + v) - \sin (u - v) = 2 \cos u \sin v.$

Using (6.2) and (6.3) in similar fashion gives us

(6.20) $\cos (u + v) + \cos (u - v) = 2 \cos u \cos v,$

(6.21) $\cos (u - v) - \cos (u + v) = 2 \sin u \sin v.$

The identities (6.18)–(6.21) are sometimes called the **product formulas.** By starting with the expression on the right side of each equation we can express a product as a sum, as shown in the following example.

EXAMPLE 1   Express each of the following as a sum.

(a) $\sin 4\theta \cos 3\theta.$ (b) $\sin 3x \sin x.$

Solution

(a)   Using (6.18) with $u = 4\theta$ and $v = 3\theta$, we obtain

$2 \sin 4\theta \cos 3\theta = \sin (4\theta + 3\theta) + \sin (4\theta - 3\theta),$

or

$\sin 4\theta \cos 3\theta = \tfrac{1}{2} \sin 7\theta + \tfrac{1}{2} \sin \theta.$

An equivalent formula could have been found by using (6.19).

(b)   Using (6.21) with $u = 3x$ and $v = x$ gives us

$2 \sin 3x \sin x = \cos (3x - x) - \cos (3x + x),$

or

$\sin 3x \sin x = \tfrac{1}{2} \cos 2x - \tfrac{1}{2} \cos 4x.$

Identities (6.18)–(6.21) may also be used to express a sum as a product. In order to obtain a form which can be applied more easily, we shall change the notation as follows. If we let

$u + v = a$ and $u - v = b,$

285

then $(u + v) + (u - v) = a + b$, which simplifies to

$$u = \frac{a + b}{2}.$$

Similarly, from $(u + v) - (u - v) = a - b$ we obtain

$$v = \frac{a - b}{2}.$$

If we now substitute for $u + v$ and $u - v$ on the left sides of (6.18)–(6.21), and for u and v on the right sides, we get

(6.22)
$$\sin a + \sin b = 2 \cos \frac{a - b}{2} \sin \frac{a + b}{2}$$
$$\sin a - \sin b = 2 \cos \frac{a + b}{2} \sin \frac{a - b}{2}$$
$$\cos a + \cos b = 2 \cos \frac{a + b}{2} \cos \frac{a - b}{2}$$
$$\cos b - \cos a = 2 \sin \frac{a + b}{2} \sin \frac{a - b}{2}$$

The identities (6.22) are sometimes referred to as the **sum** or **factoring formulas.**

EXAMPLE 2   Express $\sin 5x - \sin 3x$ as a product.

Solution

Using the second identity in (6.22) with $a = 5x$ and $b = 3x$, we have

$$\sin 5x - \sin 3x = 2 \cos \frac{5x + 3x}{2} \sin \frac{5x - 3x}{2}$$
$$= 2 \cos 4x \sin x.$$

EXAMPLE 3   Verify the identity

$$\frac{\sin 3t + \sin 5t}{\cos 3t - \cos 5t} = \cot t.$$

Solution

Using the first and fourth formulas in (6.22) gives us

$$\frac{\sin 3t + \sin 5t}{\cos 3t - \cos 5t} = \frac{2 \cos \dfrac{3t - 5t}{2} \sin \dfrac{3t + 5t}{2}}{2 \sin \dfrac{5t + 3t}{2} \sin \dfrac{5t - 3t}{2}}$$

$$= \frac{2 \cos(-t) \sin 4t}{2 \sin 4t \sin t}$$

$$= \frac{\cos(-t)}{\sin t}$$

$$= \cot t,$$

where the last step follows from (5.13) and (5.9).

EXAMPLE 4   Find the solutions of the equation

$$\cos t - \sin 2t - \cos 3t = 0.$$

Solution

By (6.22)

$$\cos t - \cos 3t = 2 \sin \frac{3t + t}{2} \sin \frac{3t - t}{2}$$

$$= 2 \sin 2t \sin t.$$

Hence the given equation is equivalent to

$$2 \sin 2t \sin t - \sin 2t = 0,$$

or to

$$\sin 2t (2 \sin t - 1) = 0.$$

Therefore the solution set is the union of the solution sets of

$$\sin 2t = 0 \quad \text{and} \quad \sin t = \tfrac{1}{2}.$$

The first equation gives us

$$2t = n\pi, \quad \text{or} \quad t = \frac{\pi}{2} n,$$

where $n \in \mathbf{Z}$. The second equation gives us the solutions

$$t = \frac{\pi}{6} + 2n\pi \quad \text{and} \quad t = \frac{5\pi}{6} + 2n\pi,$$

where $n \in \mathbf{Z}$.

The addition formulas may also be employed to derive the so-called **reduction formulas.** The latter formulas can be used to write expressions such as

$$\sin\left(\theta + n\frac{\pi}{2}\right) \quad \text{and} \quad \cos\left(\theta + n\frac{\pi}{2}\right),$$

where n is any integer, in terms of only $\sin \theta$ or $\cos \theta$. Similar formulas are true for the other trigonometric functions. For example, (6.4) illustrates the case in which $n = 1$ and $\theta = -u$. Instead of deriving general reduction formulas we shall illustrate several special cases by means of examples.

EXAMPLE 5 Express $\sin(\theta - 3\pi/2)$ and $\cos(\theta + \pi)$ in terms of a function of θ.

Solution

From (6.7) and (6.3)

$$\sin\left(\theta - \frac{3\pi}{2}\right) = \sin\theta\cos\frac{3\pi}{2} - \cos\theta\sin\frac{3\pi}{2}$$
$$= \sin\theta \cdot 0 - \cos\theta \cdot (-1)$$
$$= \cos\theta,$$

$$\cos(\theta + \pi) = \cos\theta\cos\pi - \sin\theta\sin\pi$$
$$= \cos\theta \cdot (-1) - \sin\theta \cdot 0$$
$$= -\cos\theta.$$

EXERCISES

In each of Exercises 1–8 express the product as a sum or difference.

1 $2\sin 7\theta\cos 4\theta$

2 $2\cos 4\theta\sin 3\theta$

3 $\sin 5t\sin 3t$

4 $\sin(-6x)\cos 7x$

5 $\cos 8u\cos(-6u)$

6 $\sin 2t\sin 10t$

7 $4\cos x\sin 2x$

8 $5\cos 5u\sin u$

In each of Exercises 9–16 write the expression as a product.

9 $\sin 8\theta + \sin 4\theta$

10 $\sin 4\theta - \sin 6\theta$

11 $\cos 7x - \cos 5x$

12 $\cos 4t + \cos 5t$

13 $\sin 3t - \sin 5t$

14 $\cos\theta - \cos 3\theta$

15 $\cos x + \cos 2x$

16 $\sin 8t + \sin 2t$

Verify the identities in Exercises 17–22.

17 $\dfrac{\sin 4t + \sin 6t}{\cos 4t - \cos 6t} = \cot t$

18 $\dfrac{\sin\theta + \sin 3\theta}{\cos\theta + \cos 3\theta} = \tan 2\theta$

19 $\dfrac{\sin u + \sin v}{\cos u + \cos v} = \tan\dfrac{u + v}{2}$

20 $\dfrac{\sin u - \sin v}{\cos u - \cos v} = -\cot\dfrac{u + v}{2}$

21 $\sin 2x + \sin 4x + \sin 6x = 4\cos x\cos 2x\sin 3x$

22 $\dfrac{\cos t + \cos 4t + \cos 7t}{\sin t + \sin 4t + \sin 7t} = \cot 4t$

In Exercises 23 and 24 find the solution sets of the given equations.

23 $\sin 5t + \sin 3t = 0$

24 $\sin t + \sin 3t = \sin 2t$

In Exercises 25–30 verify the given reduction formulas by using addition formulas.

25 $\sin (\theta + \pi) = -\sin \theta$

26 $\sin (\theta + 3\pi/2) = -\cos \theta$

27 $\cos (\theta - 5\pi/2) = \sin \theta$

28 $\cos (\theta - 3\pi) = -\cos \theta$

29 $\tan (\pi - \theta) = -\tan \theta$ (*Hint:* $\tan (\pi - \theta) = \sin (\pi - \theta)/\cos (\pi - \theta)$.)

30 $\tan (\theta + \pi/2) = -\cot \theta$

6 SUMMARY OF FORMULAS

In order to make the principal trigonometric identities and formulas which we have developed readily available for reference, we list them as follows.

Fundamental Identities

$$\csc u = \frac{1}{\sin u}, \quad \sec u = \frac{1}{\cos u}, \quad \cot u = \frac{1}{\tan u}$$

$$\tan u = \frac{\sin u}{\cos u}, \quad \cot u = \frac{\cos u}{\sin u}$$

$$\sin^2 u + \cos^2 u = 1$$

$$\tan^2 u + 1 = \sec^2 u$$

$$\cot^2 u + 1 = \csc^2 u$$

$$\sin (-u) = -\sin u, \quad \cos (-u) = \cos u, \quad \tan (-u) = -\tan u$$

Addition Formulas

$$\sin (u \pm v) = \sin u \cos v \pm \cos u \sin v$$

$$\cos (u \pm v) = \cos u \cos v \mp \sin u \sin v$$

$$\tan (u \pm v) = \frac{\tan u \pm \tan v}{1 \mp \tan u \tan v}$$

Double Angle Formulas

$$\sin 2u = 2 \sin u \cos u$$

$$\cos 2u = \cos^2 u - \sin^2 u$$

$$= 1 - 2 \sin^2 u = 2 \cos^2 u - 1$$

$$\tan 2u = \frac{2 \tan u}{1 - \tan^2 u}$$

Half-Angle Formulas

$$\left|\sin \frac{u}{2}\right| = \sqrt{\frac{1 - \cos u}{2}}$$

$$\left|\cos \frac{u}{2}\right| = \sqrt{\frac{1 + \cos u}{2}}$$

$$\tan \frac{u}{2} = \frac{1 - \cos u}{\sin u} = \frac{\sin u}{1 + \cos u}$$

Product Formulas

$$2 \sin u \cos v = \sin (u + v) + \sin (u - v)$$
$$2 \cos u \sin v = \sin (u + v) - \sin (u - v)$$
$$2 \cos u \cos v = \cos (u + v) + \cos (u - v)$$
$$2 \sin u \sin v = \cos (u - v) - \cos (u + v)$$

Factoring Formulas

$$\sin u \pm \sin v = 2 \cos \frac{u \mp v}{2} \sin \frac{u \pm v}{2}$$

$$\cos u + \cos v = 2 \cos \frac{u + v}{2} \cos \frac{u - v}{2}$$

$$\cos u - \cos v = 2 \sin \frac{v + u}{2} \sin \frac{v - u}{2}$$

7 THE INVERSE TRIGONOMETRIC FUNCTIONS

Inverse functions were discussed in Section 7 of Chapter Three. At that time we pointed out that if f is a one-to-one function with domain X and range Y, then for each element y in Y there is one and only one element x in X such that $f(x) = y$. In this case we can define a function g from Y to X by letting $g(y) = x$. The function g is called the *inverse function* of f and is often denoted by f^{-1}. In words, f^{-1} *reverses* the correspondence from X to Y given by f; that is

(6.23) $f^{-1}(u) = v$ if and only if $f(v) = u.$

for all u in Y and v in X.
As in (3.27) the inverse function f^{-1} of f is characterized by the equations

(6.24) $f^{-1}(f(v)) = v,$ for all v in X,
$f(f^{-1}(u)) = u,$ for all u in Y.

Since the trigonometric functions are not one-to-one, they do not have inverses. However, by restricting the domains it is possible to obtain functions which behave in the same way as the trigonometric functions (over the smaller domains) and which do possess inverse functions.

Let us first consider the sine function, with domain $\mathbf{R}$ and range the set of real numbers in the interval $[-1, 1]$. This function is not one-to-one since, for example, numbers such as $\pi/6$, $5\pi/6$, and $-7\pi/6$ have the same image, $\frac{1}{2}$. It is easy to find a subset S of $\mathbf{R}$ with the property that as x varies through S, $\sin x$ takes on each value between -1 and 1 once and only once. It is convenient to choose S as the interval $[-\pi/2, \pi/2]$. The new function obtained by restricting the domain of the sine function to $[-\pi/2, \pi/2]$ is one-to-one and hence has an inverse function. Applying (6.23) leads to the following definition.

(6.25) Definition

The **inverse sine function,** denoted by $\sin^{-1}$, is defined by

$$\sin^{-1} u = v \quad \text{if and only if} \quad \sin v = u,$$

where $-1 \leq u \leq 1$ and $-\pi/2 \leq v \leq \pi/2$.

It is also customary to refer to this function as the **arcsine function** and to use the notation $\arcsin u$ in place of $\sin^{-1} u$. The expression $\arcsin u$ is used because if $v = \arcsin u$, then $\sin v = u$, that is, v is a number (or an *arc*length) whose sine is u. Since both notations $\sin^{-1}$ and $\arcsin$ are commonly used in mathematics and its applications, we shall employ both of them in our work. Note that by (6.25) we have

$$-\pi/2 \leq \sin^{-1} u \leq \pi/2,$$

or equivalently

$$-\pi/2 \leq \arcsin u \leq \pi/2.$$

EXAMPLE 1 Sketch the graph of $y = \sin^{-1} x$.

Solution

By (6.25) the graph of $y = \sin^{-1} x$ is the same as the graph of $\sin y = x$, except that the variables are restricted as follows:

$$-\frac{\pi}{2} \leq y \leq \frac{\pi}{2} \quad \text{and} \quad -1 \leq x \leq 1.$$

Several coordinates of points on the graph are given in the following table.

y	$-\dfrac{\pi}{2}$	$-\dfrac{\pi}{3}$	$-\dfrac{\pi}{6}$	0	$\dfrac{\pi}{6}$	$\dfrac{\pi}{3}$	$\dfrac{\pi}{2}$
x	-1	$-\dfrac{\sqrt{3}}{2}$	-1	0	$\dfrac{1}{2}$	$\dfrac{\sqrt{3}}{2}$	1

The graph is sketched in Fig. 6.4.

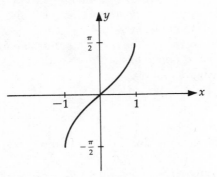

Figure 6.4 $y = \arcsin x = \sin^{-1} x$

Using (6.24) gives us the following important identities:

(6.26)
$$\sin^{-1} (\sin v) = v \quad \text{if} \quad -\frac{\pi}{2} \le v \le \frac{\pi}{2}$$
$$\sin (\sin^{-1} u) = u \quad \text{if} \quad -1 \le u \le 1.$$

Of course (6.26) may also be written in the form

$$\arcsin (\sin v) = v \quad \text{and} \quad \sin (\arcsin u) = u,$$

provided v and u are suitably restricted.

EXAMPLE 2 Find $\sin^{-1} (\sqrt{2}/2)$ and $\arcsin (-\tfrac{1}{2})$.

Solutions

If $v = \sin^{-1} (\sqrt{2}/2)$, then $\sin v = \sqrt{2}/2$ and consequently $v = \pi/4$. Note that it is essential to choose v in the interval $[-\pi/2, \pi/2]$. A number such as $3\pi/4$ is incorrect, even though $\sin (3\pi/4) = \sqrt{2}/2$. In like manner, if $v = \arcsin (-\tfrac{1}{2})$, then $\sin v = -\tfrac{1}{2}$ and hence $v = -\pi/6$.

EXAMPLE 3 Find $\sin^{-1} (\tan 3\pi/4)$.

Solution

If we let

$$v = \sin^{-1} \left(\tan \frac{3\pi}{4} \right) = \sin^{-1} (-1),$$

then by (6.25), sin $v = -1$. Consequently $v = -\pi/2$.

The other trigonometric functions may also be used to introduce inverse functions. The procedure is first to determine a convenient subset of the domain so that a one-to-one function is obtained and then to apply (6.23).

If the domain of the cosine function is restricted to the interval $[0, \pi]$ and if v increases from 0 to π, then cos v takes on each value between -1 and 1 once and only once. This leads to the next definition.

(6.27) Definition

The **inverse cosine function,** denoted by $\cos^{-1}$, is defined by

$$\cos^{-1} u = v \quad \text{if and only if} \quad \cos v = u,$$

where $-1 \le u \le 1$ and $0 \le v \le \pi$.

The inverse cosine function is also referred to as the **arccosine function** and the notation arccos u is used interchangeably with $\cos^{-1} u$.

As in (6.26), if $0 \le v \le \pi$ and $-1 \le u \le 1$, then

(6.28)
$$\cos(\cos^{-1} u) = \cos(\text{arccos } u) = u, \quad \text{and}$$
$$\cos^{-1}(\cos v) = \text{arccos}(\cos v) = v.$$

By (6.27) the graph of $y = \cos^{-1} x$ is the same as the graph of cos $y = x$, where $0 \le y \le \pi$. It is left as an exercise to verify that the graph has the appearance indicated in Fig. 6.5.

Figure 6.5 $y = \text{arccos } x = \cos^{-1} x$

EXAMPLE 4 Approximate $\cos^{-1}(-0.7951)$.

Solution

If $v = \cos^{-1}(-0.7951)$, then cos $v = -0.7951$. If v' is the reference number for v, then cos $v' = 0.7951$. From Table 3 we see that $v' \approx .6516$. According to (6.27), v must be chosen between 0 and π. Hence

$$v = \pi - v' \approx 3.1416 - .6516,$$

that is,

$$\cos^{-1}(-0.7951) \approx 2.4900.$$

If we restrict the domain of the tangent function to the open interval $(-\pi/2, \pi/2)$, then a one-to-one correspondence is obtained. We may, therefore, adopt the following definition.

(6.29)   Definition

The **inverse tangent function** or **arctangent function,** denoted by $\tan^{-1}$ or arctan, is defined by

$$\tan^{-1} u = \arctan u = v \quad \text{if and only if} \quad \tan v = u,$$

where u is any real number and $-\pi/2 < v < \pi/2$.

Note that the domain of the arctangent function is all of **R** and the range is the open interval $(-\pi/2, \pi/2)$. In Exercise 32 the student is asked to verify the graph of the inverse tangent function sketched in Fig. 6.6.

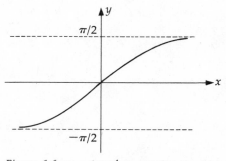

Figure 6.6 $y = \tan^{-1} x = \arctan x$

An analogous procedure may be used for the other trigonometric functions. The functions we have defined are generally referred to as the **inverse trigonometric functions.**

EXAMPLE 5 Without using tables find sec $(\arctan \frac{2}{3})$.

Solution

If $v = \arctan \frac{2}{3}$, then $\tan v = \frac{2}{3}$. Since $\sec^2 v = 1 + \tan^2 v$ and $0 < v < \pi/2$, we may write

$$\sec v = \sqrt{1 + \tan^2 v} = \sqrt{1 + (\tfrac{2}{3})^2} = \sqrt{13}/3.$$

Hence $\sec(\arctan \tfrac{2}{3}) = \sec v = \sqrt{13}/3$.

The following examples illustrate some of the manipulations that can be carried out with the inverse trigonometric functions.

EXAMPLE 4 Evaluate $\sin(\arctan \tfrac{1}{2} - \arccos \tfrac{4}{5})$.

Solution

If we let $v = \arctan \tfrac{1}{2}$ and $w = \arccos \tfrac{4}{5}$, then $\tan v = \tfrac{1}{2}$ and $\cos w = \tfrac{4}{5}$. We wish to find $\sin(v - w)$. Since v and w lie in the interval $[0, \pi/2]$, they may be considered as the radian measure of positive acute angles, and other functional values of v and w may be found by referring to suitable right triangles (see Fig. 6.7). Using (5.27), we have $\sin v = 1/\sqrt{5}$, $\cos v = 2/\sqrt{5}$, and $\sin w = 3/5$. Consequently by (6.7),

$$\sin(v - w) = \sin v \cos w - \cos v \sin w$$

$$= \frac{1}{\sqrt{5}} \frac{4}{5} - \frac{2}{\sqrt{5}} \frac{3}{5}$$

$$= \frac{-2}{5\sqrt{5}} = \frac{-2\sqrt{5}}{25}.$$

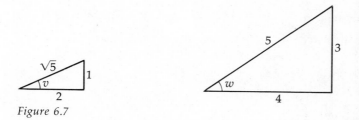

Figure 6.7

EXAMPLE 7 Write $\cos(\sin^{-1} u)$ as an algebraic expression in u.

Solution

Let $v = \sin^{-1} u$, so that $\sin v = u$. We wish to find $\cos v$. Since $-\pi/2 \le v \le \pi/2$ it follows that $\cos v \ge 0$, and hence

$$\cos v = \sqrt{1 - \sin^2 v} = \sqrt{1 - u^2}.$$

Consequently $\cos(\sin^{-1} u) = \sqrt{1 - u^2}$.

EXERCISES

In Exercises 1–6 find the exact values, and in Exercises 7–8 the approximate values, of the given numbers.

1 **a** $\sin^{-1} \sqrt{3}/2$ **b** $\sin^{-1}(-\sqrt{3}/2)$

2 **a** $\sin^{-1} 0$ **b** $\arccos 0$

3 a $\cos^{-1} \sqrt{2}/2$ **b** $\cos^{-1} (-\sqrt{2}/2)$

4 a $\arcsin (-1)$ **b** $\cos^{-1} (-1)$

5 a $\tan^{-1} \sqrt{3}$ **b** $\arctan (-\sqrt{3})$

6 a $\tan^{-1} (-1)$ **b** $\arccos \frac{1}{2}$

7 a $\sin^{-1} (-0.6494)$ **b** $\arccos (0.1132)$

8 a $\arccos (0.5225)$ **b** $\tan^{-1} (-2.2286)$

Determine the numbers in Exercises 9–20 without the use of tables.

9 $\sin [\cos^{-1} \sqrt{3}/2]$ **10** $\cos [\sin^{-1} 0]$

11 $\sin [\arccos \frac{3}{8}]$ **12** $\tan [\tan^{-1} 10]$

13 $\arcsin (\sin \sqrt{5})$ **14** $\tan^{-1} (\cos 0)$

15 $\cos [\sin^{-1} \frac{3}{5} + \tan^{-1} \frac{4}{3}]$ **16** $\sin [\arcsin \frac{1}{2} + \arccos 0]$

17 $\tan [\arctan \frac{3}{4} + \arccos \frac{3}{5}]$ **18** $\cos [2 \sin^{-1} \frac{8}{17}]$

19 $\sin [2 \arccos (-\frac{4}{5})]$ **20** $\cos [\frac{1}{2} \tan^{-1} \frac{15}{8}]$

In each of Exercises 21–24 rewrite the given expression as an algebraic expression in u.

21 $\sin (\tan^{-1} u)$ **22** $\tan (\arccos u)$

23 $\cos (\frac{1}{2} \arccos u)$ **24** $\cos (2 \tan^{-1} u)$

Verify the identities in Exercises 25–28.

25 $\sin^{-1} u + \cos^{-1} u = \pi/2$

26 $\sin^{-1} (-u) = -\sin^{-1} u$

27 $\arctan u + \arctan 1/u = \pi/2, u > 0$

28 $\arcsin \dfrac{2u}{1 + u^2} = 2 \arctan u, |u| \le 1$

29 Define $\cot^{-1}$ by restricting the domain of cot to the interval $(0, \pi)$.

30 Define $\sec^{-1}$ by restricting the domain of sec to $[0, \pi/2) \cup [\pi, 3\pi/2)$.

31 Verify the sketch in Figure 6.5.

32 Verify the sketch in Figure 6.6.

Sketch the graphs of the equations in Exercises 33–40.

33 $y = \sin^{-1} 2x$ **34** $y = \cos^{-1} (x/2)$

35 $y = \frac{1}{2} \sin^{-1} x$ **36** $y = 2 \cos^{-1} x$

37 $y = 2 \tan^{-1} x$ **38** $y = \tan^{-1} 2x$

39 $y = 2 + \tan^{-1} x$ **40** $y = \sin^{-1} (x + 1)$

A triangle that does not contain a right angle is referred to as an **oblique triangle.** Since it is always possible to divide such triangles into two right triangles, methods developed in Chapter Five may be used for solving them; however, sometimes it is cumbersome to proceed in this manner. In this and the next section, formulas are obtained which aid in simplifying solutions of oblique triangles.

If two angles and a side of a triangle are known, or if two sides and an angle opposite one of them are known, then the remaining parts of the triangle may be found by means of the formula given in (6.30). We shall use the letters A, B, C, a, b, c, α, β, and γ for parts of triangles as they were used in Section 9 of Chapter Five. Given triangle ABC, let us place angle α in standard position on a rectangular coordinate system so that B is on the positive x-axis. The case where α is obtuse is illustrated in Fig. 6.8. The type of argument we shall give may also be used if α is acute.

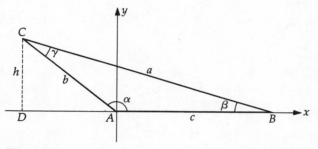

Figure 6.8

Consider the line through C parallel to the y-axis and intersecting the x-axis at point D. Suppose $d(C, D) = h$, so that the ordinate of C is h. By (5.26)

$$\sin \alpha = \frac{h}{b}, \quad \text{or} \quad h = b \sin \alpha.$$

Referring to triangle BDC and using (5.27) we obtain

$$\sin \beta = \frac{h}{a}, \quad \text{or} \quad h = a \sin \beta.$$

Consequently

$$a \sin \beta = b \sin \alpha,$$

which may be written

$$\frac{a}{\sin \alpha} = \frac{b}{\sin \beta}.$$

If α is taken in standard position so that C is on the positive x-axis, then by the same reasoning,

$$\frac{a}{\sin \alpha} = \frac{c}{\sin \gamma}.$$

The last two equalities give us the following result.

(6.30)   The Law of Sines

If ABC is any oblique triangle labeled in the usual manner, then

$$\frac{a}{\sin \alpha} = \frac{b}{\sin \beta} = \frac{c}{\sin \gamma}.$$

The next example illustrates a method of applying (6.30) to the case in which two angles and a side of a triangle are known. We shall use the rules for rounding off answers discussed in Section 9 of Chapter Five.

EXAMPLE 1   Given triangle ABC with $\alpha = 48°20'$, $\gamma = 57°30'$, and $b = 47.3$, approximate the remaining parts.

Solution

The triangle is represented in Fig. 6.9. Since the sum of the angles is 180°, we have

$$\beta = 180° - (57°30' + 48°20') = 74°10'.$$

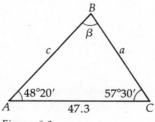

Figure 6.9

Applying (6.30), we have

$$a = \frac{b \sin \alpha}{\sin \beta} = \frac{(47.3) \sin 48°20'}{\sin 74°10'}.$$

Consulting Table 3,

$$a \approx \frac{(47.3)(.7470)}{(.9621)} \approx 36.7$$

If we wish to find a by means of logarithms we may proceed as follows:

$$\log a = \log 47.3 + \log \sin 48°20' - \log \sin 74°10'.$$

Using Tables 2 and 4,

$$
\begin{aligned}
\log 47.3 &\approx 1.6749 \\
\log \sin 48°20' &\approx \underline{9.8733 - 10} \\
&\quad 11.5482 - 10 \\
\log \sin 74°10' &\approx \underline{9.9832 - 10} \\
\log a &\approx 1.5650
\end{aligned}
$$

Hence, from Table 2, $a \approx 36.7$. Similarly,

$$c = \frac{b \sin \gamma}{\sin \beta} = \frac{(47.3) \sin 57°30'}{\sin 74°10'}$$

$$\approx \frac{(47.3)(.8434)}{(.9621)} \approx 41.5$$

To find c using logarithms we have

$$
\begin{aligned}
\log 47.3 &\approx 1.6749 \\
\log \sin 57°30' &\approx \underline{9.9260 - 10} \\
&\quad 11.6009 - 10 \\
\log \sin 74°10' &\approx \underline{9.9832 - 10} \\
\log c &\approx 1.6177
\end{aligned}
$$

Consulting Table 2 gives us $c \approx 41.5$.

Data such as that given in Example 1 always gives us a unique triangle ABC. However, if two sides and an angle opposite one of them are given, a unique triangle is not always determined. To illustrate, suppose that two numbers a and b are to be lengths of sides of a triangle ABC. In addition, suppose there is given an angle α which is to be opposite the side of length a. Let us consider the case in which α is acute. Place α in standard position on a rectangular coordinate system and consider the line segment AC of length b on the terminal side of α, as shown in Fig. 6.10.

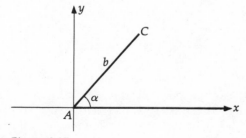

Figure 6.10

The third vertex B should be somewhere on the x-axis. Since the length a of the side opposite α is given, B may be found by striking off a circular arc of length a with center at C. There are four possible outcomes for this construction, as illustrated in Fig. 6.11, where the coordinate axes have been deleted.

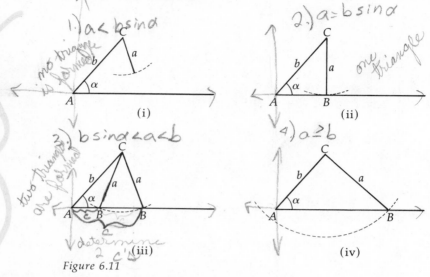

Figure 6.11

These four possibilities may be listed as follows:

(i)   The arc does not intersect the x-axis and no triangle is formed.

(ii) The arc is tangent to the x-axis and a right triangle is formed.

(iii) The arc intersects the positive x-axis in two distinct points and two triangles are formed.

(iv) The arc intersects both the positive and nonpositive parts of the x-axis and one triangle is formed.

Since the distance from C to the x-axis is $b \sin \alpha$ (Why?), we see that (i) occurs if $a < b \sin \alpha$, (ii) occurs if $a = b \sin \alpha$, (iii) occurs if $b \sin \alpha < a < b$, and (iv) occurs if $a \geq b$. It is unnecessary to memorize these facts, since in any specific problem the case that occurs will become evident when the solution is attempted. For example, in solving the equation

$$\frac{a}{\sin \alpha} = \frac{b}{\sin \beta}$$

suppose we obtain $|\sin \beta| > 1$. This will indicate that no triangle exists. If the equation $\sin \beta = 1$ is obtained, then $\beta = 90°$ and hence case (ii) occurs. On the other hand, if $|\sin \beta| < 1$, then there are two possible choices for the angle β. By checking both possibilities, it will become apparent whether (iii) or (iv) occurs.

If the measure of α is greater than 90°, then as in Fig. 6.12 a triangle exists if and only if $a > b$. Needless to say, our discussion is independent of the symbols we have used; that is, we might be given b, c, β, or a, c, γ, and so on.

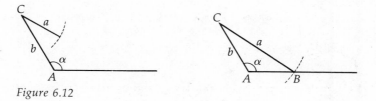

Figure 6.12

Since different possibilities may arise, the case in which two sides and an angle opposite one of them are given is sometimes called the **ambiguous case.**

EXAMPLE 2   Solve triangle ABC if $\alpha = 67°$, $c = 125$, and $a = 100$.

Solution

Using $\sin \gamma = \dfrac{c \sin \alpha}{a}$ we obtain

$$\sin \gamma = \frac{(125) \sin 67°}{100}$$
$$\approx \frac{(125)(.9205)}{100}$$
$$\approx 1.1506.$$

Since $\sin \gamma > 1$, there is no triangle with the given parts.

EXAMPLE 3   Approximate the remaining parts of triangle ABC if $a = 12.4$, $b = 8.7$, and $\beta = 36°40'$.

Solution

Using $\sin \alpha = \dfrac{a \sin \beta}{b}$ we have

$$\sin \alpha = \frac{(12.4) \sin 36°40'}{8.7}$$
$$\approx \frac{(12.4)(.5972)}{8.7} \approx .8512.$$

If a logarithmic solution is desired we write

$$\log \sin \alpha = \log 12.4 + \log \sin 36°40' - \log 8.7.$$

Using Tables 2 and 4,

$$\begin{aligned}
\log 12.4 &\approx 1.0934 \\
\log \sin 36°40' &\approx \underline{9.7761 - 10} \\
& 10.8695 - 10 \\
\log 8.7 &\approx \underline{.9395} \\
\log \sin \alpha &\approx 9.9300 - 10.
\end{aligned}$$

There are two possible angles α between $0°$ and $180°$ such that

$$\sin \alpha \approx .8512, \quad \text{or} \quad \log \sin \alpha \approx 9.9300 - 10.$$

If we let α' denote the reference angle for α, then from either Table 3 or Table 4 we obtain

$$\alpha' \approx 58°20'.$$

Consequently the two possibilities for α are

$$\alpha_1 \approx 58°20' \quad \text{and} \quad \alpha_2 \approx 121°40'.$$

Let γ_1 and γ_2 denote the third angle of the triangle corresponding to the angles α_1 and α_2, respectively. Then

$$\gamma_1 \approx 180° - (36°40' + 58°20') = 85°$$

and

$$\gamma_2 \approx 180° - (36°40' + 121°40') = 21°40'.$$

Thus there are two possible triangles which have the given parts. These are the triangles A_1BC and A_2BC shown in Fig. 6.13.

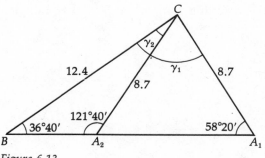

Figure 6.13

If c_1 is the side opposite γ_1 in triangle A_1BC, then

$$c_1 = \frac{a \sin \gamma_1}{\sin \alpha_1}$$

$$c_1 \approx \frac{(12.4) \sin 85°}{\sin 58°20'}$$

$$\approx \frac{(12.4)(.9962)}{.8511} \approx 14.5$$

If we wish to find c_1 using logarithms we have

$$\log c_1 \approx \log 12.4 + \log \sin 85° - \log \sin 58°20'.$$

Arranging our work in the usual way,

$$
\begin{array}{lll}
\log 12.4 & \approx & 1.0934 \\
\log \sin 85° & \approx & \underline{9.9983 - 10} \\
 & & 11.0917 - 10 \\
\log \sin 58°20' & \approx & \underline{9.9300 - 10} \\
\log c_1 & \approx & 1.1617.
\end{array}
$$

Consulting Table 2 gives us

$$c_1 \approx 14.5.$$

Hence the solution for triangle A_1BC is $\alpha_1 \approx 58°20'$, $\gamma_1 \approx 85°$, and $c_1 \approx 14.5$.

Similarly, if c_2 is the side opposite angle γ_2 in triangle A_2BC, we have

$$c_2 = \frac{a \sin \gamma_2}{\sin \alpha_2}$$

or

$$c_2 \approx \frac{(12.4) \sin 21°40'}{\sin 121°40'}$$

$$\approx \frac{(12.4)(.3692)}{.8511} \approx 5.4.$$

As before, if a logarithmic solution is desired we have

$$
\begin{array}{lll}
\log 12.4 & \approx & 1.0934 \\
\log \sin 21°40' & \approx & \underline{9.5673 - 10} \\
 & & 10.6607 - 10 \\
\log \sin 121°40' & \approx & \underline{9.9300 - 10} \\
\log c_2 & \approx & .7307.
\end{array}
$$

Using Table 2 and rounding off, we obtain $c_2 \approx 5.4$. Consequently the solution for triangle A_2BC is $\alpha_2 \approx 121°40'$, $\gamma_2 \approx 21°40'$, and $c_2 \approx 5.4$.

EXERCISES

In each of Exercises 1–12 approximate the remaining parts of triangle ABC.

1 $\alpha = 41°$, $\gamma = 77°$, $a = 10.5$

2 $\beta = 20°$, $\gamma = 31°$, $b = 210$

3 $\alpha = 27°40'$, $\beta = 52°10'$, $a = 32.4$

4 $\alpha = 42°10'$, $\gamma = 61°20'$, $b = 19.7$

5 $\beta = 50°50'$, $\gamma = 70°30'$, $c = 537$ **6** $\alpha = 7°10'$, $\beta = 11°40'$, $a = 2.19$

7 $\alpha = 65°10'$, $a = 21.3$, $b = 18.9$ **8** $\beta = 30°$, $b = 17.9$, $a = 35.8$

9 $\gamma = 53°20'$, $a = 140$, $c = 115$ **10** $\alpha = 27°30'$, $c = 52.8$, $a = 28.1$

11 $\beta = 113°10'$, $b = 248$, $c = 195$ **12** $\gamma = 81°$, $c = 11$, $b = 12$

13 It is desired to find the distance between two points A and B lying on opposite banks of a river. A line segment AC of length 240 yards is laid off and the measure of angles BAC and ACB are found to be $63°20'$ and $54°10'$, respectively. Approximate the distance from A to B.

14 In order to determine the distance between two points A and B, a surveyor chooses a point C which is 375 yards from A and 530 yards from B. If angle BAC has measure $49°30'$, approximate the required distance.

15 When the angle of elevation of the sun is $64°$, a telegraph pole that is tilted at an angle of $12°$ directly away from the sun casts a shadow 34 feet long on level ground. Approximate the length of the pole.

16 A straight road makes an angle of $15°$ with the horizontal. When the angle of elevation of the sun is $57°$, a vertical pole at the side of the road casts a shadow 75 feet long directly down the road. Approximate the length of the pole.

17 The angles of elevation of a balloon from two points A and B on level ground are $24°10'$ and $47°40'$, respectively. If A and B are 8.4 miles apart and the balloon is between A and B in the same vertical plane, approximate the height of the balloon above the ground.

18 An airport A is 480 miles due east of airport B. A pilot flew in the direction $235°$ from A to C and then in the direction $320°$ from C to B. Approximate the total distance he flew.

19 A forest ranger at an observation point A sights a fire in the direction N27°10′E. Another ranger at an observation point B, 6 miles due east of A, sights the same fire at N52°40′W. Approximately how far is the fire from each of the observation points?

20 A surveyor notes that the direction from point A to point B is S63°W and the direction from A to C is S38°W. If the distance from A to B is 239 yards and the distance from B to C is 374 yards, approximate the distance and direction from A to C.

9 THE LAW OF COSINES

If two sides and the included angle or the three sides of a triangle are given, then the Law of Sines cannot be applied directly. We may, however, use the following result.

(6.31) The Law of Cosines

If *ABC* is any triangle labeled in the usual manner, then

$$a^2 = b^2 + c^2 - 2bc \cos \alpha,$$
$$b^2 = a^2 + c^2 - 2ac \cos \beta,$$
$$c^2 = a^2 + b^2 - 2ab \cos \gamma.$$

Instead of memorizing each of the formulas given in (6.31), it is more convenient to remember the following statement, which takes all of them into account: *the square of the length of any side of a triangle equals the sum of the squares of the lengths of the other two sides minus twice the product of the lengths of the other two sides and the cosine of the angle between them.*

We shall use the distance formula to establish (6.31). We again place α in standard position on a rectangular coordinate system, as illustrated in Fig. 6.8. Although α is pictured as an obtuse angle, our development is also true if α is acute. Since the segment *AB* has length *c*, the coordinates of *B* are $(c, 0)$. By (5.26) the coordinates of *C* are $(b \cos \alpha,$ $b \sin \alpha)$. We also have $a = d(B, C)$ and hence $a^2 = [d(B, C)]^2$. Using the distance formula leads to the following equations:

$$\begin{aligned} a^2 &= (b \cos \alpha - c)^2 + (b \sin \alpha - 0)^2 \\ &= b^2 \cos^2 \alpha - 2bc \cos \alpha + c^2 + b^2 \sin^2 \alpha \\ &= b^2 (\cos^2 \alpha + \sin^2 \alpha) + c^2 - 2bc \cos \alpha \\ &= b^2 + c^2 - 2bc \cos \alpha. \end{aligned}$$

This gives us the first formula of (6.31). The second and third formulas may be obtained by placing β and γ, respectively, in standard position on a rectangular coordinate system and using a similar procedure.

EXAMPLE 1 Approximate the remaining parts of triangle *ABC* if $a = 5.0$, $c = 8.0$, and $\beta = 77°10'$.

Solution

Using the Law of Cosines, we have

$$\begin{aligned} b^2 &= (5.0)^2 + (8.0)^2 - 2(5.0)(8.0) \cos 77°10' \\ &\approx 25 + 64 - (80)(.2221) \\ &\approx 71.2. \end{aligned}$$

305

Consequently $b \approx 8.44$. We now use the Law of Sines to find γ. Thus

$$\sin \gamma \approx \frac{(8.0) \sin 77°10'}{8.44} \approx \frac{(8.0)(.9750)}{8.44} \approx .9242$$

Consulting Table 3 we see that $\gamma \approx 67°30'$. Hence

$$\alpha \approx 180° - (77°10' + 67°30') = 35°20'.$$

EXAMPLE 2 Given sides $a = 90$, $b = 70$, and $c = 40$ of triangle ABC, approximate angles α, β, and γ.

Solution

According to the first equation of (6.31), we have

$$\cos \alpha = \frac{b^2 + c^2 - a^2}{2bc}$$
$$= \frac{4900 + 1600 - 8100}{5600}$$
$$\approx -.2857.$$

From Table 3 we see that the reference angle for α is approximately $73°20'$. Hence $\alpha \approx 106°40'$.

Similarly, from the second equation of (6.31),

$$\cos \beta = \frac{a^2 + c^2 - b^2}{2ac}$$
$$= \frac{8100 + 1600 - 4900}{7200}$$
$$\approx .6667,$$

and hence $\beta \approx 48°10'$. Finally $\gamma \approx 180° - (106°40' + 48°10') = 25°10'$.

EXERCISES

In Exercises 1–10 approximate the remaining parts of triangle ABC.

1 $\alpha = 60°$, $b = 20$, $c = 30$ **2** $\gamma = 45°$, $b = 10$, $a = 15$

3 $\beta = 150°$, $a = 150$, $c = 30$ **4** $\beta = 73°50'$, $c = 14$, $a = 87$

5 $\gamma = 115°10'$, $a = 1.1$, $b = 2.1$ **6** $\alpha = 23°40'$, $c = 4.3$, $b = 70$

7 $a = 2$, $b = 3$, $c = 4$ **8** $a = 10$, $b = 15$, $c = 12$

9 $a = 25$, $b = 80$, $c = 60$ **10** $a = 20$, $b = 20$, $c = 10$

11 A parallelogram has sides of length 30 inches and 70 inches. If one of the angles has measure 65°, approximate the length of each diagonal.

12 The angle at one corner of a triangular plot of ground has measure 73°40′. If the sides that meet at this corner are 175 feet and 150 feet long, approximate the length of the third side.

13 A vertical pole 40 feet tall stands on a hillside that makes an angle of 17° with the horizontal. What is the minimal length of rope that will reach from the top of the pole to a point directly down the hill 72 feet from the base of the pole?

14 To find the distance between two points A and B, a surveyor chooses a point C which is 420 yards from A and 540 yards from B. If angle ACB has measure 63°10′, approximate the distance.

15 Two automobiles leave from the same point and travel along straight highways which differ in direction by 84°. If their speeds are 60 miles per hour and 45 miles per hour, respectively, approximately how far apart will they be at the end of 20 minutes?

16 A triangular plot of land has sides of length 420 feet, 350 feet, and 180 feet. Find the smallest angle between the sides.

17 A ship leaves point P at 1:00 P.M. and travels S35°E at the rate of 24 miles per hour. Another ship leaves P at 1:30 P.M. and travels S20°W at 18 miles per hour. Approximately how far apart are the ships at 3:00 P.M.?

18 An airplane flies 165 miles from point A in the direction 130° and then travels in the direction 245° for 80 miles. Approximately how far is the airplane from A?

19 Show that for every triangle ABC,

a $a^2 + b^2 + c^2 = 2(bc \cos \alpha + ac \cos \beta + ab \cos \gamma)$,

b $\dfrac{\cos \alpha}{a} + \dfrac{\cos \beta}{b} + \dfrac{\cos \gamma}{c} = \dfrac{a^2 + b^2 + c^2}{2abc}$.

20 Show that if ABC is a right triangle, then one part of (6.31) reduces to the Pythagorean Theorem.

10 REVIEW EXERCISES

Oral

Define or discuss each of the following:

1 Trigonometric equation

2 The solution set of a trigonometric equation

3 Trigonometric identity

4 The addition formulas

Written

Verify the identities in Exercises 1–10.

1 $(\cot^2 x + 1)(1 - \cos^2 x) = 1$

2 $\cos \theta + \sin \theta \tan \theta = \sec \theta$

3 $\dfrac{\tan^3 \phi - \cot^3 \phi}{\tan^2 \phi + \csc^2 \phi} = \tan \phi - \cot \phi$

4 $\dfrac{\sin u + \sin v}{\csc u + \csc v} = \dfrac{1 - \sin u \sin v}{-1 + \csc u \csc v}$

5 $\cos (x - 5\pi/2) = \sin x$

6 $\tan (x + 3\pi/4) = \dfrac{\tan x - 1}{\tan x + 1}$

7 $\frac{1}{4} \sin 4\beta = \sin \beta \cos^3 \beta - \cos \beta \sin^3 \beta$

8 $\tan \frac{1}{2}\theta = \csc \theta - \cot \theta$

9 $\sin 8\theta = 8 \sin \theta \cos \theta (1 - 2 \sin^2 \theta)(1 - 8 \sin^2 \theta \cos^2 \theta)$

10 $\arcsin \dfrac{2u}{1 + u^2} = 2 \arctan u,$ where $|u| \le 1$

In each of Exercises 11–22 find the solutions of the given equation which are in the interval $[0, 2\pi)$, and also find the degree measure of each solution.

11 $2 \sin^3 \theta - \sin \theta = 0$ **12** $2 \sin \alpha + \cot \alpha = \csc \alpha$

13 $\cos \theta = \cot \theta$ **14** $\sec^5 \theta - 4 \sec \theta = 0$

15 $2 \sin^3 t + \sin^2 t - 2 \sin t - 1 = 0$

16 $\sin x \tan^2 x = \sin x$

17 $\cos t + 2 \sin^2 t = 1$

18 $\cos 2x + 3 \cos x + 2 = 0$

19 $2 \sec u \sin u + 2 = 4 \sin u + \sec u$

     20 $\sin \theta - \sin 2\theta = 0$

21 $3 \cos x - 2 \cos^2 \frac{1}{2}x = 0$

22 $\csc 2\theta \sec 2\theta = 2 \csc 2\theta$

Find the exact value of Exercises 23–26 without the use of tables.

23 $\sin 75°$ **24** $\tan 285°$ **25** $\cos 195°$ **26** $\sin 3\pi/8$

If θ and ϕ are acute angles such that $\csc \theta = \frac{5}{3}$ and $\cos \phi = \frac{8}{17}$, find the numbers in Exercises 27–33.

27 $\sin (\theta - \phi)$ **28** $\cos (\theta - \phi)$ **29** $\tan (\theta - \phi)$

30 $\sin 2\phi$ **31** $\cos 2\phi$ **32** $\sin \dfrac{\theta}{2}$

33 $\tan \dfrac{\theta}{2}$

34 Express $\cos (\alpha + \beta + \gamma)$ in terms of functions of α, β, and γ.

35 Express each of the following products as a sum or difference:
   a $\sin 5t \sin 2t$;   **b** $\cos (u/3) \cos (-u/4)$;   **c** $4 \cos 3x \sin 2x$.

36 Express each of the following as a product:
   a $\sin 6u + \sin 4u$;   **b** $\cos 5\theta - \cos 6\theta$;   **c** $\sin (t/2) - \sin (t/3)$.

Find each of the numbers in Exercises 37–45 without the use of tables:

37 $\cos^{-1} (-\sqrt{3}/2)$ **38** $\sin^{-1} (\sqrt{2}/2)$

39 $\arccos (-1)$ **40** $\arctan (-\sqrt{3}/3)$

41 $\sin \arccos (-\sqrt{3}/2)$ **42** $\cos [\sin^{-1} \frac{15}{17} - \sin^{-1} \frac{8}{17}]$

43 $\cos [2 \sin^{-1} \frac{4}{5}]$ **44** $\sin [\sin^{-1} \frac{2}{3}]$

45 $\cos^{-1} (\sin 0)$

46 Sketch the graphs of the following equations.
   a $y = \cos^{-1} 3x$ **b** $y = 4 \sin^{-1} x$ **c** $y = 1 - \sin^{-1} x$

Without using tables, find the remaining parts of triangle ABC in Exercises 47–50.

47 $\alpha = 60°, \beta = 45°, b = 100$ **48** $\gamma = 30°, a = 2\sqrt{3}, c = 2$

49 $\alpha = 60°, b = 6, c = 7$ **50** $a = 2, b = 3, c = 4$

7

Systems of Equations and Inequalities

In certain types of mathematical problems it is necessary to work simultaneously with more than one equation in several variables. We then refer to the given equations as a **system of equations.** *It is usually desirable to find the solutions which are common to all equations in the system. In this chapter we shall investigate special types of systems of equations and develop methods for finding their common solutions. Of particular importance are the matrix techniques introduced for systems of linear equations. We shall also touch briefly on systems of inequalities and on linear programming.*

1 EQUIVALENT SYSTEMS OF EQUATIONS

We have often used notations such as $f(x)$ and $h(x)$ to represent expressions in a variable x. For example, we could have

$$f(x) = 3x^2 - 4x + 1$$

and

$$h(x) = 2^x + 3 \log x.$$

Although those equations define two functions, f and h, we shall now regard $f(x)$ and $h(x)$ primarily as symbols which represent the indicated expressions. If a is a real number, then as in our previous work $f(a)$ will denote the numerical value of $f(x)$ obtained by substituting a for x. For example, using the expressions above,

$$f(2) = 3(2)^2 - 4(2) + 1 = 5$$

and

$$h(10) = 2^{10} + 3 \log 10$$
$$= 1024 + 3(1) = 1027.$$

In similar fashion, it is convenient to use symbols such as $g(x, y)$ and $k(x, y)$ to denote expressions in *two* variables x and y.

For example, we might consider

$$g(x, y) = 3x^2 - 4y + 5$$

or

$$k(x, y) = y^2 \log x.$$

If a and b are real numbers, then $g(a, b)$ and $k(a, b)$ will denote the numerical values obtained by substituting a for x and b for y. To illustrate, using the above expressions we have

$$g(2, -1) = 3(2)^2 - 4(-1) + 5 = 21,$$
$$k(3, 100) = 3^2 \log 100 = 9(2) = 18.$$

If we are given an equation in two variables x and y, say

$$x^2 - 4y - 2 = 3x + 2y + 6,$$

then it is always possible to rewrite it so that 0 is on the right-hand side. For example, using the previous equation we may write

$$x^2 - 3x - 6y - 8 = 0.$$

To reiterate, every equation in two variables x and y can be written in the form

(7.1) $g(x, y) = 0,$

where $g(x, y)$ is some expression in x and y. As in Chapter 2, a **solution** of (7.1) is an ordered pair (a, b) of real numbers such that a true statement is obtained when a is substituted for x and b for y; that is,

$$g(a, b) = 0.$$

The **solution set** of the equation is the collection of all solutions. Two equations in x and y are **equivalent** if they have the same solution set.

As in our early work, the **graph** of (7.1) is the graph of its solution set. For example, we know from (3.23) that the graph of every linear equation

$$Ax + By + C = 0$$

where A, B, and C are real numbers, is a straight line.

By a **system** of two equations in two variables x and y we mean a pair of equations

$$g(x, y) = 0, \qquad k(x, y) = 0,$$

where $g(x, y)$ and $k(x, y)$ are expressions in x and y. If S and T are the solution sets of these equations, then by definition **the solution set of the system** is $S \cap T$. Thus (a, b) is a solution of the system if and only if $g(a, b) = 0$ and also $k(a, b) = 0$. Moreover, if (a, b) is in both S and T, then the corresponding point $P(a, b)$ in a rectangular coordinate system is on the graph of each equation. It follows that the points which correspond to the pairs in the solution set are precisely the points at which the graphs of the two equations intersect.

As a concrete example, consider the system

$$x^2 - y = 0, \qquad 2x - y + 3 = 0.$$

The following table exhibits some solutions of the equation $x^2 - y = 0$

x	-2	-1	0	1	2	3	4
y	4	1	0	1	4	9	16

A similar table for $2x - y + 3 = 0$ is

x	-2	-1	0	1	2	3	4
y	-1	1	3	5	7	9	11

Note that the pairs $(3, 9)$ and $(-1, 1)$ are in both solution sets and hence are in the solution set of the system. We shall prove later that these are the *only* pairs in the solution set. The graphs of the two equations are sketched in Fig. 7.1 on the same axes. As pointed out before, the two pairs in the solution set of the system represent points of intersection of the two graphs.

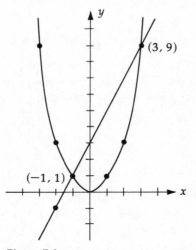

Figure 7.1

For brevity the phrase "solve a system of equations" is sometimes used instead of "find the solution set of a system of equations." We shall now introduce a technique which is often useful for solving a system of two equations in two variables.

Let us consider any system in two variables of the form

(7.2) $g(x, y) = 0,$ $k(x, y) = 0.$

Although we have used the variables x and y, other symbols could be used. Suppose that we can solve the equation $k(x, y) = 0$ for y in terms of x, say $y = f(x)$, where f is a function. Since this means that the equations $k(x, y) = 0$ and $y = f(x)$ are equivalent, it follows that the system

(7.3) $g(x, y) = 0,$ $y = f(x)$

is equivalent to (7.2). If we substitute $f(x)$ for y in the first equation we obtain the following system:

(7.4) $g(x, f(x)) = 0,$ $y = f(x).$

We shall now prove that (7.4) is equivalent to (7.2).

Suppose the ordered pair (a, b) is a solution of (7.4). Thus, if a is substituted for x and b for y, equality results, that is,

$$g(a, f(a)) = 0, \qquad b = f(a).$$

Since $b = f(a)$, we have

$$g(a, b) = 0, \qquad b = f(a).$$

This implies that (a, b) is a solution of (7.3) and hence of (7.2). We have shown, therefore, that every solution of (7.4) is also a solution of (7.2).

Conversely, if (a, b) is a solution of (7.2), then

$$g(a, b) = 0, \qquad k(a, b) = 0.$$

Since $k(x, y) = 0$ is equivalent to $y = f(x)$, it follows that

$$g(a, b) = 0, \qquad b = f(a).$$

Consequently

$$g(a, f(a)) = g(a, b) = 0$$

which implies that (a, b) is a solution of (7.4). This completes the proof.

A pair (a, b) is in the solution set of (7.4) if and only if a is in the solution set of $g(x, f(x)) = 0$ and $b = f(a)$. Accordingly, to find the solution set of system (7.2), we can begin by finding the solutions of $g(x, f(x)) = 0$. These are the x values for the pairs in the solution set. The corresponding y values may then be found from $y = f(x)$. This technique for solving system (7.2) will be called the **method of substitution.** Of course, sometimes it may be more convenient to solve for x in terms of y. In general, the steps used to solve a system of two equations in two variables by the method of substitution are as follows:

(1) Solve one of the equations for one variable in terms of the other variable.

(2) Substitute the expression obtained in step (1) in the second equation, obtaining an equation in one variable.

(3) Find the solution set of the equation obtained in step (2).

(4) Use the solutions from step (3) together with the expression obtained in step (1) to find the solutions of the system.

EXAMPLE 1 Find the solution set of the system

$$x^2 - y = 0, \qquad 2x - y + 3 = 0.$$

Solution 1

This is the same system considered earlier. Solving the first equation for y gives us $y = x^2$. Substituting for y in the second equation we obtain $2x - x^2 + 3 = 0$, or equivalently, $x^2 - 2x - 3 = 0$. Since this factors as $(x - 3)(x + 1) = 0$, it has solutions $x = 3$ and $x = -1$. These are the x values for the pairs in the solution set of the system. To get the corresponding y values, we substitute 3 and -1 for x in the equation $y = x^2$, obtaining $y = 9$ and $y = 1$ respectively. Hence the solution set of the system is $\{(3, 9), (-1, 1)\}$.

Solution 2

If we solve the second equation of the system for x in terms of y, we obtain $x = (y - 3)/2$. Substituting for x in the first equation gives us

$$\left(\frac{y-3}{2}\right)^2 - y = 0,$$

which simplifies to $y^2 - 10y + 9 = 0$. The solutions of the latter equation are $y = 9$ and $y = 1$. We now substitute in $x = (y - 3)/2$ to get the corresponding x values, obtaining $x = 3$ and $x = -1$ respectively. This gives us the same solution set as before.

EXAMPLE 2 Find the solution set of the system $x^2 + y^2 = 25$, $x^2 + y = 19$. Sketch the graph of each equation, showing the points of intersection.

Solving the second equation for y we obtain $y = 19 - x^2$. Substituting for y in the first equation leads to the following chain of equivalent equations:

$$x^2 + (19 - x^2)^2 = 25$$
$$x^4 - 37x^2 + 336 = 0$$
$$(x^2 - 16)(x^2 - 21) = 0.$$

The solutions of the last equation are 4, -4, $\sqrt{21}$ and $-\sqrt{21}$. The corresponding y values are found by substituting for x in the equation $y = 19 - x^2$. Substitution of 4 or -4 for x gives us $y = 3$, whereas substitution of $\sqrt{21}$ or $-\sqrt{21}$ gives us $y = -2$. Hence the solution set of the system is

$$\{(4, 3), (-4, 3), (\sqrt{21}, -2), (-\sqrt{21}, -2)\}.$$

The graph of $x^2 + y^2 = 25$ is a circle of radius 5 with center at the origin, and the graph of $y = 19 - x^2$ is a parabola with a vertical axis. The graphs and their points of intersection are illustrated in Fig. 7.2.

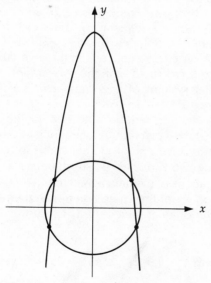

Figure 7.2

EXAMPLE 3 Find the solutions of the system $y - \log(x + 3) = 1$, $2 - y = \log x$.

Solution

Since $\log x$ appears in one of the equations, the domain of x is the set of positive real numbers. From the second equation we have $y = 2 - \log x$. Substituting in the first equation and simplifying, we obtain

$\log x + \log (x + 3) = 1.$

This leads to the following chain of equivalent equations (supply the reasons):

$$\log x(x + 3) = 1$$
$$x(x + 3) = 10$$
$$x^2 + 3x - 10 = 0$$
$$(x + 5)(x - 2) = 0.$$

Since -5 is not in the domain of x, the only solution of the last equation is 2. The corresponding value for y is $2 - \log 2$. Hence the system has one solution, $(2, 2 - \log 2)$.

We can also consider expressions in three variables x, y, and z. It is again convenient to use a notation such as $g(x, y, z)$ to denote an expression of this type. For example, we might consider

$$g(x, y, z) = x^2 y - 4z^2 + x \log z + 3^y.$$

We say that the corresponding equation $g(x, y, z) = 0$ has a **solution** (a, b, c) if $g(a, b, c) = 0$, that is, if a substitution of a, b, and c for x, y, and z, respectively, produces a true statement. We refer to (a, b, c) as an **ordered triple** of real numbers. Solution sets and equivalent equations are defined as before. A system of equations in three variables and the corresponding solution set are defined as in the two-variable case. In like manner, we can consider systems of *any* number of equations in *any* number of variables.

The method of substitution can be extended to these more complicated systems. For example, given three equations in three variables, suppose it is possible to solve one of the equations for one variable in terms of the remaining two variables. By substituting this expression in each of the other equations, we obtain a system of two equations in two variables. The solution set of the latter system can then be used to find the solution set of the original system. This is illustrated in the following example.

EXAMPLE 4 Solve the system $x - y + z = 2$, $xyz = 0$, $2y + z = 1$.

Solution

Solving the third equation for z gives us

$$z = 1 - 2y.$$

Substituting in the first two equations of the system we obtain

$$\begin{cases} x - y + (1 - 2y) = 2 \\ \quad xy(1 - 2y) = 0, \end{cases}$$

or equivalently,

(7.5) $\begin{cases} x - 3y - 1 = 0 \\ xy(1 - 2y) = 0. \end{cases}$

We now solve system (7.5). Solving the first equation for x in terms of y gives us

(7.6) $x = 3y + 1.$

Substituting $3y + 1$ for x in the second equation of (7.5) we obtain

$$(3y + 1)y(1 - 2y) = 0,$$

which has for its solutions the numbers $-\frac{1}{3}$, 0, $\frac{1}{2}$. These are the y values for the solution set of the system. To obtain the corresponding x values, we use (7.6), obtaining $x = 0$, 1, and $\frac{5}{2}$, respectively. Finally returning to $z = 1 - 2y$, we obtain the z values $\frac{5}{3}$, 1, and 0. Hence the solution set of the original system consists of the ordered triples $(0, -\frac{1}{3}, \frac{5}{3})$, $(1, 0, 1)$, and $(\frac{5}{2}, \frac{1}{2}, 0)$.

EXERCISES

In Exercises 1–28 use the method of substitution to find the solution set of each system of equations.

1 $\begin{cases} y = x^2 - 1 \\ y = x + 1 \end{cases}$ $\quad x = y - 1$

2 $\begin{cases} y = x^2 - 4 \\ 2x + y = -1 \end{cases}$

3 $\begin{cases} x = y^2 \\ y = 4x \end{cases}$

4 $\begin{cases} x = y^2 + 1 \\ x + 2y = 4 \end{cases}$

5 $\begin{cases} 2x - y = 7 \\ x - 3y = 11 \end{cases}$

6 $\begin{cases} 3x - 2y = 5 \\ 4x + y = -8 \end{cases}$

7 $\begin{cases} x - 3y = 2 \\ 2x - 6y = 5 \end{cases}$

8 $\begin{cases} 2x - 5y = 1 \\ -4x + 10y = -2 \end{cases}$

9 $\begin{cases} x^2 + y^2 = 25 \\ 3x + y = 5 \end{cases}$

10 $\begin{cases} x^2 + y^2 = 25 \\ 4x - 3y = 25 \end{cases}$

11 $\begin{cases} x^2 + y^2 = 25 \\ x - y = 7 \end{cases}$

12 $\begin{cases} x^2 + y^2 = 25 \\ x + y^2 = 19 \end{cases}$

13 $\begin{cases} x^2 - y = 0 \\ y^2 + x = 0 \end{cases}$

14 $\begin{cases} y = x^3 \\ y = 4x \end{cases}$

15 $\begin{cases} x^2 + y^2 + 4x - 2y - 5 = 0 \\ x + y = 1 \end{cases}$

16 $\begin{cases} xy = 2 \\ x - 3y = 5 \end{cases}$

317

Divide

17 $\begin{cases} y = 6/x^2 \\ y = 7 - x^2 \end{cases}$

18 $\begin{cases} y = x^2 - 4x + 5 \\ y - x = 1 \end{cases}$

19 $\begin{cases} 9x^2 + 16y^2 = 140 \\ x^2 - 4y^2 = 4 \end{cases}$

20 $\begin{cases} 25x^2 - 16y^2 = 400 \\ 9x^2 - 4y^2 = 36 \end{cases}$

21 $\begin{cases} x^2 + y^2 = 4 \\ 2x^2 - 3y^2 = 1 \end{cases}$

22 $\begin{cases} 2x^3 + 4y^3 = 1 \\ 3x^3 - 5y^3 = 8 \end{cases}$

23 $\begin{cases} x + 2y + z = 3 \\ 2x - y - 3z = -4 \\ 2x + 3y - z = 4 \end{cases}$

24 $\begin{cases} 2y = x^2 + 3z \\ y = x^2 + z - 1 \\ y^2 = yz \end{cases}$

25 $\begin{cases} \log (3r + 1) = s + 5 \\ \log (r - 2) = s + 4 \end{cases}$

26 $\begin{cases} \log u = v - 3 \\ 5 = v + \log (u - 21) \end{cases}$

27 $\begin{cases} y = 3^{2x} \\ y = 3^x + 2 \end{cases}$

28 $\begin{cases} a + 3 \cdot 2^b = 2^{2b} \\ a = 2^{b+1} + 6 \end{cases}$

Solve Exercises 29–34 by introducing several variables and using a suitable system of equations.

29 Find two positive integers whose difference is 4 and whose squares differ by 88.

30 Find two real numbers whose difference and product both equal 4.

31 The sum of the digits in a two-digit number is 12. If the digits are reversed the number is increased by 18. Find the number.

32 A chemist has two acid solutions, the first containing 10% acid and the second 30% acid. How many ounces of each should be mixed in order to obtain 10 ounces of a solution containing 25% acid?

33 The perimeter of a rectangle is 40 inches and its area is 96 square inches. Find the length and width.

34 Generalize Exercise 33 to the case where the perimeter is P and area is A. Express the length and width in terms of P and A. What restrictions are necessary?

35 Find three numbers whose sum and product are 20 and 60 respectively, and such that one of the numbers equals the sum of the other two.

36 Find the values of b for which the solution set of the system

$$\begin{cases} x^2 + y^2 = 4 \\ y = x + b \end{cases}$$

consists of (a) one real number, (b) two real numbers, (c) no real numbers. Interpret the three cases geometrically.

An equation 'of the form $ax + by + c = 0$ (or equivalently $ax + by = -c$) is called a linear equation in x and y. Similarly, a linear equation in three variables x, y, and z is an equation of the form $ax + by + cz = d$, where the coefficients are real numbers. Linear equations in any number of variables may be defined in like manner. If a large number of variables occurs, a subscript notation is usually employed. If we let $x_1, x_2, \cdots, x_n$ denote variables, where n is any positive integer, then an expression of the form

(7.7) $\qquad a_1x_1 + a_2x_2 + \cdots + a_nx_n = a$

where $a_1, a_2, \cdots, a_n$ and a are real numbers, is called a **linear equation in n variables** with real coefficients.

In modern applications of mathematics, perhaps the most common systems of equations are those in which all the equations are linear. In this section we shall only consider systems of two linear equations in two variables. Systems involving more than two variables are discussed in Section 3.

The method used to solve a system of equations in several variables is similar to that used for one equation in one variable, in the sense that the given system is replaced by a chain of equivalent systems until a stage is reached from which the solution set is easily obtained. Some general rules which allow us to transform a system of equations into an equivalent system are given below in (7.8). Since we prefer not to specify the number or types of variables in (iii) of (7.8), we shall use the notation $p = 0$ and $q = 0$ for typical equations in a system. The reader should bear in mind that the symbols p and q represent expressions in the variables under consideration. For example, if $p = x - 2y^2 + z^3$ and $q = 4x^2 + y - 5z$, the equations $p = 0$ and $q = 0$ have the form

$$x - 2y^2 + z^3 = 0, \qquad 4x^2 + y - 5z = 0.$$

(7.8) Transformations that Lead to Equivalent Systems

The following transformations do not change the solution set of a system of equations.

- (i) Interchanging the position of any two equations.
- (ii) Multiplying both sides of an equation in the system by a nonzero real number.
- (iii) Replacing an equation $q = 0$ of the system by $kp + q = 0$, where $p = 0$ is any other equation in the system and k is any real number.

Proof

It is easy to show that (i) and (ii) do not change the solution set of the system and therefore we shall omit the proofs. To prove (iii), we first note that a solution of the original system is a solution of both equations $p = 0$ and $q = 0$. Accordingly, each of the expressions p and q equals zero if the variables are replaced by appropriate real numbers, and hence the expression $kp + q$ will also equal zero. Since none of the other equations has been changed, this shows that any solution of the original system is also a solution of the transformed system obtained by replacing the equation $q = 0$ by $kp + q = 0$.

Conversely, given a solution of the transformed system, both of the expressions $kp + q$ and p equal zero if the variables are replaced by appropriate numbers. This implies, however, that $(kp + q) - kp$ equals 0. Since $(kp + q) - kp = q$, we see that q must also equal zero when the substitution is made. Thus a solution of the transformed system is also a solution of the original system. This completes the proof.

A similar situation exists if the equations do not have zero on the right-hand side. For example, the system

$$(7.9) \quad \begin{cases} p_1 = p_2 \\ q_1 = q_2 \end{cases}$$

where p_1, p_2, q_1, and q_2 represent certain expressions containing variables, is equivalent to the system

$$\begin{cases} p_1 - p_2 = 0 \\ q_1 - q_2 = 0. \end{cases}$$

By (7.8) the latter system is equivalent to

$$\begin{cases} p_1 - p_2 = 0 \\ k(p_1 - p_2) + (q_1 - q_2) = 0 \end{cases}$$

which, in turn, is equivalent to

$$(7.10) \quad \begin{cases} p_1 = p_2 \\ kp_1 + q_1 = kp_2 + q_2. \end{cases}$$

For convenience, we shall describe the process of going from the system (7.9) to the equivalent system (7.10) by the phrase "add to the second equation k times the first equation." Of course, to *add* two equations means to add corresponding sides. To *multiply* an equation by k means to multiply both sides of the equation by k.

EXAMPLE 1 Find the solution set of the system

$$\begin{cases} x + 3y = -1 \\ 2x - y = 5. \end{cases}$$

Solution

By (ii) of (7.8) we may multiply the second equation by 3. This gives us the equivalent system

$$\begin{cases} x + 3y = -1 \\ 6x - 3y = 15. \end{cases}$$

Next, by (iii) of (7.8) we may add to the second equation 1 times the first. This gives us

$$x + 3y = -1$$
$$7x \quad = 14.$$

We see from the last equation that the only possible value for x is 2. The corresponding value for y may be found by substituting for x in the first equation. This gives us the following:

$$2 + 3y = -1, \quad 3y = -3, \quad y = -1.$$

Consequently the solution set of the given equation is $\{(2, -1)\}$.

There are other methods of solution. For example, we could begin by multiplying the first equation by -2, obtaining

$$\begin{cases} -2x - 6y = 2 \\ 2x - y = 5. \end{cases}$$

If we next add the first equation to the second, we get

$$\begin{cases} -2x - 6y = 2 \\ -7y = 7. \end{cases}$$

The last equation implies that $y = -1$. Substitution for y in the first equation gives us

$$-2x - 6(-1) = 2, \quad -2x = -4, \quad x = 2,$$

and again we see that the solution set is $\{(2, -1)\}$.

The system can also be solved by using the method of substitution. The graphs of the two equations, showing the point of intersection $(2, -1)$, are sketched in Fig. 7.3.

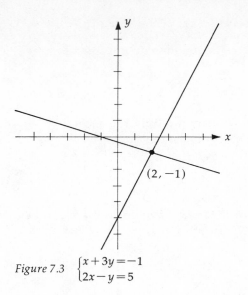

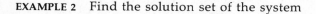

$Figure\ 7.3$ $\begin{cases} x+3y=-1 \\ 2x-y=5 \end{cases}$

EXAMPLE 2 Find the solution set of the system

$$\begin{cases} 3x + y = 6 \\ 6x + 2y = 12. \end{cases}$$

Solution

Multiplying the second equation by $\frac{1}{2}$ gives us

$$\begin{cases} 3x + y = 6 \\ 3x + y = 6. \end{cases}$$

Consequently (a, b) is a solution if and only if $3a + b = 6$, that is, $b = 6 - 3a$. It follows that the solution set may be written $\{(a, 6 - 3a) : a \in \mathbf{R}\}$. If we wish to find particular solutions we may substitute various values for a. For example, a few solutions are $(0, 6)$, $(1, 3)$, $(3, -3)$, $(-2, 12)$, $(\sqrt{2}, 6 - 3\sqrt{2})$.

EXAMPLE 3 Find the solution set of the system

$$\begin{cases} 3x + y = 6 \\ 6x + 2y = 20. \end{cases}$$

Solution

If we add to the second equation -2 times the first equation we obtain the equivalent system

$$\begin{cases} 3x + y = 6 \\ 0 = 8. \end{cases}$$

Since the last equation is never true, the solution set of the system is the empty set ∅. Of course, this means that the graphs of the given equations do not intersect (see Fig. 7.4).

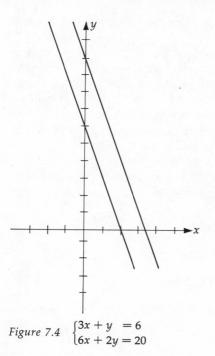

Figure 7.4 $\begin{cases} 3x + y = 6 \\ 6x + 2y = 20 \end{cases}$

Every system of two linear equations in two variables x and y has the form

(7.11) $\begin{cases} a_1x + a_2y = a \\ b_1x + b_2y = b \end{cases}$

where the coefficients are real numbers. We know from (3.23) that the graph of each equation in the system is a straight line. It follows that precisely *one* of the following three possibilities must occur:

(a) The lines are identical.
(b) The lines are parallel.
(c) The lines intersect in one and only one point.

Since the pairs in the solution set of the system (7.11) are the coordinates of the points of intersection of the graphs of the two equations, we may interpret statements (a)–(c) in the following way: if S and T denote the solution sets of the equations in (7.11), then one and only one of the following statements is true:

(a) $S = T$.
(b) $S \cap T = \varnothing$.
(c) $S \cap T$ consists of exactly one ordered pair.

If (a) occurs, then every solution of one equation in (7.11) is also a solution of the other, and we say that the equations are **dependent**.

If (b) occurs, there is no solution and the system is said to be **inconsistent**.

If (c) occurs, there is a unique solution and the system is called **consistent**.

In practice, there should be little difficulty in determining which of the three cases occurs. In case (a) we have a solution similar to that for Example 2, where one of the equations can be transformed into the other by using (7.8). In case (b) the lack of a solution is indicated by an absurdity such as the statement $0 = 8$ which appeared in Example 3. The case of the unique solution (c) will be apparent when (7.8) is applied to the system, as illustrated in Example 1.

In Section 7 we shall state conditions on the coefficients that can be used to predict which of (a), (b), or (c) occurs. Perhaps the reader can already tell what must be true of the coefficients in (7.11) if the lines are parallel.

EXERCISES

Without solving, determine which of the systems in Exercises 1–6 are equivalent to the system

$$\begin{cases} 3x + y = 1 \\ x - 2y = -3 \end{cases}$$

and justify your answers.

1 $\begin{cases} 3x + y = 1 \\ -2x + 4y = 6 \end{cases}$ **3** $\begin{cases} 2x + 3y = 4 \\ x - 2y = -3 \end{cases}$ **5** $\begin{cases} 3x + y = 1 \\ x - y = 2 \end{cases}$

2 $\begin{cases} 3x + y = 1 \\ -3x + 6y = 9 \end{cases}$ **4** $\begin{cases} 3x + y = 1 \\ 7x = -1 \end{cases}$ **6** $\begin{cases} x + \frac{1}{3}y = \frac{1}{3} \\ \frac{1}{4}x - \frac{1}{2}y = -\frac{3}{4} \end{cases}$

Find the solution sets of the systems in Exercises 7–22.

7 $\begin{cases} 3x + 2y = 5 \\ x - y = 5 \end{cases}$ **8** $\begin{cases} a - 2b = 1 \\ a + 2b = 3 \end{cases}$

9 $\begin{cases} 6r + 7s = 10 \\ 2r - 3s = 14 \end{cases}$ **10** $\begin{cases} x - 3y = -9 \\ 2x + 5y = 4 \end{cases}$

11 $\begin{cases} x = y + 2 \\ 3y = 2x + 1 \end{cases}$ **12** $\begin{cases} 5u - 3v = 1 \\ 14u + 11v = -36 \end{cases}$

13 $\begin{cases} -2x + y = 10 \\ 10x - 5y = 3 \end{cases}$ **14** $\begin{cases} 9x - 12y = 24 \\ 3x - 4y = 8 \end{cases}$

15 $\begin{cases} \frac{1}{3}u - \frac{1}{2}v = 4 \\ u + \frac{2}{3}v = -1 \end{cases}$ **16** $\begin{cases} -\frac{1}{5}x + \frac{1}{2}y = \frac{3}{2} \\ \frac{1}{3}x + \frac{1}{4}y = -\frac{17}{12} \end{cases}$

17 $\begin{cases} 12x - 3y = 6 \\ -4x + y = -2 \end{cases}$

18 $\begin{cases} 3c - 3d = 9 \\ 2c - 2d = 4 \end{cases}$

19 $\begin{cases} \dfrac{x+y}{3} - \dfrac{x-y}{4} = 1 \\ \dfrac{3x-y}{2} - \dfrac{x+2y}{3} = -2 \end{cases}$

20 $\begin{cases} 0.14x - 0.05y = 0.13 \\ 0.04x + 0.11y = 0.41 \end{cases}$

21 $\begin{cases} 1/x + 3/y = 7 \\ 4/x - 2/y = 1 \end{cases}$

(*Hint:* Let $x' = 1/x$ and $y' = 1/y$.)

22 $2/(x+1) + 3/(y-2) = 4$
$1/(x+1) + 2/(y-2) = 3$

23 Determine a and b so that the straight line with equation $ax + by = 5$ passes through the points $P_1(-1, 2)$ and $P_2(3, 1)$.

24 Generalize Exercise 23 to the case where the graph of $ax + by = 5$ passes through the points $P_1(x_1, y_1)$ and $P_2(x_2, y_2)$. (Express a and b in terms of x_1, y_1, x_2, and y_2.) Is it necessary to restrict P_1 or P_2 in any way?

Solve each of the following by employing a system of linear equations.

25 Find two positive integers whose sum is 82 and whose difference is 24.

26 A collection of nickels and dimes amounts to $2.45. If there are 34 coins in all, find the number of dimes.

27 An operator of a candy store has two types of candy, the first selling for 75 cents a pound and the second for 95 cents a pound. A customer wants a three-pound mixture of the two, which sells for $2.50. How much of each type of candy should be used?

28 An airplane made a 1300-mile trip with the wind in 2 hours and the return trip against the wind in 2 hours and 10 minutes. If the speed of the wind was the same throughout the trip, find the speed of the wind and the average speed of the airplane in still air.

29 A man receives income from two investments at simple interest rates of $6\frac{1}{2}\%$ and 8%, respectively. He has twice as much invested at $6\frac{1}{2}\%$ as at 8%. If his annual income from the two investments is $89.25, find how much is invested at each rate.

30 Two water pipes running at the same time can fill a swimming pool in 4 hours. If both pipes run for 2 hours and the first is then shut off, it takes 3 more hours for the second to fill the pool. How long does it take each pipe to fill the pool by itself?

3 SYSTEMS OF LINEAR EQUATIONS IN MORE THAN TWO VARIABLES

For systems of equations containing more than two variables we can use either the method of substitution explained in Section 1 or the techniques developed in Section 2. Let us illustrate this by solving a system in two different ways.

EXAMPLE 1 Find the solution set of the system

$$\begin{cases} x - 2y + 3z = 4 \\ 2x + y - 4z = 3 \\ -3x + 4y - z = -2. \end{cases}$$

Solution 1

Using the method of substitution we solve the first equation for x, obtaining

$$x = 4 + 2y - 3z.$$

Substituting for x in the second and third equations gives us

$$\begin{cases} 2(4 + 2y - 3z) + y - 4z = 3 \\ -3(4 + 2y - 3z) + 4y - z = -2, \end{cases}$$

which simplifies to

$$\begin{cases} y - 2z = -1 \\ y - 4z = -5. \end{cases}$$

If we now add to the second equation -1 times the first equation we obtain

$$\begin{aligned} y - 2z &= -1 \\ -2z &= -4. \end{aligned}$$

The solution of this system is $z = 2$, $y = 3$. The corresponding value of x is found by substituting for y and z in the equation $x = 4 + 2y - 3z$. This gives us $x = 4 + 2(3) - 3(2) = 4$ and hence the solution set of the system consists of one ordered triple $(4, 3, 2)$. We shall leave it to the reader to show that this triple is a solution of *each* of the given equations. Of course, other approaches to the problem could be used, such as solving the third equation for z and substituting in the first and second equations, obtaining two equations in x and y, and so on.

Solution 2

3
Systems
of Linear
Equations
in More
Then Two
Variables

We begin by eliminating x from the second and third equations. Adding to the second equation -2 times the first equation gives us the equivalent system

$$\begin{cases} x - 2y + 3z = 4 \\ 5y - 10z = -5 \\ -3x + 4y - z = -2. \end{cases}$$

Next, adding to the third equation 3 times the first equation, we obtain the equivalent system

$$\begin{cases} x - 2y + 3z = 4 \\ 5y - 10z = -5 \\ -2y + 8z = 10. \end{cases}$$

To simplify computations we multiply the second equation by $\frac{1}{5}$, obtaining

$$\begin{cases} x - 2y + 3z = 4 \\ y - 2z = -1 \\ -2y + 8z = 10. \end{cases}$$

We now eliminate y from the third equation by adding to it 2 times the second equation. This gives us the system

(7.12) $$\begin{cases} x - 2y + 3z = 4 \\ y - 2z = -1 \\ 4z = 8. \end{cases}$$

The solution set of the last system is easy to obtain. From the third equation it is clear that the only possible z value is 2. Substituting this for z in the second equation of (7.12), we get $y - 2(2) = -1$, or $y = 3$. Finally, the x value is found by substituting for y and z in the first equation. This gives us $x - 2(3) + 3(2) = 4$ and hence $x = 4$. Thus the solution set is $\{(4, 3, 2)\}$.

The system (7.12) is said to be in **echelon form**. If a system is in this form, the solution can be readily found.

The general form for a system of three linear equations in three variables is

(7.13) $$\begin{cases} a_1x + a_2y + a_3z = a \\ b_1x + b_2y + b_3z = b \\ c_1x + c_2y + c_3z = c, \end{cases}$$

where the coefficients are real numbers.

It can be shown that the system (7.13) has either (a) a *unique solution,* (b) *many solutions,* or (c) *no solutions.* As with the case of two equations in two variables, the terminology which is used to describe these cases is (a) *consistent,* (b) *dependent,* or (c) *inconsistent,* respectively.

Similar remarks and methods of solution hold for systems of four linear equations in four variables or, for that matter, systems of n linear equations in n variables, where n is any positive integer. Using the method of substitution we can always solve one of the equations for one variable in terms of the remaining variables. By substituting in each of the remaining equations we obtain a system containing one less equation in one less variable. We keep repeating this process until we obtain a form similar to (7.12), from which the solution may be easily found.

Sometimes it is necessary to work with systems in which the number of equations is not the same as the number of variables. Such systems can be attacked by using methods similar to those we have discussed.

EXAMPLE 2   Find the solution set of the system

$$\begin{cases} 2x + 3y + 4z = 1 \\ 3x + 4y + 5z = 3. \end{cases}$$

Solution

We can eliminate x from the second equation by adding to it $-\frac{3}{2}$ times the first equation. This gives us the equivalent system

$$\begin{cases} 2x + 3y + 4z = 1 \\ \quad -\frac{1}{2}y - z = \frac{3}{2}. \end{cases}$$

To clear of fractions we multiply the second equation by -2, obtaining

$$\begin{cases} 2x + 3y + 4z = 1 \\ \quad y + 2z = -3. \end{cases}$$

There are infinite number of solutions (b, c) for the second equation. Indeed, if we write $y = -3 - 2z$, then by substituting any number c for z, the corresponding value b of y is given by $b = -3 - 2c$. For each such solution (b, c) of $y + 2z = -3$ there corresponds a number a such that the triple (a, b, c) is a solution of the first equation, that is, $2a + 3b + 4c = 1$. Solving the latter equation for a we obtain

$$a = \tfrac{1}{2}(1 - 3b - 4c) = \tfrac{1}{2} - \tfrac{3}{2}b - 2c.$$

Since $b = -3 - 2c$, we have

$$a = \tfrac{1}{2} - \tfrac{3}{2}(-3 - 2c) - 2c,$$

which reduces to

$$a = 5 + c.$$

3
Systems
of Linear
Equations
in More
Than Two
Variables

Consequently the solution set of the system consists of all ordered triples of the form

$$(5 + c, -3 - 2c, c),$$

where c is any real number. These solutions may be checked by substituting $5 + c$ for x, $-3 - 2c$ for y, and c for z in the two given equations. We can obtain any number of solutions for the system by substituting specific real numbers for c. For example, if $c = 0$, we obtain $(5, -3, 0)$; if $c = 2$, we have $(7, -7, 2)$; if $c = -\tfrac{1}{2}$, we get $(\tfrac{9}{2}, -2, -\tfrac{1}{2})$; and so on.

Another method of solution for Example 2 is to rewrite the original system as

$$\begin{cases} 2x + 3y = 1 - 4z \\ 3x + 4y = 3 - 5z. \end{cases}$$

If a number c is substituted for z, we obtain a system of two equations in two variables x and y, and the solution set can be found as in Section 2. For a different approach we could solve for two *other* variables in terms of the third. This method can be extended. For example, given two equations in four variables, say x, y, z, and w, it may be possible to solve for two of the variables, say x and y, in terms of the remaining variables z and w. If we then substitute numbers c and d for z and w, respectively, and calculate the corresponding values for x and y, we obtain the solution set.

The system of linear equations (7.13) is called **homogeneous** if $a = b = c = 0$. The same terminology is used for more than three equations. A system of homogeneous equations always has the **trivial solution** obtained by substituting zero for all the variables. Nontrivial solutions sometimes exist. The procedure for finding solution sets is the same as that used for nonhomogeneous systems.

EXAMPLE 3 Find the solution set of the system

$$\begin{cases} x - y + 4z = 0 \\ 2x + y + z = 0 \\ -x - y + 2z = 0. \end{cases}$$

Solution

Adding to the second equation -2 times the first equation

and then adding the first equation to the third gives us the equivalent system

$$\begin{cases} x - y + 4z = 0 \\ \quad 3y - 9z = 0 \\ \quad -2y + 6z = 0. \end{cases}$$

Multiplying the second equation by $\frac{1}{3}$ and the third equation by $-\frac{1}{2}$ leads to the same equation, $y - 3z = 0$. Hence the given system is equivalent to

$$\begin{cases} x - y + 4z = 0 \\ \quad y - 3z = 0. \end{cases}$$

The solution may now be found as in the previous example. If we substitute any value c for z, then the second equation gives us $y = 3c$. Substituting $3c$ for y in the first equation, we obtain $x - 3c + 4c = 0$, or $x = -c$. Thus the solution set consists of all triples of the form $(-c, 3c, c)$, where c is any real number.

EXAMPLE 4 Find the solution set of the system

$$\begin{cases} x + y + z = 0 \\ x - y + z = 0 \\ x - y - z = 0. \end{cases}$$

Solution

By the usual process we obtain the equivalent system

$$\begin{cases} x + y + z = 0 \\ \quad -2y = 0 \\ \quad -2y - 2z = 0. \end{cases}$$

Multiplying the second and third equations by $-\frac{1}{2}$ gives us

$$\begin{cases} x + y + z = 0 \\ \quad y = 0 \\ \quad y + z = 0. \end{cases}$$

From the second equation of this system we see that the only possible y value is 0. Substituting this in the last equation we have $0 + z = 0$, or $z = 0$. Finally, substituting 0 for y and z in the first equation we see that x also must equal 0. Hence the only solution for the system is the trivial one, $(0, 0, 0)$.

3
Systems
of Linear
Equations
in More
Than Two
Variables

EXERCISES

Find the solution sets of the systems in Exercises 1–20.

1 $\begin{cases} x + 2y - z = -1 \\ x - y + 2z = 8 \\ 2x - y - z = 5 \end{cases}$

2 $\begin{cases} x - 2y + 2z = -6 \\ 2x + 3y - z = 10 \\ x - y - z = 4 \end{cases}$

3 $\begin{cases} 3x - 2y + z = 1 \\ x + y + z = 0 \\ y - 2z = 2 \end{cases}$

4 $\begin{cases} 4x - y + 3z = -8 \\ -x - y + 2z = 13 \\ 3x + y - z = -19 \end{cases}$

5 $\begin{cases} 2x - 3y - z = 4 \\ 3x + y - 2z = 7 \\ -4x + 6y + 2z = 1 \end{cases}$

6 $\begin{cases} x - 2y = 4 \\ 3y + z = -1 \\ 2x - 5z = 3 \end{cases}$

7 $\begin{cases} x - y - z = 0 \\ 3x + y - z = 0 \\ 2x + 4y + z = 0 \end{cases}$

8 $\begin{cases} 2x - y - z = 0 \\ z + x = 0 \\ z + 4x - y = 0 \end{cases}$

9 $\begin{cases} x + 2y + z = 0 \\ 2x - y + 2z = 0 \\ x + y + z = 0 \end{cases}$

10 $\begin{cases} x - y + 2z = 0 \\ y + 2x - z = 0 \\ 3y - 2z - x = 0 \end{cases}$

11 $\begin{cases} x + 2y - 3z = 1 \\ 2x + 3y + z = -3 \end{cases}$

12 $\begin{cases} -3x + y - 2z = 1 \\ 2x - y + z = 5 \end{cases}$

13 $\begin{cases} x + 3y - 2z + w = -3 \\ 2x - y - 3z + w = 2 \\ -x + 2y + z - 3w = -6 \\ 3x + y + 2z - 4w = 8 \end{cases}$

14 $\begin{cases} 3/x - 1/y + 4/z = -13 \\ 1/x + 2/y - 1/z = 12 \\ 4/x - 1/y + 3/z = -7 \end{cases}$

15 $\begin{cases} 2x - 3y = 4 \\ x + y = 5 \\ 3x - y = -1 \end{cases}$

16 $\begin{cases} 2x + y = 1 \\ 3x + 2y = 3 \\ -x + 2y = 7 \end{cases}$

17 $\begin{cases} 4/x + 1/y + 1/z = 5 \\ 1/x - 4/y + 2/z = 0 \\ -3/x + 1/y + 3/z = -1 \end{cases}$

18 $\begin{cases} x + 3y - 2z + w = 2 \\ 2x + 4y + 2z - 3w = -1 \\ -x + y + z + 2w = 1 \\ 2x + 5y - z - w = 1 \end{cases}$

19 $\begin{cases} 2x - y - z = 0 \\ z + x = 0 \\ z + 4x - y = 0 \end{cases}$

20 $\begin{cases} y + 2x = 0 \\ y - 5z = 1 \\ z + 3x = -2 \end{cases}$

21 A collection of nickels, dimes, and quarters amounts to $4.50. If there are 57 coins in all and there are three times as many nickels as there are dimes, find the number of quarters.

22 A three-digit number is equal to 25 times the sum of its digits. If the three digits are reversed, the resulting number exceeds the given number by 198. The tens digit is one less than the sum of the hundreds and units digits. Find the number.

23 A chemist has three solutions containing a certain acid. The first contains

10% acid, the second 30%, and the third 50%. He wishes to use all three solutions to obtain a mixture of 25 ounces containing 40% acid, using twice as much of the 50% solution as the 30% solution. How many ounces of each solution should be used?

24 A swimming pool can be filled by three pipes A, B, and C. Pipe A can fill the pool by itself in 8 hours. If pipes A and C are used together, the pool can be filled in 6 hours. If B and C are used together, it takes 10 hours. How long does it take to fill the pool if all three pipes are used?

25 Find the equation of a circle which passes through the three points $P_1(5, 7)$, $P_2(-2, 6)$, and $P_3(6, 0)$. (*Hint:* The equation of the circle has the form $x^2 + y^2 + ax + by + c = 0$.)

26 Determine a, b, and c such that the graph of the equation

$$y = ax^2 + bx + c$$

passes through the points $P_1(2, 1)$, $P_2(1, -1)$, and $P_3(-1, 4)$.

4 MATRIX METHODS

If we analyze Solution 2 of Example 1 in the preceding section, we see that the symbols used for the variables really have little effect on the solution of the problem. The *coefficients* of the variables are the important things to consider. Thus if different symbols such as r, s, and t are used for the variables in that example, we obtain the system

$$(7.14) \quad \begin{cases} r - 2s + 3t = 4 \\ 2r + s - 4t = 3 \\ -3r + 4s - t = -2. \end{cases}$$

The method of solution could then proceed exactly as before. Since this is true, it is possible to use a process which reduces the amount of labor involved. Specifically, we introduce a notation for keeping track of the coefficients in such a way that the variables do not have to be written down. Referring to (7.14), and checking that corresponding variables appear underneath one another and terms not involving variables are to the right of the equals sign, we list the numbers which are involved in the equations in the following manner:

$$(7.15) \quad \begin{bmatrix} 1 & -2 & 3 & 4 \\ 2 & 1 & -4 & 3 \\ -3 & 4 & -1 & -2 \end{bmatrix}.$$

If some variable had not appeared in one of the equations, we would have used a zero in the appropriate position. An array of numbers such as (7.15) is called a *matrix*. The *rows* of (7.15) are the numbers which appear next to one another *horizontally*. Thus the first row is 1 −2 3 4, the second row is 2 1 −4 3, and the third row is −3 4 −1 −2. The *columns* of (7.15) are the sets of numbers which form a *vertical* pattern. For example, the second column consists of the numbers −2, 1, 4, (in that order); the fourth column consists of 4, 3, −2; and so on.

Before discussing the method of solving (7.14) by means of the matrix (7.15), let us introduce a general definition of matrix. It will be convenient to use a *double subscript notation* for the symbols which represent the numbers in the matrix. Specifically, the symbol a_{ij} will denote the number which appears in row i and column j. We call i the **row subscript** and j the **column subscript** of a_{ij}.

(7.16)   Definition

If m and n are positive integers, then an $m \times n$ **matrix** over the real number system is an array of the form

$$\begin{bmatrix} a_{11} & a_{12} & a_{13} & \cdots & a_{1n} \\ a_{21} & a_{22} & a_{23} & \cdots & a_{2n} \\ a_{31} & a_{32} & a_{33} & \cdots & a_{3n} \\ \cdot & \cdot & \cdot & & \cdot \\ \cdot & \cdot & \cdot & & \cdot \\ \cdot & \cdot & \cdot & & \cdot \\ a_{m1} & a_{m2} & a_{m3} & \cdots & a_{mn} \end{bmatrix}$$

where each a_{ij} is a real number.

The symbol $m \times n$ in (7.16) is read "m by n." The matrix in (7.15) is a 3×4 matrix over the real numbers. It is also possible to consider matrices where the elements a_{ij} belong to other mathematical systems; however, we shall not do so in this book.

The rows and columns of a matrix are defined as before. Thus the matrix in (7.16) has m rows and n columns. The reader should carefully examine the notation for a matrix, observing, for example, that the symbol a_{23} appears in row 2 and column 3, whereas a_{32} appears in row 3 and column 2. Each a_{ij} is called an **element of the matrix.** The elements $a_{11}, a_{22}, a_{33}, \cdots$ are called the **main diagonal elements.** If $m \neq n$, the matrix is sometimes referred to as a **rectangular matrix.** If $m = n$, we refer to the matrix as a **square matrix of order** n.

It is possible to develop a comprehensive theory for matrices. Later in this chapter we shall define *equality* of matrices, the *sum* of two matrices, the *product* of two matrices, and the *inverse* of a matrix. All these ideas are very important and useful in mathematics and applications. In particular, matrix techniques are well suited for the types of problems which are solved by computers. We shall examine algebraic properties of matrices in Sections 10 and 11. Our main purpose in introducing matrices at this time is for use as a tool in solving systems of linear equations. This technique is well illustrated by considering some specific systems.

Let us return to the system (7.14). The matrix (7.15) is called the **matrix of the system.** In certain cases we may wish to consider only the *coefficients* of the variables of a system of linear equations. The corresponding matrix is called the **coefficient matrix.** The coefficient matrix for the system (7.14) is

$$\begin{bmatrix} 1 & -2 & 3 \\ 2 & 1 & -4 \\ -3 & 4 & -1 \end{bmatrix}.$$

In order to distinguish the matrix of the system from the coefficient matrix, the former is sometimes called the **augmented matrix.** After forming the matrix of the system, we work with the rows of the matrix *just as though they were equations.* The only items missing are the symbols for the variables, the addition signs used between terms, and the equals signs. We simply keep in mind that the numbers in the first column are the coefficients of the first variable, the numbers in the second column are the coefficients of the second variable, and so on. Our rules for transforming a matrix are formulated in such a way that they always produce a matrix of an equivalent system of equations. The key to these rules is given by (7.8). As a matter of fact, we have the following theorem for matrices, which is a perfect analogue of Theorem (7.8) for equations.

(7.17) Matrix Transformation Theorem

Given a matrix of a system of linear equations, each of the following transformations results in a matrix of an equivalent system of linear equations:

(i) Interchanging any two rows.

(ii) Multiplying all of the elements in a row by the same nonzero real number k.

(iii) Adding to the elements in a row k times the corresponding elements of any other row, where k is any real number.

For convenience we shall refer to (ii) as the process of "multiplying a row by the number k" and (iii) will be described by saying "add to a row k times any other row." Rules (i)–(iii) of (7.17) are called the **elementary row transformations** of a matrix.

Since the rules given in (7.8) produce equivalent systems of equations, it is evident that each of the rules in (7.17) results in the matrix of an equivalent system. Let us find the solution of the system (7.14) by using (7.17). The reader should compare our work with Solution 2 of Example 1 in the previous section. We shall use arrows to indicate that the form of the matrices has been changed by means of elementary row transformations, and we shall justify each transformation by stating the reason for each step. Thus

$$\begin{bmatrix} 1 & -2 & 3 & 4 \\ 2 & 1 & -4 & 3 \\ -3 & 4 & -1 & -2 \end{bmatrix} \rightarrow \begin{bmatrix} 1 & -2 & 3 & 4 \\ 0 & 5 & -10 & -5 \\ -3 & 4 & -1 & -2 \end{bmatrix}$$

Add to the second row -2 times the first row.

$$\rightarrow \begin{bmatrix} 1 & -2 & 3 & 4 \\ 0 & 5 & -10 & -5 \\ 0 & -2 & 8 & 10 \end{bmatrix}$$

Add to the third row 3 times the first row.

$$\rightarrow \begin{bmatrix} 1 & -2 & 3 & 4 \\ 0 & 1 & -2 & -1 \\ 0 & -2 & 8 & 10 \end{bmatrix}$$

Multiply row 2 by $\frac{1}{5}$.

$$\rightarrow \begin{bmatrix} 1 & -2 & 3 & 4 \\ 0 & 1 & -2 & -1 \\ 0 & 0 & 4 & 8 \end{bmatrix}$$

Add to the third row 2 times the second row.

The last matrix has zeros everywhere below the main diagonal and is said to be in **echelon form.** We now use this matrix to return to the system of equations

$$\begin{cases} r - 2s + 3t = 4 \\ s - 2t = -1 \\ 4t = 8, \end{cases}$$

which is equivalent to the given system. The solution set is then found as in (7.12).

The method used above is completely general. Beginning with *any* system of linear equations, we can use (iii) of (7.17) many times and also interchange rows if necessary to introduce zeros everywhere in the first column after the first row; that is, we can obtain a matrix of an equivalent system which has the form

$$\begin{bmatrix} a_{11} & a_{12} & a_{13} & \cdot & \cdot & \cdot \\ 0 & b_{22} & b_{23} & \cdot & \cdot & \cdot \\ 0 & c_{32} & c_{33} & \cdot & \cdot & \cdot \\ 0 & \cdot & \cdot & \cdot & \cdot & \cdot \\ \cdot & \cdot & \cdot & \cdot & \cdot & \cdot \\ \cdot & \cdot & \cdot & \cdot & \cdot & \cdot \\ \cdot & \cdot & \cdot & \cdot & \cdot & \cdot \\ 0 & \cdot & \cdot & \cdot & \cdot & \cdot \end{bmatrix}$$

where the dots indicate that possibly more numbers appear.

If this matrix has a nonzero element in the second column and after the first row, then again using (iii) of (7.17) we may transform the matrix into the form

$$\begin{bmatrix} a_{11} & a_{12} & a_{13} & \cdot & \cdot & \cdot \\ 0 & k_{22} & k_{23} & \cdot & \cdot & \cdot \\ 0 & 0 & l_{33} & \cdot & \cdot & \cdot \\ \cdot & \cdot & \cdot & \cdot & \cdot & \cdot \\ \cdot & \cdot & \cdot & \cdot & \cdot & \cdot \\ \cdot & \cdot & \cdot & \cdot & \cdot & \cdot \\ 0 & 0 & \cdot & \cdot & \cdot & \cdot \end{bmatrix}$$

where only zeros appear below k_{22} in the second column. This process is continued until we obtain the matrix of a system which is in echelon form. This will be true when all elements below the main diagonal are zero. The final matrix is then used to obtain a system of equations which is equivalent to the original system. The solution set of the latter system is found as in previous sections.

Let us illustrate this technique by solving a system of four linear equations.

EXAMPLE 1   Find the solution set of the system

$$\begin{cases} x & -2z + 2w = & 1 \\ -2x + 3y + 4z & = -1 \\ & y + z - w = & 0 \\ 3x + y - 2z - w = & 3. \end{cases}$$

Solution

Notice that we have arranged the equations so that the same variables appear in vertical columns. We shall begin with the matrix of the system and proceed to the solution.

$$\begin{bmatrix} 1 & 0 & -2 & 2 & 1 \\ -2 & 3 & 4 & 0 & -1 \\ 0 & 1 & +1 & -1 & 0 \\ 3 & 1 & -2 & -1 & 3 \end{bmatrix} \rightarrow \begin{bmatrix} 1 & 0 & -2 & 2 & 1 \\ 0 & 3 & 0 & 4 & 1 \\ 0 & 1 & 1 & -1 & 0 \\ 0 & 1 & 4 & -7 & 0 \end{bmatrix}$$

Add to the second row 2 times the first row and then add to the fourth row -3 times the first row.

$$\rightarrow \begin{bmatrix} 1 & 0 & -2 & 2 & 1 \\ 0 & 1 & 1 & -1 & 0 \\ 0 & 3 & 0 & 4 & 1 \\ 0 & 1 & 4 & -7 & 0 \end{bmatrix}$$

Interchange rows 2 and 3.

$$\rightarrow \begin{bmatrix} 1 & 0 & -2 & 2 & 1 \\ 0 & 1 & 1 & -1 & 0 \\ 0 & 0 & -3 & 7 & 1 \\ 0 & 0 & 3 & -6 & 0 \end{bmatrix}$$

Add to the third row -3 times the second row and then add to the fourth row -1 times the second row.

$$\rightarrow \begin{bmatrix} 1 & 0 & -2 & 2 & 1 \\ 0 & 1 & 1 & -1 & 0 \\ 0 & 0 & -3 & 7 & 1 \\ 0 & 0 & 0 & 1 & 1 \end{bmatrix}$$

Add row 3 to row 4.

The final matrix corresponds to the system of equations

$$\begin{cases} x & -2z + 2w = 1 \\ y + z - w = 0 \\ -3z + 7w = 1 \\ w = 1. \end{cases}$$

From the last equation we see that $w = 1$ Substituting in the third equation, we get $-3z + 7 \cdot 1 = 1$, or $z = 2$. Using the second equation, we get $y + 2 - 1 = 0$, or $y = -1$. Finally, from the first equation, we have $x - 2 \cdot 2 + 2 \cdot 1 = 1$, or $x = 3$. Hence the system has only one solution, $x = 3, y = -1, z = 2$, and $w = 1$.

EXERCISES

1–8   Use matrices to solve Exercises 1–8 of Section 3.
9–16 Use matrices to solve Exercises 7–14 of Section 2.
17–18 Use matrices to solve Exercises 13 and 14 of Section 3.

Solve the following by means of matrices.

19
$$\begin{cases} 2x + y - 3z + 4s + t = -2 \\ 3x + 2y - 2z - s + 2t = 7 \\ x - y + z - 2s - 2t = 0 \\ -x - 2y - z + 2s + t = 5 \\ 2x + 2y - 2z + s + 2t = 3. \end{cases}$$

20
$$\begin{cases} x + 2y = 1 \\ 2x - y = -1 \\ -3x + 2y = 5. \end{cases}$$

21
$$\begin{cases} x + y - z + 2w = 1 \\ 2x - y - z - w = -1 \\ -x + 2y - 2z + w = 2. \end{cases}$$

22
$$\begin{cases} 4x - y - 3z = 0 \\ x + 2y - z = 3 \\ 2x - 3y - z = -4 \\ -x + 3y + 2z = -1. \end{cases}$$

5 DETERMINANTS

Throughout this section and the next it is assumed that all matrices under discussion are *square* matrices. Associated with each such matrix A is a number called the *determinant* of A. Determinants can be used to solve systems of linear equations if the number of equations is the same as the number of variables. In this section we shall state the definition and give some basic properties of determinants. It may be difficult to see exactly why the definitions given below are used. One reason will be pointed out in Section 7, where it is shown that our definitions arise naturally when solving systems of linear equations.

The determinant of a square matrix A will be denoted by $|A|$. This notation should not be confused with the symbol used for the absolute value of a real number. To avoid any misunderstanding, the expression det A is used in some mathematics texts instead of $|A|$. We shall arrive at the general definition of $|A|$ by beginning with the case in which A has order 1 and then by increasing the order a step at a time. The concept of mathematical induction (see Chapter Ten) may be used to justify this type of definition.

If A is a square matrix of order 1, then A has only one element. Thus $A = [a_{11}]$ and we define $|A| = a_{11}$ If A is a square matrix of order 2, then we may write

$$A = \begin{bmatrix} a_{11} & a_{12} \\ a_{21} & a_{22} \end{bmatrix}$$

and the determinant of A is defined by

$$|A| = a_{11}a_{22} - a_{21}a_{12}.$$

Another notation for $|A|$ is obtained by replacing the brackets in the symbol for A with vertical bars as follows:

(7.18) $\quad |A| = \begin{vmatrix} a_{11} & a_{12} \\ a_{21} & a_{22} \end{vmatrix} = a_{11}a_{22} - a_{21}a_{12}.$

EXAMPLE 1 Find $|A|$ if $A = \begin{bmatrix} 2 & -1 \\ 4 & -3 \end{bmatrix}$.

Solution

Using (7.18),

$$|A| = \begin{vmatrix} 2 & -1 \\ 4 & -3 \end{vmatrix} = (2)(-3) - (4)(-1) = -6 + 4 = -2.$$

For matrices of higher order it is convenient to introduce additional terminology. First, if A is a matrix of order 3, then the **minor** M_{ij} of an element a_{ij} is the determinant of the matrix of order 2 obtained by deleting row i and column j of A. Thus, to determine the minor of an element we discard the row and column in which the element appears and then find the determinant of the resulting matrix. To illustrate, if

(7.19) $\quad A = \begin{bmatrix} a_{11} & a_{12} & a_{13} \\ a_{21} & a_{22} & a_{23} \\ a_{31} & a_{32} & a_{33} \end{bmatrix}$

then

$$M_{11} = \begin{vmatrix} a_{22} & a_{23} \\ a_{32} & a_{33} \end{vmatrix} = a_{22}a_{33} - a_{32}a_{23}$$

$$M_{12} = \begin{vmatrix} a_{21} & a_{23} \\ a_{31} & a_{33} \end{vmatrix} = a_{21}a_{33} - a_{31}a_{23}$$

(7.20)

$$M_{13} = \begin{vmatrix} a_{21} & a_{22} \\ a_{31} & a_{32} \end{vmatrix} = a_{21}a_{32} - a_{31}a_{22}$$

$$M_{23} = \begin{vmatrix} a_{11} & a_{12} \\ a_{31} & a_{32} \end{vmatrix} = a_{11}a_{32} - a_{31}a_{12}$$

and likewise for the other minors M_{21}, M_{22}, M_{31}, M_{32}, and M_{33}. The **cofactor** A_{ij} of the element a_{ij} is defined by

$$A_{ij} = (-1)^{i+j}M_{ij}.$$

Thus, to obtain the cofactor of a_{ij} we find the minor and multiply it by 1 or -1 depending on whether the sum of i and j is even or odd, respectively.

EXAMPLE 2 Given

$$A = \begin{bmatrix} 1 & -3 & 3 \\ 4 & 2 & 0 \\ -2 & -7 & 5 \end{bmatrix}$$

find M_{11}, M_{21}, M_{22}, A_{11}, A_{21}, and A_{22}.

Solution

By definition

$$M_{11} = \begin{vmatrix} 2 & 0 \\ -7 & 5 \end{vmatrix} = (2)(5) - (-7)(0) = 10,$$

$$M_{21} = \begin{vmatrix} -3 & 3 \\ -7 & 5 \end{vmatrix} = (-3)(5) - (-7)(3) = 6,$$

$$M_{22} = \begin{vmatrix} 1 & 3 \\ -2 & 5 \end{vmatrix} = (1)(5) - (-2)(3) = 11.$$

All that is necessary to obtain the cofactors is to prefix the corresponding minors with the proper signs. Thus

$$A_{11} = (-1)^{1+1}M_{11} = (1)(10) = 10,$$
$$A_{21} = (-1)^{2+1}M_{21} = (-1)(6) = -6,$$
$$A_{22} = (-1)^{2+2}M_{22} = (1)(11) = 11.$$

The determinant $|A|$ of a square matrix of order 3 is defined by

$$(7.21) \quad |A| = \begin{vmatrix} a_{11} & a_{12} & a_{13} \\ a_{21} & a_{22} & a_{23} \\ a_{31} & a_{32} & a_{33} \end{vmatrix} = a_{11}A_{11} + a_{12}A_{12} + a_{13}A_{13}.$$

Since $A_{11} = (-1)^{1+1}M_{11} = M_{11}$, $A_{12} = (-1)^{1+2}M_{12} = -M_{12}$, and $A_{13} = (-1)^{1+3}M_{13} = M_{13}$, definition (7.21) may also be written thus:

$$(7.22) \quad |A| = a_{11}M_{11} - a_{12}M_{12} + a_{13}M_{13}.$$

If we substitute for M_{11}, M_{12}, and M_{13} from (7.20) we obtain the following formula for $|A|$ in terms of the elements of A:

$$(7.23) \quad |A| = a_{11}a_{22}a_{33} - a_{11}a_{32}a_{23} - a_{12}a_{21}a_{33} + a_{12}a_{31}a_{23} + a_{13}a_{21}a_{32}$$
$$- a_{13}a_{31}a_{22}.$$

Formula (7.21) displays a pattern of multiplying each element in row 1 by its cofactor and then adding to find $|A|$. This is referred to as *expanding* $|A|$ *by the first row*. By actually carrying out the computations, it is not difficult to show that $|A|$ *can be expanded in similar fashion by using any row or column*. As an illustration, the expansion by the second column in (7.19) is:

$$|A| = a_{12}A_{12} + a_{22}A_{22} + a_{32}A_{32}$$
$$= a_{12} \left(- \begin{vmatrix} a_{21} & a_{23} \\ a_{31} & a_{32} \end{vmatrix} \right) + a_{22} \left(+ \begin{vmatrix} a_{11} & a_{13} \\ a_{31} & a_{33} \end{vmatrix} \right) + a_{32} \left(- \begin{vmatrix} a_{11} & a_{13} \\ a_{21} & a_{23} \end{vmatrix} \right).$$

If we use (7.18) for the determinants in parentheses, multiply as indicated, and rearrange the terms in the sum, then the expression for $|A|$ given in (7.23) will result. Similarly, the expansion of (7.19) by the third row is

$$|A| = a_{31}A_{31} + a_{32}A_{32} + a_{33}A_{33}$$
$$= a_{31} \left(+ \begin{vmatrix} a_{12} & a_{13} \\ a_{22} & a_{23} \end{vmatrix} \right) + a_{32} \left(- \begin{vmatrix} a_{11} & a_{13} \\ a_{21} & a_{23} \end{vmatrix} \right) + a_{33} \left(+ \begin{vmatrix} a_{11} & a_{12} \\ a_{21} & a_{22} \end{vmatrix} \right).$$

Once again it can be shown that this equals (7.23).

EXAMPLE 3 Find $|A|$ if

$$A = \begin{bmatrix} -1 & 3 & 1 \\ 2 & 5 & 0 \\ 3 & 1 & -2 \end{bmatrix}.$$

Solution

Expanding $|A|$ by the second row gives us

$$|A| = (2)A_{21} + (5)A_{22} + (0)A_{23}.$$

$$A_{21} = (-1)^3 M_{21} = -\begin{vmatrix} 3 & 1 \\ 1 & -2 \end{vmatrix} = -[(3)(-2) - (1)(1)] = 7,$$

$$A_{22} = (-1)^4 M_{22} = \begin{vmatrix} -1 & 1 \\ 3 & -2 \end{vmatrix} = [(-1)(-2) - (3)(1)] = -1.$$

Consequently

$$|A| = (2)(7) + (5)(-1) + (0)A_{23} = 14 - 5 + 0 = 9.$$

If A is a matrix of order 4, we define the minor M_{ij} of the element a_{ij} as the determinant of the matrix of order 3 obtained by deleting row i and column j of A. The cofactor A_{ij} is again given by $(-1)^{i+j} M_{ij}$. In a manner analogous to (7.21) we define

$$|A| = a_{11}A_{11} + a_{12}A_{12} + a_{13}A_{13} + a_{14}A_{14}.$$

This is called *the expansion of $|A|$ by the first row*. In terms of minors, the formula may be written as

$$|A| = a_{11}M_{11} - a_{12}M_{12} + a_{13}M_{13} - a_{14}M_{14}.$$

It can be shown that the same number is obtained if $|A|$ is expanded by any other row or column.

The method of defining determinants of matrices of arbitrary order n should now be apparent. If a_{ij} is an element of a matrix of order $n > 1$, then the **minor** M_{ij} is defined as the determinant of the matrix of order $n - 1$ obtained by deleting row i and column j. The **cofactor** A_{ij} is defined as $(-1)^{i+j} M_{ij}$. We then define the determinant $|A|$ of a matrix A of order n as the expansion by the first row, that is

$$(7.24) \quad |A| = a_{11}A_{11} + a_{12}A_{12} + \cdots + a_{1n}A_{1n}$$

or, in terms of minors,

$$|A| = a_{11}M_{11} - a_{12}M_{12} + \cdots + a_{1n}(-1)^{1+n}M_{1n}.$$

As was the case with matrices of small order, the number $|A|$ may be found by using *any* row or column. Specifically, we have the following theorem.

(7.25) Expansion Theorem for Determinants

If A is a square matrix of order $n > 1$, then the determinant $|A|$ may be found by multiplying the elements of any row (or column) by their respective cofactors and adding the resulting products.

The proof of (7 25) is difficult and may be found in texts on matrix theory. The theorem is quite useful if many zeros appear in a row or column, as illustrated in the following example.

EXAMPLE 4 Find $|A|$ if

$$A = \begin{bmatrix} 1 & 0 & 2 & 5 \\ -2 & 1 & 5 & 0 \\ 0 & 0 & -3 & 0 \\ 0 & -1 & 0 & 3 \end{bmatrix}.$$

Solution

Note that all but one of the elements in the third row is zero. Hence if we expand $|A|$ by the third row there will be at most one nonzero term. Specifically,

$$|A| = 0 \cdot A_{31} + 0 \cdot A_{32} + (-3)A_{33} + 0 \cdot A_{34} = -3A_{33},$$

where

$$A_{33} = \begin{vmatrix} 1 & 0 & 5 \\ -2 & 1 & 0 \\ 0 & -1 & 3 \end{vmatrix}.$$

Expanding A_{33} by column 1, we obtain

$$A_{33} = 1 \begin{vmatrix} 1 & 0 \\ -1 & 3 \end{vmatrix} + (-2)\left(- \begin{vmatrix} 0 & 5 \\ -1 & 3 \end{vmatrix}\right) + 0 \begin{vmatrix} 0 & 5 \\ 1 & 0 \end{vmatrix}$$
$$= 3 + 10 + 0 = 13,$$

and therefore,

$$|A| = -3A_{33} = (-3)(13) = -39.$$

In general, if all but one element a in some row (or column) of A is zero and if the determinant $|A|$ is expanded by that row (or column), then all terms drop out except the product of the element a with its cofactor. We will make important use of this fact in the next section.

If *every* element in a row (or column) of a matrix A is zero, then upon expanding $|A|$ by that row (or column) we obtain the number 0. This provides the following result.

(7.26) Theorem

If every element of a row (or column) of a square matrix A is zero, then $|A| = 0$.

In each of Exercises 1–4 find all the minors and cofactors of the elements in the given matrix.

1 $\begin{bmatrix} 3 & 1 & 0 \\ -1 & 4 & 2 \\ 5 & 0 & -8 \end{bmatrix}$

2 $\begin{bmatrix} 4 & -3 & 7 \\ 2 & 0 & -1 \\ -2 & 5 & 1 \end{bmatrix}$

3 $\begin{bmatrix} 3 & 1 \\ -2 & 0 \end{bmatrix}$

4 $\begin{bmatrix} 1 & 2 \\ 3 & 4 \end{bmatrix}$

5–8 Find the determinants of the matrices given in Exercises 1–4.

In Exercises 9–20 find the determinants of the given matrices.

9 $\begin{bmatrix} 2 & -3 \\ -1 & -2 \end{bmatrix}$

10 $\begin{bmatrix} 3 & 0 \\ 1 & 5 \end{bmatrix}$

11 $\begin{bmatrix} a & b \\ -a & -b \end{bmatrix}$

12 $\begin{bmatrix} a & b \\ b & a \end{bmatrix}$

13 $\begin{bmatrix} 2 & 0 & 8 \\ -3 & 1 & -2 \\ 5 & 2 & 4 \end{bmatrix}$

14 $\begin{bmatrix} 1 & 4 & 7 \\ 2 & 5 & 8 \\ 3 & 6 & 9 \end{bmatrix}$

15 $\begin{bmatrix} 2 & -1 & 2 \\ 3 & 1 & -1 \\ 1 & 5 & -3 \end{bmatrix}$

16 $\begin{bmatrix} -1 & 0 & 5 \\ 6 & 9 & -3 \\ -2 & 1 & 0 \end{bmatrix}$

17 $\begin{bmatrix} 4 & 0 & -1 & 0 \\ 0 & 3 & 0 & -2 \\ -5 & 0 & 6 & 0 \\ 0 & 1 & 0 & 4 \end{bmatrix}$

18 $\begin{bmatrix} 6 & -1 & 0 & -2 \\ 0 & 4 & 0 & 1 \\ 1 & 1 & -2 & 2 \\ 0 & 5 & 0 & 3 \end{bmatrix}$

19 $\begin{bmatrix} 0 & 0 & c & 0 \\ a & 0 & 0 & 0 \\ 0 & 0 & 0 & d \\ 0 & b & 0 & 0 \end{bmatrix}$

20 $\begin{bmatrix} 1 & 1 & 1 & 1 \\ 0 & 2 & 2 & 2 \\ 0 & 0 & 3 & 3 \\ 0 & 0 & 0 & 4 \end{bmatrix}$

Verify the identities in Exercises 21–28.

21 $\begin{vmatrix} a & b \\ c & d \end{vmatrix} = - \begin{vmatrix} c & d \\ a & b \end{vmatrix}$

22 $\begin{vmatrix} a & b \\ c & d \end{vmatrix} = - \begin{vmatrix} b & a \\ d & c \end{vmatrix}$

23 $\begin{vmatrix} a & kb \\ c & kd \end{vmatrix} = k \begin{vmatrix} a & b \\ c & d \end{vmatrix}$

24 $\begin{vmatrix} a & b \\ kc & kd \end{vmatrix} = k \begin{vmatrix} a & b \\ c & d \end{vmatrix}$

25 $\begin{vmatrix} a & b \\ c & d \end{vmatrix} = \begin{vmatrix} a & b \\ ka+c & kb+d \end{vmatrix}$

26 $\begin{vmatrix} a & b \\ c & d \end{vmatrix} = \begin{vmatrix} a & ka+b \\ c & kc+d \end{vmatrix}$

27 $\begin{vmatrix} a & b \\ c & d \end{vmatrix} + \begin{vmatrix} a & e \\ c & f \end{vmatrix} = \begin{vmatrix} a & b+e \\ c & d+f \end{vmatrix}$

28 $\begin{vmatrix} a & b \\ c & d \end{vmatrix} + \begin{vmatrix} a & b \\ e & f \end{vmatrix} = \begin{vmatrix} a & b \\ c+e & d+f \end{vmatrix}$

29 Prove that if a matrix A of order 2 has two identical rows or columns, then $|A| = 0$.

30 Verify (7.23) for the matrix

$$A = \begin{bmatrix} a_{11} & a_{12} & a_{13} \\ a_{21} & a_{22} & a_{23} \\ a_{31} & a_{32} & a_{33} \end{bmatrix}.$$

6 PROPERTIES OF DETERMINANTS

The method of evaluating a determinant by means of the Expansion Theorem (7.25) is not very efficient for matrices of high order. For example, if a determinant of a matrix of order 10 is expanded by any row, a sum of 10 terms is obtained, where each term contains the determinant of a matrix of order 9 (that is, a cofactor of the original matrix). If any of the latter determinants is expanded by a row (or column), a sum of 9 terms is obtained, each containing the determinant of a matrix of order 8. Hence, at this stage there are 90 determinants of matrices of order 8 to evaluate! This process could be continued until only determinants of matrices of order 2 remain. Unless many elements of the original matrix are zero, it is an enormous task to carry out all of the computations.

We shall now consider some rules which make the process of evaluating determinants simpler. These rules are used mainly for introducing zeros into the determinant. They may also be used to change the determinant to echelon form, that is, a form in which the elements below the main diagonal elements a_{ii} are all zero. The transformations on rows given in (7.27) below are the same as the elementary row transformations of a matrix given in (7.17). However, for determinants, we may also employ similar transformations on columns.

(7.27)   Row and Column Transformations of a Determinant

Let A be a matrix of order n.

   (i)   If a matrix B is obtained from A by interchanging two rows (or columns), then $|B| = -|A|$.
   (ii) If B is obtained from A by multiplying every element of one row (or column) of A by a real number k, then $|B| = k|A|$.
   (iii) If B is obtained from A by adding to any row (or column) of A, k times another row (or column), where k is any real number, then $|B| = |A|$.

When using (7.27) to justify manipulations with determinants, we shall refer to the rows (or columns) of the *determinant* in the obvious way. For example, property (iii) may be phrased: "adding to any row (or column) of a determinant k times another row (or column) does not affect the value of the determinant."

We shall not give the general proof of (7.27). For the case of matrices of orders 2 or 3 the theorem can be proved by evaluating $|B|$. For example, given a matrix of order 3 as in (7.19), suppose B is the matrix obtained by adding to row 2 the product of k times row 1, that is,

$$B = \begin{bmatrix} a_{11} & a_{12} & a_{13} \\ ka_{11} + a_{21} & ka_{12} + a_{22} & ka_{13} + a_{23} \\ a_{31} & a_{32} & a_{33} \end{bmatrix}.$$

To evaluate $|B|$ we expand by row 2, obtaining

$$|B| = (ka_{11} + a_{21}) \left(- \begin{vmatrix} a_{12} & a_{13} \\ a_{32} & a_{33} \end{vmatrix} \right) + (ka_{12} + a_{22}) \begin{vmatrix} a_{11} & a_{13} \\ a_{31} & a_{33} \end{vmatrix}$$
$$+ (ka_{13} + a_{23}) \left(- \begin{vmatrix} a_{11} & a_{12} \\ a_{31} & a_{32} \end{vmatrix} \right).$$

The determinants which appear in this equation are those associated with the cofactors A_{21}, A_{22}, and A_{23} of the original matrix. Hence

$$|B| = (ka_{11} + a_{21})A_{21} + (ka_{12} + a_{22})A_{22} + (ka_{13} + a_{23})A_{23},$$

which may also be written in the form

$$|B| = k(a_{11}A_{21} + a_{12}A_{22} + a_{13}A_{23}) + (a_{21}A_{21} + a_{22}A_{22} + a_{23}A_{23}).$$

By (7.25) the second expression in parentheses equals $|A|$. By actually carrying out the computations it can be shown that the first expression in parentheses is zero. Consequently

$$|B| = 0 + |A| = |A|.$$

Statement (i) of (7.27) is often phrased thus: "interchanging two rows (or columns) changes the sign of a determinant." As illustrations we have

$$\begin{vmatrix} 2 & 0 & 1 \\ 6 & 4 & 3 \\ 0 & 3 & 5 \end{vmatrix} = - \begin{vmatrix} 6 & 4 & 3 \\ 2 & 0 & 1 \\ 0 & 3 & 5 \end{vmatrix}$$

$$\begin{vmatrix} 2 & 0 & 1 \\ 6 & 4 & 3 \\ 0 & 3 & 5 \end{vmatrix} = - \begin{vmatrix} 1 & 0 & 2 \\ 3 & 4 & 6 \\ 5 & 3 & 0 \end{vmatrix},$$

where in the first case, rows 1 and 2 were interchanged, and in the second case we interchanged columns 1 and 3.

As an illustration of (iii) of (7.27),

$$\begin{vmatrix} 1 & -3 & 4 \\ 2 & -1 & 0 \\ 3 & 1 & 6 \end{vmatrix} = \begin{vmatrix} 1 & -3 & 4 \\ 0 & 5 & -8 \\ 3 & 1 & 6 \end{vmatrix},$$

where we have added to the second row -2 times the first row. This type of manipulation is very important for evaluating determinants of large order as will be shown below in Example 1. In like manner,

$$\begin{vmatrix} 1 & -3 & 4 \\ 2 & -1 & 0 \\ 3 & 1 & 6 \end{vmatrix} = \begin{vmatrix} -5 & -3 & 4 \\ 0 & -1 & 0 \\ 5 & 1 & 6 \end{vmatrix},$$

where we have added to the first column 2 times the second column. Rule (ii) of (7.27) will be illustrated in Examples 2 and 3.

(7.28) Theorem

If two rows (or columns) of a square matrix A are identical, then $|A| = 0$.

Proof

If B is the matrix obtained from A by interchanging the two identical rows (or columns), then B and A are the same and consequently $|B| = |A|$. However, by (i) of (7.27), $|B| = -|A|$ and hence $-|A| = |A|$. This implies that $|A| = 0$.

We shall illustrate the use of the previous theorems by means of several examples.

EXAMPLE 1 Find $|A|$ if

$$A = \begin{bmatrix} 2 & 3 & 0 & 4 \\ 0 & 5 & -1 & 6 \\ 1 & 0 & -2 & 3 \\ -3 & 2 & 0 & -5 \end{bmatrix}$$

Solution

We plan on using (iii) of (7.27) to introduce zero everywhere except in one position in some row or column. To do this, it is convenient to work with an element of the matrix which equals 1 or -1, since this enables us to avoid the use of fractions. If no such element appears in the original matrix, it is always possible to introduce the number 1 by using (iii) or (ii) of (7.27). In this example there is no such problem, since 1 appears in row 3 and -1 in row 2. Let us work with the element 1 and introduce zero everywhere else in the first column, as shown below:

$$|A| = \begin{vmatrix} 0 & 3 & 4 & -2 \\ 0 & 5 & -1 & 6 \\ 1 & 0 & -2 & 3 \\ 0 & 2 & -6 & 4 \end{vmatrix}$$

Add to the first row -2 times the third row and then add to the fourth row 3 times the third row.

$$= 1 \cdot \begin{vmatrix} 3 & 4 & -2 \\ 5 & -1 & 6 \\ 2 & -6 & 4 \end{vmatrix}$$

Expand by column 1.

$$= \begin{vmatrix} 23 & 4 & 22 \\ 0 & -1 & 0 \\ -28 & -6 & -32 \end{vmatrix}$$

Add to the first column 5 times the second column and then add to the third column 6 times the second column.

$$= (-1) \begin{vmatrix} 23 & 22 \\ -28 & -32 \end{vmatrix}$$

Expand by row 2.

$$= (-1)[(23)(-32) - (-28)(22)]$$
$$= 120.$$

By (7.18).

Part (ii) of (7.27) is useful for finding factors of determinants. To illustrate, for a determinant of a matrix of order 3, we have the following:

$$\begin{vmatrix} a_{11} & a_{12} & a_{13} \\ ka_{21} & ka_{22} & ka_{23} \\ a_{31} & a_{32} & a_{33} \end{vmatrix} = k \begin{vmatrix} a_{11} & a_{12} & a_{13} \\ a_{21} & a_{22} & a_{23} \\ a_{31} & a_{32} & a_{33} \end{vmatrix}.$$

Similar formulas hold if k is a common factor of the elements of some other row or column. When (7.27) is used in this way, we often use the phrase "k is a common factor in the row (or column)."

EXAMPLE 2 Find $|A|$ if

$$A = \begin{bmatrix} 14 & -6 & 4 \\ 4 & -5 & 12 \\ -21 & 9 & -6 \end{bmatrix}.$$

Solution

$$|A| = 2 \begin{vmatrix} 7 & -3 & 2 \\ 4 & -5 & 12 \\ -21 & 9 & -6 \end{vmatrix}$$

2 is a common factor in row 1.

$$= (2)(-3) \begin{vmatrix} 7 & -3 & 2 \\ 4 & -5 & 12 \\ 7 & -3 & 2 \end{vmatrix}$$

-3 is a common factor in row 3.

$$= 0$$

By (7.28).

EXAMPLE 3 Without expanding, show that $a - b$ is a factor of

$$\begin{vmatrix} 1 & 1 & 1 \\ a & b & c \\ a^2 & b^2 & c^2 \end{vmatrix}.$$

Solution

$$\begin{vmatrix} 1 & 1 & 1 \\ a & b & c \\ a^2 & b^2 & c^2 \end{vmatrix} = \begin{vmatrix} 0 & 1 & 1 \\ a-b & b & c \\ a^2-b^2 & b^2 & c^2 \end{vmatrix}$$

Add to the first column -1 times the second column.

$$= (a-b) \begin{vmatrix} 0 & 1 & 1 \\ 1 & b & c \\ a+b & b^2 & c^2 \end{vmatrix}$$

$a-b$ is a common factor of column 1.

EXERCISES

Without expanding, explain why the statements in Exercises 1–14 are true.

1 $\begin{vmatrix} 1 & 2 & 0 \\ 0 & 1 & 2 \\ 2 & 0 & 1 \end{vmatrix} = - \begin{vmatrix} 1 & 2 & 0 \\ 2 & 0 & 1 \\ 0 & 1 & 2 \end{vmatrix}$

2 $\begin{vmatrix} 1 & 2 & 1 \\ 0 & 0 & 2 \\ 2 & 1 & 0 \end{vmatrix} = - \begin{vmatrix} 1 & 2 & 1 \\ 2 & 0 & 0 \\ 0 & 1 & 2 \end{vmatrix}$

3 $\begin{vmatrix} 1 & 1 & 0 \\ 2 & 1 & 2 \\ 1 & 2 & 1 \end{vmatrix} = \begin{vmatrix} 1 & 1 & 0 \\ 2 & 1 & 2 \\ 0 & 1 & 1 \end{vmatrix}$

4 $\begin{vmatrix} -3 & 1 & 2 \\ 4 & -2 & 6 \\ 1 & 1 & 5 \end{vmatrix} = 2 \begin{vmatrix} -3 & 1 & 2 \\ 2 & -1 & 3 \\ 1 & 1 & 5 \end{vmatrix}$

5 $\begin{vmatrix} 1 & 3 & -2 \\ 2 & 1 & 3 \\ 1 & 3 & -2 \end{vmatrix} = 0$

6 $\begin{vmatrix} 1 & -1 & 1 \\ -1 & 3 & -1 \\ 2 & 4 & 2 \end{vmatrix} = 0$

7 $\begin{vmatrix} 1 & 0 & 2 \\ -1 & 1 & 0 \\ 0 & 1 & -1 \end{vmatrix} = - \begin{vmatrix} 1 & 0 & 2 \\ 0 & 1 & -1 \\ -1 & 1 & 0 \end{vmatrix}$

8 $\begin{vmatrix} 1 & 0 & 4 \\ 2 & 1 & -2 \\ 0 & 5 & 6 \end{vmatrix} = 2 \begin{vmatrix} 1 & 0 & 2 \\ 2 & 1 & -1 \\ 0 & 5 & 3 \end{vmatrix}$

9 $\begin{vmatrix} 1 & 3 \\ -4 & 7 \end{vmatrix} = - \begin{vmatrix} 1 & 3 \\ 4 & -7 \end{vmatrix}$

10 $\begin{vmatrix} 1 & 3 \\ -4 & 7 \end{vmatrix} = - \begin{vmatrix} 3 & 1 \\ 7 & -4 \end{vmatrix}$

11 $\begin{vmatrix} 0 & 0 & 2 \\ -2 & 0 & 0 \\ 0 & 0 & 1 \end{vmatrix} = 0$

12 $\begin{vmatrix} 1 & 0 & 1 \\ 0 & 0 & 0 \\ 1 & 0 & -1 \end{vmatrix} = 0$

13 $\begin{vmatrix} 1 & 0 & -2 \\ 1 & -1 & 0 \\ 2 & 0 & -2 \end{vmatrix} = \begin{vmatrix} 1 & 0 & 0 \\ 1 & -1 & 2 \\ 2 & 0 & 2 \end{vmatrix}$

14 $\begin{vmatrix} 1 & 0 & 3 \\ 0 & -1 & 2 \\ 1 & 2 & 0 \end{vmatrix} = \begin{vmatrix} 1 & 0 & 3 \\ 0 & -1 & 2 \\ 0 & 2 & -3 \end{vmatrix}$

In each of Exercises 15–22 find the determinant of the matrix after introducing zeros as in Example 1.

15 $\begin{bmatrix} 2 & 0 & -2 \\ -3 & 1 & 0 \\ 1 & 2 & -1 \end{bmatrix}$

16 $\begin{bmatrix} 3 & 5 & -1 \\ -2 & 2 & 1 \\ 2 & -3 & 4 \end{bmatrix}$

17 $\begin{bmatrix} 4 & -2 & 3 \\ 9 & 4 & 5 \\ -3 & 2 & 7 \end{bmatrix}$

18 $\begin{bmatrix} 3 & -4 & 3 \\ -2 & 2 & 6 \\ 2 & 8 & 5 \end{bmatrix}$

19 $\begin{bmatrix} 0 & 2 & 1 & 4 \\ -2 & 1 & 0 & 5 \\ -1 & 0 & 1 & -2 \\ 3 & 1 & 0 & 2 \end{bmatrix}$

20 $\begin{bmatrix} 2 & 0 & -3 & 0 \\ 2 & -1 & 1 & 0 \\ 3 & 0 & 1 & 1 \\ -2 & 1 & 0 & 5 \end{bmatrix}$

21 $\begin{bmatrix} 1 & 0 & -1 & 0 & -3 \\ 2 & 1 & 0 & 0 & 3 \\ 0 & -2 & 0 & 1 & 2 \\ -1 & 4 & 2 & 0 & 0 \\ 0 & 0 & 1 & 3 & 0 \end{bmatrix}$

22 $\begin{bmatrix} 0 & 0 & 1 & 2 & -5 \\ 3 & 0 & -2 & 1 & 3 \\ 0 & 0 & 4 & -1 & 0 \\ -2 & 0 & 0 & 1 & 0 \\ 0 & -1 & 0 & 2 & 1 \end{bmatrix}$

23 Prove that

$$\begin{vmatrix} 1 & 1 & 1 \\ a & b & c \\ a^2 & b^2 & c^2 \end{vmatrix} = (a - b)(b - c)(c - a).$$

(*Hint:* See Example 3.)

24 Prove that

$$\begin{vmatrix} 1 & 1 & 1 \\ a & b & c \\ a^3 & b^3 & c^3 \end{vmatrix} = (a - b)(b - c)(c - a)(a + b + c).$$

25 If A is a matrix of order 4 of the form

$$A = \begin{bmatrix} a_{11} & a_{12} & a_{13} & a_{14} \\ 0 & a_{22} & a_{23} & a_{24} \\ 0 & 0 & a_{33} & a_{34} \\ 0 & 0 & 0 & a_{44} \end{bmatrix},$$

show that $|A| = a_{11}a_{22}a_{33}a_{44}$. Generalize this result to matrices of order n.

26 If

$$A = \begin{bmatrix} a & b & 0 & 0 \\ c & d & 0 & 0 \\ 0 & 0 & e & f \\ 0 & 0 & g & h \end{bmatrix},$$

prove that

$$|A| = \begin{vmatrix} a & b \\ c & d \end{vmatrix} \begin{vmatrix} e & f \\ g & h \end{vmatrix}$$

27 If $P_1(x_1, y_1)$ and $P_2(x_2, y_2)$ are distinct points in a Cartesian plane, prove that the equation

$$\begin{vmatrix} 1 & x & y \\ 1 & x_1 & y_1 \\ 1 & x_2 & y_2 \end{vmatrix} = 0$$

is an equation of the straight line through P_1 and P_2.

28 Use Exercise 27 to find an equation of the line passing through the points $P_1(3, -5)$ and $P_2(-2, 8)$.

7 CRAMER'S RULE

Determinants arise naturally in the study of solutions of systems of linear equations. To illustrate, let us consider the case of two linear equations in two unknowns:

$$(7.29) \qquad \begin{cases} a_{11}x + a_{12}y = k_1 \\ a_{21}x + a_{22}y = k_2 \end{cases}$$

where at least one nonzero coefficient appears in each equation. We may as well assume that $a_{11} \neq 0$, for otherwise $a_{12} \neq 0$ and we could regard y as the "first" variable instead of x. We shall use the matrix method to obtain the matrix of an equivalent system. Transforming the matrix of the system by means of (7.17), we obtain

$$\begin{bmatrix} a_{11} & a_{12} & k_1 \\ a_{21} & a_{22} & k_2 \end{bmatrix}$$

$$\rightarrow \begin{bmatrix} a_{11} & a_{12} & k_1 \\ 0 & a_{22} - \left(\dfrac{a_{12}a_{21}}{a_{11}}\right) & k_2 - \left(\dfrac{a_{21}k_1}{a_{11}}\right) \end{bmatrix} \quad \begin{array}{l}\text{Add to the second row} \\ -a_{21}/a_{11} \text{ times the first row.}\end{array}$$

$$\rightarrow \begin{bmatrix} a_{11} & a_{12} & k_1 \\ 0 & (a_{11}a_{22} - a_{12}a_{21}) & (a_{11}k_2 - a_{21}k_1) \end{bmatrix} \quad \text{Multiply row 2 by } a_{11}.$$

Thus the given system is equivalent to

$$a_{11}x + a_{12}y = k_1$$

$$(a_{11}a_{22} - a_{12}a_{21})y = a_{11}k_2 - a_{21}k_1$$

Notice that the numbers in the second equation may be written in determinant form as follows:

$$(7.30) \qquad \begin{cases} a_{11}x + a_{12}y = k_1 \\ \begin{vmatrix} a_{11} & a_{12} \\ a_{21} & a_{22} \end{vmatrix} y = \begin{vmatrix} a_{11} & k_1 \\ a_{21} & k_2 \end{vmatrix} . \end{cases}$$

The following results now follow from the discussion in Section 2.

(a)   If

$$\begin{vmatrix} a_{11} & a_{12} \\ a_{21} & a_{22} \end{vmatrix} = 0 \quad \text{and} \quad \begin{vmatrix} a_{11} & k_1 \\ a_{21} & k_2 \end{vmatrix} = 0,$$

then the equations are dependent.

(b)   If

$$\begin{vmatrix} a_{11} & a_{12} \\ a_{21} & a_{22} \end{vmatrix} = 0 \quad \text{and} \quad \begin{vmatrix} a_{11} & k_1 \\ a_{21} & k_2 \end{vmatrix} \neq 0,$$

then the equations are inconsistent.

(c)   If

$$\begin{vmatrix} a_{11} & a_{12} \\ a_{21} & a_{22} \end{vmatrix} \neq 0,$$

then the equations are consistent.

If (c) occurs, then we can solve the second equation of (7.30) for y, obtaining

$$(7.31) \quad y = \frac{\begin{vmatrix} a_{11} & k_1 \\ a_{21} & k_2 \end{vmatrix}}{\begin{vmatrix} a_{11} & a_{12} \\ a_{21} & a_{22} \end{vmatrix}}$$

The corresponding value for x may be found by substituting for y in the first equation. It can be shown that this leads to

$$(7.32) \quad x = \frac{\begin{vmatrix} k_1 & a_{12} \\ k_2 & a_{22} \end{vmatrix}}{\begin{vmatrix} a_{11} & a_{12} \\ a_{21} & a_{22} \end{vmatrix}}$$

This proves that *if the determinant of the coefficient matrix of (7 29) is not zero, then the system has a unique solution given by (7.31) and (7 32).* The last two formulas constitute what is known as **Cramer's Rule** for the solution of a system of two linear equations in two variables.

There is an easy way to remember **Cramer's rule.** Let

$$D = \begin{bmatrix} a_{11} & a_{12} \\ a_{21} & a_{22} \end{bmatrix}$$

be the coefficient matrix of the system and let D_x denote the matrix obtained from D by replacing the coefficients a_{11}, a_{21} of x by the numbers

k_1, k_2, respectively. Similarly, let D_y denote the matrix obtained from D by replacing the coefficients a_{12}, a_{22} of y by the numbers k_1, k_2, respectively. Thus

$$D_x = \begin{bmatrix} k_1 & a_{12} \\ k_2 & a_{22} \end{bmatrix}, \qquad D_y = \begin{bmatrix} a_{11} & k_1 \\ a_{21} & k_2 \end{bmatrix}.$$

According to (7.32) and (7.31), if $|D| \neq 0$, the solution (x, y) is given by

(7.33) $$x = \frac{|D_x|}{|D|}, \qquad y = \frac{|D_y|}{|D|}.$$

EXAMPLE 1 Use Cramer's Rule to solve the system

$$\begin{cases} 2x - 3y = -4 \\ 5x + 7y = 1. \end{cases}$$

Solution

The determinant of the coefficient matrix is

$$|D| = \begin{vmatrix} 2 & -3 \\ 5 & 7 \end{vmatrix} = 29.$$

Using the notation introduced above, we have

$$|D_x| = \begin{vmatrix} -4 & -3 \\ 1 & 7 \end{vmatrix} = -25, \qquad |D_y| = \begin{vmatrix} 2 & -4 \\ 5 & 1 \end{vmatrix} = 22$$

Hence by (7.33),

$$x = \frac{|D_x|}{|D|} = \frac{-25}{29}, \qquad y = \frac{|D_y|}{|D|} = \frac{22}{29}.$$

Thus the system has the unique solution $(-\frac{25}{29}, \frac{22}{29})$.

Let us briefly consider the case in which each of the equations in (7.29) is homogeneous, that is, $k_1 = 0$ and $k_2 = 0$. In this event the determinants $|D_x|$ and $|D_y|$ in (7.33) are both zero (why?). Consequently, if

$$|D| = \begin{vmatrix} a_{11} & a_{12} \\ a_{21} & a_{22} \end{vmatrix} \neq 0,$$

then $x = 0/|D| = 0$ and $y = 0/|D| = 0$; that is, the only solution is the trivial one $(0, 0)$. This proves that *if a system of homogeneous equations*

$$\begin{cases} a_{11}x + a_{12}y = 0 \\ a_{21}x + a_{22}y = 0 \end{cases}$$

$$\begin{vmatrix} a_{11} & a_{12} \\ a_{21} & a_{22} \end{vmatrix} = 0.$$

It can be shown, conversely, that if the determinant of the coefficient matrix is zero, then a homogeneous system has a nontrivial solution.

The previous discussion can be extended to systems of n linear equations in n variables. It is possible to show that such a system has a unique solution if and only if the determinant of the coefficient matrix is different from zero. If the system is homogeneous, then nontrivial solutions exist if and only if the determinant of the coefficient matrix is zero.

Cramer's Rule can be extended to systems of n linear equations in n variables $x_1, x_2, \cdots, x_n$, where each equation is written as in (7.7). To solve such a system, let D denote the coefficient matrix and let D_{x_i} denote the matrix obtained by replacing the coefficients of x_i, as in D by the column of numbers $k_1, \cdots, k_n$ which appears to the right of the equals signs in the system. It can be shown that if $|D| \neq 0$, then the system has a unique solution given by

$$(7.34) \quad x_1 = \frac{|D_{x_1}|}{|D|}, \quad x_2 = \frac{|D_{x_2}|}{|D|}, \cdots, \quad x_n = \frac{|D_{x_n}|}{|D|}.$$

If $|D| = 0$, the equations are dependent or inconsistent, depending on whether all the D_{x_i} are zero or at least one of them is not zero.

EXAMPLE 2 Use Cramer's Rule to solve the system

$$\begin{cases} x & - 2z = 3 \\ & -y + 3z = 1 \\ 2x & + 5z = 0. \end{cases}$$

Solution

We shall merely list the various determinants which are used, leaving the reader to check the answers:

$$|D| = \begin{vmatrix} 1 & 0 & -2 \\ 0 & -1 & 3 \\ 2 & 0 & 5 \end{vmatrix} = -9, \quad |D_x| = \begin{vmatrix} 3 & 0 & -2 \\ 1 & -1 & 3 \\ 0 & 0 & 5 \end{vmatrix} = -15,$$

$$|D_y| = \begin{vmatrix} 1 & 3 & -2 \\ 0 & 1 & 3 \\ 2 & 0 & 5 \end{vmatrix} = 27, \quad |D_z| = \begin{vmatrix} 1 & 0 & 3 \\ 0 & -1 & 1 \\ 2 & 0 & 0 \end{vmatrix} = 6.$$

By (7.34) the solution is

$$x = \frac{|D_x|}{|D|} = \frac{-15}{-9} = \frac{5}{3}, \quad y = \frac{|D_y|}{|D|} = \frac{27}{-9} = -3, \quad z = \frac{|D_z|}{|D|} = \frac{6}{-9} = \frac{-2}{3}.$$

In general, Cramer's Rule is difficult to apply if there are a large number of equations, since many determinants must be evaluated. Note also that Cramer's Rule cannot be used if $|D| = 0$, or if the number of equations is not the same as the number of variables. In these cases the matrix method or the method of substitution should be employed.

EXERCISES

1–8 Use Cramer's Rule to solve Exercises 7–14 of Section 2.

9–14 Use Cramer's Rule to solve Exercises 1–6 of Section 3.

15 Use Cramer's Rule to solve Exercise 13 of Section 3.

16 Use Cramer's Rule to solve Example 1 of Section 4.

8 SYSTEMS OF INEQUALITIES

Our previous work with inequalities was restricted to inequalities in one variable. The notion of inequalities in several variables can be developed in a manner similar to our work with equations in several variables. For example, if p, q, and r are expressions in two variables x and y, then statements of the form $p < q$, $p \le q$, $q \ge p > r$, and so on, are called **inequalities in x and y.** As with equations, a **solution** of an inequality is defined as an ordered pair (a, b) which produces a true statement when a and b are substituted for x and y, respectively. The **solution set** of an inequality is the collection of all solutions. The **graph of an inequality** is the graph of its solution set. Two inequalities are **equivalent** if they have the same solution sets. It is possible to prove the analogue of (1.12) in the present situation; that is, an inequality in two variables can be simplified by adding an expression in x and y to both sides or by multiplying both sides by such an expression, provided we are careful about signs. Similar definitions and remarks hold for inequalities in more than two variables. We shall restrict our discussion, however, to the case of inequalities in two variables.

EXAMPLE 1 Find the solution set and sketch the graph of the inequality $3x - 3 < 5x - y$.

Solution

Adding the expression $y + 3 - 3x$ to both sides, we obtain the equivalent inequality $y < 2x + 3$. Hence the solution set may be written as

$$\{(x, y) : y < 2x + 3\},$$

where it is understood that x and y are real numbers.

There is a close relationship between the graph of the inequality $y < 2x + 3$ and the graph of the equation $y = 2x + 3$. The graph of the equation is the straight line sketched in Fig. 7.5. For each real number a, the point on

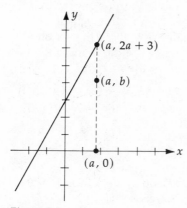

Figure 7.5

the line with this abscissa has coordinates $(a, 2a + 3)$. A point $P(a, b)$ belongs to the graph of the *inequality* if and only if $b < 2a + 3$, that is, if and only if the point $P(a, b)$ lies directly below the point with coordinates $(a, 2a + 3)$ as shown in Fig. 7.5. It follows that the graph of the inequality $y < 2x + 3$ consists of all points in the plane which lie below the line $y = 2x + 3$. In Fig. 7.6 we have shaded a portion of the graph. Dashes used for the line indicate that it is not part of the graph of the inequality. A region of this type is called a **half-plane.** More precisely, when the line is *not* included, we refer to such a region as an **open half-plane.** If the line *is* included, as would be the case for the graph of the inequality $y \le 2x + 3$, then the region is called a **closed half-plane.**

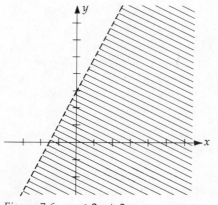

Figure 7.6 $y < 2x + 3$

By an argument similar to that used in Example 1, it can be shown that the graph of the inequality $y > 2x + 3$ is the open half-plane which lies *above* the line with equation $y = 2x + 3$.

If an inequality involves only polynomials of the first degree in x and y, as was the case in Example 1, it is called a **linear inequality.**

The procedure used in Example 1 can be generalized to inequalities of the form $y < f(x)$, where f is a function. Specifically, it can be shown that the graph of the inequality $y < f(x)$ is the set of points which lie *below* the graph of the equation $y = f(x)$. Similarly, the graph of $y > f(x)$ is the set of points which lie *above* the graph of $y = f(x)$.

EXAMPLE 2 Find the solution set and sketch the graph of the inequality $x(x + 1) - 2y > 3(x - y)$.

Solution

The given inequality is equivalent to $x^2 + x - 2y > 3x - 3y$. Adding $3y - x^2 - x$ to both sides we obtain $y > 2x - x^2$ and hence the solution set is given by $\{(x, y) : y > 2x - x^2\}$. To find the graph, we begin by sketching the graph of $y = 2x - x^2$ with dashes as illustrated in Fig. 7.7, and then we shade the region above the graph.

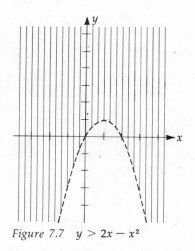

Figure 7.7 $y > 2x - x^2$

It can be shown that the graph of an inequality of the form $x < g(y)$, *where g is a* function, is the set of points to the *left* of the graph of the equation $x = g(y)$. Similarly, the graph of $x > g(y)$ is the set of points to the *right* of the graph of $x = g(y)$. If $\geq$ or $\leq$ is used, then we also include the graphs of the equation $x = g(y)$.

EXAMPLE 3 Sketch the graph of $x \geq y^2$.

Solution

We first sketch the graph of the equation $x = y^2$. The graph of the inequality consists of all points on this parabola together with the points in the region to the right of this curve (see Fig. 7.8).

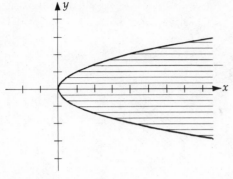

Figure 7.8 $x \geq y^2$

It is sometimes necessary to work simultaneously with several inequalities in two variables. In this case we refer to the given inequalities as a **system of inequalities.** The **solution set of a system** of inequalities is, by definition, the intersection of the solution sets of all the inequalities in the system. It should be clear how to define **equivalent systems** and the **graph of a system** of inequalities.

The following examples illustrate a method for solving systems of inequalities.

EXAMPLE 4   Find the solution set and sketch the graph of the system

$$\begin{cases} x + y \leq 4 \\ 2x - y \leq 4. \end{cases}$$

Solution

The given system is equivalent to

$$\begin{cases} y \leq 4 - x \\ y \geq 2x - 4. \end{cases}$$

We begin by sketching the graphs of the lines $y = 4 - x$ and $y = 2x - 4$. These lines intersect at the point $(\frac{8}{3}, \frac{4}{3})$ shown in Fig. 7.9. The graph of $y \leq 4 - x$ includes the points on the graph of $y = 4 - x$ together with the points which lie below this line. The graph of $y \geq 2x - 4$ includes the points on the graph of $y = 2x - 4$ together with the points which lie above this line. A portion of each of these regions is shown in Fig. 7.9. The graph of the system consists of the points that are in *both* regions. This corresponds to the cross-hatched region shown in Fig. 7.9. We may denote the solution set of the system symbolically by

$$\{(x, y) : y \leq 4 - x\} \cap \{(x, y) : y \geq 2x - 4\}.$$

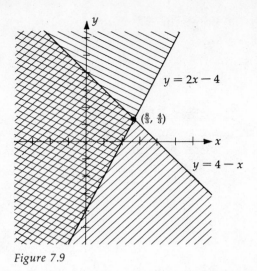

$y = 2x - 4$

$(\frac{8}{3}, \frac{4}{3})$

$y = 4 - x$

Figure 7.9

EXAMPLE 5 Sketch the graph of the system

$$\begin{cases} x + y \leq 4 \\ 2x - y \leq 4 \\ x \geq 0 \\ y \geq 0. \end{cases}$$

Solution

The first two inequalities are the same as those considered in Example 4 and hence the points on the graph of the present system must lie within the region shown in Fig. 7.9. In addition, the third and fourth inequalities in the system tell us that the points must lie in the first quadrant. This gives us the region shown in Fig. 7.10.

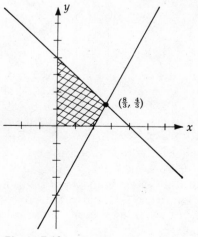

$(\frac{8}{3}, \frac{4}{3})$

Figure 7.10

EXAMPLE 6 Sketch the graph of the system

$$\begin{cases} x^2 + y^2 \le 1 \\ (x-1)^2 + y^2 \le 1. \end{cases}$$

Solution

The graph of the equation $x^2 + y^2 = 1$ is a unit circle with center at the origin, and the graph of $(x-1)^2 + y^2 = 1$ is a unit circle with center at the point $C(1, 0)$. By the distance formula, the graphs of the given inequalities are the regions within and on these circles. The graph of the system is the intersection of these regions, as illustrated in Fig. 7.11.

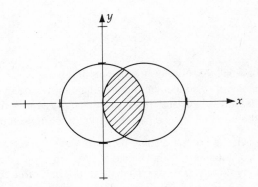

Figure 7.11

EXERCISES

In each of Exercises 1–10 find the solution set and sketch the graph of the given inequality.

1 $y > 2x + 1$

2 $3x + 2y < 6$

3 $2y + 1 \le 3x + y$

4 $2y + 3(x + 1) > 4x$

5 $y + 3 < x^2$

6 $y - x^2 \ge 4$

7 $y \ge x^2 + 2$

8 $x > y^2 + 1$

9 $y - 1 < x^3$

10 $yx^2 \le 1$

In each of Exercises 11–24 sketch the graph of the given system.

11 $\begin{cases} 3y + 2x > 6 \\ x < y \end{cases}$

12 $\begin{cases} 2y \le x + 4 \\ 3x + 2y + 12 < 0 \end{cases}$

13 $\begin{cases} 2x + y - 4 < 0 \\ 2 - y - 2x < 0 \end{cases}$

14 $\begin{cases} 2y - x > 1 \\ 2y + 1 < x \end{cases}$

15 $\begin{cases} y + 4 \ge 3x \\ x + y \le 6 \\ x \ge 0 \\ y \ge 0 \end{cases}$

16 $\begin{cases} 2y - x \le 12 \\ x \le y + 3 \\ x \ge 1 \\ y \ge 1 \end{cases}$

17 $\begin{cases} x^2 + y^2 \le 1 \\ y - x \ge 1 \end{cases}$

18 $\begin{cases} y \le 2x + 2 \\ 2y + 5x \le 13 \\ y + x + 1 \ge 0 \end{cases}$

19 $\begin{cases} y + x^2 \le 4 \\ y + x + 2 \ge 0 \end{cases}$

20 $\begin{cases} x^2 + y^2 \le 4 \\ x^2 \ge 3y \end{cases}$

21 $\begin{cases} y < 3^x \\ y > 2^x \\ x \ge 0 \end{cases}$

22 $\begin{cases} y \le \log x \\ y \ge 1 - x \end{cases}$

23 $\begin{cases} y - x^3 \le 1 \\ y - 2^{-x} \ge 0 \\ x - 4 \le 0 \end{cases}$

24 $\begin{cases} x^2 + y^2 \le 4 \\ x^2 + y^2 \ge 1 \end{cases}$

25 . The manager of a baseball team wishes to buy bats and balls, costing $3.50 and $2.50, respectively. If the maximum amount he can spend is $40 and he wants to buy at least two balls and three bats, write a system of inequalities describing all the possibilities and sketch the graph of the solution set.

26 An office worker wishes to purchase some 8-cent postage stamps and also some 10-cent stamps, totaling not more than $7.00. Moreover, he wants at least twice as many 8-cent stamps as 10-cent stamps and more than ten 8-cent stamps. Write a system of inequalities describing all the possibilities and sketch the graph of the solution set.

9 LINEAR PROGRAMMING

In applications problems sometimes arise which require solving a system of linear inequalities. A typical problem is that of finding maximum and minimum values of certain expressions which are subject to various constraints. The branch of mathematics which deals with problems of this type is referred to as *linear programming*. This area of mathematics has become very important in business and economics. The logical development of the theorems and techniques of linear programming would take us beyond the objectives of our present work. We shall therefore limit ourselves to an illustration of one type of problem.

EXAMPLE A manufacturer of a certain product has two warehouses W_1 and W_2. There are 80 units of his product stored at W_1 and 70 units at W_2. Two customers X and Y order 35 units and 60 units, respectively. The shipping cost from each warehouse to X and Y is determined according to the table at the top of p. 361. How should the order be filled so as to minimize the total shipping cost?

Warehouse	Customer	Shipping cost per unit
W_1	X	$ 8
W_1	Y	12
W_2	X	10
W_2	Y	13

Solution

If we let x denote the number of units to be sent to X from W_1, then $35 - x$ units must be sent from W_2 to X. Similarly, if y denotes the number of units to be sent from W_1 to Y, then $60 - y$ units must be sent from W_2 to y. We wish to determine values for x and y which make the total shipping costs minimal. We first note that since x and y are between 35 and 60, respectively, the pair (x, y) must be in the solution set of the following system of inequalities:

(7.35) $$0 \leq x \leq 35, \qquad 0 \leq y \leq 60.$$

The graph of the solution set is the rectangular region shown in Fig. 7.12.

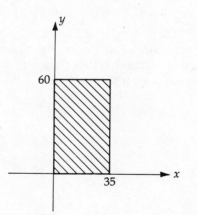

Figure 7.12

There are further constraints on x and y which make it possible to reduce the size of the region above. Since the total number of units shipped from W_1 cannot exceed 80 and the total shipped from W_2 cannot exceed 70, the pair (x, y) must also be in the solution set of the system

$$\begin{cases} x + y \leq 80 \\ (35 - x) + (60 - y) \leq 70. \end{cases}$$

This system is equivalent to

(7.36) $\begin{cases} x + y \leq 80 \\ x + y \geq 25. \end{cases}$

The graph of the solution set of (7.36) is the region between the parallel lines $x + y = 80$ and $x + y = 25$ (see Fig. 7.13).

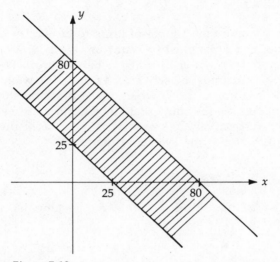

Figure 7.13

The pair (x, y) which we seek must lie in the intersection of the solution sets of (7.35) and (7.36). The corresponding graph is the polygonal region shown in Fig. 7.14.

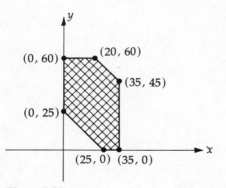

Figure 7.14

Now let C denote the total cost (in dollars) of shipping the merchandise to X and Y. We see from the table that the cost of shipping the 35 units to X is $8x + 10(35 - x)$, whereas the cost of shipping the 60 units to Y is $12y + 13(60 - y)$. Hence the total cost is

$$C = (8x + 350 - 10x) + (12y + 780 - 13y),$$

or

$$C = 1130 - 2x - y.$$

For each point (x, y) of the region shown in Fig. 7.14, we obtain a value for C. For example, at $(20, 40)$,

$$C = 1130 - 40 - 40 = 1050;$$

at $(10, 50)$,

$$C = 1130 - 20 - 50 = 1060;$$

and so on.

Since x and y must be integers, there are only a finite number of possible values for C. By checking each possibility we could find the pair (x, y) which produces the smallest cost. However, since there are a very large number of pairs, the task of checking each one would be very tedious. This is where the theory developed in linear programming is helpful. It can be shown that if we are interested in the value C of a linear expression $ax + by + c$, where a, b and c are real numbers and where each pair (x, y) corresponds to a point in a plane polygonal region which is *convex*, in the sense that for every pair of points A, B in the region, all points on the line segment AB are in the region, then C takes on its maximum and minimum values at the *vertices* of the polygon. This means that in order to determine the minimum (or maximum) values of C we need only check the vertices $(0, 25)$, $(0, 60)$, $(20, 60)$, $(35, 45)$, $(35, 0)$, and $(25, 0)$. These values are arranged in tabular form below.

Vertex	$1130 - 2x - y = C$
$(0, 25)$	$1130 - 2(0) - 25 = 1105$
$(0, 60)$	$1130 - 2(0) - 60 = 1070$
$(20, 60)$	$1130 - 2(20) - 60 = 1030$
$(35, 45)$	$1130 - 2(35) - 45 = 1015$
$(35, 0)$	$1130 - 2(35) - 0 = 1060$
$(25, 0)$	$1130 - 2(25) - 0 = 1080$

According to our remarks, the minimal shipping cost \$1015 occurs if $x = 35$ and $y = 45$. This means that the manufacturer should ship all of the units to X from W_1. In addition, the manufacturer should ship 45 units to Y from W_1 and 15 units to Y from W_2. Note that the *maximum* shipping cost will occur when $x = 0$ and $y = 25$, that is, when all 35 units are shipped to X from W_2 and when Y receives 25 units from W_1 and 35 units from W_2.

The above illustration is an elementary problem of linear programming which can be solved by rather crude methods. The much more complicated problems that occur are usually solved by employing techniques which are adapted for solutions by computers.

EXERCISES

1   If, in the example of this section, the shipping costs are $12 per unit from W_1 to X, $10 per unit from W_2 to X, $16 per unit from W_1 to Y, and $12 per unit from W_2 to Y, determine how the order should be filled so as to minimize shipping costs.

2   A coffee company purchases mixed lots of coffee beans and then grades them into premium, regular, and unusable beans. The company needs at least 280 tons of premium-grade and 200 tons of regular-grade coffee beans. The company can purchase ungraded coffee from two suppliers A and B in any amount desired. Samples from the two suppliers contain the following percentages of premium, regular, and unusable beans:

Supplier	Premium	Regular	Unusable
A	20%	50%	30%
B	40%	20%	40%

If A charges $125 per ton and B charges $200 per ton, how much should the company purchase from each supplier to fulfill its needs at minimum cost?

3   A farmer has 100 acres available for planting two crops A and B. The seed for crop A costs $4 per acre and the seed for crop B costs $6 per acre. The total cost of labor will amount to $20 per acre for crop A and $10 per acre for crop B. The expected income from crop A is $110 per acre, whereas from crop B it is $150 per acre. If the farmer does not wish to spend more than $480 for seed and $1400 for labor, how many acres of each crop should be planted in order to obtain the maximum profit?

4   A firm manufactures two products A and B. For each product it is necessary to employ two different machines X and Y. To manufacture product A, machine X must be used for $\frac{1}{2}$ hour and machine Y for 1 hour. To manufacture product B, machine X must be used for 2 hours and machine Y for 2 hours. The profit on product A is $20 per unit and the profit on B is $50 per unit.

If machine X can be used for 8 hours per day and machine Y for 12 hours per day, determine how many units of each product should be manufactured each day in order to maximize the profit.

5 Three substances X, Y, and Z each contain four ingredients A, B, C, and D. The percentage of each ingredient and the cost in cents per ounce of the substances are given in the table below.

Substance	A	B	C	D	Cost/ Ounce
X	20%	10%	25%	45%	25¢
Y	20%	40%	15%	25%	35¢
Z	10%	20%	25%	45%	50¢

If the cost is to be minimum, how many ounces of each substance should be combined in order to obtain a mixture of 20 ounces containing at least 14% A, 16% B, and 20% C? What combination would make the cost greatest?

6 A man plans to operate a stand at a one-day fair in which he will sell bags of peanuts and bags of candy. He has $100 available to purchase his stock, which will cost 10¢ per bag of peanuts and 20¢ per bag of candy. He intends to sell the peanuts at 15¢ and the candy at 26¢ per bag. His stand can accommodate up to 500 bags of peanuts and 400 bags of candy. From past experience he knows that he will sell no more than a total of 700 bags. Find the number of bags of each that he should have available in order to maximize his profit. What is the maximum profit?

10 THE ALGEBRA OF MATRICES

It is possible to develop a comprehensive theory for matrices which has many mathematical and scientific applications. In this section we shall discuss several basic algebraic properties of matrices which serve as the starting point for such a theory. Our work will be restricted to matrices whose elements are real numbers.

In order to conserve space it is sometimes convenient to denote an $m \times n$ matrix A of the type displayed in (7.16) by the symbol (a_{ij}). If (b_{ij}) denotes another $m \times n$ matrix B, then we say that A and B are **equal** and we write $A = B$, if and only if $a_{ij} = b_{ij}$ for every i and j. For example

$$\begin{bmatrix} 1 & 0 & 5 \\ \sqrt[3]{8} & 3^2 & -2 \end{bmatrix} = \begin{bmatrix} (-1)^2 & 0 & \sqrt{25} \\ 2 & 9 & -2 \end{bmatrix}.$$

If $A = (a_{ij})$ and $B = (b_{ij})$ are $m \times n$ matrices, then their **sum** $A + B$ is defined as the $m \times n$ matrix $C = (c_{ij})$, where $c_{ij} = a_{ij} + b_{ij}$ for all i and j. Thus, to add two matrices we add the elements which appear in corresponding positions in each matrix. Note that two matrices can be added only if they have the same number of rows and the same number of columns. Using the parentheses notation,

$$(a_{ij}) + (b_{ij}) = (a_{ij} + b_{ij}).$$

Although we have used the symbol $+$ in two different ways, there is little chance for confusion, since whenever $+$ appears between symbols for matrices it refers to matrix addition, and when $+$ is used between real numbers it denotes their sum. An example of the sum of two 3×2 matrices is

$$\begin{bmatrix} 4 & -5 \\ 0 & 4 \\ -6 & 1 \end{bmatrix} + \begin{bmatrix} 3 & 2 \\ 7 & -4 \\ -2 & 1 \end{bmatrix} = \begin{bmatrix} 7 & -3 \\ 7 & 0 \\ -8 & 2 \end{bmatrix}.$$

It is not difficult to prove that addition of matrices is both a commutative and associative operation, that is,

$$A + B = B + A, \qquad A + (B + C) = (A + B) + C$$

for all $m \times n$ matrices A, B, and C.

The $m \times n$ **zero matrix,** denoted by O, is the matrix with m rows and n columns in which every element is 0. It is an identity element relative to addition, since $A + O = A$ for every $m \times n$ matrix A. For example,

$$\begin{bmatrix} a_{11} & a_{12} \\ a_{21} & a_{22} \\ a_{31} & a_{32} \end{bmatrix} + \begin{bmatrix} 0 & 0 \\ 0 & 0 \\ 0 & 0 \end{bmatrix} = \begin{bmatrix} a_{11} & a_{12} \\ a_{21} & a_{22} \\ a_{31} & a_{32} \end{bmatrix}.$$

The **additive inverse** $-A$ of the matrix $A = (a_{ij})$ is, by definition, the matrix $(-a_{ij})$ obtained by changing the sign of each element of A. For example,

$$-\begin{bmatrix} 2 & -3 & 4 \\ -1 & 0 & 5 \end{bmatrix} = \begin{bmatrix} -2 & 3 & -4 \\ 1 & 0 & -5 \end{bmatrix}.$$

It follows that for every $m \times n$ matrix A,

$$A + (-A) = 0.$$

Subtraction of two $m \times n$ matrices is defined by

$$A - B = A + (-B).$$

Using the parentheses notation for matrices, this implies that

$$(a_{ij}) - (b_{ij}) = (a_{ij}) + (-b_{ij}) = (a_{ij} - b_{ij})$$

Thus, to subtract two matrices we merely subtract the elements in the same positions.

The **product** of a real number c and an $m \times n$ matrix $A = (a_{ij})$ is defined by

$$cA = (ca_{ij})$$

that is, we multiply each element of A by c. For example,

$$3 \begin{bmatrix} 4 & -1 \\ 2 & 3 \end{bmatrix} = \begin{bmatrix} 12 & -3 \\ 6 & 9 \end{bmatrix}.$$

It should be noted that this is *not* the same as rule (ii) of (7.27) for determinants.

The following results may be established, where A and B are $m \times n$ matrices, and c and d are any real numbers.

$$c(A + B) = cA + cB,$$
$$(7.37) \quad (c + d)A = cA + dA,$$
$$(cd)A = c(dA).$$

Those formulas are used in advanced courses in mathematics.

The following definition of the product of two matrices may appear unusual to the beginning student. Although there are many applications which justify the form of the definition, we shall not discuss them in this text. In order to define the product AB of two matrices A and B, *the number of columns of A must be the same as the number of rows of B.* Suppose that $A = (a_{ij})$ is $m \times n$ and $B = (b_{ij})$ is $n \times p$. To determine the element c_{ij} of the product, we single out row i of A and column j of B as illustrated in (7.38).

$$(7.38) \quad \begin{bmatrix} a_{11} & a_{12} & \cdots & a_{1n} \\ & & & \\ & & & \\ \hline a_{i1} & a_{i2} & \cdots & a_{in} \\ & & & \\ & & & \\ a_{m1} & a_{m2} & \cdots & a_{mn} \end{bmatrix} \begin{bmatrix} b_{11} & \cdots & b_{1j} & \cdots & b_{1p} \\ b_{21} & \cdots & b_{2j} & \cdots & b_{2p} \\ & & & & \\ & & & & \\ b_{n1} & \cdots & b_{nj} & \cdots & b_{np} \end{bmatrix}$$

Next we multiply pairs of elements and then add them, according to this formula:

$$(7.39) \quad c_{ij} = a_{i1}b_{1j} + a_{i2}b_{2j} + \cdots + a_{in}b_{nj}.$$

For example, the element c_{11} in the first row and the first column of AB is given by $c_{11} = a_{11}b_{11} + a_{12}b_{21} + \cdots + a_{1n}b_{n1}$. The element c_{12} in the first row and second column of AB is $c_{12} = a_{11}b_{12} + a_{12}b_{22} + \cdots + a_{1n}b_{n2}$, and so on. By definition, the product AB has the same number of rows as A and the same number of columns as B. In particular, if A is $m \times n$ and B is $n \times p$, then AB is $m \times p$. This is illustrated by the following product of a 2×3 matrix and a 3×4 matrix:

$$\begin{bmatrix} 1 & 2 & -3 \\ 4 & 0 & -2 \end{bmatrix} \begin{bmatrix} 5 & -4 & 2 & 0 \\ -1 & 6 & 3 & 1 \\ 7 & 0 & 4 & 8 \end{bmatrix} = \begin{bmatrix} -18 & 8 & -4 & -20 \\ 6 & -16 & 0 & -16 \end{bmatrix}.$$

Here are some typical computations of the elements c_{ij} in the product:

$$c_{11} = (1)(5) + (2)(-1) + (-3)(7) = 5 - 2 - 21 = -18$$
$$c_{13} = (1)(2) + (2)(3) + (-3)(4) = 2 + 6 - 12 = -4$$
$$c_{23} = (4)(2) + (0)(3) + (-2)(4) = 8 + 0 - 8 = 0$$
$$c_{24} = (4)(0) + (0)(1) + (-2)(8) = 0 + 0 - 16 = -16.$$

The reader should check the remaining elements.

The product operation for matrices is not commutative. Indeed, if A is 2×3 and B is 3×4, then AB may be found, but BA is undefined since the number of columns of B is different from the number of rows of A. Even if AB and BA are both defined, it is often true that these products are different. This is illustrated in the next example, along with the fact that the product of two nonzero matrices may equal a zero matrix.

EXAMPLE 1 If $A = \begin{bmatrix} 2 & 2 \\ -1 & -1 \end{bmatrix}$ and $B = \begin{bmatrix} 1 & 2 \\ 1 & 2 \end{bmatrix}$, show that $AB \neq BA$.

Solution

Using the definition of product we obtain

$$AB = \begin{bmatrix} 4 & 8 \\ -2 & -4 \end{bmatrix} \quad \text{and} \quad BA = \begin{bmatrix} 0 & 0 \\ 0 & 0 \end{bmatrix}.$$

Hence $AB \neq BA$. Note that BA is a zero matrix.

It is possible to show that matrix multiplication is associative in the sense that $A(BC) = (AB)C$, provided the indicated products are defined. This will be the case when A is $m \times n$, B is $n \times p$, and C is $p \times q$. The distributive laws also hold if the matrices involved have the proper number of rows and columns. Thus if A_1 and A_2 are $m \times n$ matrices and if B_1 and B_2 are $n \times p$ matrices, then

$$A_1(B_1 + B_2) = A_1 B_1 + A_1 B_2$$

and

$$(A_1 + A_2)B_1 = A_1 B_1 + A_2 B_1.$$

As a special case, if all matrices are square, of order n, then the associative and distributive laws are true. We shall not give proofs for these theorems.

EXERCISES

In Exercises 1–8 find $A + B$, $A - B$, $3A$, and $-2B$.

1 $A = \begin{bmatrix} 2 & -1 \\ 3 & 4 \end{bmatrix}$, $B = \begin{bmatrix} 1 & 3 \\ -2 & 5 \end{bmatrix}$ **2** $A = \begin{bmatrix} 4 & 7 \\ 0 & 1 \end{bmatrix}$, $B = \begin{bmatrix} -2 & -1 \\ 8 & 3 \end{bmatrix}$

3 $A = \begin{bmatrix} 4 & 0 \\ -7 & 2 \\ 1 & -5 \end{bmatrix}$, $B = \begin{bmatrix} 1 & -3 \\ 6 & 1 \\ 0 & 2 \end{bmatrix}$

4 $A = \begin{bmatrix} 1 & 0 & -2 \\ 4 & 6 & -1 \end{bmatrix}$, $B = \begin{bmatrix} -2 & 1 & 7 \\ -5 & 2 & 1 \end{bmatrix}$

5 $A = \begin{bmatrix} 1 & 0 & 5 \end{bmatrix}$, $B = \begin{bmatrix} -3 & 6 & 2 \end{bmatrix}$

6 $A = \begin{bmatrix} 7 \\ -6 \end{bmatrix}$, $B = \begin{bmatrix} -8 \\ 2 \end{bmatrix}$

7 $A = \begin{bmatrix} 1 & 0 & 8 & 7 \\ 2 & -1 & -3 & 0 \end{bmatrix}$, $B = \begin{bmatrix} -1 & 0 & -8 & -7 \\ 5 & 1 & 0 & -1 \end{bmatrix}$

8 $A = \begin{bmatrix} 8 \end{bmatrix}$, $B = \begin{bmatrix} -3 \end{bmatrix}$

In Exercises 9–16 find AB and BA.

9 $A = \begin{bmatrix} 3 & 1 \\ -2 & 4 \end{bmatrix}$, $B = \begin{bmatrix} 2 & 5 \\ -1 & 6 \end{bmatrix}$ **10** $A = \begin{bmatrix} 2 & -1 \\ 0 & 5 \end{bmatrix}$, $B = \begin{bmatrix} 1 & 3 \\ -2 & 0 \end{bmatrix}$

11 $A = \begin{bmatrix} 1 & 0 & -2 \\ 0 & 5 & -1 \\ 3 & 1 & 4 \end{bmatrix}$, $B = \begin{bmatrix} -2 & -1 & 4 \\ 1 & 0 & 1 \\ 3 & 2 & -3 \end{bmatrix}$

12 $A = \begin{bmatrix} 3 & 0 & 0 \\ 0 & 2 & 0 \\ 0 & 0 & -1 \end{bmatrix}$, $B = \begin{bmatrix} 2 & 0 & 0 \\ 0 & -4 & 0 \\ 0 & 0 & 5 \end{bmatrix}$

13 $A = \begin{bmatrix} 5 & -2 & 7 \\ -6 & 1 & 3 \end{bmatrix}$, $B = \begin{bmatrix} 1 & 4 \\ -1 & 0 \\ 3 & -2 \end{bmatrix}$

14 $A = \begin{bmatrix} 1 & -1 & 2 & 3 \\ 0 & 2 & -4 & 1 \\ 5 & -3 & 0 & 1 \end{bmatrix}$, $B = \begin{bmatrix} 3 & -5 & 1 \\ 1 & 0 & -1 \\ 0 & 1 & 4 \\ -1 & 2 & 0 \end{bmatrix}$

15 $A = \begin{bmatrix} 2 & -1 & 5 \end{bmatrix}$, $B = \begin{bmatrix} 3 \\ 2 \\ 1 \end{bmatrix}$ **16** $A = \begin{bmatrix} 3 & 5 \end{bmatrix}$, $B = \begin{bmatrix} -2 \\ 1 \end{bmatrix}$

In Exercises 17–20 find AB.

17 $A = \begin{bmatrix} 2 & -3 \\ 0 & 1 \\ -1 & 4 \end{bmatrix}$, $B = \begin{bmatrix} 5 \\ 3 \end{bmatrix}$ **18** $A = \begin{bmatrix} 8 \\ -7 \\ 5 \end{bmatrix}$, $B = [4 \quad 6]$

19 $A = \begin{bmatrix} 1 & 0 & -2 & 3 \\ 0 & -5 & 1 & 7 \end{bmatrix}$, $B = \begin{bmatrix} 2 & 0 & 1 \\ 1 & -3 & -1 \\ 0 & 4 & 0 \\ -1 & 0 & 5 \end{bmatrix}$

20 $A = \begin{bmatrix} 5 & 2 & -7 \\ -1 & 6 & 0 \end{bmatrix}$, $B = \begin{bmatrix} 2 & -1 & 0 & 1 \\ 1 & 2 & 5 & -3 \\ 0 & 0 & 0 & 1 \end{bmatrix}$

21 If $A = \begin{bmatrix} 3 & 4 \\ 0 & 1 \end{bmatrix}$ and $B = \begin{bmatrix} 1 & 0 \\ -2 & -3 \end{bmatrix}$, show that $(A + B)(A - B) \neq A^2 - B^2$, where $A^2 = AA$ and $B^2 = BB$.

22 If A and B are the matrices of Exercise 21, show that $(A + B)(A + B) \neq A^2 + 2AB + B^2$.

23 If A and B are the matrices of Exercise 21 and $C = \begin{bmatrix} 2 & 4 \\ 5 & 1 \end{bmatrix}$, prove that $A(B + C) = AB + AC$.

24 If A, B, and C are the matrices of Exercise 23 prove that $A(BC) = (AB)C$.

Prove the identities given in Exercises 25–28, where $A = \begin{bmatrix} a_{11} & a_{12} \\ a_{21} & a_{22} \end{bmatrix}$, $B = \begin{bmatrix} b_{11} & b_{12} \\ b_{21} & b_{22} \end{bmatrix}$, $C = \begin{bmatrix} c_{11} & c_{12} \\ c_{21} & c_{22} \end{bmatrix}$, and c, d are real numbers.

25 $c(A + B) = cA + cB$ **26** $(c + d)A = cA + dA$

27 $A(B + C) = AB + AC$ **28** $A(BC) = (AB)C$

11 INVERSES OF MATRICES

Throughout this section we shall concentrate on square matrices. The symbol I_n will be used to denote the square matrix of order n which has 1 in each position on the main diagonal and 0 elsewhere. For example,

$$I_2 = \begin{bmatrix} 1 & 0 \\ 0 & 1 \end{bmatrix}, \qquad I_3 = \begin{bmatrix} 1 & 0 & 0 \\ 0 & 1 & 0 \\ 0 & 0 & 1 \end{bmatrix},$$

and so on. It can be shown that if A is any square matrix of order n, then

$$AI_n = A = I_nA.$$

For this reason I_n is called an **identity matrix of order** n. To illustrate, if $A = (a_{ij})$ is of order 2, then a direct calculation shows that

$$\begin{bmatrix} a_{11} & a_{12} \\ a_{21} & a_{22} \end{bmatrix}\begin{bmatrix} 1 & 0 \\ 0 & 1 \end{bmatrix} = \begin{bmatrix} a_{11} & a_{12} \\ a_{21} & a_{22} \end{bmatrix} = \begin{bmatrix} 1 & 0 \\ 0 & 1 \end{bmatrix}\begin{bmatrix} a_{11} & a_{12} \\ a_{21} & a_{22} \end{bmatrix}.$$

Some, but not all, $n \times n$ matrices A have **multiplicative inverses** in the sense that there is a matrix B such that $AB = I_n = BA$. If A has a multiplicative inverse we denote it by A^{-1} and write

$$AA^{-1} = I_n = A^{-1}A.$$

The symbol A^{-1} is read "A inverse." In matrix theory it is *not* acceptable to use the symbol $1/A$ in place of A^{-1}.

In order to find A^{-1}, when it exists, it is convenient to employ the concept of the **transpose** A' of the matrix A. By definition, if $A = (a_{ij})$, then

$$A' = (a'_{ij}), \text{ where } a'_{ij} = a_{ji}.$$

In words, to find A' we interchange the rows and columns of A. For example,

$$\begin{bmatrix} 1 & 2 & 3 \\ 4 & 5 & 6 \\ 7 & 8 & 9 \end{bmatrix}' = \begin{bmatrix} 1 & 4 & 7 \\ 2 & 5 & 8 \\ 3 & 6 & 9 \end{bmatrix}.$$

Transposes may also be defined for rectangular matrices.

If $A = (a_{ij})$ is an $n \times n$ matrix, then by (7.25) the determinant $|A|$ may be found by expanding by any row. If row i is used this leads to the formula

(7.40) $\quad |A| = a_{i1}A_{i1} + a_{i2}A_{i2} + \cdots + a_{in}A_{in}$

where A_{ij} denotes the cofactor of a_{ij}. It can also be shown that if, with the elements of row i in (7.40), we use the cofactors of the elements from row k, where $k \neq i$, then 0 is obtained; that is

(7.41) $\quad a_{i1}A_{k1} + a_{i2}A_{k2} + \cdots + a_{in}A_{kn} = 0.$

The proof of (7.41) may be found in texts on the theory of matrices.

Let us now consider the $n \times n$ matrix (A_{ij}) obtained from the $n \times n$ matrix (a_{ij}) by replacing each element a_{ij} by its cofactor A_{ij}. The transpose $(A_{ij})'$ is called the **adjoint** of A and is denoted by adj A. Thus, by definition,

(7.42) $\quad$ adj $A = (A_{ij})'.$

By employing (7.40) and (7.41) it can be shown that the product A adj A is the $n \times n$ matrix with $|A|$ in each position on the main diagonal and 0 elsewhere, that is,

$$(7.43) \quad A \text{ adj } A = \begin{bmatrix} |A| & 0 & 0 & \cdots & 0 \\ 0 & |A| & 0 & \cdots & 0 \\ \cdot & & \cdot & & \cdot \\ \cdot & & \cdot & & \cdot \\ \cdot & & \cdot & & \cdot \\ 0 & 0 & 0 & \cdots & |A| \end{bmatrix} = |A| I_n.$$

We shall not give a general proof of this identity since the notation becomes very cumbersome. As a special case, if A has order 2, then

$$A \text{ adj } A = \begin{bmatrix} a_{11} & a_{12} \\ a_{21} & a_{22} \end{bmatrix} \begin{bmatrix} A_{11} & A_{21} \\ A_{12} & A_{22} \end{bmatrix}$$

$$= \begin{bmatrix} a_{11}A_{11} + a_{12}A_{12} & a_{11}A_{21} + a_{12}A_{22} \\ a_{21}A_{11} + a_{22}A_{12} & a_{21}A_{21} + a_{22}A_{22} \end{bmatrix}.$$

This simplifies to $\begin{bmatrix} |A| & 0 \\ 0 & |A| \end{bmatrix}$ since the element $a_{11}A_{11} + a_{12}A_{12}$ is the expansion of $|A|$ by the first row and $a_{21}A_{21} + a_{22}A_{22}$ is the expansion by the second row, whereas the other elements are of the form (7.41).

It can also be shown that for any matrix A of order n,

$$(7.44) \quad (\text{adj } A)A = |A| I_n.$$

Consequently, if $|A| \neq 0$, it follows from (7.43) and (7.44) that

$$A \left(\frac{1}{|A|} \text{ adj } A \right) = I_n = \left(\frac{1}{|A|} \text{ adj } A \right) A.$$

This implies that the multiplicative inverse A^{-1} of A is given by

$$(7.45) \quad A^{-1} = \frac{1}{|A|} \text{ adj } A.$$

EXAMPLE 1 Find A^{-1} if $A = \begin{bmatrix} 3 & 5 \\ 1 & 4 \end{bmatrix}$.

Solution

The minor of each element of A is the determinant of a matrix of order 1, that is, of a matrix having only one element. By definition, the determinant of such a matrix is the element of the matrix. Thus the cofactors of A are given by $A_{11} = (-1)^{1+1} \cdot 4 = 4$, $A_{12} = (-1)^{1+2} \cdot 1 = -1$, $A_{21} = (-1)^{2+1} \cdot 5 = -5$, and $A_{22} = (-1)^{2+2} \cdot 3 = 3$. By (7.42) adj A is found by replacing each element of A by its cofactor and then finding the transpose of the resulting matrix. Thus

$$\text{adj } A = (A_{ij})' = \begin{bmatrix} 4 & -1 \\ -5 & 3 \end{bmatrix}' = \begin{bmatrix} 4 & -5 \\ -1 & 3 \end{bmatrix}$$

Since $|A| = (3)(4) - (1)(5) = 7$, we have by (7.45),

$$A^{-1} = \frac{1}{|A|} \text{ adj } A = \frac{1}{7} \begin{bmatrix} 4 & -5 \\ -1 & 3 \end{bmatrix},$$

which may also be written in the form

$$A^{-1} = \begin{bmatrix} \frac{4}{7} & -\frac{5}{7} \\ -\frac{1}{7} & \frac{3}{7} \end{bmatrix}.$$

The reader should verify that $AA^{-1} = I_2 = A^{-1}A$.

EXAMPLE 2 Find A^{-1} if $A = \begin{bmatrix} -1 & 3 & 1 \\ 2 & 5 & 0 \\ 3 & 1 & -2 \end{bmatrix}$.

Solution

The matrix A was considered in Example 3 of Section 5 where we found that $|A| = 9$, $A_{21} = 7$, and $A_{22} = -1$. The remaining cofactors are

$$A_{11} = (-1)^{1+1} \begin{vmatrix} 5 & 0 \\ 1 & -2 \end{vmatrix} = -10, \qquad A_{12} = (-1)^{1+2} \begin{vmatrix} 2 & 0 \\ 3 & -2 \end{vmatrix} = 4,$$

$$A_{13} = (-1)^{1+3} \begin{vmatrix} 2 & 5 \\ 3 & 1 \end{vmatrix} = 13, \qquad A_{23} = (-1)^{2+3} \begin{vmatrix} -1 & 3 \\ 3 & 1 \end{vmatrix} = 10,$$

$$A_{31} = (-1)^{3+1} \begin{vmatrix} 3 & 1 \\ 5 & 0 \end{vmatrix} = -5, \qquad A_{32} = (-1)^{3+2} \begin{vmatrix} -1 & 1 \\ 2 & 0 \end{vmatrix} = 2,$$

$$A_{33} = (-1)^{3+3} \begin{vmatrix} -1 & 3 \\ 2 & 5 \end{vmatrix} = -11.$$

The adjoint of A is given by

$$\text{adj } A = (A_{ij})' = \begin{bmatrix} -10 & 4 & -13 \\ 7 & -1 & 10 \\ -5 & 2 & -11 \end{bmatrix}' = \begin{bmatrix} -10 & 7 & -5 \\ 4 & -1 & 2 \\ -13 & 10 & -11 \end{bmatrix}.$$

Using (7.45) we obtain

$$A^{-1} = \frac{1}{|A|} \text{ adj } A = \frac{1}{9} \begin{bmatrix} -10 & 7 & -5 \\ 4 & -1 & 2 \\ -13 & 10 & -11 \end{bmatrix}.$$

The reader should check the fact that $AA^{-1} = I_3 = A^{-1}A$.

373

In general, it is very tedious to use (7.45) when the order of A is large. There are more efficient methods for finding A^{-1}, but they will not be discussed in this text. We have shown that if $|A| \neq 0$, then A^{-1} exists. The converse is also true, namely if A^{-1} exists, then $|A| \neq 0$. The proof of this fact requires results not discussed in this book.

There are many uses for inverses of matrices. One application concerns solutions of systems of linear equations. Let us consider a general system of two linear equations in two unknowns x and y, as given in (7.29). We may express the system by means of matrices as follows:

$$(7.46) \quad \begin{bmatrix} a_{11}x + a_{12}y \\ a_{21}x + a_{22}y \end{bmatrix} = \begin{bmatrix} k_1 \\ k_2 \end{bmatrix}.$$

If we let

$$A = \begin{bmatrix} a_{11} & a_{12} \\ a_{21} & a_{22} \end{bmatrix}, \qquad X = \begin{bmatrix} x \\ y \end{bmatrix}, \qquad \text{and} \qquad B = \begin{bmatrix} k_1 \\ k_2 \end{bmatrix},$$

then (7.46) may be written in the form

$$(7.47) \quad AX = B.$$

If A^{-1} exists (or equivalently, if $|A| \neq 0$), then multiplying both sides of (7.47) by A^{-1} gives us $A^{-1}AX = A^{-1}B$. Since $A^{-1}A = I_n$ and $I_n X = X$, this leads to

$$X = A^{-1}B,$$

from which the solution (x, y) may be found. This technique may be extended to systems of n linear equations in n unknowns.

EXAMPLE 3 Solve the following system of equations:

$$\begin{cases} -x + 3y + z = 1 \\ 2x + 5y = 3 \\ 3x + y - 2z = -2. \end{cases}$$

Solution

If we let

$$A = \begin{bmatrix} -1 & 3 & 1 \\ 2 & 5 & 0 \\ 3 & 1 & -2 \end{bmatrix}, \qquad X = \begin{bmatrix} x \\ y \\ z \end{bmatrix}, \qquad \text{and} \qquad B = \begin{bmatrix} 1 \\ 3 \\ -2 \end{bmatrix},$$

then as in (7.46) and (7.47), the given system may be written in terms of matrices as $AX = B$. From the preceding discussion this implies that $X = A^{-1}B$. The matrix A^{-1} was found in Example 2. Substituting for X, A^{-1}, and B in the last equation gives us

$$\begin{bmatrix} x \\ y \\ z \end{bmatrix} = \tfrac{1}{9} \begin{bmatrix} -10 & 7 & -5 \\ 4 & -1 & 2 \\ -13 & 10 & -11 \end{bmatrix} \begin{bmatrix} 1 \\ 3 \\ -2 \end{bmatrix} = \tfrac{1}{9} \begin{bmatrix} 21 \\ -3 \\ 39 \end{bmatrix} = \begin{bmatrix} \tfrac{7}{3} \\ -\tfrac{1}{3} \\ \tfrac{13}{3} \end{bmatrix}.$$

It follows that $x = \tfrac{7}{3}$, $y = -\tfrac{1}{3}$, and $z = \tfrac{13}{3}$. Hence the ordered triple $(\tfrac{7}{3}, -\tfrac{1}{3}, \tfrac{13}{3})$ is the solution of the given system.

EXERCISES

In Exercises 1–10 find the inverse of the given matrix, if possible.

1 $\begin{bmatrix} 3 & -1 \\ -4 & 2 \end{bmatrix}$ **2** $\begin{bmatrix} 2 & 1 \\ 3 & 4 \end{bmatrix}$ **3** $\begin{bmatrix} 3 & 2 \\ 6 & 4 \end{bmatrix}$

4 $\begin{bmatrix} 1 & -1 \\ -1 & 1 \end{bmatrix}$ **5** $\begin{bmatrix} 2 & 0 & 0 \\ 0 & 1 & -3 \\ 0 & 2 & 4 \end{bmatrix}$ **6** $\begin{bmatrix} 2 & 0 & 2 \\ 0 & -1 & 0 \\ 3 & 0 & 3 \end{bmatrix}$

7 $\begin{bmatrix} 1 & -1 & 0 \\ 2 & 3 & -2 \\ -3 & 0 & 4 \end{bmatrix}$ **8** $\begin{bmatrix} 3 & 0 & -1 \\ 1 & 5 & 0 \\ -4 & 3 & 2 \end{bmatrix}$

9 $\begin{bmatrix} 1 & 0 & -2 & 0 \\ 2 & 3 & 0 & -1 \\ 0 & -1 & 2 & 1 \\ 1 & 0 & 0 & -1 \end{bmatrix}$ **10** $\begin{bmatrix} 1 & 2 & 1 & 3 \\ 0 & 1 & 1 & 2 \\ 0 & 0 & -1 & -2 \\ 0 & 0 & 0 & 1 \end{bmatrix}$

11 State conditions on a, b, c, and d which guarantee that the matrix $\begin{bmatrix} a & b \\ c & d \end{bmatrix}$ has an inverse, and find a formula for the inverse when it exists.

12 If $abc \neq 0$, find the inverse of $\begin{bmatrix} a & 0 & 0 \\ 0 & b & 0 \\ 0 & 0 & c \end{bmatrix}$.

13 If $A = \begin{bmatrix} a_{11} & a_{12} & a_{13} \\ a_{21} & a_{22} & a_{23} \\ a_{31} & a_{32} & a_{33} \end{bmatrix}$, prove that $AI_3 = A = I_3A$.

14 Prove that $AI_4 = A = I_4A$ for every square matrix A of order 4.

Solve the following systems by the method of Example 3. (Refer to inverses of matrices found in Exercises 1–8.)

15 $\begin{array}{l} 3x - y = 1 \\ -4x + 2y = 3 \end{array}$ **16** $\begin{array}{l} 2x + y = 5 \\ 3x + 4y = -2 \end{array}$

17 $$\begin{aligned} x - y &= 2 \\ 2x + 3y - 2z &= 1 \\ -3x + 4z &= -1 \end{aligned}$$

18 $$\begin{aligned} 3x - z &= 1 \\ x + 5y &= -3 \\ -4x + 3y + 2z &= 0 \end{aligned}$$

19 If A and B are matrices of order 2, prove that $|AB| = |A||B|$. Use this fact to prove that if A^{-1} exists, then $|A| \neq 0$.

20 If two $n \times n$ matrices A and B have inverses A^{-1} and B^{-1}, prove that AB has an inverse $B^{-1}A^{-1}$.

12 REVIEW EXERCISES

Oral

Define or discuss each of the following.

1 System of equations

2 Solution set of a system of equations

3 Equivalent systems of equations

4 System of linear equations

5 Homogeneous system of linear equations

6 An $m \times n$ matrix

7 A square matrix of order n

8 The coefficient matrix of a system of linear equations; the augmented matrix

9 Elementary row transformations

10 Minor

11 Cofactor

12 Determinant

13 Properties of determinants

14 Cramer's Rule

15 The sum and product of two matrices

16 Zero matrix

17 Identity matrix

18 Transpose of a matrix

19 Adjoint of a matrix

20 Inverse of a matrix

Find the solution sets of the systems in Exercises 1–12. Work each exercise twice, using different methods.

1 $\begin{cases} y - x^2 = 1 \\ y + 2x = 4 \end{cases}$

2 $\begin{cases} x^2 + y^2 = 25 \\ x + 2y = -10 \end{cases}$

3 $\begin{cases} y = 2^x + 2 \\ y = 4^x \end{cases}$

4 $\begin{cases} y = 4x^2 \\ y = 10 - x^2 \end{cases}$

5 $\begin{cases} 4x - 3y = 19 \\ 5x - 2y = 8 \end{cases}$

6 $\begin{cases} 2x + y = 4 \\ 4x + 2y = 2 \end{cases}$

7 $\begin{cases} 2x + 3y - z = -4 \\ x - y + 4z = 23 \\ -3x + y + 3z = 8 \end{cases}$

8 $\begin{cases} x + 2y - z = 0 \\ 3x + 4y - z = 0 \\ 2x + 4y - 2z = 0 \end{cases}$

9 $\begin{cases} x + 2y - z = 1 \\ 3x + y + 2z = 1 \\ -2x - 4y + 2z = 3 \end{cases}$

10 $\begin{cases} 3x + y - z + w = 4 \\ x + 2y + z - w = -3 \\ -2x - y + z + 3w = 5 \\ x + y - 2z + 2w = 4 \end{cases}$

11 $\begin{cases} 2x + y + 6 > 0 \\ 4x - y + 3 < 0 \\ y + x - 3 < 0 \end{cases}$

12 $\begin{cases} x^2 - y < 0 \\ y - 2x < 5 \\ xy < 0 \end{cases}$

Find the determinants of the matrices in Exercises 13–18.

13 $[-5]$

14 $\begin{bmatrix} 4 & -3 \\ -2 & -6 \end{bmatrix}$

15 $\begin{bmatrix} 1 & 2 & -1 \\ 3 & 0 & 4 \\ 2 & -1 & 1 \end{bmatrix}$

16 $\begin{bmatrix} 3 & 0 & 1 & 4 \\ -1 & 2 & 3 & -5 \\ 2 & 1 & 0 & -3 \\ -4 & 7 & 1 & -1 \end{bmatrix}$

17 $\begin{bmatrix} 1 & 0 & -1 & 0 & 1 \\ 0 & 1 & 0 & -1 & 0 \\ -1 & 0 & 1 & 1 & 0 \\ 0 & 0 & -1 & 0 & 1 \\ 1 & -1 & 0 & 1 & -1 \end{bmatrix}$

18 The $n \times n$ matrix (a_{ij}), where $a_{ij} = 0$ if $i \neq j$

19 Without expanding, show that

$$\begin{vmatrix} 1 & x & y+z \\ 1 & y & x+z \\ 1 & z & x+y \end{vmatrix} = 0.$$

Find the inverses of the matrices in Exercises 20 and 21.

20 $\begin{bmatrix} 2 & 1 & 5 \\ -1 & 2 & 0 \\ 0 & 1 & 3 \end{bmatrix}$

21
$$\begin{bmatrix} 1 & 2 & 0 & 0 \\ 2 & 3 & 0 & 0 \\ 0 & 0 & 2 & -1 \\ 0 & 0 & 1 & -1 \end{bmatrix}$$

Express each of the following as a single matrix.

22 $\begin{bmatrix} 1 & 2 & 0 \\ -2 & 0 & 1 \end{bmatrix}\begin{bmatrix} 1 & 2 & 1 \\ 0 & -1 & 0 \\ 3 & 2 & 3 \end{bmatrix}$

23 $\begin{bmatrix} 1 & 2 \\ 3 & 4 \end{bmatrix}\begin{bmatrix} 5 \\ 6 \end{bmatrix}$

24 $\begin{bmatrix} 1 & 0 \\ -1 & 5 \\ 2 & 1 \end{bmatrix}\begin{bmatrix} 0 & -1 & 4 \\ 1 & 6 & 2 \end{bmatrix}$

25 $\begin{bmatrix} 0 & -1 & 4 \\ 1 & 6 & 2 \end{bmatrix}\begin{bmatrix} 1 & 0 \\ -1 & 5 \\ 2 & 1 \end{bmatrix}$

26 $3\begin{bmatrix} 2 & -1 & 0 \\ 4 & 2 & -5 \end{bmatrix} - 2\begin{bmatrix} -3 & 1 & -2 \\ 0 & 4 & 1 \end{bmatrix}$

27 $\begin{bmatrix} c & 0 \\ 0 & c \end{bmatrix}\begin{bmatrix} 1 & 2 \\ 3 & 4 \end{bmatrix}$

28 $\begin{bmatrix} c & 0 \\ 0 & d \end{bmatrix}\begin{bmatrix} 1 & 2 \\ 3 & 4 \end{bmatrix}$

29 $\begin{bmatrix} 1 & 2 \\ 3 & 4 \end{bmatrix} \text{adj} \begin{bmatrix} 1 & 2 \\ 3 & 4 \end{bmatrix}$

30 $\begin{bmatrix} 2 & 3 & 5 \\ 1 & 6 & 4 \\ 8 & 2 & 5 \end{bmatrix}\begin{bmatrix} 2 & 3 & 5 \\ 1 & 6 & 4 \\ 8 & 2 & 5 \end{bmatrix}^{-1}$

8

Complex Numbers

*Although real numbers are adequate for many mathematical and scientific problems, there is a serious defect in the system when it comes to solving some equations. Indeed, since the square of a real number cannot be negative, the solution set of an equation such as $x^2 = -5$ is empty if x is restricted to **R**. For many applications we need a mathematical system which contains the real numbers and which has the additional property that equations such as $x^2 = -5$ do have solutions. Fortunately, it is possible to construct such a system: the system of complex numbers discussed in this chapter.*

1 DEFINITION OF COMPLEX NUMBERS

Let us consider the problem of inventing a new mathematical system **C** which contains the real number system **R** and which has certain other properties. If **C** is to be used to find solutions of equations, then it must possess *operations,* that is, rules which may be applied to every pair of its elements to obtain another element. As a matter of fact, we would like to define operations called *addition* and *multiplication* in such a way that if we restrict the elements to the subset **R**, then the operations behave in the same way as addition and multiplication of real numbers. Since we are extending the notions of addition and multiplication to the set **C**, we shall continue to use the symbols "+" and "·" for those operations.

In order to gain some insight into the construction of **C**, let us begin by taking an intuitive approach. We first note that if **R** is to be a subset of **C** and if we want equations of the form $x^2 = -k$, where k is a positive real number, to have solutions, then in particular when $k = 1$, it is necessary for **C** to contain some element i such that $i^2 = -1$. If $b \in \mathbf{R}$, then also $b \in \mathbf{C}$, and since **C** is to be closed relative to multiplication, we must have $bi \in \mathbf{C}$. Moreover if

379

$a \in \mathbf{R}$ and if $\mathbf{C}$ is to be closed relative to addition, then $a + bi \in \mathbf{C}$. Thus $\mathbf{C}$ must contain elements of the form $a + bi$, $c + di$, and so on, where a, b, c, and d are real numbers and $i^2 = -1$. If we want the field properties (1.3)–(1.7) to be valid, it is evident that these elements must obey the following rules:

(8.1) $(a + bi) + (c + di) = (a + c) + (b + d)i$,

(8.2) $(a + bi)(c + di) = (ac - bd) + (ad + bc)i$,

where we have merely applied associative, commutative, and distributive properties to the symbols, together with the additional property $i^2 = -1$.

The preceding discussion indicates the manner in which elements must behave if a system of the required type is to exist. Moreover, our discussion provides a key to the actual construction of $\mathbf{C}$. Thus we begin by *defining* a **complex number** as any symbol of the form $a + bi$, where a and b are real numbers. The real number a is called the **real part** of the complex number and bi is called the **imaginary part.** At the outset the letter i is given no specific meaning and the $+$ sign which appears in $a + bi$ is not to be interpreted as the symbol for addition, but only as part of the notation for a complex number. As above, $\mathbf{C}$ will denote the set of all complex numbers. Two complex numbers $a + bi$ and $c + di$ are said to be **equal** and we write

(8.3) $a + bi = c + di$ if and only if $a = c$ and $b = d$.

Next we *define* addition and multiplication of complex numbers by means of (8.1) and (8.2). It should be observed that the $+$ sign is used in three different ways in (8.1). First, by our previous remarks, it is part of the symbol for a complex number. Secondly, it is used to denote addition of the complex numbers $a + bi$ and $c + di$. Thirdly, it is the addition sign for real numbers, as in the expressions $a + c$ and $b + d$ on the right-hand side of (8.1). The need for remembering this threefold use of $+$ will disappear after we agree on the notational conventions which follow.

Let us consider the subset $\mathbf{R}'$ of $\mathbf{C}$ consisting of all complex numbers of the form $a + 0i$, that is

$$\mathbf{R}' = \{a + 0i : a \in \mathbf{R}\}.$$

By associating a with $a + 0i$ we obtain a one-to-one correspondence between the sets $\mathbf{R}$ and $\mathbf{R}'$. Applying (8.1) and (8.2) to the elements of $\mathbf{R}'$ (that is, letting $b = d = 0$) we obtain

$$(a + 0i) + (c + 0i) = (a + c) + 0i$$
$$(a + 0i)(c + 0i) = ac + 0i.$$

This shows that in order to add (or multiply) two elements of $\mathbf{R}'$ we merely add (or multiply) the real parts, *disregarding* the imaginary parts.

Hence, as far as properties of addition and multiplication are concerned, the only difference between **R** and **R'** is the notation for the elements. Accordingly, we shall use the symbol a in place of $a + 0i$. For example, an element such as 3 of **R** (or **C**) is considered the same as the element $3 + 0i$ of **C**. It is also convenient to abbreviate the complex number $0 + bi$ by the symbol bi. Applying (8.1) gives us

$$(a + 0i) + (0 + bi) = (a + 0) + (0 + b)i = a + bi.$$

This indicates that $a + bi$ may be thought of as the sum of the two complex numbers a and bi (that is, $a + 0i$ and $0 + bi$). With these agreements on notation, all the $+$ signs in (8.1) may be regarded as addition of complex numbers.

EXAMPLE 1   Express each of the following in the form $a + bi$, where a and b are real numbers.

(a) $(3 + 4i) + (2 + 5i)$ and $(3 + 4i)(2 + 5i)$.
(b) $(3 + 4i)^2$.

Solutions

(a)   Applying (8.1) and (8.2) we have

$$(3 + 4i) + (2 + 5i) = (3 + 2) + (4 + 5)i = 5 + 9i,$$
$$(3 + 4i)(2 + 5i) = (3 \cdot 2 - 4 \cdot 5) + (3 \cdot 5 + 4 \cdot 2)i$$
$$= -14 + 23i.$$

(b)   Exponents are defined in **C** exactly as they are in **R**. Thus

$$(3 + 4i)^2 = (3 + 4i)(3 + 4i)$$
$$= (3 \cdot 3 - 4 \cdot 4) + (3 \cdot 4 + 4 \cdot 3)i$$
$$= -7 + 24i.$$

It is not difficult to show that complex numbers obey all the field properties (1.3)–(1.7). In particular the commutative and associative laws for addition and multiplication are true. The identity element relative to addition is 0 (or equivalently $0 + 0i$), since

$$(a + bi) + 0 = (a + bi) + (0 + 0i)$$
$$= (a + 0) + (b + 0)i$$
$$= a + bi.$$

As usual we refer to 0 as **zero** or the **zero element.** It follows from (8.2) that the *product* of zero and any complex number is zero. We may also use (8.2) to prove that 1 (that is, $1 + 0i$) is the identity element relative to multiplication.

If $(-a) + (-b)i$ is added to $a + bi$ we obtain 0. This implies that $(-a) + (-b)i$ is the additive inverse of $a + bi$, that is,

$$-(a + bi) = (-a) + (-b)i.$$

We shall postpone the discussion of multiplicative inverses until the next section.

Subtraction on **C** is defined using additive inverses as follows:

$$(a + bi) - (c + di) = (a + bi) + [-(c + di)].$$

Since $-(c + di) = (-c) + (-d)i$ it follows that

$$(a + bi) - (c + di) = (a - c) + (b - d)i.$$

The special case of this formula with $b = c = 0$ gives us

$$(a + 0i) - (0 + di) = (a - 0) + (0 - d)i,$$

which may be written in the form

$$a - di = a + (-d)i.$$

The preceding formula is useful when the real number associated with the imaginary part of a complex number is negative.

If c, d, and k are real numbers, then by (8.2) and our agreement on notation,

$$k(c + di) = (k + 0i)(c + di) = (kc - 0d) + (kd + 0c)i,$$

that is,

(8.4) $k(c + di) = kc + (kd)i.$

One illustration of that formula is $3(5 + 2i) = 15 + 6i$. The special case of (8.4) with $k = -1$ gives us

$$(-1)(c + di) = (-c) + (-d)i = -(c + di).$$

Hence, as in **R**, the additive inverse of a complex number may be found by multiplying it by -1.

The complex number $0 + 1i$ (or equivalently $1i$) will be denoted by i. Applying (8.4) with $k = b$, $c = 0$, and $d = 1$ gives us

$$b(0 + 1i) = b \cdot 0 + (b \cdot 1)i = 0 + bi = bi.$$

Thus the symbol bi which has been used throughout this section may be regarded as the *product* of b and i.

Finally, using (8.2) with $a = c = 0$ and $b = d = 1$ we obtain

$$\begin{aligned}
i^2 &= (0 + 1i)(0 + 1i) \\
&= (0 \cdot 0 - 1 \cdot 1) + (0 \cdot 1 + 1 \cdot 0)i \\
&= -1 + 0i, \\
&= -1.
\end{aligned}$$

which gives us this important rule: $i^2 = -1$.

If we collect all of the formulas and remarks made in this section it becomes evident that when working with complex numbers *we may treat all symbols just as though they represented real numbers with exactly one exception: wherever the symbol i^2 appears it may be replaced by* -1. Consequently manipulations can be carried out without referring to (8.1) or (8.2), which is what we had in mind from the very beginning of our discussion! We shall use this technique in the solution of the next example. If, as in Example 2, we are asked to write an expression in the form $a + bi$, we shall also accept the form $a - di$, since we have seen that it equals $a + (-d)i$.

EXAMPLE 2   Write each of the following in the form $a + bi$.

   (a) $4(2 + 5i) - (3 - 4i)$
   (b) $(4 - 3i)(2 + i)$
   (c) $i(3 - 2i)^2$
   (d) i^{51}

Solutions

   (a) $4(2 + 5i) - (3 - 4i) = 8 + 20i - 3 + 4i = 5 + 24i.$
   (b) $(4 - 3i)(2 + i) = 8 - 6i + 4i - 3i^2 = 11 - 2i.$
   (c) $i(3 - 2i)^2 = i(9 - 12i + 4i^2) = i(5 - 12i) = 5i - 12i^2 = 12 + 5i.$
   (d)   Taking successive powers of i, we obtain $i^1 = i$, $i^2 = -1$, $i^3 = -i$, $i^4 = 1$, and then the cycle starts over: $i^5 = i$, $i^6 = i^2 = -1$, and so on. In particular, $i^{51} = i^{48}i^3 = (i^4)^{12}i^3 = (1)^{12}i^3 = i^3 = -i.$

Finally, it should be pointed out that there is another way to define complex numbers. Observe that each symbol $a + bi$ determines a unique ordered pair (a, b) of real numbers. Conversely, every ordered pair (a, b) can be used to obtain a symbol $a + bi$. In this way we obtain a one-to-one correspondence between the symbols $a + bi$ and ordered pairs (a, b). This fact suggests using ordered pairs of real numbers to define the system **C**. If we do this, then (8.1) and (8.2) can be used to motivate definitions for addition and multiplication. Specifically, we define **C** as the set of all ordered pairs of real numbers subject to these two laws:

$$(a, b) + (c, d) = (a + c, b + d),$$
$$(a, b)(c, d) = (ac - bd, ad + bc).$$

Notice the manner in which (8.1) and (8.2) were used to help formulate the definition. We merely replaced symbols such as $a + bi$ by (a, b) and translated the rules accordingly. It may then be shown that the field properties (1.3)–(1.7) are true for ordered pairs. By a suitable change in notation we can obtain the $a + bi$ form for complex numbers introduced in this section.

EXERCISES

In each of Exercises 1–36 write the expression in the form $a + bi$.

1 $(2 + 5i) + (-7 + 3i)$ **2** $(3 - 6i) + (1 - 4i)$

3 $(-6 + 2i) + (3 - i)$ **4** $(4 + 8i) + (-6 - 5i)$

5 $(18 + 11i) - (8 + 14i)$ **6** $(1 - 5i) - (8 + 3i)$

7 $-(-3 + 6i) + (-5 + 5i)$ **8** $-(8 - 9i) - (-4 - 2i)$

9 $9 - (5 - 9i)$ **10** $-8 + (3 + 8i)$

11 $3i - (4 + 9i)$ **12** $(7 + 3i) - 8i$

13 $(3 + 2i)(-4 + 5i)$ **14** $(5 - 6i)(2 + i)$

15 $(-6 + i)(-5 + i)$ **16** $(2 + 5i)(2 - 5i)$

17 $(4 + 3i)(4 - 3i)$ **18** $8i(12 + 7i)$

19 $-8i(6 - 7i)$ **20** $(5i)(-3i)$

21 $8(3 - 10i)$ **22** $-4(-2 + 11i)$

23 $(-\sqrt{3}i)(\sqrt{6}i)$ **24** $(1 + i)(1 - i)$

25 $(\sqrt{5} + \sqrt{2}i)(\sqrt{5} - \sqrt{2}i)$ **26** $(-8 + 3i)^2$

27 $(5 + 6i)^2$ **28** $5i(1 + 2i)^2$

29 $i(5 + 2i)(3 - i)$ **30** $(1 + i)^4$

31 $\left(-\dfrac{1}{2} + \dfrac{\sqrt{3}}{2}i\right)^3$ **32** $\left(-\dfrac{1}{2} - \dfrac{\sqrt{3}}{2}i\right)^3$

33 i^{34} **34** i^{19}

35 i^{137} **36** $(-i)^{30}$

In Exercises 37–40 solve for x and y.

37 $3x + 5i = -7 + yi$ **38** $8 - 9yi = 4x + 2i$

39 $i(x - yi) = x + 5 + 2yi$

40 $(x + 2y) + (4x - 3y)i = (2x - 1) + (y - 6)i$

2 CONJUGATES AND INVERSES

The complex number $a - bi$ is called the **conjugate** of the complex number $a + bi$. Since

(8.5) $(a + bi)(a - bi) = a^2 + b^2,$

we see that the product of a complex number and its conjugate is a real number. If $a^2 + b^2 \neq 0$, then multiplying both sides of (8.5) by

$1/(a^2 + b^2)$ and rearranging terms on the left-hand side gives us

$$\left(\frac{1}{a^2 + b^2}\right)(a - bi)(a + bi) = 1.$$

Hence, if $a + bi \neq 0$, then the complex number $[1/(a^2 + b^2)](a - bi)$ is the multiplicative inverse of $a + bi$. As usual, this inverse is written as either $1/(a + bi)$ or $(a + bi)^{-1}$.

If $c + di \neq 0$, we define the **quotient**

(8.6) $\dfrac{a + bi}{c + di}$

to be the product of $a + bi$ and $1/(c + di)$. The rules for quotients of real numbers can be extended to quotients of complex numbers. We can write the quotient (8.6) in the form $u + vi$, where u and v are real numbers, by multiplying numerator and denominator by the conjugate $c - di$ of the denominator. Thus

$$\frac{a + bi}{c + di} = \frac{a + bi}{c + di} \cdot \frac{c - di}{c - di}$$

$$= \frac{(ac + bd) + (bc - ad)i}{c^2 + d^2}$$

$$= \left(\frac{ac + bd}{c^2 + d^2}\right) + \left(\frac{bc - ad}{c^2 + d^2}\right)i.$$

This technique may also be used to find the multiplicative inverse of $(a + bi)$. Specifically, we multiply numerator and denominator of $1/(a + bi)$ by $a - bi$ as follows:

$$\frac{1}{a + bi} = \frac{1}{a + bi} \cdot \frac{a - bi}{a - bi}$$

$$= \frac{a - bi}{a^2 + b^2} = \frac{1}{a^2 + b^2}(a - bi),$$

which is the same result we obtained previously.

EXAMPLE 1   Express each of the following in the form $a + bi$.

(a) $1/(9 + 2i)$.   (b) $(7 - i)/(3 - 5i)$.

Solutions

(a) $\dfrac{1}{9 + 2i} = \dfrac{1}{9 + 2i} \cdot \dfrac{9 - 2i}{9 - 2i} = \dfrac{9 - 2i}{81 + 4} = \dfrac{9}{85} - \dfrac{2}{85}i.$

(b) $\dfrac{7 - i}{3 - 5i} = \dfrac{7 - i}{3 - 5i} \cdot \dfrac{3 + 5i}{3 + 5i}$

$= \dfrac{21 - 3i + 35i - 5i^2}{9 - 25i^2}$

$= \dfrac{26 + 32i}{34} = \dfrac{13}{17} + \dfrac{16}{17}i.$

Conjugates of complex numbers have several interesting and useful properties. To simplify the notation, if $z = a + bi$ is a complex number, then its conjugate will be denoted by $\bar{z}$, that is, $\bar{z} = a - bi$.

(8.7) Theorem on Conjugates

If z and w are complex numbers, then

    (i) $\overline{z + w} = \bar{z} + \bar{w}$,

    (ii) $\overline{z \cdot w} = \bar{z} \cdot \bar{w}$,

    (iii) $\overline{z^n} = \bar{z}^n$, for all positive integers n,

    (iv) $\bar{z} = z$ if and only if z is real.

Proof

Let $z = a + bi$ and $w = c + di$, where a, b, c, and d are real numbers. Since $z + w = (a + c) + (b + d)i$, we have, by definition of conjugate and properties of addition of complex numbers,

$$\overline{z + w} = (a + c) - (b + d)i$$
$$= (a - bi) + (c - di)$$
$$= \bar{z} + \bar{w}.$$

That proves (i).

By (8.2) $z \cdot w = (ac - bd) + (ad + bc)i$ and hence the conjugate is

$$\overline{z \cdot w} = (ac - bd) - (ad + bc)i = (a - bi)(c - di) = \bar{z}\bar{w}.$$

This proves (ii).

If we set $w = z$ in (ii), then $\overline{z \cdot z} = \bar{z} \cdot \bar{z}$, that is, $\overline{z^2} = \bar{z}^2$. We may then write $\overline{z^3} = \overline{z^2 \cdot z} = \overline{z^2} \cdot \bar{z} = \bar{z}^2 \cdot \bar{z} = \bar{z}^3$. Continuing in this manner it appears that $\overline{z^n} = \bar{z}^n$, for all positive integers n. A complete proof of (iii) requires the method of mathematical induction discussed in Chapter Ten.

Finally, let us prove (iv). If $z = a + bi$ is real, then $b = 0$ and $\bar{z} = a - 0i = a + 0i = z$. Conversely, if $\bar{z} = z$, then $a - bi = a + bi$ and by (8.3), $-b = b$. This implies that $b = 0$; that is, z is real.

It is not difficult to extend (i) and (ii) of (8.7). Thus if z, w, and u are complex numbers, then applying (i) twice we have

$$\overline{(z + w) + u} = \overline{z + w} + \bar{u} = \bar{z} + \bar{w} + \bar{u}.$$

The analogous result holds for more than three complex numbers. This fact may be stated thus: "the conjugate of a sum of complex numbers equals the sum of the conjugates." A similar result is true for products.

An important application of (8.7) will be made in the next chapter. We shall conclude this section by extending the notion of absolute value to complex numbers. Taking the principal square root of both sides of equation (8.5) we have

$$\sqrt{(a + bi)(a - bi)} = \sqrt{a^2 + b^2} \cdot$$

If $b = 0$, the right side of the equation reduces to $\sqrt{a^2}$, which equals $|a|$ (see 1.27). We shall use this fact as our motivation for the following definition.

(8.8)   Definition

The **absolute value** of a complex number $a + bi$ is denoted by $|a + bi|$ and is defined to be the nonnegative real number $\sqrt{a^2 + b^2}$.

The preceding definition extends the concept of absolute value to **C**, since if $a + bi$ is real (that is, $b = 0$), then (8.8) agrees with our definition in **R**.

EXAMPLE 2 Find (a) $|2 - 6i|$,   (b) $|3i|$.

   Solutions

   (a) $|2 - 6i| = \sqrt{4 + 36} = \sqrt{40} = 2\sqrt{10}$,
   (b) $|3i| = \sqrt{0 + 9} = 3$.

We have made considerable use of the field properties (1.3)–(1.7). Notice, however, that inequalities have not been mentioned, because it is impossible to define positive elements in **C** which have the same properties as the positive elements in **R**. In particular we never say that a complex number with nonzero imaginary part is less than, or greater than, another complex number.

EXERCISES

In each of Exercises 1–28 express the given number in the form $a + bi$.

1 $1/(3 + 2i)$

2 $1/(7 + 4i)$

3 $2/(6 - 5i)$

4 $-3/(5 - 11i)$

5 $(7 - 2i)/(5 + 3i)$

6 $(1 + 2i)/(-4 + 3i)$

7 $(4 + 5i)/(1 - 4i)$

8 $(2 - 5i)/(-3 - i)$

9 $(19 - 8i)/i$

10 $(10 + 11i)/(-5i)$

11 $\dfrac{2 + 3i}{1 - i} + \dfrac{6 - 2i}{5 + 2i}$

12 $\dfrac{1 - 7i}{4 + 3i} - \dfrac{2 - 5i}{i}$

13 $3 + \dfrac{2 - i}{1 + 2i}$

14 $\dfrac{1}{6 - i} + 4i$

15 $\dfrac{1}{(2+i)^3}$

16 $\dfrac{(1+i)^3}{(1-i)}$

17 $\dfrac{i^2-9}{i-3}$

18 $\dfrac{9i^2-16}{3i+4}$

19 $(1/4i)^3$

20 $1/(2+3i)^2$

21 $|4+3i|$

22 $|4-7i|$

23 $|-5-9i|$

24 $|1+i|$

25 $|7i|$

26 $|i^3|$

27 $|i^{1000}|$

28 $|-10i|$

In each of Exercises 29–32 find all complex numbers z that satisfy the given equation, and express z in the form $a+bi$.

29 $3z+2i=iz+4$

30 $5iz-2=7i+3z$

31 $(z-4i)^2=(z+3i)^2$

32 $z(z+2i)=(z-i)(z+5i)$

If $z=a+bi$ and $w=c+di$, verify the identities in Exercises 33–40.

33 $\overline{z}=z$

34 $\overline{z-w}=\overline{z}-\overline{w}$

35 $|z|=\sqrt{z\overline{z}}$

36 $|-z|=|z|$

37 $|\overline{z}|=|z|$

38 $|z|=0$ if and only if $z=0$

39 $|zw|=|z|\,|w|$

40 $|z/w|=|z|/|w|,\quad w\neq 0$

3 COMPLEX ROOTS OF EQUATIONS

It is easy to see that if k is a positive real number, then equations of the form $x^2=-k$ do have solutions in **C**. As a matter of fact, one solution is $\sqrt{k}i$, since $(\sqrt{k}i)^2=(\sqrt{k})^2i^2=k(-1)=-k$. Similarly, $-\sqrt{k}i$ is also a solution. Moreover, these are the only solutions, for if a complex number z is in the solution set, then $z^2+k=0$ and hence

$$(z+\sqrt{k}i)(z-\sqrt{k}i)=0,$$

which implies that either $z=-\sqrt{k}i$ or $z=\sqrt{k}i$.

The next definition is motivated by the fact that $(\sqrt{k}i)^2=-k$.

(8.9) **Definition**

If k is a positive real number, then the **principal square root** of $-k$ is denoted by $\sqrt{-k}$ and is defined to be the complex number $\sqrt{k}i$.

As illustrations of (8.9) we have

$$\sqrt{-9}=\sqrt{9}i=3i,\quad \sqrt{-5}=\sqrt{5}i,\quad \sqrt{-1}=\sqrt{1}i=i.$$

Care must be taken in using the radical sign if the radicand is negative. For example, the formula $\sqrt{a}\,\sqrt{b} = \sqrt{ab}$ established for positive real numbers is not true when a and b are both negative. To illustrate,

$$\sqrt{-3}\,\sqrt{-3} = (\sqrt{3}\,i)(\sqrt{3}\,i) = (\sqrt{3})^2 i^2 = 3(-1) = -3,$$

whereas

$$\sqrt{(-3)(-3)} = \sqrt{9} = 3.$$

Hence

$$\sqrt{-3}\,\sqrt{-3} \neq \sqrt{(-3)(-3)}.$$

However, if only *one* of a or b is negative, then we may show that $\sqrt{a}\,\sqrt{b} = \sqrt{ab}$. In general, we shall not apply laws of radicals if radicands are negative. Instead, we shall change the form of radicals, using (8.9) before performing any operations.

EXAMPLE 1   Express $(5 - \sqrt{-3})(-1 + \sqrt{-4})$ in the form $a + bi$.

Solution

$$
\begin{aligned}
(5 - \sqrt{-3})(-1 + \sqrt{-4}) &= (5 - \sqrt{3}\,i)(-1 + 2i)\\
&= -5 - 2\sqrt{3}\,i^2 + 10i + \sqrt{3}\,i\\
&= (-5 + 2\sqrt{3}) + (10 + \sqrt{3})i.
\end{aligned}
$$

In Chapter Two we saw that if a, b, and c are real numbers such that $b^2 - 4ac \geq 0$ and $a \neq 0$, then the solution set of the quadratic equation

$$ax^2 + bx + c = 0$$

consists of

(8.10) $\dfrac{-b + \sqrt{b^2 - 4ac}}{2a}$ and $\dfrac{-b - \sqrt{b^2 - 4ac}}{2a}$.

We may now extend this fact to the case where $b^2 - 4ac < 0$. Indeed, the same manipulations used to obtain the quadratic formula, together with the developments in this chapter, show that if $b^2 - 4ac < 0$, then the solution set of $ax^2 + bx + c = 0$ consists of the two *complex* numbers given by (8.10). Notice that these numbers are conjugates of one another. This completes the theory of solutions of quadratic equations in one variable with real coefficients. The solution set is *always* given by the quadratic formula (2.3).

EXAMPLE 2 Find the solutions of the equation $5x^2 + 2x + 1 = 0$.

Solution

By the quadratic formula we have

$$x = \frac{-2 \pm \sqrt{4-20}}{10} = \frac{-2 \pm \sqrt{-16}}{10} = \frac{-2 \pm 4i}{10}.$$

Hence the solutions of the equation are $-\frac{1}{5} + (\frac{2}{5})i$ and $-\frac{1}{5} - (\frac{2}{5})i$.

EXAMPLE 3 Find the solution set of the equation $x^3 - 1 = 0$.

Solution

By (1.35) the given equation may be written as

$$(x-1)(x^2 + x + 1) = 0.$$

Setting each factor equal to zero and solving the resulting equations, we obtain the solution set

$$\left\{ 1, \; \frac{-1 \pm \sqrt{1-4}}{2} \right\},$$

which may be written as

$$\left\{ 1, \; -\frac{1}{2} + \frac{\sqrt{3}}{2}i, \; -\frac{1}{2} - \frac{\sqrt{3}}{2}i \right\}.$$

Those three numbers are called the **cube roots of unity**. It can be shown that if n is any positive integer, then the equation $x^n - 1 = 0$ has n distinct complex roots. They are called the n**th roots of unity**.

It is easy to form a quadratic equation having complex roots z and w. We merely write

$$(x - z)(x - w) = 0,$$

or equivalently,

$$x^2 - (z + w)x + zw = 0.$$

If z and w are both real or if they are complex conjugates of one another, then this quadratic equation has real coefficients. This is true because the sum or product of a complex number $a + bi$ and its conjugate $a - bi$ is a real number (see (8.5) and Exercise 39).

EXAMPLE 4 Find a quadratic equation which has roots $3 + 2i$ and $3 - 2i$.

Solution

By the preceding remarks the equation is given by $(x - 3 - 2i)(x - 3 + 2i) = 0$, which simplifies to $x^2 - 6x + 13 = 0$.

Quadratic equations with complex coefficients may also be considered. It can be shown that the solutions are again given by the quadratic formula. Since in this case $b^2 - 4ac$ may be complex, the solution of such an equation may involve finding the square root of a complex number. We shall not discuss the general theory of roots of complex numbers here. In special cases it is possible to determine such roots directly, as is illustrated in the following example.

EXAMPLE 5 Find the complex solutions of the equation $x^2 = 3 + 4i$.

Solution

A complex number $a + bi$ is a solution if and only if $(a + bi)^2 = 3 + 4i$, that is, if and only if

$$(a^2 - b^2) + 2abi = 3 + 4i.$$

By (8.3) the latter equation is true if and only if

$$\begin{cases} a^2 - b^2 = 3 \\ \quad 2ab = 4. \end{cases}$$

We may solve this system of equations by the method of substitution. From the second equation we obtain $b = 2/a$. Substituting for b in the first equation gives us

$$a^2 - \frac{4}{a^2} = 3,$$

which simplifies to

$$a^4 - 3a^2 - 4 = 0.$$

Factoring we have

$$(a^2 - 4)(a^2 + 1) = 0.$$

Since a is real, $a^2 + 1 \neq 0$ and hence $a^2 - 4 = 0$. Consequently, $a = 2$ or $a = -2$. The corresponding values for b, obtained from $b = 2/a$, are 1 and -1. Hence the solutions of the given equation are the complex numbers $2 + i$ and $-2 - i$, which are the square roots of $3 + 4i$.

EXERCISES

In each of Exercises 1–12 express the given number in the form $a + bi$.

1 $\sqrt{-7}\ \sqrt{-7}$

2 $(-\sqrt{7})(\sqrt{-7})$

3 $(4 + \sqrt{-9}) - (5 - \sqrt{-4})$

4 $(6 + \sqrt{-1}) + (3 - \sqrt{-16})$

5 $(5 + \sqrt{-3})(5 - \sqrt{-3})$

6 $(\sqrt{-25} + 3)(\sqrt{-36} - 5)$

7 $\sqrt{-3}(\sqrt{3} + \sqrt{-3})$

8 $\sqrt{-2}\ (2 - \sqrt{-8})$

9 $(\sqrt{-2})^3$

10 $\sqrt{(-2)^3}$

11 $\dfrac{1 + \sqrt{-9}}{5 + \sqrt{-4}}$

12 $\dfrac{1}{2 + \sqrt{-3}}$

In Exercises 13–26 find the solution sets of the given equations.

13 $x^2 - 4x + 13 = 0$

14 $x^2 - 6x + 18 = 0$

15 $x^2 + 3x + 4 = 0$

16 $x^2 + 8x + 41 = 0$

17 $3x^2 + x + 2 = 0$

18 $-4x^2 + 3x - 5 = 0$

19 $x^3 - 27 = 0$

20 $x^3 + 125 = 0$

21 $x^6 - 64 = 0$

22 $x^4 = 1$

23 $4x^4 + 19x^2 + 12 = 0$

24 $9x^4 + 80x^2 - 9 = 0$

25 $x^3 + x^2 + x = 0$

26 $x^3 - x^2 + x - 1 = 0$

In each of Exercises 27–36 find a quadratic equation with the given roots.

27 $2 + 5i,\ 2 - 5i$

28 $4 + i,\ 4 - i$

29 $-1 + i,\ -1 - i$

30 $\tfrac{2}{3} + \tfrac{1}{2}i,\ \tfrac{2}{3} - \tfrac{1}{2}i$

31 $2i,\ 3i$

32 $2 + i,\ 1 + 2i$

33 $2 - 3i,\ 2 + 3i$

34 $5i,\ -5i$

35 $i,\ 1/i$

36 $i^3,\ i^5$

Use the method of Example 5 to solve the equations in Exercises 37 and 38. Write solutions in the form $a + bi$.

37 $x^2 = i$

38 $x^2 = 8 + 6i$

39 Prove that if z is any complex number, then $z + \bar{z}$ is a real number. Is $z - \bar{z}$ necessarily a real number?

40 Prove that the sum and product of the roots of the equation $ax^2 + bx + c = 0$ are $-b/a$ and c/a, respectively.

Each complex number $a + bi$ determines a unique ordered pair (a, b) of real numbers. The corresponding point $P(a, b)$ in a coordinate plane will be called the **geometric representation of** $a + bi$. To emphasize that we are assigning complex numbers to points in a plane, the point $P(a, b)$ will be labeled $a + bi$. A coordinate plane with a complex number assigned to each point is referred to as the **complex plane** instead of the xy-plane. Also, according to this scheme, the x-axis is called the **real axis** and the y-axis the **imaginary axis.** In Fig. 8.1 we have indicated the geometric representation of several complex numbers.

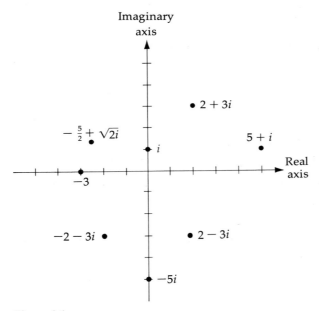

Figure 8.1

If $a + bi$ is an arbitrary complex number, then by (3.3) the distance from the origin to the point $P(a, b)$ is $\sqrt{a^2 + b^2}$, which by (8.8) is the same as $|a + bi|$. Hence a geometric interpretation of the absolute value of a complex number is the distance from the origin to the point corresponding to the number. This again justifies our definition of $|a + bi|$, since as we saw in Chapter One, the absolute value of a *real* number equals the distance from the origin of a coordinate line to the point corresponding to the real number. It is worth noting that the points which correspond to all of the complex numbers having a fixed absolute value lie on a circle with center at the origin in the complex plane. For example, the points corresponding to the complex numbers z with $|z| = 1$ lie on a unit circle.

The geometric representation of a complex number leads to a method of representation that involves trigonometric functions. Let us consider a nonzero complex number $z = a + bi$ and its geometric representation $P(a, b)$. Let θ be any angle in standard position whose terminal side lies on the segment OP and let $r = |z|$, that is, $r = d(O, P) = \sqrt{a^2 + b^2}$. An illustration of this situation is shown in Fig. 8.2.

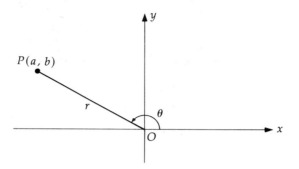

Figure 8.2 $z = a + bi = r\,(\cos\theta + i\,\sin\theta)$

Using (5.26), we may write $a = r\cos\theta$ and $b = r\sin\theta$. Since $z = a + bi$, this gives us

$$z = (r\cos\theta) + (r\sin\theta)i,$$

which may be written in the form

(8.11) $z = r(\cos\theta + i\,\sin\theta).$

If $z = 0$, then $r = 0$ and hence (8.11) may be used to represent the complex number 0 where θ is *any* angle. The form given in (8.11) is called the **trigonometric form** or **polar form** for the complex number $z = a + bi$. The trigonometric form for z is not unique since there are an infinite number of different choices for the angle θ. When the trigonometric form is used, the absolute value r of z is often referred to as the **modulus** of z and the angle θ associated with z is called the **argument** (or **amplitude**) of z.

EXAMPLE 1 Express each of the following complex numbers in trigonometric form:

(a) $-4 + 4i$,   (b) $2\sqrt{3} - 2i$,   (c) $2 + 7i$.

Solutions

The three complex numbers are represented geometrically in Fig. 8.3. Using (8.11) we obtain

(a) $-4 + 4i = 4\sqrt{2}\left[\cos\dfrac{3\pi}{4} + i\,\sin\dfrac{3\pi}{4}\right],$

(b) $2\sqrt{3} - 2i = 4\left[\cos\left(-\dfrac{\pi}{6}\right) + i\,\sin\left(-\dfrac{\pi}{6}\right)\right],$

(c) $2 + 7i = \sqrt{53}\,[\cos(\arctan\tfrac{7}{2}) + i\,\sin(\arctan\tfrac{7}{2})].$

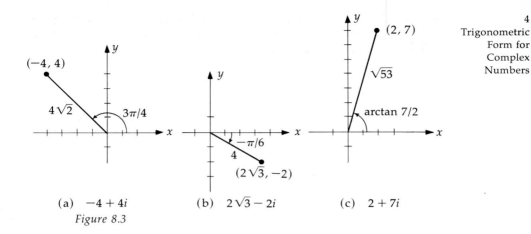

(a) $-4 + 4i$

(b) $2\sqrt{3} - 2i$

(c) $2 + 7i$

Figure 8.3

If complex numbers are expressed in trigonometric form, then multiplications and divisions may be carried out in the simple manner indicated by the next theorem.

(8.12)   Theorem

If the trigonometric forms for two complex numbers z_1 and z_2 are given by $z_1 = r_1(\cos \theta_1 + i \sin \theta_1)$ and $z_2 = r_2(\cos \theta_2 + i \sin \theta_2)$, then

(i) $z_1 z_2 = r_1 r_2 [\cos (\theta_1 + \theta_2) + i \sin (\theta_1 + \theta_2)],$

(ii) $z_1/z_2 = (r_1/r_2)[\cos (\theta_1 - \theta_2) + i \sin (\theta_1 - \theta_2)], \quad z_2 \neq 0.$

Proof

We shall prove (i) and leave (ii) as an exercise. Thus

$$z_1 z_2 = r_1(\cos \theta_1 + i \sin \theta_1) \cdot r_2(\cos \theta_2 + i \sin \theta_2)$$
$$= r_1 r_2 \{[\cos \theta_1 \cos \theta_2 - \sin \theta_1 \sin \theta_2]$$
$$+ i [\sin \theta_1 \cos \theta_2 + \cos \theta_1 \sin \theta_2]\}.$$

Applying the addition formulas for $\cos (\theta_1 + \theta_2)$ and $\sin (\theta_1 + \theta_2)$ gives us (i).

Part (i) of (8.12) states that the modulus of a product of two complex numbers is the product of their moduli and the argument is the sum of their arguments. An analogous statement can be made for (ii).

EXAMPLE 2   Use trigonometric forms to find $z_1 z_2$ and z_1/z_2 if $z_1 = 2\sqrt{3} - 2i$ and $z_2 = -1 + \sqrt{3}i$. Check by using the methods of Sections 1 and 2.

Solution

From Example 1 we have

$$z_1 = 2\sqrt{3} - 2i = 4\left[\cos\left(-\frac{\pi}{6}\right) + i\sin\left(-\frac{\pi}{6}\right)\right].$$

Similarly the reader may verify that

$$z_2 = -1 + \sqrt{3}\,i = 2\left[\cos\frac{2\pi}{3} + i\sin\frac{2\pi}{3}\right].$$

Applying (i) of (8.12), we have

$$z_1 z_2 = 8\left[\cos\left(-\frac{\pi}{6} + \frac{2\pi}{3}\right) + i\sin\left(-\frac{\pi}{6} + \frac{2\pi}{3}\right)\right]$$

$$= 8\left[\cos\frac{\pi}{2} + i\sin\frac{\pi}{2}\right]$$

$$= 8i.$$

As a check, using the methods of Section 1 we have

$$z_1 z_2 = (2\sqrt{3} - 2i)(-1 + \sqrt{3}\,i)$$
$$= (-2\sqrt{3} + 2\sqrt{3}) + i(2 + 6)$$
$$= 8i,$$

which is in agreement with our previous answer.

Applying (ii) of (8.12) we obtain

$$\frac{z_1}{z_2} = 2\left[\cos\left(-\frac{\pi}{6} - \frac{2\pi}{3}\right) + i\sin\left(-\frac{\pi}{6} - \frac{2\pi}{3}\right)\right]$$

$$= 2\left[\cos\left(-\frac{5\pi}{6}\right) + i\sin\left(-\frac{5\pi}{6}\right)\right]$$

$$= 2\left[-\frac{\sqrt{3}}{2} + i\left(-\frac{1}{2}\right)\right]$$

$$= -\sqrt{3} - i.$$

Using the methods of Section 2 gives us

$$\frac{z_1}{z_2} = \frac{2\sqrt{3} - 2i}{-1 + \sqrt{3}\,i} \cdot \frac{-1 - \sqrt{3}\,i}{-1 - \sqrt{3}\,i}$$

$$= \frac{(-2\sqrt{3} - 2\sqrt{3}) + i(2 - 6)}{4}$$

$$= -\sqrt{3} - i.$$

EXERCISES

In each of Exercises 1–12 represent the given complex number geometrically.

1 $3 + 5i$ **2** $-3 + 2i$

3 $2 - 6i$ **4** $-8 - 5i$

5 $-(2-7i)$	**6** $(2+i)^2$
7 $i(4+9i)$	**8** $(-i)(5-3i)$
9 $(1-i)^2$	**10** $(-4)(-2+i)$
11 $1/i$	**12** $1/(1+i)$

Change the complex numbers in Exercises 13–32 to trigonometric form.

13 $-1+i$	**14** $1+\sqrt{3}i$
15 $4\sqrt{3}-4i$	**16** $-5-5i$
17 $-100i$	**18** 37
19 $(-10)(\sqrt{3}+i)$	**20** $-5i$
21 -15	**22** $2i(\sqrt{3}-i)$
23 $1+2i$	**24** $-3+4i$
25 $-3-3i$	**26** $4-4i$
27 $7i$	**28** -3
29 $-4+4\sqrt{3}i$	**30** $-5\sqrt{3}+5i$
31 9	**32** 0

In Exercises 33–38 find z_1z_2 and z_1/z_2 by changing to trigonometric form and then using (8.12). Check by using the methods of Section 2.

33 $z_1=1-i,\ z_2=1+i$	**34** $z_1=-\sqrt{3}+i,\ z_2=\sqrt{3}-i$
35 $z_1=-2\sqrt{3}-2i,\ z_2=3i$	**36** $z_1=-4+4i,\ z_2=-2i$
37 $z_1=-2,\ z_2=-3$	**38** $z_1=i,\ z_2=-i$

39 Extend (i) of (8.12) to the case of three complex numbers. Generalize to n complex numbers.

40 Prove (ii) of (8.12).

5 DE MOIVRE'S THEOREM AND nTH ROOTS OF COMPLEX NUMBERS

If z is a complex number and n is a positive integer, then a complex number w is called an nth *root* of z if $w^n = z$. In this section we shall show that every nonzero complex number has n distinct nth roots. Since $\mathbf{R} \subseteq \mathbf{C}$, it will also follow that every nonzero real number has n distinct nth roots. If a is a positive real number and $n = 2$, then we already know that these roots are $\sqrt{a}$ and $-\sqrt{a}$.

If in (8.12) we let z_1 and z_2 both equal $z = r(\cos\theta + i\sin\theta)$, we obtain

$$z^2 = r^2(\cos 2\theta + i\sin 2\theta).$$

If we now apply (8.12) to z and z^2, then

$$z^2 \cdot z = (r^2 \cdot r)[\cos (2\theta + \theta) + i \sin (2\theta + \theta)],$$

or

$$z^3 = r^3(\cos 3\theta + i \sin 3\theta).$$

Next, applying (8.12) to z^3 and z produces

$$z^4 = r^4(\cos 4\theta + i \sin 4\theta).$$

Continuing this process, it appears that *for every positive integer n,*

(8.13) $[r(\cos \theta + i \sin \theta)]^n = r^n[\cos n\theta + i \sin n\theta].$

A complete proof of this formula may be given by using the method of mathematical induction to be discussed in Chapter Ten. It is possible to show that (8.13) is also true when n is any negative integer. The theorem which gives us (8.13) is referred to as *De Moivre's Theorem,* in honor of the French mathematician Abraham De Moivre (1667–1754).

EXAMPLE 1   Find $(1 + i)^{20}$.

Solution

Introducing the trigonometric form, we have

$$1 + i = \sqrt{2} \left(\cos \frac{\pi}{4} + i \sin \frac{\pi}{4} \right).$$

Next by (8.13),

$$(1 + i)^{20} = (2^{1/2})^{20} \left[\cos 20 \left(\frac{\pi}{4} \right) + i \sin 20 \left(\frac{\pi}{4} \right) \right]$$
$$= 2^{10}(\cos 5\pi + i \sin 5\pi)$$
$$= -1024.$$

If a nonzero complex number z has an nth root w, then $w^n = z$. If the trigonometric forms for these numbers are

$$w = s(\cos \alpha + i \sin \alpha) \quad \text{and} \quad z = r(\cos \theta + i \sin \theta),$$

then by (8.13) we get

(8.14) $s^n[\cos n\alpha + i \sin n\alpha] = r(\cos \theta + i \sin \theta).$

If two complex numbers are equal, then so are their absolute values. Consequently $s^n = r$, and since s and r are nonnegative, $s = \sqrt[n]{r}$.

Substituting s^n for r in (8.14) and dividing both sides by s^n we obtain

$$\cos n\alpha + i \sin n\alpha = \cos \theta + i \sin \theta.$$

By (8.3) this implies that

$$\cos n\alpha = \cos \theta \quad \text{and} \quad \sin n\alpha = \sin \theta.$$

Since both the sine and cosine functions have period 2π, the last two equations are true if and only if $n\alpha$ and θ differ by a multiple of 2π, that is, for some integer k,

$$n\alpha = \theta + 2\pi k,$$

and hence

$$\alpha = \frac{\theta + 2\pi k}{n}.$$

Substituting in the trigonometric form for w, we obtain

$$(8.15) \quad w = \sqrt[n]{r} \left[\cos \left(\frac{\theta + 2\pi k}{n} \right) + i \sin \left(\frac{\theta + 2\pi k}{n} \right) \right].$$

If we substitute $k = 0, 1, \cdots, n - 1$ successively in (8.15), there result n distinct values for w and hence n distinct nth roots of z. No other value of k will produce a new nth root. For example, if $k = n$ we obtain the angle $(\theta + 2\pi n)/n$, or $\theta/n + 2\pi$, which gives us the same nth root as $k = 0$. Similarly, $k = n + 1$ yields the same nth root as $k = 1$, and so on. The same is true for negative values of k. Our discussion also shows that the numbers given by (8.15) are the only possible nth roots of z. We have therefore proved the following theorem.

(8.16) Theorem

If $z = r(\cos \theta + i \sin \theta)$ is any nonzero complex number and if n is any positive integer, then z has precisely n distinct nth roots. Moreover, these roots are given by

$$\sqrt[n]{r} \left[\cos \left(\frac{\theta + 2\pi k}{n} \right) + i \sin \left(\frac{\theta + 2\pi k}{n} \right) \right],$$

where $k = 0, 1, \cdots, n - 1$.

Note that the nth roots of z all have modulus $\sqrt[n]{r}$ and hence they lie on a circle of radius $\sqrt[n]{r}$ with center at O. Moreover, they are equispaced on this circle since the difference in the arguments of successive nth roots is $2\pi/n$.

It is sometimes convenient to use degree measure for θ. In this event the formula in (8.16) becomes

(8.17) $\sqrt[n]{r}\left[\cos\left(\dfrac{\theta+k\cdot 360°}{n}\right)+i\sin\left(\dfrac{\theta+k\cdot 360°}{n}\right)\right],$

where $k=0,1,\cdots,n-1$.

EXAMPLE 2   Find the four fourth roots of $-8(1+i\sqrt{3})$.

Solution

The geometric representation of the given number is shown in Fig. 8.4. Introducing trigonometric form we have

$$-8(1+i\sqrt{3})=16[\cos 240°+i\sin 240°].$$

By (8.17) the fourth roots are given by

$$2\left[\cos\left(\dfrac{240°+k\cdot 360°}{4}\right)+i\sin\left(\dfrac{240°+k\cdot 360°}{4}\right)\right],$$

where $k=0,1,2,$ and 3. The formula may be rewritten as

$$2[\cos(60°+k\cdot 90°)+i\sin(60°+k\cdot 90°)].$$

Substituting $k=0,1,2,$ and 3 for k yields the following fourth roots:

$$\begin{aligned}
2(\cos 60°\ +i\sin 60°)\ &=1+\sqrt{3}i,\\
2(\cos 150°+i\sin 150°)&=-\sqrt{3}+i,\\
2(\cos 240°+i\sin 240°)&=-1-\sqrt{3}i,\\
2(\cos 330°+i\sin 330°)&=\sqrt{3}-i.
\end{aligned}$$

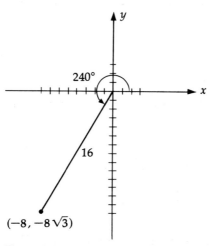

Figure 8.4

EXAMPLE 3 Find the six sixth roots of -1.

Solution

Writing $-1 = 1(\cos \pi + i \sin \pi)$ and using (8.16), we find that the sixth roots of -1 are given by

$$\cos \left(\frac{\pi + 2\pi k}{6} \right) + i \sin \left(\frac{\pi + 2\pi k}{6} \right),$$

or

$$\cos \left(\frac{\pi}{6} + \frac{\pi}{3} k \right) + i \sin \left(\frac{\pi}{6} + \frac{\pi}{3} k \right).$$

Substituting 0, 1, 2, 3, 4, and 5 for k gives us these sixth roots:

$$\begin{aligned}
\cos \ \pi/6 \ + i \sin \ \pi/6 \ &= \ \sqrt{3}/2 + (1/2)i, \\
\cos \ \pi/2 \ + i \sin \ \pi/2 \ &= i, \\
\cos \ 5\pi/6 + i \sin \ 5\pi/6 &= -\sqrt{3}/2 + (1/2)i, \\
\cos \ 7\pi/6 + i \sin \ 7\pi/6 &= -\sqrt{3}/2 - (1/2)i, \\
\cos \ 3\pi/2 + i \sin \ 3\pi/2 &= -i, \\
\cos \ 11\pi/6 + i \sin \ 11\pi/6 &= \sqrt{3}/2 - (1/2)i.
\end{aligned}$$

The special case in which $z = 1$ is of particular interest. The n distinct nth roots of 1 are called the **nth roots of unity.**

EXAMPLE 4 Find the three third roots of unity.

Solution

Writing $1 = \cos 0 + i \sin 0$ we obtain from (8.16) the three roots

$$\cos \frac{2\pi k}{3} + i \sin \frac{2\pi k}{3},$$

where $k = 0, 1,$ and 2. Substituting for k we obtain

$$\cos 0 \ + i \sin 0 \ = 1,$$

$$\cos \frac{2\pi}{3} + i \sin \frac{2\pi}{3} = -\tfrac{1}{2} + (\sqrt{3}/2)i,$$

$$\cos \frac{4\pi}{3} + i \sin \frac{4\pi}{3} = -\tfrac{1}{2} - (\sqrt{3}/2)i.$$

The reader should compare this solution with Example 3 of Section 3.

EXERCISES

In Exercises 1–10 use De Moivre's Theorem to express the given numbers in the form $a + bi$, where a and b are real numbers.

1 $(1 + i)^{10}$

2 $(2 + 2i)^4$

3 $(-1 + i)^5$

4 $(1 - i)^6$

5 $(\sqrt{3} - i)^4$

6 $(-1 + \sqrt{3}i)^7$

7 $\left(\dfrac{\sqrt{2}}{2} + \dfrac{\sqrt{2}}{2}i\right)^{20}$

8 $\left(\dfrac{\sqrt{2}}{2} - \dfrac{\sqrt{2}}{2}i\right)^{10}$

9 $\left(-\dfrac{\sqrt{3}}{2} + \dfrac{1}{2}i\right)^{15}$

10 $\left(-\dfrac{1}{2} - \dfrac{\sqrt{3}}{2}i\right)^{25}$

11 Find the two square roots of $1 + \sqrt{3}i$.

12 Find the two square roots of $-4i$.

13 Find the four fourth roots of $-1 + \sqrt{3}i$.

14 Find the four fourth roots of $-8 - 8\sqrt{3}i$.

15 Find the three cube roots of $-8i$.

16 Find the three cube roots of i.

17 Find the six sixth roots of unity.

18 Find the eight eighth roots of unity.

In each of Exercises 19–24 find the solution set of the given equation.

19 $x^4 - 1 = 0$

20 $x^6 + 729 = 0$

21 $x^6 + 64 = 0$

22 $x^3 + 27i = 0$

23 $x^6 + 8i = 0$

24 $x^5 - 1 = 0$

6 VECTOR REPRESENTATION OF COMPLEX NUMBERS

Another method of representing complex numbers, called the **vector method,** is useful in physical applications. Vectors may be approached from either a geometric or an algebraic point of view. Since each approach complements the other, it is best if they are interwoven. The algebraic point of view leads to simpler, more precise proofs of results, whereas the geometric approach provides a better physical and intuitive feeling for vectors. Although we begin geometrically, we shall bring in algebraic ideas shortly thereafter.

Previously we assigned positive directions to certain lines such as the x and y-axes. In a similar manner, a **directed line segment** is a line segment to which a positive direction has been assigned. To represent such a segment geometrically, an arrowhead is placed on one end to indicate the positive direction. Another name for a directed line segment is a **geometric vector,** or simply, a **vector.** We shall use boldface letters such as **v** and **w**, to denote vectors. If A and B are the endpoints of a vector **v** and if the positive direction is from A to B, as shown in Fig. 8.5, then A is called the **initial point** and B the **terminal point** of **v**. By definition, the **magnitude** of **v** is the length of the line segment AB.

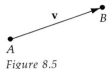

Figure 8.5

Consider a nonzero complex number $z = a + bi$ and let **z** denote the geometric vector from the origin of a rectangular coordinate system to the point with coordinates (a, b). We call **z** the **vector representation** of the complex number. Note that the magnitude of **z** is equal to the absolute value $\sqrt{a^2 + b^2}$ of z. Let $w = c + di$ be another nonzero complex number with vector representation **w**. If $u = z + w$, then $u = (a + c) + (b + d)i$, and hence the vector **u**, which represents u, is the directed line segment from the origin to the point with coordinates $(a + c, b + d)$, as shown in Fig. 8.6.

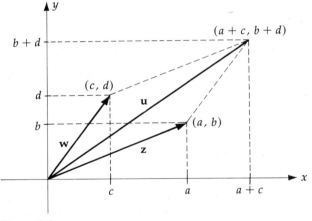

Figure 8.6

If the vectors are not collinear, then a geometric argument will show that the latter point is the fourth vertex of a parallelogram with adjacent sides **z** and **w**. If **z** and **w** lie on the same straight line through the origin, then **u** will also lie on this straight line. The vector **u** is called the **sum** of **z** and **w**, written $u = z + w$. We have defined an operation of **addition** on the set of all nonzero geometric vectors with initial point at the origin. This definition can be extended to include the so-called **zero vector** **0**, whose geometric representation is the point $(0, 0)$, by defining $z + 0 = z = 0 + z$ for all vectors **z**.

Many physical concepts can be represented by means of vectors. A common example is a *force* acting at a point, where the magnitude and direction of the force are indicated by an appropriate vector. Velocity and acceleration are examples of forces. Another example is the pressure of a liquid acting on a submerged object. It can be shown that if two forces act at a point, then there is a single force, called the **resultant,** which produces the same effect as the two taken together. Moreover, if the forces are represented by geometric vectors **z** and **w** in a Cartesian plane with initial points at the origin, then it can be shown experimentally that the resultant is precisely the vector **z** + **w** we have defined previously.

The vector representation of complex numbers suggests a method of treating vectors in a plane algebraically rather than geometrically. If vectors are considered as originating at the origin of a rectangular coordinate system, then every vector is completely determined by its terminal point $P(a, b)$. Indeed, there is a one-to-one correspondence between such geometric vectors and ordered pairs of real numbers. This enables us to regard a vector in a Cartesian plane as an ordered pair of real numbers instead of as a directed line segment. To avoid confusion with the notation used for open intervals or for points in a plane, we shall use the symbol $\langle a, b \rangle$ for an ordered pair which represents a vector. We shall then also refer to $\langle a, b \rangle$ as a vector and denote it by a boldface letter. The numbers a and b are called the **components** of the vector $\langle a, b \rangle$. Addition of such ordered pairs is defined so that it agrees with our definition of vector addition, which in turn is the same as that of addition of complex numbers. Specifically, we define

(8.18) $\langle a, b \rangle + \langle c, d \rangle = \langle a + c, b + d \rangle$.

Thus, to add vectors we add their corresponding components.

In applications it is often necessary to "multiply" a vector $\langle a, b \rangle$ by a real number k (called a **scalar**). Since $k(a + bi) = ka + (kb)i$, it is natural to define

(8.19) $k \langle a, b \rangle = \langle ka, kb \rangle$.

The geometric counterpart of (8.19) consists of lengthening or shortening a vector or, if $k < 0$, producing a vector having a direction opposite to that of the given vector (see Exercise 15).

EXAMPLE 1 If $\mathbf{a} = \langle 3, -2 \rangle$ and $\mathbf{b} = \langle -6, 7 \rangle$, find $\mathbf{a} + \mathbf{b}$, $4\mathbf{a}$, and $2a + 3b$.

Solution

By (8.18) and (8.19) we have

$$\mathbf{a} + \mathbf{b} = \langle 3, -2 \rangle + \langle -6, 7 \rangle = \langle -3, 5 \rangle$$
$$4\mathbf{a} = 4\langle 3, -2 \rangle = \langle 12, -8 \rangle$$
$$2\mathbf{a} + 3\mathbf{b} = \langle 6, -4 \rangle + \langle -18, 21 \rangle = \langle -12, 17 \rangle.$$

The zero vector **0** corresponds to $\langle 0, 0 \rangle$. Also, if $\mathbf{a} = \langle a, b \rangle$, then we define $-\mathbf{a} = \langle -a, -b \rangle$. Using these definitions we may establish the following properties, where **a**, **b**, and **c** denote arbitrary vectors.

(8.20)
$$\mathbf{a} + \mathbf{b} = \mathbf{b} + \mathbf{a}$$
$$\mathbf{a} + (\mathbf{b} + \mathbf{c}) = (\mathbf{a} + \mathbf{b}) + \mathbf{c}$$
$$\mathbf{a} + \mathbf{0} = \mathbf{a}$$
$$\mathbf{a} + (-\mathbf{a}) = \mathbf{0}$$

The proof of (8.20) follows readily from the definition of vector addition and properties of real numbers. For example, if $\mathbf{a} = \langle a_1, a_2 \rangle$ and $\mathbf{b} = \langle b_1, b_2 \rangle$, then since $a_1 + b_1 = b_1 + a_1$ and $a_2 + b_2 = b_2 + a_2$,

$$\mathbf{a} + \mathbf{b} = \langle a_1 + b_1, a_2 + b_2 \rangle$$
$$= \langle b_1 + a_1, b_2 + a_2 \rangle$$
$$= \mathbf{b} + \mathbf{a}.$$

The remainder of the proof is left as an exercise. The reader should also give geometric interpretations for (8.20).

The operation of *subtraction* of vectors, denoted by "−", is defined as follows:

(8.21) $\mathbf{a} - \mathbf{b} = \mathbf{a} + (-\mathbf{b})$.

If we use the ordered pair notation for **a** and **b**, then since $-\mathbf{b} = \langle -b_1, -b_2 \rangle$, it follows from (8.21) and (8.18) that

$$\mathbf{a} - \mathbf{b} = \langle a_1, a_2 \rangle + \langle -b_1, -b_2 \rangle$$
$$= \langle a_1 - b_1, a_2 - b_2 \rangle.$$

Thus, to find $\mathbf{a} - \mathbf{b}$ we merely subtract the components of **b** from the corresponding components of **a**.

EXAMPLE 2   If $\mathbf{a} = \langle 2, -4 \rangle$ and $\mathbf{b} = \langle 6, 7 \rangle$, find $3\mathbf{a} - 5\mathbf{b}$.

Solution

Using (8.19) and the preceding formula for subtraction we obtain

$$3\mathbf{a} - 5\mathbf{b} = \langle 6, -12 \rangle - \langle 30, 35 \rangle = \langle -24, -47 \rangle.$$

The following properties of scalar multiples of vectors can be proved for any vectors **a**, **b** and real numbers c, d.

(8.22)
$$c(\mathbf{a} + \mathbf{b}) = c\mathbf{a} + c\mathbf{b}.$$
$$(c + d)\mathbf{a} = c\mathbf{a} + d\mathbf{a}.$$
$$(cd)\mathbf{a} = c(d\mathbf{a}) = d(c\mathbf{a}).$$
$$1\mathbf{a} = \mathbf{a}.$$
$$0\mathbf{a} = \mathbf{0} = c\mathbf{0}.$$

We shall prove the first property and leave proofs of the others as exercises. Letting $\mathbf{a} = \langle a_1, a_2 \rangle$ and $\mathbf{b} = \langle b_1, b_2 \rangle$ we have

$$c(\mathbf{a} + \mathbf{b}) = c\langle a_1 + b_1, a_2 + b_2 \rangle$$
$$= \langle ca_1 + cb_1, ca_2 + cb_2 \rangle$$
$$= \langle ca_1, ca_2 \rangle + \langle cb_1, cb_2 \rangle$$
$$= c\mathbf{a} + c\mathbf{b}.$$

By employing strictly algebraic techniques we could develop an entire theory of vectors in a plane. As indicated above, theorems can be proved by using properties of real numbers. There are many advantages to this algebraic approach. We have already mentioned the simplicity of proofs. Perhaps more important, the algebraic approach lends itself to generalization more readily than does the geometric approach. For example, we can develop a theory of vectors in *three* dimensions by considering **ordered triples** $\langle a_1, a_2, a_3 \rangle$ of real numbers and by defining

$$\langle a_1, a_2, a_3 \rangle = \langle b_1, b_2, b_3 \rangle \quad \text{if and only if}$$
$$a_1 = b_1, \; a_2 = b_2, \; a_3 = b_3,$$
(8.23) $\quad \langle a_1, a_2, a_3 \rangle + \langle b_1, b_2, b_3 \rangle = \langle a_1 + b_1, a_2 + b_2, a_3 + b_3 \rangle,$
$$k\langle a_1, a_2, a_3 \rangle = \langle ka_1, ka_2, ka_3 \rangle, \quad k \in \mathbf{R}.$$

Similarly we could go on to study ordered 4-tuples $\langle a_1, a_2, a_3, a_4 \rangle$ of real numbers. Using definitions similar to (8.23) leads to a system of **vectors in four dimensions.** This theory is of major importance in modern physics. Actually there is no need to stop at 4-tuples. We could study **vectors in n dimensions,** where n is any positive integer. These vectors may be defined as **ordered n-tuples** $\langle a_1, a_2, \cdots, a_n \rangle$ of real numbers subject to laws similar to (8.23). There are many important applications for such vectors. The reader is referred to books on linear algebra or matrix theory for a discussion of vectors in many dimensions.

EXERCISES

In each of Exercises 1–4 find $\mathbf{a} + \mathbf{b}$, $\mathbf{a} - \mathbf{b}$, $4\mathbf{a} + 5\mathbf{b}$, $4\mathbf{a} - 5\mathbf{b}$, and the magnitudes of $\mathbf{a}$ and $\mathbf{b}$.

1   $\mathbf{a} = \langle 2, -3 \rangle$, $\mathbf{b} = \langle 1, 4 \rangle$ 2 $\mathbf{a} = \langle -2, 6 \rangle$, $\mathbf{b} = \langle 2, 3 \rangle$

3   $\mathbf{a} = -\langle 7, -2 \rangle$, $\mathbf{b} = 4\langle -2, 1 \rangle$ 4 $\mathbf{a} = 2\langle 5, -4 \rangle$, $\mathbf{b} = -\langle 6, 0 \rangle$

5   Complete the proof of (8.20) and give geometric interpretations for each property.

6   Complete the proof of (8.22) and give geometric interpretations for the first three properties.

Prove each of the properties in Exercises 7–14, where **a** and **b** are arbitrary vectors and c is any real number.

7 $(-1)\mathbf{a} = -\mathbf{a}$

8 $(-c)\mathbf{a} = -c\mathbf{a}$

9 $-(\mathbf{a} + \mathbf{b}) = -\mathbf{a} - \mathbf{b}$

10 $c(\mathbf{a} - \mathbf{b}) = c\mathbf{a} - c\mathbf{b}$

11 If $\mathbf{a} + \mathbf{b} = \mathbf{0}$, then $\mathbf{b} = -\mathbf{a}$

12 If $\mathbf{a} + \mathbf{b} = \mathbf{a}$, then $\mathbf{b} = \mathbf{0}$

13 If $c\mathbf{a} = \mathbf{0}$ and $c \neq 0$, then $\mathbf{a} = \mathbf{0}$

14 If $c\mathbf{a} = \mathbf{0}$ and $\mathbf{a} \neq \mathbf{0}$, then $c = 0$

15 If a complex number $a + bi$ is represented by a geometric vector **v**, prove each of the following.

   a The magnitude of $2\mathbf{v}$ is twice the magnitude of **v**.

   b The magnitude of $\frac{1}{2}\mathbf{v}$ is one-half the magnitude of **v**.

   c The magnitude of $-2\mathbf{v}$ is twice the magnitude of **v**.

   d If k is any real number, then the magnitude of $k\mathbf{v}$ is $|k|$ times the magnitude of **v**.

16 If **v** and **w** are geometric representations for complex numbers, give a geometric interpretation for $\mathbf{v} - \mathbf{w}$.

17 Prove (8.20) and (8.22) for vectors in three dimensions.

18 If the vectors **i**, **j**, and **k** in three dimensions are defined by $\mathbf{i} = \langle 1, 0, 0 \rangle$, $\mathbf{j} = \langle 0, 1, 0 \rangle$, and $\mathbf{k} = \langle 0, 0, 1 \rangle$, show that any vector $\mathbf{v} = \langle a_1, a_2, a_3 \rangle$ can be written in one and only one way in the form $\mathbf{v} = a_1\mathbf{i} + a_2\mathbf{j} + a_3\mathbf{k}$.

19 Using the notation of Exercise 18, if $\mathbf{v} = a_1\mathbf{i} + a_2\mathbf{j} + a_3\mathbf{k}$ and $\mathbf{u} = b_1\mathbf{i} + b_2\mathbf{j} + b_3\mathbf{k}$ are vectors in three dimensions, their **inner product** $\mathbf{v} \cdot \mathbf{u}$ is defined by $\mathbf{v} \cdot \mathbf{u} = a_1b_1 + a_2b_2 + a_3b_3$. Prove each of the following:

   a $\mathbf{v} \cdot \mathbf{u} = \mathbf{u} \cdot \mathbf{v}$

   b $(c\mathbf{v}) \cdot \mathbf{u} = c(\mathbf{v} \cdot \mathbf{u})$, for every real number c

   c $\mathbf{v} \cdot (\mathbf{u} + \mathbf{w}) = \mathbf{v} \cdot \mathbf{u} + \mathbf{v} \cdot \mathbf{w}$, where $\mathbf{w} = c_1\mathbf{i} + c_2\mathbf{j} + c_3\mathbf{k}$

20 If **v** and **u** are as in Exercise 19, their **vector product** $\mathbf{v} \times \mathbf{u}$ is defined by

$$\mathbf{v} \times \mathbf{u} = \begin{vmatrix} a_2 & a_3 \\ b_2 & b_3 \end{vmatrix} \mathbf{i} + \begin{vmatrix} a_3 & a_1 \\ b_3 & b_1 \end{vmatrix} \mathbf{j} + \begin{vmatrix} a_1 & a_2 \\ b_1 & b_2 \end{vmatrix} \mathbf{k}.$$

Prove each of the following.

   a $\mathbf{v} \times \mathbf{u} = -(\mathbf{u} \times \mathbf{v})$

   b $\mathbf{v} \times (\mathbf{u} + \mathbf{w}) = \mathbf{v} \times \mathbf{u} + \mathbf{v} \times \mathbf{w}$

   c $(\mathbf{v} \times \mathbf{u}) \cdot \mathbf{w} = \begin{vmatrix} a_1 & a_2 & a_3 \\ b_1 & b_2 & b_3 \\ c_1 & c_2 & c_3 \end{vmatrix}$, where $\mathbf{w} = c_1\mathbf{i} + c_2\mathbf{j} + c_3\mathbf{k}$

7 REVIEW EXERCISES

Oral

Define or discuss the following.

1   The system of complex numbers

2   The conjugate of a complex number

3   The absolute value of a complex number

4   The *n*th roots of unity

5   Geometric representation of a complex number

6   The complex plane

7   Trigonometric form for complex numbers

8   De Moivre's Theorem

9   Vector representation of complex numbers

10   Properties of vectors

Written

Express each number in Exercises 1–20 in the form $a + bi$ where a and b are real numbers.

1   $(3 + 2i) + (-5 + 7i)$ 2 $-(-2 + 5i) + (-3 + 3i)$

3   $(3 + 2i)(-2 + 5i)$ 4 $(-4 + i)(-7 + 2i)$

5   $5i(6 + 4i)$ 6 $(-2 + 3i)(-2 - 3i)$

7   $(\sqrt{6} + 2i)(\sqrt{6} - 2i)$ 8 $(4 + 7i)^2$

9   $i(2 + i)(3 - 4i)$ 10 $(5 - 2i)/(7 + 3i)$

11   $(2 + 7i)/(3 - 4i)$ 12 $(17 - 5i)/2i$

13   $1/(9 - 2i)$ 14 $(4i)/(-2i)$

15   $|3 - 5i|$ 16 $|-7 + 2i|$

17   $5i - i(4 + 2i)$ 18 $(1 + i)^2 - (4i + 6)$

19   $1/i - 2/(4 - i)$ 20 i^{999}

Find the solutions of the equations in Exercises 21–24.

21   $3x^2 - x + 1 = 0$ 22 $4x^2 - 5x + 2 = 0$

23   $4x^4 + 19x^2 + 12 = 0$ 24 $x^4 + 13x^3 + 36x^2 = 0$

In each of Exercises 25–28 find a quadratic equation with the given roots.

25   $3 - 5i,\ 3 + 5i$ 26 $1 - i,\ i$

27   $-4 - i,\ -4 + i$ 28 $10i,\ -10i$

Change the complex numbers in Exercises 29–34 to trigonometric form.

29 $-3 - 3i$

30 $1 - \sqrt{3}i$

31 -8

32 $-3i$

33 $-6\sqrt{3} + 6i$

34 $7i$

In each of Exercises 35–38 use De Moivre's Theorem to write the given number in the form $a + bi$, where a and b are real numbers.

35 $(-\sqrt{3} + i)^5$

36 $[\sqrt{2}/2 + (\sqrt{2}/2)i]^{30}$

37 $(1 - i)^6$

38 $(2 + 2\sqrt{3}i)^{10}$

39 Find the three cube roots of -27.

40 Find the solution set of the equation $x^5 - 32 = 0$.

41 Given the vectors $\mathbf{a} = \langle 3, -4 \rangle$ and $\mathbf{b} = \langle -2, 6 \rangle$, find $\mathbf{a} + \mathbf{b}$, $\mathbf{a} - \mathbf{b}$, $3\mathbf{a} - 2\mathbf{b}$, and the magnitudes of $\mathbf{a}$, $\mathbf{b}$, and $\mathbf{a} + \mathbf{b}$.

42 If $\mathbf{v}$ and $\mathbf{w}$ are geometric representations of complex numbers, give a geometric proof that the magnitude of $\mathbf{v} + \mathbf{w}$ is not greater than the sum of the magnitudes of $\mathbf{v}$ and $\mathbf{w}$.

9

Polynomials

The concept of polynomial was introduced in Chapter One. At that time we were interested primarily in basic manipulations such as adding, multiplying, and factoring. In this chapter we shall explore the theory of polynomials more deeply and discuss equations which are determined by polynomials of any degree.

1 THE ALGEBRA OF POLYNOMIALS

As in (3.16) every polynomial in a variable x, with real coefficients, can be written in the form

$$(9.1) \qquad a_n x^n + a_{n-1} x^{n-1} + \cdots + a_1 x + a_0,$$

where n is a nonnegative integer and the coefficients $a_0, a_1, \cdots, a_n$ are real numbers. Recall that the coefficient a_n of the highest power of x in (9.1) is known as the **leading coefficient,** and if $a_n \neq 0$ the polynomial has **degree n.** If all the coefficients of a polynomial are zero it is called the **zero polynomial** and is denoted by 0. It is customary not to assign a degree to the zero polynomial.

According to the definition of degree, if a is a nonzero real number, then a is a polynomial of degree 0. Such polynomials (together with the zero polynomial) are called **constant polynomials.** A polynomial of the form $a_1 x + a_0$, $a_1 \neq 0$, is of degree 1 and is called a **linear polynomial.** Polynomials of degrees 2, 3, 4, and 5 are called **quadratic, cubic, quartic,** and **quintic polynomials** respectively.

If we allow the coefficients a_i to be complex numbers, then (9.1) is called a polynomial in x with *complex* coefficients.

Much of the theory to follow will hold for polynomials of that type or for polynomials in which all coefficients are real. Therefore the letter F will be used to denote either the system of real numbers or the system of complex numbers, and $F[x]$ will denote the set of all polynomials in x with coefficients in F.

Although x is a variable which may have real or complex values, there are parts of mathematics where it has other connotations. For example, in certain theories one may wish to let x represent a square matrix. Consequently it is often convenient to regard x merely as a symbol with no specific meaning attached to it. If this is done, then polynomials may be applied to a variety of systems, and the meaning of x is determined by the particular system being studied. When x is used in this way, it is often referred to as an **indeterminate** rather than a variable. To develop a theory of polynomials from this point of view we could consider all *formal* expressions of the type given in (9.1), where the coefficients are chosen from some system. By defining operations on these expressions in a suitable way we can obtain a theory which is applicable to the various situations where polynomials are needed. We shall not use the latter approach. Instead, we shall continue to regard x as a variable and the coefficients as real (or complex) numbers. However, the development of polynomials by means of indeterminates is very similar to the discussion given in this section.

As a notational aid, let us employ symbols such as $f(x)$ and $g(x)$ to denote polynomials in x. This is a natural notation, since each polynomial determines a function f, where the value $f(c)$ of the function at the number c is the result obtained by substituting c for x in the given polynomial. If $f(c) = 0$, then c is called a **zero,** or **root,** of the polynomial $f(x)$.

By inserting terms with zero coefficients if necessary, we can assume that when discussing a finite set of polynomials, the same powers of x appear. As an illustration, given the polynomials $f(x) = 4x^3 - 3x + 2$ and $g(x) = x^2 + 5$, we can rewrite them as $f(x) = 4x^3 + 0x^2 - 3x + 2$ and $g(x) = 0x^3 + x^2 + 0x + 5$.

Let us consider any two polynomials $f(x)$ and $g(x)$ in $F[x]$. By the preceding remarks we may write

$$f(x) = a_n x^n + a_{n-1} x^{n-1} + \cdots + a_1 x + a_0,$$
$$g(x) = b_n x^n + b_{n-1} x^{n-1} + \cdots + b_1 x + b_0,$$

where $a_i, b_i \in F$ and n is a nonnegative integer. The polynomials $f(x)$ and $g(x)$ are **equal,** and we write $f(x) = g(x)$, if and only if coefficients of like powers of x are the same, that is, $a_0 = b_0, a_1 = b_1, \cdots, a_n = b_n$. To *add* $f(x)$ and $g(x)$, we add corresponding coefficients. Thus

$$f(x) + g(x) = (a_n + b_n)x^n + \cdots + (a_1 + b_1)x + (a_0 + b_0).$$

Many of the field properties $(1.3) - (1.7)$ are true in $F[x]$. The commutative law $f(x) + g(x) = g(x) + f(x)$ follows easily since for each k, $a_k + b_k = b_k + a_k$. Similarly, the associative law is a consequence of the fact that the associative law is valid for the coefficients. The zero polynomial

$$0 = 0x^n + 0x^{n-1} + \cdots + 0x + 0$$

is the additive identity, since $f(x) + 0 = f(x)$ for every polynomial $f(x)$. The additive inverse $-f(x)$ of $f(x)$ is given by

$$-f(x) = (-a_n)x^n + \cdots + (-a_1)x + (-a_0),$$

since the sum of $f(x)$ and $-f(x)$ is the zero polynomial. The rule for subtracting the polynomials $f(x)$ and $g(x)$ is

$$f(x) - g(x) = (a_n - b_n)x^n + \cdots + (a_1 - b_1)x + (a_0 - b_0).$$

Multiplication in $F[x]$ is carried out in the usual way, using properties of real numbers. For this operation there is no advantage in assuming that the highest exponent of x is the same for each polynomial. If $f(x)$ has degree n and $g(x)$ has degree m, we may write

$$f(x) = a_n x^n + a_{n-1}x^{n-1} + \cdots + a_1 x + a_0, \qquad a_n \neq 0,$$
$$g(x) = b_m x^m + b_{m-1}x^{m-1} + \cdots + b_1 x + b_0, \qquad b_m \neq 0,$$

and hence

(9.2) $\quad f(x)g(x) = a_n b_m x^{n+m} + (a_n b_{m-1} + a_{n-1}b_m)x^{n+m-1}$
$$+ \cdots + (a_1 b_0 + a_0 b_1)x + a_0 b_0.$$

Note that for each expression $a_i b_j$ occurring in any term of (9.2), the sum $i + j$ of the subscripts is equal to the exponent of x for that term. In general, *the coefficient of x^k in (9.2) is*

$$a_k b_0 + a_{k-1}b_1 + a_{k-2}b_2 + \cdots + a_1 b_{k-1} + a_0 b_k.$$

In particular, the coefficient of x^2 is $a_2 b_0 + a_1 b_1 + a_0 b_2$; the coefficient of x^3 is $a_3 b_0 + a_2 b_1 + a_1 b_2 + a_0 b_3$; and so on. As a special case of (9.2), if $f(x) = a_0$ is a constant polynomial, then

$$a_0 g(x) = a_0 b_m x^m + a_0 b_{m-1}x^{m-1} + \cdots + a_0 b_1 x + a_0 b_0;$$

that is, to find the product $a_0 g(x)$ we multiply each coefficient of $g(x)$ by a_0. Of course, this is simply a generalization of the distributive law (1.5).

If $a_n \neq 0$ and $b_m \neq 0$, then $a_n b_m \neq 0$ and it follows from (9.2) that $f(x)g(x)$ has degree $n + m$. This provides the following useful result.

(9.3)   Theorem

The degree of the product of two nonzero polynomials equals the sum of the degrees of the two polynomials.

As an illustration of (9.3), if $f(x) = 5x^3 - x + 1$ and $g(x) = 2x^2 + 3$, then the term of highest degree in $f(x)g(x)$ is $10x^5$. Hence the degree of the product is 5, which equals the sum of the degrees of the two given polynomials.

Because of the complexity of (9.2), it is rather tedious to check the field properties (1.3)–(1.7) which deal with multiplication. Suffice it to say, it can be proved that multiplication in $F[x]$ is commutative and associative and that the distributive laws are valid. Evidently, the zero degree polynomial 1 is the identity element relative to multiplication. The only field property which is *not* true in $F[x]$ is the part of (1.7) about multiplicative inverses, for consider a polynomial $f(x)$ of degree 1 or higher. Now *if* there were a polynomial $g(x)$ such that $f(x)g(x) = 1$, then applying (9.3) the degree of the polynomial $f(x)g(x)$ would be at least 1, whereas the degree of the constant polynomial 1 is 0. Since equal polynomials must have the same degree, we have reached a contradiction. Consequently no such $g(x)$ can exist.

(9.4)   Theorem

If $f(x)$ and $g(x)$ are polynomials such that $f(x)g(x) = 0$, then either $f(x) = 0$ or $g(x) = 0$.

Proof

We shall give an indirect proof. Suppose that $f(x)g(x) = 0$ and that *both* $f(x)$ and $g(x)$ are different from the zero polynomial. If $f(x)$ and $g(x)$ have degrees n and m, respectively, then by (9.3), $f(x)g(x)$ has degree $n + m$. On the other hand, the zero polynomial 0 has *no* degree. This contradiction establishes (9.4).

(9.5)   Cancellation Law for Polynomials

Suppose $f(x)$, $g(x)$, and $h(x)$ are polynomials with $h(x) \neq 0$. If $f(x)h(x) = g(x)h(x)$, then $f(x) = g(x)$.

Proof

If $f(x)h(x) = g(x)h(x)$, then

$$[f(x) - g(x)]h(x) = 0.$$

Since $h(x) \neq 0$, it follows from (9.4) that $f(x) - g(x) = 0$ and the theorem is proved.

Examples of addition, subtraction, and multiplication of polynomials were given in Chapter One. By way of review, the next set of exercises contains several more problems of this type, together with other exercises of a more theoretical nature.

EXERCISES

In Exercises 1–6 find $f(x) + g(x)$, $f(x) - g(x)$, and $f(x)g(x)$.

1 $f(x) = 2x^3 - x + 5$, $g(x) = x^2 + x + 2$

2 $f(x) = 4x^2 - 1$, $g(x) = x^3 + 6x + 1$

3 $f(x) = 7x^4 + x^2 - 1$, $g(x) = 7x^4 - x^3 + 4x$

4 $f(x) = -x^5 + 5$, $g(x) = x^5 - 5$

5 $f(x) = 3$, $g(x) = x$

6 $f(x) = 0$, $g(x) = 2x^2 - 7x + 4$

7 Illustrate (9.3) with the polynomials $2x^4 - 3x + 7$ and $3x^3 + 8x^2$.

8 If $f(x)$ and $g(x)$ are polynomials in x of degree 5, does it follow that $f(x) + g(x)$ has degree 5? Explain.

9 Give examples of polynomials $f(x)$ and $g(x)$ of degree 3 such that the degree of $f(x) + g(x)$ is   **a** 3,   **b** 2,   **c** 1,   **d** 0.

10 If $f(x)$, $g(x)$, and $h(x)$ are polynomials such that $f(x)g(x)h(x) = 0$, prove that at least one of the polynomials is 0.

11 If $f(x)$ and $g(x)$ are polynomials of degrees n and m, respectively, and $n > m$, prove that the degree of $f(x) + g(x)$ is n. What can be said if $n = m$?

12 Use (9.3) to prove that the degree of the product of three nonzero polynomials equals the sum of the degrees of the polynomials.

13 Let $\mathbf{Z}[x]$ denote the subset of $F[x]$ consisting of polynomials with coefficients in the set $\mathbf{Z}$ of integers. Which of the field properties (1.3)–(1.7) are true in $\mathbf{Z}[x]$?

14 Determine which of (9.3), (9.4), and (9.5) are true for the set $\mathbf{Z}[x]$ defined in Exercise 13.

15 Prove that the only polynomials in $F[x]$ that have multiplicative inverses are polynomials of degree 0.

16 Let S denote the subset of $F[x]$ consisting of the polynomials (9.1) with $a_0 = 0$. Prove that S is closed relative to addition, subtraction, and multiplication. Which of the field properties (1.3)–(1.7) are true in S?

2 PROPERTIES OF DIVISION

If S denotes a set of numbers, then a polynomial with coefficients in S is said to be **prime,** or **irreducible over** S, if it cannot be written as a product of two polynomials of positive degree with coefficients in S. Note that a polynomial may be irreducible over one set but not over another. For example, $x^2 - 2$ is irreducible over the rational numbers; however, if real coefficients are allowed then it *can* be expressed as the product

$(x - \sqrt{2})(x + \sqrt{2})$ of two polynomials of degree 1. Similarly, $x^2 + 1$ is irreducible over the real numbers but not over the complex numbers, since $x^2 + 1 = (x + i)(x - i)$. Every polynomial $ax + b$ of degree 1 in $F[x]$ is irreducible over F, for if $ax + b = f(x)g(x)$ where $f(x)$, $g(x) \in F[x]$, then by (9.3) the sum of the degrees of $f(x)$ and $g(x)$ is 1, which implies that one of the latter polynomials has degree 0.

If a polynomial $g(x)$ is a factor of a polynomial $f(x)$ we often say that $f(x)$ is **divisible** by $g(x)$. For example $x^2 - 25$ is divisible by $x - 5$ and by $x + 5$. A division process may also be introduced if $g(x)$ is *not* a factor of $f(x)$. The process here is similar to that in the set of integers. For example, the number 24 has positive factors 1, 2, 3, 4, 6, 8, 12, and 24; that is, 24 is *divisible* by those numbers. A different situation exists if we divide 24 by nonfactors such as 5, 7, 15 and 32. In these cases a process called *long division* is used which yields a *quotient* and *remainder*. The reader undoubtedly remembers the process from elementary arithmetic where, for example, the work involved in dividing 4126 by 23 might be arranged as follows:

$$
\begin{array}{r}
179 \\
23\overline{)4126} \\
23 \\
\overline{182} \\
161 \\
\overline{216} \\
207 \\
\overline{9}
\end{array}
$$

Here the number 179 is called the *quotient* and 9 the *remainder*. To complete the terminology, 23 is called the *divisor* and 4126 the *dividend*. The remainder should always be less than the divisor, for otherwise the quotient can be increased. The above result is often written as follows:

$$\frac{4126}{23} = 179 + \frac{9}{23}.$$

If we multiply by 23, we get

$$4126 = 23 \cdot 179 + 9.$$

The above form is very useful for theoretical purposes and can be generalized to arbitrary integers. Specifically, it can be shown that if a and b are integers with $b > 0$, then there exist unique integers q and r such that

$$a = bq + r,$$

where $0 \leq r < b$. The integer q is called the **quotient** and r the **remainder** in the division of a by b.

A similar discussion can be given for polynomials. For example, the polynomial $x^4 - 16$ is divisible by $x - 2$, $x + 2$, $x^2 - 4$, and $x^2 + 4$; but $x^2 + 3x + 1$ is not a factor of $x^4 - 16$. However, by another process called *long division,* we write

$$
\begin{array}{r}
x^2 - 3x + 8 \\
x^2 + 3x + 1\overline{\smash{\big)}\,x^4 \qquad\qquad\qquad - 16} \\
\underline{x^4 + 3x^3 + x^2} \\
-3x^3 - x^2 \\
\underline{-3x^3 - 9x^2 - 3x} \\
8x^2 + 3x - 16 \\
\underline{8x^2 + 24x + 8} \\
-21x - 24
\end{array}
$$

which yields the quotient $x^2 - 3x + 8$ and the remainder $-21x - 24$. In this division we proceed as indicated until we arrive at a polynomial (the remainder) which is either 0 or has smaller degree than the divisor. We shall assume familiarity with this process and not attempt to justify it here. As with integers, the result of this division process is often written as follows:

$$
\frac{x^4 - 16}{x^2 + 3x + 1} = (x^2 - 3x + 8) + \left(\frac{-21x - 24}{x^2 + 3x + 1}\right),
$$

or multiplying by $x^2 + 3x + 1$,

$$
x^4 - 16 = (x^2 + 3x + 1)(x^2 - 3x + 8) + (-21x - 24).
$$

This has the same general form $a = bq + r$ that was given above for integers.

The preceding example illustrates the following theorem, which we state without proof.

(9.6) Division Algorithm for Polynomials

If $f(x)$ and $g(x)$ are polynomials in $F[x]$ and if $g(x) \neq 0$, then there exist unique polynomials $q(x)$ and $r(x)$ such that

$$
f(x) = g(x)q(x) + r(x),
$$

where either $r(x) = 0$ or the degree of $r(x)$ is less than the degree of $g(x)$. The polynomial $q(x)$ is called the **quotient** and $r(x)$ the **remainder** in the division of $f(x)$ by $g(x)$.

An interesting special case of (9.6) occurs if $f(x)$ is divided by a linear polynomial of the form $x - c$, where $c \in F$. If the remainder is not 0, then by (9.6) it must have smaller degree than the divisor $x - c$. This implies that the remainder has degree less than 1, that is, degree 0. This in turn means that the remainder is a nonzero element of F. Consequently we have

(9.7) $f(x) = (x - c)q(x) + d,$

where $d \in F$ (possibly $d = 0$). If c is substituted for x in (9.7), we obtain

$f(c) = (c - c)q(c) + d,$

which reduces to $f(c) = d$. Substituting for d in (9.7) gives us

(9.8) $f(x) = (x - c)q(x) + f(c).$

We have therefore proved the following theorem.

(9.9)   Remainder Theorem

If a polynomial $f(x)$ in $F[x]$ is divided by $x - c$, where $c \in F$, then the remainder is $f(c)$.

EXAMPLE 1   Verify the Remainder Theorem if $f(x) = x^3 - 3x^2 + x + 5$ and $c = 2$.

Solution

We first note that $f(2) = 2^3 - 3 \cdot 2^2 + 2 + 5 = 3$. Hence according to (9.9), when $f(x)$ is divided by $x - 2$, the remainder should be 3. By long division we get

$$
\begin{array}{r}
x^2 - x - 1 \\
x - 2\overline{\smash{\big)}\ x^3 - 3x^2 + x + 5} \\
\underline{x^3 - 2x^2} \\
-x^2 + x \\
\underline{-x^2 + 2x} \\
-x + 5 \\
\underline{-x + 2} \\
3
\end{array}
$$

which is what we wished to show.

The following important result is a consequence of (9.9).

(9.10)   Factor Theorem

A polynomial $f(x)$ has a factor $x - c$ if and only if $f(c) = 0$.

Proof

From (9.8) we have $f(x) = (x - c)q(x) + f(c)$. If $f(c) = 0$, then $f(x) = (x - c)q(x)$; that is, $x - c$ is a factor of $f(x)$. Conversely, if

$x - c$ is a factor, then the remainder upon division of $f(x)$ by $x - c$ must be 0 and hence by the Remainder Theorem, $f(c) = 0$.

The Factor Theorem is useful for finding factors of polynomials, as illustrated in the next example.

EXAMPLE 2 Show that $x - 2$ is a factor of the polynomial $f(x) = x^3 - 4x^2 + 3x + 2$.

Solution

Since $f(2) = 8 - 16 + 6 + 2 = 0$, it follows from the Factor Theorem that $x - 2$ is a factor of $f(x)$. Of course, another method of solution would be to divide $f(x)$ by $x - 2$ and show that the remainder is 0. The quotient in this division would be another factor of $f(x)$.

EXERCISES

In Exercises 1–6 express the given polynomial as a product of polynomials which are (a) irreducible over **R**; (b) irreducible over **C**.

1 $2x^2 - 9x - 5$ 2 $12x^2 + 32x + 5$

3 $x^4 - 16$ 4 $x^2 + x + 1$

5 $8x^3 - 27$ 6 $2x^4 - x^3 - 2x^2 + x$

In Exercises 7–12 find the quotient $q(x)$ and the remainder $r(x)$ if $f(x)$ is divided by $g(x)$.

7 $f(x) = x^4 - 2x^3 + x + 4$, $g(x) = x^2 + 3x - 5$

8 $f(x) = 2x^3 + x^2 - 5$, $g(x) = x^2 - 6x$

9 $f(x) = 5x^3 - 7x$, $g(x) = 2x^2 + 3$

10 $f(x) = -3x^4 - x^3 + x^2 - 4x + 3$, $g(x) = 2x^3 + 4x - 1$

11 $f(x) = 9x^3 + 4x - 1$, $g(x) = x^4 + 7x^2 - 5$

12 $f(x) = 3x - 10$, $g(x) = 7x^2 + 4x - 13$

In Exercises 13–16 use the Remainder Theorem to find $f(c)$. Check by substituting c for x.

13 $f(x) = x^4 - 5x^3 + 2x^2 - x + 2$, $c = 3$

14 $f(x) = 2x^3 - x^2 - 6x - 3$, $c = 4$

15 $f(x) = 4x^3 - 7x + 9$, $c = -2$

16 $f(x) = x^3 - 8x + 4$, $c = i$

17 Determine k so that $f(x) = x^3 - kx^2 + 3x + 7k$ is divisible by $x + 2$.

18 Determine all values of k such that $f(x) = k^2x^4 - 3kx^2 + 1$ is divisible by $x - 1$.

19 Use the Factor Theorem to show that $x - 3$ is a factor of $f(x) = x^4 - 2x^3 + x^2 - 8x - 12$.

20 Show that $x + 2$ is a factor of $f(x) = x^{10} - 1024$.

21 Prove that $f(x) = x^4 + 2x^2 + 1$ has no factors of the form $x - c$, where c is a real number.

22 Find the remainder if the polynomial $5x^{100} - 6x^{75} + 4x^{50} + 3x^{25} + 2$ is divided by $x + 1$.

23 Use the Factor Theorem to prove that $x - y$ is a factor of $x^n - y^n$, for all positive integers n. If n is even, show that $x + y$ is also a factor of $x^n - y^n$.

24 If n is an odd positive integer, prove that $x + y$ is a factor of $x^n + y^n$.

25 Given the complex number i, show that for every integer n, i^n equals just one of $1, -1, i,$ or $-i$. (*Hint:* Write $n = 4q + r, 0 \leq r < 4$, and use the laws of exponents.)

26 Prove that a polynomial $f(x)$ in $F[x]$ has a root in F if and only if $f(x)$ has a first degree polynomial in $F[x]$ as a factor.

3 SYNTHETIC DIVISION

When applying the Remainder Theorem it is necessary to divide by polynomials of the form $x - c$. The process referred to as *synthetic division* simplifies the work if divisors are of that form. We shall illustrate the process by means of examples.

   If the polynomial $3x^4 - 8x^3 + 9x + 5$ is divided by $x - 2$ in the usual way, we obtain

$$
\begin{array}{r}
3x^3 - 2x^2 - 4x + 1 \\
x - 2 \overline{\smash{\big)}\ 3x^4 - 8x^3 + 0x^2 + 9x + 5} \\
\underline{3x^4 - 6x^3} \\
-2x^3 + 0x^2 \\
\underline{-2x^3 + 4x^2} \\
-4x^2 + 9x \\
\underline{-4x^2 + 8x} \\
x + 5 \\
\underline{x - 2} \\
7
\end{array}
$$

where the term $0x^2$ has been inserted in the dividend so that all powers of x are accounted for. Since this technique of long division seems to involve a great deal of labor for so simple a problem, we look for a means of simplifying the notation. After arranging the terms which involve like powers of x in vertical columns as above, it is seen that the repeated expressions $3x^4$, $-2x^3$, $-4x^2$, and x may be deleted without too much chance of confusion. Also, it appears unnecessary to "bring down" the terms $0x^2$, $9x$, and 5 from the dividend as indicated. With the elimination of those repetitions, our work takes on this form:

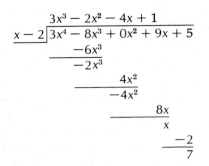

$$
\begin{array}{r}
3x^3 - 2x^2 - 4x + 1 \\
x - 2\overline{)\,3x^4 - 8x^3 + 0x^2 + 9x + 5} \\
-6x^3 \\
\hline
-2x^3 \\
4x^2 \\
\hline
-4x^2 \\
8x \\
\hline
x \\
-2 \\
\hline
7
\end{array}
$$

If we take care to keep like powers of x under one another and we account for missing terms by means of zero coefficients as above, some labor can be saved by omitting the symbol x. Doing this in the preceding display, we obtain the following:

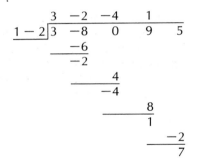

$$
\begin{array}{r}
3 \quad -2 \quad -4 \quad 1 \\
1 - 2\overline{)\,3 \quad -8 \quad 0 \quad 9 \quad 5} \\
-6 \\
\hline
-2 \\
4 \\
\hline
-4 \\
8 \\
\hline
1 \\
-2 \\
\hline
7
\end{array}
$$

Since the divisor is a polynomial of the form $x - c$, the two coefficients in the far left position are always $1 - c$. With this in mind we shall discard the coefficient 1. Moreover, to make our notation more compact let us move the numbers up in the following way:

$$
\begin{array}{r}
3 \quad -2 \quad -4 \quad 1 \\
-2\overline{)\,3 \quad -8 \quad \ 0 \quad \ 9 \quad \ 5} \\
-6 \quad \ \ 4 \quad \ 8 \quad -2 \\
\hline
-2 \quad -4 \quad \ 1 \quad \ 7
\end{array}
$$

If we now insert the leading coefficient 3 in the first position of the last row, the first four numbers of that row are the coefficients $3, -2, -4$, and 1 of the quotient, and the final number 7 is the remainder. Since there is no need to write the coefficients of the quotient two times, we discard the first row in our scheme, obtaining

$$\begin{array}{r|rrrrr}
-2 & 3 & -8 & 0 & 9 & 5 \\
& & -6 & 4 & 8 & -2 \\
\hline
& 3 & -2 & -4 & 1 & 7
\end{array}$$
(9.11)

where the top line has also been deleted since there is no longer any need for it.

There is a simple way of interpreting (9.11). Note that every number in the second row can be obtained by multiplying the number in the third row of the *preceding* column by -2. Moreover, each number in the third row can be found by subtracting the number above it in the second row from the corresponding number in the first row. This suggests a procedure for carrying out (9.11) without actually thinking of the division process. After arranging the terms of the polynomial in decreasing powers of x, we write the coefficients in a row, supplying 0 for any missing term. Next we write $-c$ (in the above case -2) to the left of this row, as indicated in (9.11). Next we bring down the leading coefficient 3 to the third row. Then we multiply that number by -2 to obtain the first number, -6, in the second row. We subtract -6 from -8 to obtain the second number -2, in the third row, and then we multiply by $-c$ (in our case -2) to obtain the third number, 4, in the second row. Again we subtract to get the third number, -4, in the third row. This process is continued until the final number in the third row (the remainder) is obtained.

It is possible to avoid the subtractions performed above if the number c is used in place of $-c$ in the far left position of the first row. In this event, when the process above is used the signs of the elements in the second row are changed, and hence to find elements of the third row, we *add* the number above it in the second row to the corresponding number in the first row. With this change (9.11) becomes

$$\begin{array}{r|rrrrr}
2 & 3 & -8 & 0 & 9 & 5 \\
& & 6 & -4 & -8 & 2 \\
\hline
& 3 & -2 & -4 & 1 & 7
\end{array}$$

The latter scheme is called the process of **synthetic division.**

EXAMPLE 1 Use synthetic division to find the quotient and remainder if $2x^4 + 5x^3 - 2x - 8$ is divided by $x + 3$.

Solution

Since we are to divide by $x + 3$, the c in the expression $x - c$ is -3. Hence the synthetic division takes this form:

$$\begin{array}{r|rrrrr}
-3 & 2 & 5 & 0 & -2 & -8 \\
& & -6 & 3 & -9 & 33 \\
\hline
& 2 & -1 & 3 & -11 & 25
\end{array}$$

The first four numbers in the third row are the coefficients of the quotient and the last number is the remainder. Hence the quotient is $2x^3 - x^2 + 3x - 11$ and the remainder is 25.

Synthetic division can be used to find values of polynomial functions, as illustrated in the next example.

EXAMPLE 2 If $f(x) = 3x^5 - 38x^3 + 5x^2 - 1$, use synthetic division to find $f(4)$.

Solution

By the Remainder Theorem $f(4)$ is the remainder when $f(x)$ is divided by $x - 4$. Dividing synthetically we have

$$
\begin{array}{r|rrrrrr}
4 & 3 & 0 & -38 & 5 & 0 & -1 \\
  &   & 12 & 48 & 40 & 180 & 720 \\
\hline
  & 3 & 12 & 10 & 45 & 180 & 719
\end{array}
$$

Consequently, $f(4) = 719$.

Synthetic division may be employed to help find zeros of polynomials. By the method illustrated in the preceding example, $f(c) = 0$ if and only if the remainder in the synthetic division by $x - c$ is 0.

EXAMPLE 3 Show that -11 is a zero of the polynomial $f(x) = x^3 + 8x^2 - 29x + 44$.

Solution

Dividing synthetically by $x + 11$, we have

$$
\begin{array}{r|rrrr}
-11 & 1 & 8 & -29 & 44 \\
    &   & -11 & 33 & -44 \\
\hline
    & 1 & -3 & 4 & 0
\end{array}
$$

Thus $f(-11) = 0$.

The preceding example shows that -11 is in the solution set of the equation $x^3 + 8x^2 - 29x + 44 = 0$. In Section 5 we shall use synthetic division in this way to find solutions of equations.

EXERCISES

In each of Exercises 1–10 use synthetic division to find the quotient and remainder if the first polynomial is divided by the second.

1 $4x^3 - 2x^2 + x - 7;\ x - 2$

2 $2x^3 + 3x^2 - x + 5;\ x + 3$

3 $x^3 + 10x - 9;\ x + 4$

4 $8x^3 - 5x^2 + 10;\ x - 5$

5 $3x^5 - x^2 + 3;\ x + 2$

6 $-3x^4 + 7x - 2;\ x - 3$

7 $2x^4 - 3x^2 + 1; \; x - \frac{1}{2}$ **8** $(\frac{1}{3})x^3 - (\frac{2}{9})x^2 + (\frac{1}{27})x + 1; \; x - \frac{1}{3}$

9 $x^4 + 2ix^3 - ix + 5; \; x - i$ **10** $x^3 - x^2; \; x - 1 + i$

Use synthetic division to solve Exercises 11–16.

11 If $f(x) = x^4 - 6x^3 + 3x^2 + 2x - 7,$ find $f(2)$ and $f(-2)$.

12 If $f(x) = 0.4x^3 - 0.02x + 0.018,$ find $f(-0.2)$ and $f(0.2)$.

13 If $f(x) = x^6 - x^5 + x^4 - x^3 + x^2 - x + 1,$ find $f(3)$.

14 If $f(x) = 2x^5 + x^3 - 3x^2 + 4,$ find $f(1/2)$.

15 If $f(x) = x^2 + 5x - 3,$ find $f(2 + i)$ and $f(2 - i)$.

16 If $f(x) = x^3 - 2x^2 + 6,$ find $f(1 + 2i)$ and $f(1 - 2i)$.

In the following exercises use synthetic division to show that c is a zero of $f(x)$.

17 $f(x) = 2x^4 - 6x^3 + 4x^2 - 17x + 15, \quad c = 3$

18 $f(x) = 3x^3 + 9x^2 - 11x + 4, \quad c = -4$

19 $f(x) = 2x^4 - 5x^3 - x^2 + 3x + 1, \quad c = -\frac{1}{2}$

20 $f(x) = 3x^5 - 2x^4 + 6x^3 - 4x^2 - 9x + 6, \quad c = \frac{2}{3}$

21 $f(x) = 2x^4 + 3x^3 - 3x^2 + 3x - 5 = 0, \quad c = -i$

22 $f(x) = 3x^3 - 5x^2 + 4x - 2, \quad c = 1 + i$

4 FACTORIZATION THEORY

The Factor Theorem (9.10) indicates that there is a close relationship between the zeros of a polynomial $f(x)$ and the factors of $f(x)$. Indeed, if a number c can be found such that $f(c) = 0$, then a factor $x - c$ is immediately obtained. Unfortunately, except in special cases, zeros of polynomials are very difficult to find. For example, given the polynomial $f(x) = x^5 - 3x^4 + 4x^3 + 4x - 10$, there are no obvious zeros. Moreover, there is no device such as the quadratic formula which can be used to produce the zeros. In spite of the practical difficulty of determining zeros of polynomials, it is possible to make some headway concerning the *theory* of such zeros. The next result is basic for the development of this theory.

(9.12) Fundamental Theorem of Algebra

If a polynomial $f(x)$ has complex coefficients and has degree greater than 0, then $f(x)$ has at least one complex zero.

Since the proof of this theorem requires advanced mathematical methods, it is impossible to give it here. We shall, however, accept its validity and use it in our work. Note that as a special case, if all the coefficients of $f(x)$ are real, then $f(x)$ has at least one complex zero. We should also remark that if $a + bi$ is a complex zero of a polynomial, it may happen that $b = 0$, in which case we refer to the number as a **real zero.**

If (9.12) is combined with the Factor Theorem, the following useful result is obtained.

(9.13) Corollary

Every polynomial that has complex coefficients and degree greater than 0 has a factor of the form $x - c$, where c is a complex number.

The preceding corollary enables us, at least in theory, to express a polynomial as a product of polynomials of degree 1. In order to see why this is true let us suppose $f(x)$ is a polynomial of degree $n > 0$ in $F[x]$. Applying (9.13) we have

(9.14) $f(x) = (x - c_1)f_1(x),$

where c_1 is a complex number and $f_1(x) \in F[x]$. By (9.3) the sum of the degrees of $x - c_1$ and $f_1(x)$ is n, and since $x - c_1$ has degree 1, the polynomial $f_1(x)$ must have degree $n - 1$. If $n - 1 = 0$, then $f_1(x)$ is a nonzero element a of F and we have $f(x) = a(x - c_1)$. However, if $n - 1 > 0$, then using (9.13) again, there exists a complex number c_2 such that

$$f_1(x) = (x - c_2)\,f_2(x),$$

where $f_2(x)$ has degree $n - 2$. Substituting in (9.14), we have

$$f(x) = (x - c_1)(x - c_2)\,f_2(x).$$

If the degree of $f_2(x)$ is 0, then

$$f(x) = a(x - c_1)(x - c_2),$$

where $a \in F$ and the factorization is complete. Otherwise, using (9.13), we continue the process, obtaining a factor $x - c_3$ of $f_2(x)$. Since the degrees of the polynomials $f_i(x)$ decrease by one at each step, and since the sum of the degrees of the polynomials in the factorization must equal n, it follows that after n steps we reach a polynomial $f_n(x)$ of degree 0. Thus $f_n(x) = a$ for some $a \in F$, and we have

(9.15) $f(x) = a(x - c_1)(x - c_2) \cdots (x - c_n),$

where $c_1, c_2, \cdots, c_n$ are complex numbers and each c_i is a zero of $f(x)$. Evidently, the leading coefficient of the polynomial on the right in (9.15) is a. Since two polynomials are equal if and only if corresponding coefficients are equal, it follows that a is the leading coefficient of $f(x)$. This proves the following theorem.

(9.16) Theorem

If $f(x)$ is a polynomial of degree $n > 0$, with complex coefficients, then there exist n complex numbers $c_1, c_2, \cdots, c_n$ such that

$$f(x) = a(x - c_1)(x - c_2) \cdots (x - c_n),$$

where a is the leading coefficient of $f(x)$.

(9.17) Corollary

If a polynomial has complex coefficients and degree n greater than 0, then $f(x)$ has at most n different complex zeros.

Proof

We shall give an indirect proof. Suppose $f(x)$ has *more* than n different complex zeros. Let us choose $n + 1$ of these zeros and label them $c_1, c_2, \cdots, c_n$, and c. We may use the c_i as in the proof of (9.16) to obtain the factorization (9.15). Substituting c for x and using the fact that $f(c) = 0$, we obtain

$$0 = a(c - c_1)(c - c_2) \cdots (c - c_n).$$

However, each factor on the right side is different from zero because $c \neq c_i$ for every i. Since the product of nonzero complex numbers cannot equal zero, we have a contradiction.

It is relatively easy to show that the zeros $c_1, c_2, \cdots, c_n$ in (9.16) are unique; for suppose there are n complex numbers $d_1, d_2, \cdots, d_n$ such that

$$a(x - d_1)(x - d_2) \cdots (x - d_n) = a(x - c_1)(x - c_2) \cdots (x - c_n).$$

If, for any i, we substitute d_i for x, then the left side reduces to 0 and hence

$$a(d_i - c_1)(d_i - c_2) \cdots (d_i - c_n) = 0.$$

Since a product of complex numbers equals 0 if and only if one of the factors is 0, it follows that d_i equals one of the c_j. By the same type of

425

argument we can show that each c_j equals one of the d_i. Thus the complex numbers $d_1, d_2, \cdots, d_n$ are the same as $c_1, c_2, \cdots, c_n$.

EXAMPLE 1 Find a polynomial $f(x)$ of degree 3 that has zeros $2, -1$, and i.

Solution

By the Factor Theorem, $f(x)$ has factors $x - 2$, $x + 1$, and $x - i$. No other factors of degree 1 exist, since by the Factor Theorem another linear factor $x - c$ would produce a fourth zero of $f(x)$ in violation of (9.17). Hence $f(x)$ has the form

$$f(x) = a(x - 2)(x + 1)(x - i),$$

where $a \in F$. Any nonzero value can be assigned to a. If we let $a = 1$, we obtain

$$f(x) = x^3 - (1 + i)x^2 + (-2 + i)x + 2i.$$

The complex numbers $c_1, c_2, \cdots, c_n$ in (9.16) are not necessarily all different. To illustrate, the polynomial $f(x) = x^3 + x^2 - 5x + 3$ has the factorization

$$f(x) = (x + 3)(x - 1)(x - 1).$$

If a factor $x - c$ occurs k times in the factorization (9.16), then c is called a **zero of multiplicity** k of $f(x)$. In the preceding illustration, 1 is a zero of multiplicity 2 and -3 is a zero of multiplicity 1.

If a zero of multiplicity k is counted as k zeros, then (9.16) tells us that a polynomial $f(x)$ of degree $n > 0$ has *at least* n zeros (not necessarily all different). Combining this statement with (9.17) gives us the following basic result.

(9.18) Theorem

If a polynomial $f(x)$ has complex coefficients and degree n greater than 0, and if a zero of multiplicity k is counted k times, then $f(x)$ has precisely n complex zeros.

EXAMPLE 2 Express $f(x) = x^5 - 4x^4 + 13x^3$ as a product of linear factors and list the five zeros of $f(x)$.

Solution

We begin by writing

$$f(x) = x^3(x^2 - 4x + 13).$$

By the quadratic formula (2.3), the zeros of the polynomial $x^2 - 4x + 13$ are given by

$$\frac{4 \pm \sqrt{16-52}}{2} = \frac{4 \pm \sqrt{-36}}{2} = 2 \pm 3i.$$

Hence by the Factor Theorem, $x^2 - 4x + 13$ has factors $x - (2 + 3i)$ and $x - (2 - 3i)$. This gives us the desired factorization

$$f(x) = x \cdot x \cdot x \cdot (x - 2 - 3i)(x - 2 + 3i).$$

Since $x - 0$ occurs as a factor three times, the number 0 is a root of multiplicity three, and the five zeros of $f(x)$ are 0, 0, 0, $2 + 3i$, and $2 - 3i$.

EXERCISES

In each of Exercises 1–4 find a polynomial of degree 3 with the indicated zeros.

1 $5, i, -i$

2 $i, -2i, 4i$

3 $3 + 2i, 3 - 2i, 3$

4 $\sqrt{2}, \sqrt{3}, \sqrt{6}$

5 Find a polynomial of degree 4 such that both 4 and -3 are zeros of multiplicity two.

6 Find a polynomial of degree 4 such that 2 is a zero of multiplicity 3 and -2 is a zero of multiplicity 1.

7 Find a polynomial of degree 7 such that 1 is a zero of multiplicity 3 and 0 is a zero of multiplicity 4.

8 Find a polynomial of degree 5 such that -1 is a zero of multiplicity 2 and 0 is a zero of multiplicity 3.

In Exercises 9–16 find the zeros of the polynomials and state the multiplicity of each zero.

9 $f(x) = (x + 5)^2(x - 1)$

10 $f(x) = (x + 4)^3(x^2 + 4)$

11 $f(x) = x^5 + x^4 - 5x^3$

12 $f(x) = (x^2 - 25)^2$

13 $f(x) = (x^2 - 9)^2(x^2 + 9)$

14 $f(x) = (x^2 - 7x + 6)^2$

15 $f(x) = (x^2 + 2x - 3)(x + 2)^4$

16 $f(x) = (5x - 8)^5$

17 Show that 3 is a zero of multiplicity two of the polynomial $f(x) = x^4 - 6x^3 + 13x^2 - 24x + 36$, and express $f(x)$ as a product of linear factors.

18 Show that 2 is a zero of multiplicity two of the polynomial $f(x) = x^4 - 6x^3 + 9x^2 + 4x - 12$, and express $f(x)$ as a product of linear factors.

19 Show that 1 is a zero of multiplicity three of $f(x) = x^4 + x^3 - 9x^2 + 11x - 4$, and find the other zero.

20 Show that -1 is a zero of multiplicity four of $f(x) = x^6 + 4x^5 + x^4 - 16x^3 - 29x^2 - 20x - 5$, and find the other zeros.

21 Let $f(x)$ and $g(x)$ be polynomials of degree not greater than n, where n is a positive integer. Show that if $f(x)$ and $g(x)$ are equal in value for more than n distinct values of x, then $f(x)$ and $g(x)$ are identical, that is, coefficients of like powers are the same. (*Hint:* Write $f(x)$ and $g(x)$ as in Section 1 and consider $h(x) = f(x) - g(x) = (a_n - b_n)x^n + \cdots + (a_0 - b_0)$. Then show that $h(x)$ has more than n distinct zeros and conclude from (9.17) that $a_i = b_i$, for all i.)

22 Determine real numbers A and B such that $A(2x + 3) + B(x - 7) = 3x - 2$ is an identity. (*Hint:* First write the equation in the form $(2A + B)x + (3A - 7B) = 3x - 2$. Next, by Exercise 21, $2A + B = 3$ and $3A - 7B = -2$. Now solve for A and B.)

Determine real numbers A, B, C, D such that the following equations are identities (see Exercise 22).

23 $A(3x - 5) + B(2x - 1) + Cx^2 = 6 - 5x$

24 $A(x - 3) + (Bx + C)(4x - 1) = 8x + 5$

25 $A(2x + 1) + (Bx + C)(2x - 1) = (x + 1)(4x + 3)$

26 $A(x - 2)(x^2 + 1) + B(x^2 + 1) + (Cx + D)(x - 2)^2 = 3x - 5$

5 ZEROS OF POLYNOMIALS WITH REAL COEFFICIENTS

In this section we shall concentrate on polynomials with real coefficients. An interesting fact about such polynomials is illustrated in Example 2 of the preceding section, where the two complex zeros of $x^5 - 4x^4 + 13x^3$ were conjugates of one another. This is no accident, since the following general result is true.

(9.19) Theorem

If $f(x)$ is a polynomial of degree $n > 0$, with real coefficients, and if z is a complex zero of $f(x)$, then the conjugate $\bar{z}$ is also a zero of $f(x)$.

Proof

We may write

$$f(x) = a_n x^n + a_{n-1} x^{n-1} + \cdots + a_1 x + a_0$$

where the a_i are real numbers and $a_n \neq 0$. If $f(z) = 0$, then

$$a_n z^n + a_{n-1} z^{n-1} + \cdots + a_1 z + a_0 = 0.$$

If two complex numbers are equal, then so are their conjugates. Consequently the conjugates of each side of the latter equation are equal, that is,

$$\overline{a_n z^n + a_{n-1} z^{n-1} + \cdots + a_1 z + a_0} = \overline{0} = 0.$$

(The fact that $\overline{0} = 0$ follows from (iv) of (8.7).) As pointed out in the preceding chapter, the conjugate of a sum of complex numbers equals the sum of the conjugates, and hence

(9.20) $$\overline{a_n z^n} + \overline{a_{n-1} z^{n-1}} + \cdots + \overline{a_1 z} + \overline{a_0} = 0.$$

Using (8.7) we have for each i,

$$\overline{a_i z^i} = \overline{a_i} \cdot \overline{z^i} = \overline{a_i} \cdot \overline{z}^i = a_i \overline{z}^i.$$

Hence (9.20) may be written as follows:

$$a_n \overline{z}^n + a_{n-1} \overline{z}^{n-1} + \cdots + a_1 \overline{z} + a_0 = 0.$$

The last equation states that $f(\overline{z}) = 0$, which completes the proof.

EXAMPLE 1 Find a polynomial $f(x)$ of degree 4 that has real coefficients and zeros $2 + i$ and $-3i$.

Solution

By (9.19), $f(x)$ must also have zeros $2 - i$ and $3i$, and hence by the Factor Theorem, $f(x)$ has factors $x - (2 + i)$, $x - (2 - i)$, $x - (-3i)$, and $x - (3i)$. Multiplying those factors gives us a polynomial of the required type. Thus

$$\begin{aligned} f(x) &= [x - (2 + i)][x - (2 - i)](x - 3i)(x + 3i) \\ &= (x^2 - 4x + 5)(x^2 + 9) \\ &= x^4 - 4x^3 + 14x^2 - 36x + 45. \end{aligned}$$

If a polynomial with real coefficients is factored as in (9.16), some of the factors $x - c_i$ may have a complex coefficient c_i. However, it is possible to obtain a factorization into polynomials with real coefficients, as stated in the next theorem.

(9.21)   Theorem

Every polynomial with real coefficients and positive degree n can be expressed as a product of linear and quadratic polynomials with real coefficients, where the quadratic factors have no real zeros.

Proof

By (9.18), $f(x)$ has precisely n complex zeros $c_1, c_2, \cdots, c_n$, and as in (9.16), we obtain the factorization

$$f(x) = a(x - c_1)(x - c_2) \cdots (x - c_n).$$

Of course, some of the c_i may be real, that is, their imaginary parts may be zero. In such cases we obtain the linear factors referred to in the statement of the theorem. If a zero c_i is not real, then by (9.19) the conjugate $\bar{c}_i$ is also a zero of $f(x)$ and hence must be one of the numbers in the set $c_1, c_2, \cdots, c_n$. This implies that both $x - c_i$ and $x - \bar{c}_i$ appear in the factorization of $f(x)$. If those factors are multiplied, we obtain $x^2 - (c_i + \bar{c}_i)x + c_i\bar{c}_i$, which has *real* coefficients (Why?). Thus the complex zeros of $f(x)$ and their conjugates give rise to quadratic polynomials which are irreducible over **R**. This completes the proof.

EXAMPLE 2 Express $x^4 - 2x^2 - 3$ (a) in the form (9.15); (b) as a product of linear and quadratic polynomials with real coefficients.

Solution

(a) The zeros of the given polynomial can be found by solving the equation $x^4 - 2x^2 - 3 = 0$, which may be regarded as quadratic in x^2. Solving for x^2 by means of the quadratic formula we obtain

$$x^2 = \frac{2 \pm \sqrt{4 + 12}}{2} = \frac{2 \pm 4}{2},$$

or $x^2 = 3$ and $x^2 = -1$. Hence the zeros are $\sqrt{3}, - \sqrt{3}, i$, and $-i$, and so we obtain the factorization

$$x^4 - 2x^2 - 3 = (x - \sqrt{3})(x + \sqrt{3})(x - i)(x + i).$$

(b) Multiplying the last two factors in the preceding factorization gives us

$$x^4 - 2x^2 - 3 = (x - \sqrt{3})(x + \sqrt{3})(x^2 + 1),$$

which is of the form stated in (9.21).

The solution of this example could also have been obtained by immediately factoring the original expression without first finding the zeros. Thus

$$x^4 - 2x^2 - 3 = (x^2 - 3)(x^2 + 1)$$
$$= (x + \sqrt{3})(x - \sqrt{3})(x + i)(x - i).$$

We have already pointed out that it is generally very difficult to find zeros of polynomials of high degree. Most of our previous results have been primarily of theoretical value, since they tell us that zeros and factorizations exist but do not indicate how to find them. However, if the coefficients are all integers or rational numbers, there is a method for finding the rational zeros, if they exist. The method is a consequence of the following theorem.

(9.22) **Theorem on Rational Zeros**

Let $f(x) = a_n x^n + a_{n-1} x^{n-1} + \cdots + a_1 x + a_0$ be a polynomial with integral coefficients. If c/d is a rational zero of $f(x)$, where c and d have no common prime factors and $c > 0$, then c is a factor of a_0 and d is a factor of a_n.

Proof

Let us show that c is a factor of a_0. If $c = 1$, the theorem follows at once, since 1 is a factor of *any* number. Now suppose $c \neq 1$. In this case $c/d \neq 1$, for if $c/d = 1$, we obtain $c = d$, and since c and d have no prime factor in common, this implies that $c = d = 1$, a contradiction. Hence in the following discussion we have $c \neq 1$ and $c \neq d$.

Since $f(c/d) = 0$,

$$a_n(c^n/d^n) + a_{n-1}(c^{n-1}/d^{n-1}) + \cdots + a_1(c/d) + a_0 = 0.$$

Multiplying by d^n and then adding $-a_0 d^n$ to both sides, we obtain

$$a_n c^n + a_{n-1} c^{n-1} d + \cdots + a_1 c d^{n-1} = -a_0 d^n,$$

or

$$c(a_n c^{n-1} + a_{n-1} c^{n-2} d + \cdots + a_1 d^{n-1}) = -a_0 d^n.$$

This shows that c is a factor of the integer $a_0 d^n$. Hence if c is factored into primes, say $c = p_1 p_2 \cdots p_k$, then each p_i is also a factor of $a_0 d^n$. However, by hypothesis, none of the p_i is a factor of d. This implies that each p_i is a factor of a_0, that is, c is a factor of a_0. A similar argument may be used to prove that d is a factor of a_n.

The technique of using (9.22) for finding rational solutions of equations with integral coefficients is illustrated in the following example.

EXAMPLE 3 Find all rational solutions of the equation $3x^4 + 14x^3 + 14x^2 - 8x - 8 = 0$.

Solution

The problem is equivalent to finding the rational zeros of the indicated polynomial. According to (9.22), if c/d is a rational zero and $c > 0$, then c is a divisor of -8 and d is a divisor of 3. Hence the possible choices for c are 1, 2, 4, 8, and the choices for d are ± 1 and ± 3. Consequently, any rational roots are included among the numbers $\pm 1, \pm 2, \pm 4, \pm 8, \pm\frac{1}{3}, \pm\frac{2}{3}, \pm\frac{4}{3}, \pm\frac{8}{3}$. Of these sixteen possibilities, not more than four can be zeros by (9.17). It is necessary to check to see which, if any, are zeros. The method of synthetic division is recommended for doing this. We have

$$\begin{array}{r|rrrrr} -2 & 3 & 14 & 14 & -8 & -8 \\ & & -6 & -16 & 4 & 8 \\ \hline & 3 & 8 & -2 & -4 & 0 \end{array}$$

which shows that -2 is a zero. Moreover, the synthetic division provides the coefficients of the quotient in the division of the polynomial by $x + 2$. Hence we have the following factorization of the given polynomial:

$$(x + 2)(3x^3 + 8x^2 - 2x - 4).$$

The remaining solutions of the equation must be zeros of the second factor, and therefore we may use the latter polynomial to check for solutions. Dividing by $x + \frac{2}{3}$ synthetically gives us

$$\begin{array}{r|rrrr} -\frac{2}{3} & 3 & 8 & -2 & -4 \\ & & -2 & -4 & 4 \\ \hline & 3 & 6 & -6 & 0 \end{array}$$

and so we see that $-\frac{2}{3}$ is a zero.

The remaining zeros are solutions of the equation $3x^2 + 6x - 6 = 0$, or equivalently, $x^2 + 2x - 2 = 0$. By the quadratic formula this equation has solutions $-1 + \sqrt{3}$ and $-1 - \sqrt{3}$. Hence the given polynomial has two rational roots, -2 and $-\frac{2}{3}$, and two irrational roots.

Theorem (9.22) may also be applied to equations with rational coefficients. We merely multiply both sides of such an equation by the least common denominator of all the coefficients to obtain an equation with integral coefficients and then proceed as above.

EXAMPLE 4 Find all rational solutions of the equation

$$(\tfrac{2}{3})x^4 + (\tfrac{1}{2})x^3 - (\tfrac{5}{4})x^2 - x - (\tfrac{1}{6}) = 0.$$

Multiplying both sides of the equation by 12 produces the equivalent equation

$$8x^4 + 6x^3 - 15x^2 - 12x - 2 = 0.$$

According to (9.22), if c/d is a rational solution, then the choices for c are 1 and 2 and the choices for d are ±1, ±2, ±4, and ±8. Hence the only possible rational roots are ±1, ±2, $\pm\frac{1}{2}$, $\pm\frac{1}{4}$, and $\pm\frac{1}{8}$. By trial we have

$$
\begin{array}{r|rrrrr}
-\frac{1}{2} & 8 & 6 & -15 & -12 & -2 \\
& & -4 & -1 & 8 & 2 \\
\hline
& 8 & 2 & -16 & -4 & 0
\end{array}
$$

and hence $-\frac{1}{2}$ is a solution. Using synthetic division on the coefficients of the quotient, we obtain

$$
\begin{array}{r|rrrr}
-\frac{1}{4} & 8 & 2 & -16 & -4 \\
& & -2 & 0 & 4 \\
\hline
& 8 & 0 & -16 & 0
\end{array}
$$

and consequently $-\frac{1}{4}$ is a solution. The last synthetic division gave us the quotient $8x^2 - 16$. Setting this equal to zero and solving, we obtain $x = \pm \sqrt{2}$. Thus the given equation has rational solutions $-\frac{1}{2}$, $-\frac{1}{4}$ and irrational solutions $\sqrt{2}, -\sqrt{2}$.

The discussion in this section gives no information about finding the irrational or complex zeros of polynomials. The examples we have worked are not typical of problems encountered in applications. Indeed, it is not unusual for a polynomial with rational coefficients to have *no* rational zeros. Except in the simplest cases, the best that can be accomplished is to find decimal approximations to the irrational zeros. There exist methods which may be used to approximate some of these irrational zeros to any degree of accuracy. A standard way is to use a technique studied in calculus called Newton's Method. In practice, computers have, to a large extent, taken over the task of approximating irrational solutions of equations.

If only rough approximations to the real solutions of an equation $f(x) = 0$ are required, then graphical methods are available. For example, we could sketch the graph of $y = f(x)$ and estimate where $y = 0$; that is, we could approximate the x-intercepts. Needless to say, the accuracy of the approximation depends on the care with which the graph is sketched.

EXERCISES

1 Find a polynomial that has real coefficients, degree 2, and root $2 - 3i$.

2 Find a polynomial that has real coefficients, degree 3, and roots $4i$ and -3.

3 Find a polynomial that has real coefficients, degree 4, and roots $2 + i$ and $1 + 2i$.

4 Find a polynomial that has real coefficients, degree 4, and roots i and $3 - \sqrt{2}\, i$.

In Exercises 5–16 find all rational solutions of the given equations. If possible find all the roots of the equations.

5 $2x^3 - 3x^2 - 17x + 30 = 0$ **6** $2x^3 - 3x^2 - 7x - 6 = 0$

7 $6x^3 + 11x^2 - 4x - 4 = 0$ **8** $6x^3 + 11x^2 - 57x - 20 = 0$

9 $3x^3 + 8x^2 - x - 20 = 0$ **10** $12x^3 - x^2 + 7x + 2 = 0.$

11 $x^4 + x^3 - 5x^2 - 15x - 18 = 0$ **12** $8x^4 + 2x^3 - 7x^2 + 2x - 15 = 0$

13 $(\frac{9}{4})x^4 - (\frac{15}{4})x^3 - 20x^2 + (\frac{11}{2})x + \frac{1}{2} = 0$

14 $(\frac{1}{3})x^4 - x^3 - x^2 + (\frac{13}{3})x - 2 = 0$

15 $x^4 + 3x^3 - 30x^2 - 6x + 56 = 0$

16 $x^4 - 3x^3 - 43x^2 + 9x + 20 = 0$

Prove that the equations in Exercises 17–20 have no rational roots.

17 $2x^4 + 2x^3 + 9x^2 - x - 5 = 0$ **18** $3x^4 - 9x^3 - 2x^2 - 15x - 5 = 0$

19 $x^5 - 3x^3 + 4x^2 + x - 2 = 0$ **20** $2x^5 + 3x^3 + 7 = 0$

21 If n is an odd positive integer, prove that a polynomial of degree n with real coefficients has at least one real zero.

22 Show that (9.19) is not necessarily true if $f(x)$ has complex coefficients.

23 Complete the proof of (9.22) by showing that d is a factor of a_n.

24 If a polynomial of the form $x^n + a_{n-1}x^{n-1} + \cdots + a_1 x + a_0$, where each a_i is an integer, has a rational root r, show that r is an integer and is a factor of a_0.

6 REVIEW EXERCISES

Oral

Define or discuss each of the following.

1 Polynomial

2 Degree of a polynomial

Written

In Exercises 1–4 find $f(x) + g(x)$, $f(x) - g(x)$, and $f(x) \cdot g(x)$.

1 $f(x) = 3x^2 + x - 5$, $g(x) = x^2 + 1$

2 $f(x) = 2x^3 + x - 1$, $g(x) = 2x^3 - x^2$

3 $f(x) = x - x^2$, $g(x) = 1$

4 $f(x) = 0$, $g(x) = 2x + 1$

5–8 State the degrees of the polynomials $f(x)$ and $g(x)$ given in Exercises 1–4. Also state the degrees of $f(x) + g(x)$, $f(x) - g(x)$, and $f(x) \cdot g(x)$.

Express each of the polynomials in Exercises 9–12 as a product of polynomials which are (a) irreducible over **R**. (b) irreducible over **C**.

9 $81 - 16x^4$

10 $x^3 + 125$

11 $3x^4 - 2x^3 + 12x^2 - 8x$

12 $x^4 - 64$

In each of Exercises 13–16 find the quotient and remainder if $f(x)$ is divided by $g(x)$:

13 $f(x) = 4x^5 - 2x^3 + x + 7$, $g(x) = x^3 + 2x - 3$

14 $f(x) = 2x^2 + 5x - 3$, $g(x) = x^3$

15 $f(x) = 6x - 5$, $g(x) = 2x + 3$

16 $f(x) = 2x^3 + x^2 - 2x + 1$, $g(x) = x^3$

17 If $f(x) = -2x^4 - 5x^3 + 4x^2 - 9$, **a** use the Remainder Theorem to find $f(2)$, **b** prove that $x + 3$ is a factor of $f(x)$.

In Exercises 18–19 use synthetic division to find the quotient and remainder if $f(x)$ is divided by $g(x)$:

18 $f(x) = 3x^5 - x^2 + 3$, $g(x) = x + 2$

19 $f(x) = x^3 - 4ix^2 + 2x + i$, $g(x) = x + i$

20 Find polynomials of degree 3 with the following zeros:

    a 1, 2, 3   **b** $-2 + 5i, -2 - 5i, -1$ **c** $2i, -i, 1$

21 Find a polynomial of degree 5 such that:

    a -2 is a zero of multiplicity 3, and 0 is a zero of multiplicity 2.

    b 0 is a zero of multiplicity 3, and -2 is a zero of multiplicity 2.

22 Show that -2 is a zero of multiplicity 3 of the polynomial $x^5 + 8x^4 + 21x^3 + 14x^2 - 20x - 24$, and express this polynomial as a product of linear factors.

Find all rational solutions of the equations in Exercises 23–25.

23 $x^4 - 3x^3 - 15x^2 + 19x + 30 = 0$

24 $30x^3 + 107x^2 - 5x - 42 = 0$

25 $x^4 + 2x^3 + 6x^2 + x + 10 = 0$

26 Determine real numbers A, B, C, and D such that the following equation is an identity:

$$A(x - 2)^3 + B(x + 1)(x - 2)^2 + C(x + 1)(x - 2) + D(x + 1)$$
$$= 3x^3 - 18x^2 + 29x - 4$$

10

Sequences

In this chapter we shall discuss a number of topics connected with properties of positive integers. The method of proof called **mathematical induction** *is very important in all branches of mathematics. The concept of* **sequences** *also plays a major role in mathematics and applications. Of special interest to us will be* **arithmetic** *and* **geometric** **sequences.** *The final two sections contain a discussion of counting processes that arise frequently in mathematics and in everyday life. Among these are the notions of* **permutations** *and* **combinations.**

1 MATHEMATICAL INDUCTION

In several earlier parts of our work we stated that the method of mathematical induction was required in order to complete a proof. This method may be used to show that certain statements or formulas are true for all positive integers. For example, if n is a positive integer, let P_n denote the statement

$$(xy)^n = x^n y^n,$$

where x and y are real numbers. Thus P_1 represents the statement $(xy)^1 = x^1 y^1$, P_2 denotes $(xy)^2 = x^2 y^2$, P_3 is $(xy)^3 = x^3 y^3$, and so on. It is easy to show that P_1, P_2, and P_3 are *true* statements. However, since the set of positive integers is infinite, it is impossible to check the validity of P_n for *every* positive integer n. In order to give a proof, the method provided by (10.2) is required. This method is based on the following fundamental axiom for the set **N** of positive integers.

(10.1) Axiom of Mathematical Induction

If a subset S of **N** has the following two properties:

(i) $1 \in S$,

(ii)   whenever $k \in S$, then also $k + 1 \in S$,

then $S = \mathbf{N}$.

There should be little reluctance about accepting (10.1). If S is a set of positive integers satisfying property (ii), then whenever S contains an arbitrary positive integer k, it must also contain the next positive integer, $k + 1$. If S also satisfies property (i), then S contains 1 and hence by (ii), S contains $1 + 1$, or 2. Applying (ii) again, we see that S contains $2 + 1$, or 3. Once again, S must contain $3 + 1$, or 4. If we continue in this manner, it can be argued that if n is any *specific* positive integer, then $n \in S$, since we can proceed a step at a time as above, eventually reaching n. Although this argument does not *prove* (10.1), it certainly makes it plausible.

We shall use (10.1) to establish the following fundamental principle.

(10.2)   Principle of Mathematical Induction

If with each positive integer n there is associated a statement P_n, then all the statements P_n are true provided the following two conditions hold:

(i) P_1 is true,

(ii)   whenever k is a positive integer such that P_k is true, then P_{k+1} is also true.

Proof

Assume that (i) and (ii) of (10.2) hold and let

$$S = \{n : P_n \text{ is true}\}.$$

By assumption, P_1 is true and consequently $1 \in S$. Thus S satisfies property (i) of (10.1). Whenever $k \in S$, then by the definition of S, P_k is true and hence from assumption (ii) of (10.2), P_{k+1} is also true. This means that $k + 1 \in S$. We have shown that whenever $k \in S$, then also $k + 1 \in S$. Consequently, property (ii) of (10.1) is true. Hence by (10.1), $S = \mathbf{N}$, that is, P_n is true for every positive integer n.

There are other variations of the principle of mathematical induction. One variation appears in (10.10). In most of our work the statement P_n will usually be given in the form of an equation involving the arbitrary positive integer n, as in our illustration $(xy)^n = x^n y^n$.

When applying (10.2), the following two steps should always be followed:

Step (i)   Prove that P_1 is true.

Step (ii)   Assume that P_k is true and prove that P_{k+1} is true.

Step (ii) is usually the most confusing for the beginning student. We do not *prove* that P_k is true (except for $k = 1$). Instead, we show that if P_k is true, *then* the next statement P_{k+1} is true. This is all that is necessary according to (10.2). The assumption that P_k is true is referred to as the **induction hypothesis.**

Many interesting formulas about positive integers can be established by using mathematical induction, two of which are illustrated in Examples 1 and 2. Others appear in the Exercises.

EXAMPLE 1   Prove that for every positive integer n, the sum of the first n positive integers is $n(n + 1)/2$.

Solution

For any positive integer n, let P_n denote the statement

(10.4) $$1 + 2 + 3 + \cdots + n = \frac{n(n + 1)}{2},$$

where by convention, when $n \leq 4$, the left side is adjusted so that there are precisely n terms in the sum. We wish to show that P_n is true for every n. Although it is instructive to check (10.4) for several values of n, it is unnecessary to do so. We need only follow steps (i) and (ii) of (10.3).

(i)   If we substitute $n = 1$ in (10.4), then by convention the left side collapses to 1 and the right side is $\dfrac{1(1 + 1)}{2}$, which also equals 1. This proves that P_1 is true.

(ii)   *Assume* that P_k is true. Thus the induction hypothesis is

(10.5) $$1 + 2 + 3 + \cdots + k = \frac{k(k + 1)}{2}.$$

Our goal is to prove that P_{k+1} is true, that is,

(10.6) $$1 + 2 + 3 \cdots + (k+1) = \frac{(k+1)[(k+1) + 1]}{2}.$$

By the induction hypothesis we already have a formula for the sum of the first k positive integers. Hence a formula for the sum of the first $k + 1$ positive integers may be found simply by adding $(k + 1)$ to both sides of (10.5). Doing this and simplifying, we obtain

$$1 + 2 + 3 + \cdots + k + (k+1) = \frac{k(k+1)}{2} + (k+1)$$

$$= \frac{k(k+1) + 2(k+1)}{2}$$

$$= \frac{k^2 + 3k + 2}{2}$$

$$= \frac{(k+1)(k+2)}{2},$$

$$= \frac{(k+1)[(k+1)+1]}{2}.$$

This shows that P_{k+1} is true; so the proof by mathematical induction is complete.

EXAMPLE 2 Prove that for each positive integer n,

$$1^2 + 3^2 + \cdots + (2n-1)^2 = \frac{n(2n-1)(2n+1)}{3}.$$

Solution

For each positive integer n, let P_n denote the given statement. Note that this is a formula for the sum of the squares of the first n odd positive integers. We again follow the two-step procedure in (10.3).

(i) Substituting 1 for n in P_n we obtain

$$1^2 = \frac{1 \cdot (2-1)(2+1)}{3} = \frac{3}{3} = 1,$$

which shows that P_1 is true.

(ii) *Assume* that P_k is true. Thus the induction hypothesis is

$$1^2 + 3^2 + \cdots + (2k-1)^2 = \frac{k(2k-1)(2k+1)}{3}.$$

We wish to prove that P_{k+1} is true; that is,

$$1^2 + 3^2 + \cdots + [2(k+1) - 1]^2$$

$$= \frac{(k+1)[2(k+1) - 1][2(k+1) + 1]}{3}.$$

The latter equation simplifies to

(10.7) $$1^2 + 3^2 + \cdots + (2k+1)^2 = \frac{(k+1)(2k+1)(2k+3)}{3}.$$

Observe that the second from the last term on the left-hand side of this equation is $(2k - 1)^2$ (Why?). In a manner similar to the solution of Example 1, we may obtain (10.7) by adding $(2k + 1)^2$ to both sides of the equation stated in the induction hypothesis. This gives us

$$1^2 + 3^2 + \cdots + (2k - 1)^2 + (2k + 1)^2$$
$$= \frac{k(2k - 1)(2k + 1)}{3} + (2k + 1)^2.$$

We leave it as an exercise for the reader to show that the right side of the preceding equation may be written in the form of the right side of (10.7). This proves that P_{k+1} is true, and hence by (10.3), P_n is true for every n.

It was mentioned in Chapter One that the laws of exponents can be proved by mathematical induction. In order to apply (10.3), we shall use the following definition of exponents.

(10.8)   Definition

If x is any real number, then

(i) $x^1 = x$;

(ii)   whenever k is a positive integer for which x^k is defined, let $x^{k+1} = x^k \cdot x$.

A definition such as (10.8) is called a **recursive definition.** In general, if a concept is defined for every positive integer n in such a way that the case corresponding to $n = 1$ is given, and if it is also stated how any case after the first is obtained from the preceding one, then the definition is a recursive definition. For example, by (i) of (10.8) we have $x^1 = x$. Next, applying (ii) of (10.8) we obtain

$$x^2 = x^{1+1} = x^1 \cdot x = x \cdot x.$$

Since x^2 is now defined we may employ (ii) again (with $k = 2$), obtaining

$$x^3 = x^{2+1} = x^2 \cdot x = (x \cdot x) \cdot x.$$

This defines x^3, and hence (ii) of (10.8) may be used again to obtain x^4. Thus

$$x^4 = x^{3+1} = x^3 \cdot x = [(x \cdot x) \cdot x] \cdot x.$$

Observe that this agrees with our previous formulation of x^n as a product of x by itself n times. It can be shown (by mathematical induction) that (10.8) defines x^n for every positive integer n.

EXAMPLE 3 If x is a real number, prove that $x^m \cdot x^n = x^{m+n}$ for all positive integers m and n.

Solution

Let m be an arbitrary positive integer. For each positive integer n, let P_n denote the statement

(10.9) $x^m \cdot x^n = x^{m+n}$.

We shall use (10.3) to prove that P_n is true for all positive integers n.

(i) To show that P_1 is true we may use (i) and (ii) of (10.8) as follows:

$$x^m \cdot x^1 = x^m \cdot x$$
$$= x^{m+1}$$

This is formula (10.9) with $n = 1$ and hence P_1 is true.

(ii) Assume that P_k is true. Thus the induction hypothesis is

$$x^m \cdot x^k = x^{m+k}.$$

We wish to prove that P_{k+1} is true, that is,

$$x^m \cdot x^{k+1} = x^{m+(k+1)}.$$

The proof may be arranged as follows, where reasons are stated to the right of each step.

$$
\begin{aligned}
x^m \cdot x^{k+1} &= x^m \cdot (x^k \cdot x) && \text{(ii) of (10.8)} \\
&= (x^m \cdot x^k) \cdot x && \text{(associative law in } \mathbf{R}) \\
&= x^{m+k} \cdot x && \text{(induction hypothesis)} \\
&= x^{(m+k)+1} && \text{(ii) of (10.8)} \\
&= x^{m+(k+1)} && \text{(associative law for integers).}
\end{aligned}
$$

By (10.3) the proof by induction is complete.

Consider a positive integer j and suppose that with each integer $n \geq j$ there is associated a statement P_n. For example, if $j = 6$, then the statements are numbered P_6, P_7, P_8, and so on. The principle of mathematical induction may be extended to cover this situation. Just as before, two steps are used. Specifically, to prove that the statements S_n are true for $n \geq j$, we use the following two steps.

(10.10) (i') Prove that S_j is true,

(ii') *Assume* that S_k is true for $k \geq j$ and *prove* that S_{k+1} is true.

EXAMPLE 4 Let a be a nonzero real number such that $a > -1$. Prove that $(1 + a)^n > 1 + na$, for all integers $n \geq 2$.

Solution

For each positive integer n, let P_n denote the inequality $(1 + a)^n > 1 + na$. Note that P_1 is *false*, since $(1 + a)^1 = 1 + 1 \cdot a$. However, we can show that P_n is true for $n \geq 2$ by using (10.10) with $j = 2$.

(i') We first note that $(1 + a)^2 = 1 + 2a + a^2$. Since $a \neq 0$, we have $a^2 > 0$ and therefore $1 + 2a + a^2 > 1 + 2a$. This gives us $(1 + a)^2 > 1 + 2a$, and hence P_2 is true.

(ii') Assume that P_k is true. Thus the induction hypothesis is

$$(1 + a)^k > 1 + ka.$$

We wish to show that P_{k+1} is true, that is,

$$(1 + a)^{k+1} > 1 + (k + 1)a.$$

Since $a > -1$, we have $a + 1 > 0$ and, consequently, multiplying both sides of the induction hypothesis by $1 + a$ will not change the inequality sign. Hence

$$(1 + a)^k(1 + a) > (1 + ka)(1 + a),$$

which may be rewritten as

$$(1 + a)^{k+1} > 1 + ka + a + ka^2,$$

or as

$$(1 + a)^{k+1} > 1 + (k + 1)a + ka^2.$$

Since $ka^2 > 0$, we have

$$1 + (k + 1)a + ka^2 > 1 + (k + 1)a$$

and therefore

$$(1 + a)^{k+1} > 1 + (k + 1)a.$$

Thus P_{k+1} is true and the proof is complete.

EXERCISES

In each of Exercises 1–12, prove that the given formula is true for every positive integer n.

1 $2 + 4 + 6 + \cdots + 2n = n(n + 1)$

2 $1 + 4 + 7 + \cdots + (3n - 2) = \dfrac{n(3n - 1)}{2}$

3 $1 + 3 + 5 + \cdots + (2n - 1) = n^2$

4 $3 + 9 + 15 + \cdots + (6n - 3) = 3n^2$

5 $1^2 + 2^2 + 3^2 + \cdots + n^2 = \dfrac{n(n + 1)(2n + 1)}{6}$

6 $1^3 + 2^3 + 3^3 + \cdots + n^3 = \left[\dfrac{n(n + 1)}{2}\right]^2$

7 $\dfrac{1}{1 \cdot 2} + \dfrac{1}{2 \cdot 3} + \dfrac{1}{3 \cdot 4} + \cdots + \dfrac{1}{n(n + 1)} = \dfrac{n}{n + 1}$

8 $\dfrac{1}{1 \cdot 2 \cdot 3} + \dfrac{1}{2 \cdot 3 \cdot 4} + \dfrac{1}{3 \cdot 4 \cdot 5} + \cdots + \dfrac{1}{n(n + 1)(n + 2)} =$

$$\dfrac{n(n + 3)}{4(n + 1)(n + 2)}$$

9 $3 + 3^2 + 3^3 + \cdots + 3^n = \tfrac{3}{2}(3^n - 1)$

10 $1^3 + 3^3 + 5^3 + \cdots + (2n - 1)^3 = n^2(2n^2 - 1)$

11 $n < 2^n$ **12** $1 + 2n \le 3^n$

13 Use mathematical induction to prove that if a is any real number greater than 1, then $a^n > 1$ for every positive integer n.

14 If $a \ne 1$, prove that

$$1 + a + a^2 + \cdots + a^{n-1} = \dfrac{a^n - 1}{a - 1},$$

for every positive integer n.

15 If a and b are real numbers, use mathematical induction to prove that $(ab)^n = a^n b^n$ for every positive integer n.

16 If a is a real number, prove that $(a^m)^n = a^{mn}$ for all positive integers m and n.

17 Use mathematical induction to prove that $a - b$ is a factor of $a^n - b^n$ for every positive integer n. (*Hint:* $a^{k+1} - b^{k+1} = a^k(a - b) + (a^k - b^k)b$.)

18 Prove that $a + b$ is a factor of $a^{2n-1} + b^{2n-1}$ for every positive integer n.

19 If z is a complex number and $\bar{z}$ is its conjugate, prove that $\overline{z^n} = \bar{z}^n$ for every positive integer n (see (8.7)).

20 Prove that for every positive integer n, if $z_1, z_2, \cdots, z_n$ are complex numbers, then $\overline{z_1 z_2 \cdots z_n} = \bar{z}_1 \bar{z}_2 \cdots \bar{z}_n$.

If X and Y are sets, then a function f from X to Y is a correspondence that associates with each element x of X a unique element $f(x)$ of Y (see (3.10)). In our previous work the domain X was usually an interval of real numbers. In this section we shall consider a different class of functions.

(10.11) Definition

An **infinite sequence** is a function whose domain is the set of positive integers.

For convenience we sometimes refer to infinite sequences merely as **sequences.** In this book the *range* of an infinite sequence will be a subset of the set of real numbers. In parts of advanced mathematics it is not unusual to take the range as a subset of the complex numbers.

If f is an infinite sequence, then to each positive integer n there corresponds a real number $f(n)$. These numbers in the range of f are usually listed by writing

(10.12) $f(1), f(2), f(3), \cdots, f(n), \cdots$

where the three dots at the end indicate that the sequence does not terminate. The number $f(1)$ is called the **first term** of the sequence, $f(2)$ the **second term** and, in general, $f(n)$ the **nth term** of the sequence.

It is customary to use a subscript notation for the terms of an infinite sequence instead of the functional notation and write (10.12) as

(10.13) $a_1, a_2, a_3, \cdots, a_n, \cdots$

where it is understood that a_n denotes the real number $f(n)$ for each positive integer n. In this way we obtain an infinite collection of real numbers which is *ordered* in the sense that there is a first element, a second element, a third element, a forty-fifth element, and so on. Although infinite sequences are functions, we shall also refer to (10.13) as an infinite sequence, meaning the sequence f defined by $f(n) = a_n$ for every positive integer n.

To specify an infinite sequence it is not sufficient to state the numbers in the range. The *order* in which these numbers appear is also important. For example, the sequence whose first five terms are

$$1, \tfrac{1}{2}, \tfrac{1}{3}, \tfrac{1}{4}, \tfrac{1}{5}, \cdots$$

is different from the sequence which begins

$$1, \tfrac{1}{3}, \tfrac{1}{2}, \tfrac{1}{4}, \tfrac{1}{5}, \cdots$$

even if all the terms after the third are the same in both sequences. In general, the sequence (10.13) is said to be *equal* to the sequence

$$b_1, b_2, b_3, \cdots, b_n, \cdots$$

if and only if $a_i = b_i$ for every positive integer i. To avoid any misunderstanding about ordering, it is customary to specify the nth term for an arbitrary positive integer n. This is often done by means of a formula, as in the following example.

EXAMPLE 1 List the first four terms and the tenth term of the sequence whose nth term is

    (a) $a_n = n/(n+1)$,
    (b) $a_n = 2 + (.1)^n$,
    (c) $a_n = (-1)^{n+1}n^2/(3n-1)$,
    (d) $a_n = 4$.

Solutions

To find the first four terms we substitute, successively, $n = 1, 2, 3,$ and 4 in the formula for a_n. The tenth term is found by substituting 10 for n. Doing this and simplifying gives us the following:

	First four terms	*Tenth term*
(a)	$\frac{1}{2}, \frac{2}{3}, \frac{3}{4}, \frac{4}{5}$	$\frac{10}{11}$
(b)	$2.1, 2.01, 2.001, 2.0001$	2.0000000001
(c)	$\frac{1}{2}, -\frac{4}{5}, \frac{9}{8}, -\frac{16}{11}$	$-\frac{100}{29}$
(d)	$4, 4, 4, 4$	4

It is not essential that a formula for a_n be given. Indeed, sometimes this is impossible, as illustrated in the next example.

EXAMPLE 2 List the first seven terms of the sequence whose nth term a_n is the nth positive prime number.

Solution

The first seven primes are 2, 3, 5, 7, 11, 13, and 17. Hence $a_1 = 2$, $a_2 = 3$, $a_3 = 5$, $a_4 = 7$, $a_5 = 11$, $a_6 = 13$, and $a_7 = 17$. No one has ever found a formula which yields the nth prime number. Our sequence is perfectly legitimate, however, since a_n is uniquely determined for every positive integer n.

An infinite sequence is sometimes defined by stating a recursive definition for the terms, as in the following example.

$$a_1 = 3 \quad \text{and} \quad a_{k+1} = 2a_k \quad \text{for } k \geq 1.$$

Solution

The sequence is defined recursively since the first term is
given and, moreover, whenever a term a_k of the sequence is
known, then the next term a_{k+1} can be found. Thus

$$a_1 = 3,$$
$$a_2 = 2a_1 = 2 \cdot 3 = 6,$$
$$a_3 = 2a_2 = 2 \cdot 2 \cdot 3 = 2^2 \cdot 3 = 12,$$
$$a_4 = 2a_3 = 2 \cdot 2 \cdot 2 \cdot 3 = 2^3 \cdot 3 = 24.$$

We have written the terms as products so as to gain some
insight into the nature of the nth term. Continuing, we ob-
tain $a_5 = 2^4 \cdot 3$ and $a_6 = 2^5 \cdot 3$; and it appears that $a_n = 2^{n-1} \cdot 3$, for every positive integer n. We shall prove that
this guess is correct by means of mathematical induction.
If we let P_n denote the statement $a_n = 2^{n-1} \cdot 3$, then P_1 is
true since $a_1 = 2^0 \cdot 3 = 3$. Next *assume* that P_k is true, that is,
$a_k = 2^{k-1} \cdot 3$. We then have

$$
\begin{array}{ll}
a_{k+1} = 2a_k & \text{(definition of } a_{k+1}) \\
= 2 \cdot 2^{k-1} \cdot 3 & \text{(induction hypothesis)} \\
= 2^k \cdot 3 & \text{(a law of exponents).}
\end{array}
$$

This shows that P_{k+1} is true. Hence $a_n = 2^{n-1} \cdot 3$ for every
positive integer n.

As a final remark, it is important to observe that if only the first
few terms of an infinite sequence are known, then it is impossible to
predict additional terms. For example, if we were given 3, 6, 9, $\cdot \cdot \cdot$
and asked to find the fourth term, we could not proceed without further
information. The infinite sequence with nth term

$$a_n = 3n + (1-n)^3(2-n)^2(3-n)$$

has for its first four terms 3, 6, 9, 120. It is possible to describe
sequences where the first three terms are 3, 6, and 9 and the fourth term
is *any* given number. This shows that when working with infinite
sequences it is essential to have specific information about the nth term
or to know a general scheme for obtaining each term from the preceding one.

EXERCISES

In Exercises 1–20, find the first five terms and the eighth term of the sequence which has the given nth term.

1 $a_n = 10 - 4n$

2 $a_n = \dfrac{4}{3n - 2}$

3 $a_n = \dfrac{2n - 3}{n^2 + 2}$

4 $a_n = 1 + \dfrac{1}{n}$

5 $a_n = 5$

6 $a_n = (n - 1)(n - 2)$

7 $a_n = 1 + (-0.1)^n$

8 $a_n = 1 + (0.1)^n$

9 $a_n = (-1)^{n-1} \dfrac{n + 1}{2n}$

10 $a_n = (-1)^n \dfrac{2 - 3n}{\sqrt{n + 1}}$

11 $a_n = 1 + (-1)^{n+1}$

12 $a_n = (-1)^{n+1} + (0.1)^{n-1}$

13 $a_n = \dfrac{2^n}{n^2 + 9}$

14 $a_n = \pi$

15 a_n is the square of the nth prime.

16 a_n is the nth prime greater than 50.

17 a_n is the largest prime less than $10n$.

18 a_n is the smallest prime greater than $15n$.

19 a_n is the number of decimal places in $(0.1)^n$.

20 a_n is the number of positive integers less than n^3.

Find the first five terms of the infinite sequences defined recursively in Exercises 21–28.

21 $a_1 = 3,\ a_{k+1} = 5a_k - 2$

22 $a_1 = 2,\ a_{k+1} = 10 - 3a_k$

23 $a_1 = -2,\ a_{k+1} = a_k^2$

24 $a_1 = 32,\ a_{k+1} = a_k/2$

25 $a_1 = 1,\ a_{k+1} = ka_k$

26 $a_1 = 2,\ a_{k+1} = 1/a_k$

27 $a_1 = 2,\ a_{k+1} = (a_k)^k$

28 $a_1 = 3,\ a_{k+1} = (a_k)^{1/k}$

29 A test question lists the first four terms of a sequence as 2, 4, 6, and 8 and asks for the fifth term. Show that the fifth term can be any real number a by finding the nth term of a sequence which has for its first five terms 2, 4, 6, 8, and a.

30 The number of bacteria in a certain culture doubles every day. If the initial number of bacteria is 500, how many are present after one day? Two days? Three days? Find a formula for the number of bacteria present after n days.

It is often desirable to find the sum of many terms of an infinite sequence. For ease in expressing such sums we use the **summation notation** described below.

Given an infinite sequence

(10.14) $a_1, a_2, a_3, \cdots, a_n, \cdots$

the symbol $\sum\limits_{i=1}^{m} a_i$ represents the sum of the first m terms, that is,

(10.15) $\sum\limits_{i=1}^{m} a_i = a_1 + a_2 + a_3 + \cdots + a_m.$

The Greek capital letter Σ (sigma) indicates a sum and the symbol a_i represents the ith term. The letter i is called the **index of summation** or the **summation variable,** and the numbers 1 and m indicate the extreme values of the summation variable.

EXAMPLE 1 Find $\sum\limits_{i=1}^{4} i^2(i-3).$

Solution

In this case, $a_i = i^2(i-3)$. To find the indicated sum we merely substitute, in succession, the integers 1, 2, 3, and 4 for i and add the resulting terms. Thus

$$\sum_{i=1}^{4} i^2(i-3) = 1^2(1-3) + 2^2(2-3) + 3^2(3-3) + 4^2(4-3)$$
$$= (-2) + (-4) + 0 + 16 = 10.$$

The letter used for the summation variable is arbitrary. For example,

$$\sum_{j=1}^{m} a_j = a_1 + a_2 + a_3 + \cdots + a_m$$

which is the same as (10.15). Other symbols can be used similarly. As a numerical example, the sum in Example 1·can be written as $\sum\limits_{k=1}^{4} k^2(k-3).$

If n is a positive integer, then the sum of the first n terms of the infinite sequence (10.14) will be denoted by S_n. Thus

(10.16) $S_n = \sum\limits_{i=1}^{n} a_i = a_1 + a_2 + \cdots + a_n.$

The definition of S_n can be given recursively by writing

(10.17)
$$S_1 = a_1$$
$$S_{k+1} = S_k + a_{k+1} \quad \text{for } k \geq 1.$$

Thus we have

$$S_1 = a_1$$
$$S_2 = S_1 + a_2 = a_1 + a_2$$
$$S_3 = S_2 + a_3 = a_1 + a_2 + a_3$$

and so on, which is in agreement with (10.16). The number S_n is called the **nth partial sum** of the sequence (10.14); and the infinite sequence

(10.18) $S_1, S_2, S_3, \cdots, S_n, \cdots$

is called the **sequence of partial sums** associated with (10.14). The sequence (10.18) is very important in calculus.

EXAMPLE 2   Find the first four terms and the nth term of the sequence of partial sums associated with the sequence 1, 2, 3, $\cdots$, n, $\cdots$ of positive integers.

Solution

The first four terms of the partial sum sequence are $S_1 = 1$, $S_2 = 1 + 2 = 3$, $S_3 = 1 + 2 + 3 = 6$, and $S_4 = 1 + 2 + 3 + 4 = 10$. From Example 1 of Section 1 we see that

$$S_n = 1 + 2 + 3 + \cdots + n = \frac{n(n+1)}{2}.$$

If a_i is the same for all positive integers i, say $a_i = c$, where c is a real number, then

$$\sum_{i=1}^{n} a_i = a_1 + a_2 + a_3 + \cdots + a_n$$
$$= c + c + c + \cdots + c$$
$$= nc.$$

It is with this in mind that we state the following rule:

(10.19) $\displaystyle\sum_{i=1}^{n} c = nc.$

The above formula could also be established by mathematical induction (see Exercise 30).

The domain of the summation variable does not have to begin at 1. For example, the following is self-explanatory:

$$\sum_{i=4}^{8} a_i = a_4 + a_5 + a_6 + a_7 + a_8.$$

As another variation, if the first term of an infinite sequence is a_0, as in

$$a_0, a_1, a_2, \cdots, a_n, \cdots,$$

then sums of the form

$$\sum_{i=0}^{n} a_i = a_0 + a_1 + a_2 + \cdots + a_n$$

may be considered. Note that this is the sum of the first $n + 1$ terms of the given sequence.

EXAMPLE 3 Find $\displaystyle\sum_{i=0}^{3} \frac{2^i}{(i+1)}$.

Solution

$$\sum_{i=0}^{3} \frac{2^i}{(i+1)} = \frac{2^0}{(0+1)} + \frac{2^1}{(1+1)} + \frac{2^2}{(2+1)} + \frac{2^3}{(3+1)}$$

$$= 1 + 1 + \frac{4}{3} + 2$$

$$= \frac{16}{3}.$$

The summation notation can be used to denote polynomials compactly. For example, in place of

$$f(x) = a_0 + a_1 x + a_2 x^2 + \cdots + a_n x^n$$

we may write

$$f(x) = \sum_{i=0}^{n} a_i x^i.$$

We conclude this section with the following useful theorem about sums.

(10.20) Theorem

If $a_1, a_2, \cdots, a_n, \cdots$ and $b_1, b_2, \cdots, b_n, \cdots$ are infinite sequences, then for every positive integer n,

(1) $\displaystyle\sum_{i=1}^{n} (a_i + b_i) = \sum_{i=1}^{n} a_i + \sum_{i=1}^{n} b_i$

(2) $\displaystyle\sum_{i=1}^{n} (a_i - b_i) = \sum_{i=1}^{n} a_i - \sum_{i=1}^{n} b_i$

(3) $\displaystyle\sum_{i=1}^{n} ca_i = c \left(\sum_{i=1}^{n} a_i \right),$ for any real number c.

Proof

Although the theorem can be proved by mathematical induction, we shall use an argument that makes the truth of the formulas transparent. We begin as follows:

$$\sum_{i=1}^{n} (a_i + b_i) = (a_1 + b_1) + (a_2 + b_2)$$
$$+ (a_3 + b_3) + \cdots + (a_n + b_n).$$

Using the commutative and associative laws many times, we may rearrange the terms on the right to produce

$$\sum_{i=1}^{n} (a_i + b_i) = (a_1 + a_2 + a_3 + \cdots + a_n)$$
$$+ (b_1 + b_2 + b_3 + \cdots + b_n).$$

Expressing the right side in summation notation gives us formula (1).

For formula (3) we have

$$\sum_{i=1}^{n} (ca_i) = ca_1 + ca_2 + ca_3 + \cdots + ca_n$$
$$= c(a_1 + a_2 + a_3 + \cdots + a_n)$$
$$= c \left(\sum_{i=1}^{n} a_i \right).$$

The proof of (2) is left as an exercise.

EXERCISES

Find the given number in each of Exercises 1–16.

1 $\displaystyle\sum_{k=1}^{5} (3k - 10)$

2 $\displaystyle\sum_{k=1}^{6} (9 - 2k)$

3 $\displaystyle\sum_{k=1}^{4} (k^2 + 1)$

4 $\displaystyle\sum_{k=1}^{10} [1 + (-1)^k]$

5 $\displaystyle\sum_{k=0}^{5} k(k-1)$

6 $\displaystyle\sum_{k=0}^{4} (k-2)(k-3)$

7 $\displaystyle\sum_{k=3}^{6} \frac{k+1}{k-2}$

8 $\displaystyle\sum_{k=1}^{6} \frac{5}{k}$

9 $\displaystyle\sum_{k=1}^{7} (-2)^{k-1}$

10 $\displaystyle\sum_{k=0}^{4} 2(3^{k})$

11 $\displaystyle\sum_{k=1}^{50} 10$

12 $\displaystyle\sum_{k=1}^{1000} 2$

13 $\displaystyle\sum_{k=1}^{n} (k^2 + 3k + 5)$ (*Hint:* Use (10.20) to write the sum as $\displaystyle\sum_{k=1}^{n} k^2 +$
$3 \displaystyle\sum_{k=1}^{n} k + \sum_{k=1}^{n} 5$. Next employ Exercise 5 and Example 1 of Section 1, together with (10.19).)

14 $\displaystyle\sum_{k=1}^{n} (3k^2 - 2k + 1)$

15 $\displaystyle\sum_{k=1}^{n} (2k - 3)^2$

16 $\displaystyle\sum_{k=1}^{n} (k^3 + 2k^2 - k + 4)$ (*Hint:* See Exercise 6 of Section 1.)

Use summation notation to express the sums in Exercises 17–26.

17 $1 + 3 + 5 + 7 + 9$

18 $1 + 4 + 7 + 10 + 13$

19 $\frac{1}{4} + \frac{2}{7} + \frac{3}{10} + \frac{4}{13}$

20 $\frac{1}{6} + \frac{2}{11} + \frac{3}{16} + \frac{4}{21}$

21 $1 - \dfrac{x^2}{2} + \dfrac{x^4}{4} - \dfrac{x^6}{6} + \cdots + (-1)^{n+1} \dfrac{x^{2n}}{2n}$

22 $2 - 4 + 8 - 16 + 32 - 64$

23 $1 - \frac{1}{2} + \frac{1}{3} - \frac{1}{4} + \frac{1}{5} - \frac{1}{6} + \frac{1}{7}$

24 $1 + x + \dfrac{x^2}{2} + \dfrac{x^3}{3} + \cdots + \dfrac{x^n}{n}$

25 $\dfrac{1}{1 \cdot 2} + \dfrac{1}{2 \cdot 3} + \dfrac{1}{3 \cdot 4} + \cdots + \dfrac{1}{99 \cdot 100}$

26 $\frac{1}{2} + \frac{2}{3} + \frac{3}{4} + \cdots + \frac{99}{100}$

27 Prove (2) of (10.20).

28 Extend (1) of (10.20) to $\displaystyle\sum_{i=1}^{n} (a_i + b_i + c_i)$.

29 Prove (10.20) by mathematical induction.

30 Prove (10.19) by mathematical induction.

4 ARITHMETIC SEQUENCES

In this section and the next we shall concentrate on two special types of sequences. The first type may be defined as follows.

(10.21) Definition

An **arithmetic sequence** is a sequence such that successive terms differ by the same real number.

Arithmetic sequences are also called **arithmetic progressions.** By definition, a sequence

$$a_1, a_2, a_3, \cdots, a_n, \cdots$$

is arithmetic if and only if there is a real number d such that

(10.22) $\quad a_{k+1} - a_k = d$

for every positive integer k. The number d is called the **common difference** associated with the arithmetic sequence.

EXAMPLE 1 Show that the sequence

$$1, 4, 7, 10, \cdots, 3n - 2, \cdots$$

is arithmetic and find the common difference.

Solution

Since $a_n = 3n - 2$ it follows that for every positive integer k,

$$\begin{aligned} a_{k+1} - a_k &= [3(k + 1) - 2] - (3k - 2) \\ &= 3k + 3 - 2 - 3k + 2 = 3. \end{aligned}$$

Hence by definition (10.21), the given sequence is arithmetic with common difference 3.

Given an arithmetic sequence we have by (10.22),

$$a_{k+1} = a_k + d$$

for every positive integer k. This provides a recursive formula for obtaining successive terms. Beginning with any real number a_1, we can generate an arithmetic sequence with common difference d simply by adding d to a_1, then to $a_1 + d$, and so on, obtaining

(10.23) $\quad a_1, \quad a_1 + d, \quad a_1 + 2d, \quad a_1 + 3d, \quad a_1 + 4d, \cdots.$

It is evident that the nth term a_n of (10.23) is given by

(10.24) $a_n = a_1 + (n-1)d$.

This formula can be proved by mathematical induction.

EXAMPLE 2 Find the fifteenth term of the arithmetic sequence whose first three terms are 20, 16.5, and 13.

Solution

The common difference is -3.5 (Why?). Substituting $a_1 = 20$, $d = -3.5$, and $n = 15$ in (10.24), we obtain

$a_{15} = 20 + (15-1)(-3.5) = 20 - 49 = -29$.

EXAMPLE 3 If the fourth term of an arithmetic sequence is 5 and the ninth term is 20, find the sixth term.

Solution

Substituting $n = 4$ and $n = 9$ in (10.24) and using the fact that $a_4 = 5$ and $a_9 = 20$, we get the following system of linear equations in the variables a_1 and d:

$$\begin{cases} 5 = a_1 + (4-1)d \\ 20 = a_1 + (9-1)d. \end{cases}$$

This system has the unique solution $d = 3$ and $a_1 = -4$ (Verify!). Substituting in (10.24), we obtain

$a_6 = (-4) + (6-1)(3) = 11$.

(10.25) Theorem

If $a_1, a_2, \cdots, a_n, \cdots$ is an arithmetic sequence with common difference d, then the nth partial sum S_n is given by either

$$S_n = \frac{n}{2}[2a_1 + (n-1)d]$$

or

$$S_n = \frac{n}{2}(a_1 + a_n).$$

Proof

Using (10.23) and (10.24) we have

$$S_n = a_1 + (a_1 + d) + (a_1 + 2d) + \cdots + [a_1 + (n-1)d],$$

or in summation notation,

$$S_n = \sum_{i=1}^{n} [a_1 + (i-1)d].$$

This may be simplified as follows:

$$S_n = \sum_{i=1}^{n} a_1 + \sum_{i=1}^{n} (i-1)d \qquad \text{((i) of (10.20))}$$

$$= na_1 + d \sum_{i=1}^{n} (i-1) \qquad \text{((10.19) and (iii) of (10.20).)}$$

However, $\sum_{i=1}^{n} (i-1) = 0 + 1 + 2 + \cdots + (n-1)$, which is the sum of the first $n-1$ positive integers. From Example 1 of Section 1 (with $n-1$ in place of n), that sum equals

$$\frac{(n-1)n}{2}.$$

Substituting in the last equation for S_n and factoring gives us

$$S_n = na_1 + d \frac{n(n-1)}{2}$$

$$= \frac{n}{2} [2a_1 + (n-1)d].$$

Since $a_n = a_1 + (n-1)d$, we get

$$S_n = \frac{n}{2} (a_1 + a_n).$$

EXAMPLE 4 Find the sum of all the even integers from 2 through 100.

Solution

The given problem is equivalent to finding the sum of the first fifty terms of the arithmetic sequence 2, 4, 6, $\cdots$, $2n$, $\cdots$. Substituting $n = 50$, $a_1 = 2$, and $a_{50} = 100$ in the second formula of (10.25), we obtain

$$S_{50} = \frac{50}{2} (2 + 100) = 2550.$$

In order to check our work we use the first formula in (10.25):

$$S_{50} = \frac{50}{2} [2 \cdot 2 + (50-1)2] = 25[4 + 98] = 2550.$$

The **arithmetic mean** of two numbers a and b is defined as $(a + b)/2$, which is also called the **average** of a and b. Note that the numbers

$$a, \quad \frac{a + b}{2}, \quad b$$

are terms of an arithmetic sequence. This concept may be generalized as follows. If $c_1, c_2, \cdots, c_k$ are real numbers such that

$$a, c_1, c_2, \cdots, c_k, b$$

is an arithmetic sequence, then $c_1, c_2, \cdots, c_k$ are called the k **arithmetic means** of the numbers a and b. The process of determining such numbers is referred to as *inserting k arithmetic means between a and b.*

EXAMPLE 5 Insert three arithmetic means between 2 and 9.

> **Solution**
>
> We wish to find three numbers c_1, c_2, and c_3 such that $2, c_1, c_2, c_3, 9$ is an arithmetic sequence. The common difference d may be found by using (10.24) with $n = 5$, $a_5 = 9$, and $a_1 = 2$. This gives us
>
> $$9 = 2 + (5 - 1)d, \quad \text{or} \quad d = \tfrac{7}{4}.$$
>
> The three arithmetic means are
>
> $c_1 = a_1 + d = 2 + \tfrac{7}{4} = \tfrac{15}{4},$
> $c_2 = c_1 + d = \tfrac{15}{4} + \tfrac{7}{4} = \tfrac{22}{4} = \tfrac{11}{2},$
> $c_3 = c_2 + d = \tfrac{11}{2} + \tfrac{7}{4} = \tfrac{29}{4}.$

EXERCISES

In each of Exercises 1–8 find the fifth term, the tenth term, and the nth term of the given arithmetic sequence.

1 1, 5, 9, 13, $\cdots$

2 15, 12, 9, 6, $\cdots$

3 2, 1.7, 1.4, 1.1, $\cdots$

4 $-5, -3.5, -2, -0.5, \cdots$

5 $-7.5, -3.4, 0.7, 4.8, \cdots$

6 $x - 7, x - 2, x + 3, x + 8, \cdots$

7 $\log 2, \log 4, \log 8, \log 16, \cdots$

8 $\log 10, 0, \log (.1), \log (.01), \cdots$

9 Find the fourteenth term of the arithmetic sequence whose first two terms are 8.4 and 6.8.

10 Find the eleventh term of the arithmetic sequence whose first two terms are $1 + \sqrt{3}$ and 2.

11 The sixth and seventh terms of an arithmetic sequence are 1.5 and 4. Find the first term.

12 Given an arithmetic sequence with $a_3 = 4$ and $a_{20} = 40$, find a_{15}.

In Exercises 13–16 find the sum of the indicated arithmetic sequence.

13 $a_1 = 50$, $d = -2$, $n = 60$ **14** $a_1 = 1$, $d = 0.1$, $n = 20$

15 $a_1 = -6$, $a_{10} = 18$, $n = 10$ **16** $a_7 = \frac{5}{3}$, $d = -\frac{2}{3}$, $n = 12$

Find the sums in Exercises 17–20.

17 $\sum_{k=1}^{18} (3k - 2)$ **18** $\sum_{k=1}^{13} (5 - 2k)$

19 $\sum_{k=0}^{20} (-1 + k/2)$ **20** $\sum_{k=0}^{10} (3 - k/4)$

21 How many integers between 32 and 395 are divisible by 6? Find their sum.

22 How many negative integers greater than -500 are divisible by 33? Find their sum.

23 How many terms are in an arithmetic sequence with first term -2, common difference $\frac{1}{4}$, and sum 21?

24 How many terms are in an arithmetic sequence with sixth term -3, common difference 0.2, and sum -33?

25 Insert five arithmetic means between 2 and 10.

26 Insert three arithmetic means between 3 and -5.

27 A pile of logs has 24 logs in the first layer, 23 in the second, 22 in the third, and so on. The last layer contains 10 logs. Find the total number of logs in the pile.

28 A seating section in a certain athletic stadium has 30 seats in the first row, 32 seats in the second, 34 in the third, and so on, until the tenth row is reached, after which there are 10 more rows, each containing 50 seats. Find the total number of seats in the section.

29 A man wishes to construct a ladder with nine rungs which diminish uniformly from 24 inches at the base to 18 inches at the top. Determine the lengths of the seven intermediate rungs.

30 A boy on a bicycle coasts down a hill, covering 4 feet the first second and in each succeeding second 5 feet more than in the preceding second. If he reaches the bottom of the hill in 11 seconds, find the total distance traveled.

5 GEOMETRIC SEQUENCES

Another important type of infinite sequence is defined as follows.

(10.26) Definition

A **geometric sequence** is a sequence such that the quotient of any term after the first by the preceding term is the same nonzero real number.

Geometric sequences are also called **geometric progressions.** By definition, a sequence

$$a_1, a_2, \cdots, a_n, \cdots$$

is geometric if and only if there is a real number $r \neq 0$ such that

(10.27) $\dfrac{a_{k+1}}{a_k} = r$

for every positive integer k. The number r is called the **common ratio** associated with the geometric sequence. We see from (10.27) that the indicated sequence is geometric (with common ratio r) if and only if

$$a_{k+1} = a_k r$$

for every positive integer k. As with arithmetic sequences, that provides a recursive method for obtaining terms. Beginning with any nonzero real number a_1, we *multiply* by the number r successively, obtaining

(10.28) $a_1, a_1 r, a_1 r^2, a_1 r^3, \cdots$.

It appears that the nth term a_n of (10.28) is given by

(10.29) $a_n = a_1 r^{n-1}$.

That can be proved by mathematical induction.

EXAMPLE 1 Find the first five terms and the tenth term of the geometric sequence having first term 3 and common ratio $-\frac{1}{2}$.

Solution

If we let $a_1 = 3$ and $r = -\frac{1}{2}$, then by (10.28) the first five terms are

$$3, -\tfrac{3}{2}, \tfrac{3}{4}, -\tfrac{3}{8}, \tfrac{3}{16}.$$

Using (10.29) with $n = 10$, we obtain

$$a_{10} = 3(-\tfrac{1}{2})^9 = -\tfrac{3}{512}.$$

EXAMPLE 2 If the third term of a geometric sequence is 5 and the ·sixth term is -40, find the eighth term.

Solution

We are given $a_3 = 5$ and $a_6 = -40$. Substituting $n = 3$ and $n = 6$ in formula (10.29) leads to the following system of equations:

$$\begin{cases} 5 = a_1 r^2 \\ -40 = a_1 r^5. \end{cases}$$

Since $r \neq 0$, the first equation is equivalent to $a_1 = 5/r^2$. Substituting for a_1 in the second equation we obtain

$$-40 = \left(\frac{5}{r^2}\right) \cdot r^5 = 5r^3,$$

and hence $r = -2$. If we now substitute -2 for r in the equation $5 = a_1 r^2$, we obtain $a_1 = \frac{5}{4}$. Finally, applying (10.29) with $n = 8$ gives us

$$a_8 = \left(\tfrac{5}{4}\right)(-2)^7 = -160.$$

Let us find a formula for S_n, the nth partial sum of a geometric sequence. Using (10.28) and (10.29), we have

(10.30) $\quad S_n = a_1 + a_1 r + a_1 r^2 + \cdots + a_1 r^{n-2} + a_1 r^{n-1}.$

If $r = 1$, this gives us $S_n = na_1$. Next suppose that $r \neq 1$. Multiplying both sides of (10.30) by r, we obtain

$$rS_n = a_1 r + a_1 r^2 + a_1 r^3 + \cdots + a_1 r^{n-1} + a_1 r^n.$$

If we subtract the preceding equation from (10.30), then many terms on the right side drop out, leaving

$$S_n - rS_n = a_1 - a_1 r^n,$$

or

$$(1 - r)S_n = a_1(1 - r^n).$$

Since $r \neq 1$ we have $1 - r \neq 0$, and dividing both sides by $1 - r$ gives us

$$S_n = a_1 \cdot \frac{(1 - r^n)}{1 - r}.$$

We have proved the following theorem.

(10.31) Theorem

The nth partial sum of a geometric sequence with first term a_1 and common ratio $r \neq 1$ is

$$S_n = a_1 \frac{(1 - r^n)}{1 - r}.$$

EXAMPLE 3   Find the sum of the first five terms of the geometric sequence which begins as follows:

$1, 0.3, 0.09, 0.027, \cdot \cdot \cdot .$

Solution

We let $a_1 = 1$, $r = .3$, and $n = 5$ in (10.31), obtaining

$$S_5 = 1 \left(\frac{1 - (.3)^5}{1 - .3} \right),$$

which reduces to $S_5 = 1.4251$.

EXAMPLE 4   A man wishes to save money by setting aside 1 cent the first day, 2 cents the second day, 4 cents the third day and so on, doubling the amount each day. If this is continued, how much must be set aside on the fifteenth day? Assuming he does not run out of money, what is the total amount saved at the end of 30 days?

Solution

The amount (in cents) set aside on successive days forms a geometric sequence

$1, 2, 4, 8, \cdot \cdot \cdot$

with first term 1 and common ratio 2. The amount needed for the fifteenth day is found by using (10.29) with $a_1 = 1$ and $n = 15$, which gives us $1 \cdot 2^{14}$, or \$163.84. To find the total amount set aside after 30 days, we use (10.31) with $n = 30$. Thus

$$S_{30} = 1 \frac{(1 - 2^{30})}{1 - 2},$$

which simplifies to \$10,737,418.23.

If a and b are positive real numbers, then a positive number c is called the **geometric mean** of a and b if a, c, and b are successive terms of a geometric sequence. If we denote the common ratio by r, then by (10.27) we must have

$$r = \frac{c}{a} = \frac{b}{c}.$$

This implies that $c^2 = ab$ or $c = \sqrt{ab}$. We have proved that *the geometric mean of the positive numbers a and b is $\sqrt{ab}$*. As a generalization of this concept, k positive numbers $c_1, c_2, \cdots, c_k$ are the *k* **geometric means** of a and b if $a, c_1, c_2, \cdots, c_k, b$ are terms of a geometric sequence.

Given the geometric series (10.28) with $r \neq 1$, we may use (10.31) to get

$$(10.32) \quad S_n = \frac{a_1}{1-r} - \frac{a_1}{1-r} r^n.$$

Although we shall not do it here, it can be shown that if $|r| < 1$, then r^n *approaches the number 0 as n increases* without bound; that is, we can make r^n as close as we wish to 0 by taking n sufficiently large. It follows from (10.32) that S_n approaches $a_1/(1-r)$ as n increases without bound. The number $a_1/(1-r)$ is called *the sum of the infinite geometric series*

$$a_1 + a_1 r + a_1 r^2 + \cdots + a_1 r^{n-1} + \cdots .$$

This gives us the following result.

(10.33) Theorem

If $|r| < 1$, then the infinite geometric series

$$a_1 + a_1 r + a_1 r^2 + \cdots + a_1 r^{n-1} + \cdots$$

has the sum

$$\frac{a_1}{1-r}.$$

Theorem (10.33) can be used to prove the fact mentioned in Chapter One that every real number represented by a repeating decimal is rational. This is illustrated in the next example.

EXAMPLE 5 Find the rational number which corresponds to the infinite repeating decimal $5.4\overline{27}$, where the bar means that the block of digits underneath is to be repeated indefinitely.

Solution

From the decimal expression $5.4272727 \cdots$ we obtain the infinite series

$$5.4 + .027 + .00027 + .0000027 + \cdots .$$

The part of the expression after the first term is

$.027 + .00027 + .0000027 + \cdots ,$

which has the form given in (10.33) with $a_1 = .027$ and $r = .01$. Hence the sum of this infinite geometric series is

$$\frac{.027}{1 - .01} = \frac{.027}{.99} = \frac{27}{990} = \frac{3}{110}.$$

Thus it appears that the desired number is $5.4 + 3/110$, or $597/110$. A check by long division shows that $597/110$ does equal the given repeating decimal.

If the terms of an infinite sequence are alternately positive and negative and if we consider the expression

$$a_1 + (-a_2) + a_3 + (-a_4) + \cdots + [(-1)^{n+1}a_n] + \cdots$$

where all the a_i are positive real numbers, then this expression is referred to as an *alternating infinite series* and we write it in the form

$$a_1 - a_2 + a_3 - a_4 + \cdots + (-1)^{n+1}a_n + \cdots .$$

Illustrations of alternating infinite series can be obtained by using infinite geometric series with negative common ratio.

EXERCISES

In each of Exercises 1–6 find the fifth term, the eighth term, and the nth term of the given geometric sequence.

1 $4, 2, 1, \frac{1}{2}, \cdots$

2 $2, 0.6, 0.18, 0.054, \cdots$

3 $7, -0.7, 0.07, -0.007, \cdots$

4 $1, -\sqrt{2}, 2, -\sqrt{8}, \cdots$

5 $1, -x^3, x^6, -x^9, \cdots$

6 $5, 5^{x+1}, 5^{2x+1}, 5^{3x+1}, \cdots$

7 Find the sixth term of the geometric sequence whose first two terms are 8 and 12.

8 Find the seventh term of the geometric sequence that has 3 and $-\sqrt{2}$ for its second and third terms respectively.

9 In a certain geometric sequence $a_5 = \frac{1}{16}$ and $r = \frac{3}{2}$. Find a_1 and S_5.

10 Given a geometric sequence in which $a_4 = 4$ and $a_7 = 12$, find r and a_{10}.

11 Insert three geometric means between 8 and 2.

12 Insert two geometric means between 8 and 3.

Find the sum in each of Exercises 13–16.

13 $\displaystyle\sum_{k=1}^{10} 2^k$

14 $\displaystyle\sum_{k=1}^{9} (-\sqrt{2})^k$

15 $\displaystyle\sum_{k=0}^{8} (-\tfrac{1}{2})^{k+1}$

16 $\displaystyle\sum_{k=1}^{6} (3^{-k})$

17 A vacuum pump removes one-half of the air in a container at each stroke. After 10 strokes, what percentage of the original amount of air remains in the container?

18 The yearly depreciation of a certain machine is 25% of its value at the beginning of the year. If the original cost of the machine is $2000, find its value after 5 years.

19 A culture of bacteria increases 20% every hour. If the original culture contains 1000 bacteria, find a formula for the number of bacteria present after t hours. How many bacteria are in the culture at the end of 10 hours?

20 If an amount P of money is deposited in a savings account which pays interest at a rate of r percent per year compounded quarterly and if the principal and accumulated interest are left in the account, find a formula for the total amount in the account after n years.

Find the sums of the infinite geometric series in Exercises 21–26, whenever they exist.

21 $1 - \tfrac{1}{2} + \tfrac{1}{4} - \tfrac{1}{8} + \cdots$

22 $4 + \tfrac{4}{3} + \tfrac{4}{9} + \tfrac{4}{27} + \cdots$

23 $0.017 + 0.00017 + 0.0000017 + \cdots$

24 $0.1 - 0.01 + 0.001 - 0.0001 + \cdots$

25 $\sqrt{5} + 5 + \sqrt{125} + 25 + \cdots$

26 $125 - 50 + 20 - 8 + \cdots$

In each of Exercises 27–32 find the rational number represented by the given repeating decimal.

27 $0.\overline{21}$

28 $0.0\overline{73}$

29 $0.6\overline{104}$

30 $7.2\overline{56}$

31 $4.3\overline{87}$

32 $38.\overline{628}$

33 A rubber ball is dropped from a height of 10 feet. If it rebounds approximately one-half the distance after each fall, use an infinite geometric series to approximate the total distance the ball travels before coming to rest.

34 The bob of a pendulum swings through an arc 24 inches long on its first swing. If each successive swing is approximately five-sixths the length of the preceding swing, use an infinite geometric series to approximate the total distance it travels before coming to rest.

It is often necessary to work with expressions of the form $(a + b)^n$, where a and b are mathematical expressions of some type and n is a large positive integer. There exists a general formula for *expanding* $(a + b)^n$, that is, for expressing it as a sum. Since $a + b$ is a binomial, the theorem which gives us the formula is called the *Binomial Theorem*. In order to obtain the formula let us first consider cases where n is a small positive integer. By examining the patterns which emerge, we shall make an educated guess about the nature of the general formula. Finally we shall prove by means of mathematical induction that our guess is correct.

If we actually perform the multiplications, the following expansions of $(a + b)^n$ are obtained for the cases $n = 2$, 3, 4, and 5:

$$(a + b)^2 = a^2 + 2ab + b^2,$$
$$(a + b)^3 = a^3 + 3a^2b + 3ab^2 + b^3,$$
$$(a + b)^4 = a^4 + 4a^3b + 6a^2b^2 + 4ab^3 + b^4,$$
$$(a + b)^5 = a^5 + 5a^4b + 10a^3b^2 + 10a^2b^3 + 5ab^4 + b^5.$$

Let us now make some observations regarding these expansions of $(a + b)^n$. We see that there are always $n + 1$ terms, the first being a^n and the last b^n. Each intermediate term contains a product of the form $a^i b^j$, where $i + j = n$. Moreover, as we move from one term to the next the exponent associated with a decreases by 1 whereas the exponent associated with b increases by 1. The pattern for the coefficients is rather interesting. If we start at one end of the expansion and consider successive terms, the coefficients match those obtained by starting at the other end of the expansion and proceeding in the reverse direction.

It requires some ingenuity to find a formula for the general term in the expansion. The second term of the special cases above for $(a + b)^n$ is always $na^{n-1}b$. The third term is seen to be

$$\frac{n(n - 1)}{2} a^{n-2}b^2.$$

If there are more than three terms, the fourth term is

(10.34) $\quad \dfrac{n(n - 1)(n - 2)}{3 \cdot 2} a^{n-3}b^3.$

It appears that for each term, if we take the product of the coefficient and the exponent of a and then divide by the number of the term, we obtain the coefficient of the next term in the expansion. For example, applying this rule to (10.34) gives us the following fifth term (when it exists):

(10.35) $\quad \dfrac{n(n - 1)(n - 2)(n - 3)}{4 \cdot 3 \cdot 2} a^{n-4}b^4.$

The denominators in (10.34) and (10.35) may be abbreviated by employing the **factorial notation.** If n is any positive integer, then the symbol $n!$ (read "n factorial") is defined by

(10.36) $n! = n(n-1)(n-2) \cdots 1,$

where there are n factors on the right side. As special cases we have $1! = 1$, $2! = 2 \cdot 1$, $3! = 3 \cdot 2 \cdot 1 = 6$, $4! = 4 \cdot 3 \cdot 2 \cdot 1 = 24$, $5! = 5 \cdot 4 \cdot 3 \cdot 2 \cdot 1 = 120$, etc. To insure certain formulas will be true for all *non-negative* integers, we define $0! = 1$. The factorial notation may also be defined recursively by writing $1! = 1$, and for any positive integer k, $(k+1)! = (k+1) \cdot k!$.

If we compare the coefficients with the exponents in (10.34) and (10.35), we are led to believe that for a positive integer r, the $(r+1)$st term in the expansion of $(a+b)^n$ is given by

(10.37) $\dfrac{n(n-1)(n-2) \cdots (n-r+1)}{r!} a^{n-r}b^r.$

Note that if $r = n$, then the coefficient reduces to $n!/n!$ and we obtain b^n, the last term in the expansion. If $r = n - 1$, then we obtain nab^{n-1}, which is the second from the last term, and so on. Hence we *conjecture* that the following formula is true for every positive integer n and all complex numbers a and b.

(10.38) $(a+b)^n = a^n + na^{n-1}b + \dfrac{n(n-1)}{2!} a^{n-2}b^2 + \cdots$

$$+ \frac{n(n-1)(n-2) \cdots (n-r+1)}{r!} a^{n-r}b^r$$

$$+ \cdots + nab^{n-1} + b^n.$$

In order to prove (10.38) we shall use mathematical induction as follows. For each positive integer n, let P_n denote the statement (10.38).

(i) If $n = 1$, then (10.38) reduces to $(a+b)^1 = a^1 + b^1$. Consequently P_1 is true.

(ii) Assume P_k is true. Thus the induction hypothesis is

(10.39) $(a+b)^k = a^k + ka^{k-1}b + \dfrac{k(k-1)}{2!} a^{k-2}b^2 + \cdots$

$$+ \frac{k(k-1)(k-2) \cdots (k-r+2)}{(r-1)!} a^{k-r+1}b^{r-1}$$

$$+ \frac{k(k-1)(k-2) \cdots (k-r+1)}{r!} a^{k-r}b^r$$

$$+ \cdots + kab^{k-1} + b^k$$

where we have shown both the rth and the $(r+1)$st terms in the expansion.

If we multiply both sides of equation (10.39) by $(a + b)$, we obtain

$$(a + b)^{k+1} = \left[a^{k+1} + ka^k b + \frac{k(k-1)}{2!} a^{k-1}b^2 + \cdots \right.$$

$$\left. + \frac{k(k-1) \cdots (k-r+1)}{r!} a^{k-r+1}b^r + \cdots + ab^k \right]$$

$$+ \left[a^k b + ka^{k-1}b^2 + \cdots + \frac{k(k-1) \cdots (k-r+2)}{(r-1)!} a^{k-r+1}b^r \right.$$

$$\left. + \cdots + kab^k + b^{k+1} \right]$$

where the terms in the first pair of brackets result from multiplying the right side of (10.39) by a and the terms in the second pair of brackets result from multiplying by b. Rearranging and combining terms, we have

$$(a + b)^{k+1} = a^{k+1} + (k+1)a^k b + \left[\frac{k(k-1)}{2!} + k \right] a^{k-1}b^2 + \cdots$$

$$+ \left[\frac{k(k-1) \cdots (k-r+1)}{r!} \right.$$

$$\left. + \frac{k(k-1) \cdots (k-r+2)}{(r-1)!} \right] a^{k-r+1}b^r$$

$$+ \cdots + (1+k)ab^k + b^{k+1}.$$

It is left to the reader to show that if the coefficients are simplified, then we obtain (10.38) with $k + 1$ substituted for n. Thus P_{k+1} is true and therefore (10.38) holds for every positive integer n.

EXAMPLE 1   Find the binomial expansion of $(2x + 3y^2)^4$.

Solution

Using (10.38) with $a = 2x$, $b = 3y^2$, and $n = 4$, we obtain

$$(2x + 3y^2)^4 = (2x)^4 + 4(2x)^3(3y^2)$$

$$+ \frac{4 \cdot 3}{2!} (2x)^2(2y^2)^2$$

$$+ \frac{4 \cdot 3 \cdot 2}{3!} (2x)(3y^2)^3$$

$$+ \frac{4 \cdot 3 \cdot 2 \cdot 1}{4!} (3y^2)^4,$$

which simplifies to

$$(2x + 3y^2)^4 = 16x^4 + 96x^3y^2 + 216x^2y^4 + 216xy^6 + 81y^8.$$

467

The symbol $\binom{n}{r}$ is often used to denote the coefficient of $a^{n-r}b^r$ in (10.37); that is,

(10.40) $\quad \binom{n}{r} = \dfrac{n(n-1) \cdots (n-r+1)}{r!}$

where r may be assigned any integral value between 0 and n. If we substitute n for r in (10.40) we obtain $\binom{n}{n} = \dfrac{n!}{n!} = 1$. It is also convenient to define $\binom{n}{0} = 1$. Formula (10.40) may then be used if r is any integer such that $0 \le r \le n$. Using this notation, (10.38) takes on the form

(10.41) $\quad (a+b)^n = \binom{n}{0} a^n b^0 + \binom{n}{1} a^{n-1}b + \binom{n}{2} a^{n-2}b^2$

$$+ \cdots + \binom{n}{r} a^{n-r}b^r$$

$$+ \cdots + \binom{n}{n-1} a^{n-(n-1)}b^{n-1} + \binom{n}{n} a^0 b^n.$$

If we use summation notation the preceding formula may be written compactly as

$$(a+b)^n = \sum_{r=0}^{n} \binom{n}{r} a^{n-r}b^r.$$

The numbers $\binom{n}{r}$ are called **binomial coefficients.** As a special case of (10.41), when $n=4$ we obtain

$$(a+b)^4 = \binom{4}{0} a^4 b^0 + \binom{4}{1} a^3 b + \binom{4}{2} a^2 b^2 + \binom{4}{3} ab^3 + \binom{4}{4} a^0 b^4.$$

By (10.40) this reduces to the expansion of $(a+b)^4$ given at the beginning of this section.

The next example illustrates the fact that if one of a or b is negative, then the terms of the expansion are alternately positive and negative.

EXAMPLE 2 Expand $\left(\dfrac{1}{x} - 2\sqrt{x}\right)^5$.

Solution

Letting $a = 1/x$, $b = -2\sqrt{x}$, and $n = 5$ in (10.38), we obtain

$$\left(\frac{1}{x} - 2\sqrt{x}\right)^5 = \left(\frac{1}{x}\right)^5 + 5\left(\frac{1}{x}\right)^4 (-2\sqrt{x})$$

$$+ \frac{5 \cdot 4}{2!} \left(\frac{1}{x}\right)^3 (-2\sqrt{x})^2$$

$$+ \frac{5 \cdot 4 \cdot 3}{3!} \left(\frac{1}{x}\right)^2 (-2\sqrt{x})^3$$

$$+ \frac{5 \cdot 4 \cdot 3 \cdot 2}{4!} \left(\frac{1}{x}\right) (-2\sqrt{x})^4$$

$$+ \frac{5 \cdot 4 \cdot 3 \cdot 2 \cdot 1}{5!} (-2\sqrt{x})^5.$$

This simplifies to

$$\left(\frac{1}{x} - 2\sqrt{x}\right)^5 = \frac{1}{x^5} - \frac{10}{x^{7/2}} + \frac{40}{x^2} - \frac{80}{x^{1/2}} + 80x - 32x^{5/2}.$$

For certain problems it is only required to find a specific term in the expansion of $(a + b)^n$. To work problems of this type we first find the exponent r that is to be assigned to b. Notice that by (10.38) or (10.41) *the exponent of b is always one less than the number of the term.* Once r is found, the exponent of a is $n - r$. Referring to (10.37), we see that the coefficient of the term involving $a^{n-r}b^r$ is of the form $p/r!$, *where p is the product of r factors and where the first factor is n and each factor is one less than the preceding factor.*

EXAMPLE 3 Find the fifth term in the expansion of $(4x^2 + 3/y)^{13}$.

Solution

We let $a = 4x^2$ and $b = 3/y$. The exponent of b in the fifth term is 4 and hence the exponent of a is 9. From the discussion of the preceding paragraph we have

$$\frac{13 \cdot 12 \cdot 11 \cdot 10}{4!} (4x^2)^9 (3/y)^4.$$

EXERCISES

In each of Exercises 1–12, expand and simplify the given expression.

1 $(a + b)^6$

2 $(a + b)^9$

3 $(a - b)^7$

4 $(a - b)^8$

5 $(5x - 3y)^4$

6 $(4s - t)^5$

7 $(3u + 2v^2)^5$

8 $(c^3 - d/2)^4$

9 $(x^{-3} - 2x)^6$ **10** $(y^{1/2} + y^{-1/2})^6$

11 $(1 + x)^{10}$ **12** $(1 - x)^{10}$

13 Find the first four terms in the binomial expansion $(a^{3/5} + 2a^{2/5})^{25}$.

14 Find the first three terms and the last three terms in the binomial expansion of $(z^2 + 2z^{-1})^{30}$.

15 Find the last two terms in the expansion of $(s^{-1} - 2s)^{15}$.

16 Find the last three terms in the expansion of $(2t - r^2)^{12}$.

Solve Exercises 17–28 without expanding completely.

17 Find the fifth term in the expansion of $(5x^2 - \frac{1}{2}\sqrt{y})^8$.

18 Find the sixth term in the expansion of $\left(\dfrac{4}{c^2} - \dfrac{c}{2}\right)^7$.

19 Find the seventh term in the expansion of $(2u + \frac{1}{2}v)^{10}$.

20 Find the fourth term in the expansion of $(a^3 + 2b^2)^6$.

21 Find the middle term in the expansion of $(a^{1/3} - b^{1/3})^{12}$.

22 Find the two middle terms in the expansion of $(xy + 1/z)^7$.

23 Find the term which does not contain x in the expansion of $(4x - 1/x)^{10}$.

24 Find the term involving x^8 in the expansion of $(3x^2 + 5y)^6$.

25 Find the term containing y^6 in the expansion of $(x^2 + 2y^3)^4$.

26 Find the term containing b^9 in the expansion of $(3a - 2b^3)^4$.

27 Find the term containing x^3 in the expansion of $(\sqrt{x} - \sqrt{y})^{10}$.

28 In the expansion of $(xy + 4y^{-3})^8$ find the term which does not contain y.

29 Use the first four terms in the binomial expansion of $(1 + 0.02)^{10}$ to approximate $(1.02)^{10}$. Also approximate $(1.02)^{10}$ by using logarithms and then compare answers.

30 Use the first four terms in the binomial expansion of $(1 - 0.01)^4$ to approximate $(0.99)^4$. Compare with the answer obtained by approximating $(0.99)^4$ through use of logarithms.

7 PERMUTATIONS

Suppose that four teams are involved in a tournament in which first, second, third, and fourth places will be determined. For identification purposes, we label the teams a, b, c, and d. Let us find the number of different ways that first and second place can be decided. It is convenient to use a **tree diagram** as in Fig. 10.1. Beginning at the word "start," the four possibilities for first place are listed. From each of these an arrow points to a possible second-place finisher. In this problem the second-place position is occupied by a team other than the one in the first-place position and hence *three* arrows are drawn from each first position. The final standings list the possible outcomes, from left to right. These are found by following all the different paths (*branches* of the tree) which lead from the word "start" to the second-place team. The total number of outcomes is 12, which we note is the product of the number of choices for first place and the number of choices for second place (after the first has been determined).

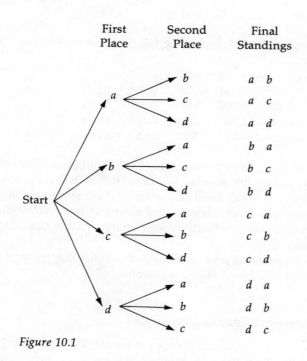

Figure 10.1

Let us now find the total number of ways that first, second, third, and fourth positions can be filled. To sketch a tree diagram we may begin by drawing arrows from the word "start" to each possible first-place finisher *a, b, c,* or *d*. Then we draw arrows from those to possible second-place finishers, as was done in Fig. 10.1. Next, from each second-place position we draw arrows indicating the possible third-place positions. Finally, we draw arrows to the fourth-place team. If we consider only the case where team *a* finishes in first place, we have the diagram shown in Fig. 10.2. Note that there are six possible final standings in which team *a* occupies first place. In a complete tree diagram there would also be three other branches of this type corresponding to first place for *b, c,* and *d,* respectively.

First Place	Second Place	Third Place	Fourth Place	Final Standings

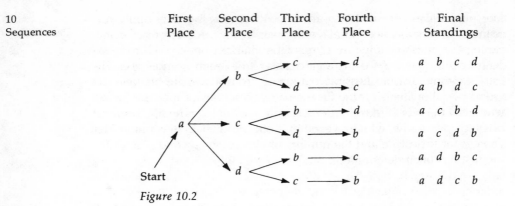

Figure 10.2

A complete diagram would display the following 24 possibilities for the final standings:

abcd, abdc, acbd, acdb, adbc, adcb,
bacd, badc, bcad, bcda, bdac, bdca,
cabd, cadb, cbad, cbda, cdab, cdba
dabc, dacb, dbac, dbca, dcab, dcba.

Note that the number 24 is the product of the number of ways (4) that first place may occur, the number of ways (3) that second place may occur (after first place has been determined), the number of possible outcomes (2) for third place (after second place has been decided), and the number of ways (1) that fourth place can occur (after the first three places have been taken).

   The discussion above illustrates the following general rule, which we accept as a basic axiom of counting.

(10.42) Fundamental Counting Principle

Let E_1, E_2, $\cdots$, E_k be a sequence of k events. If for each i, E_i can occur in m_i ways, then the total number of ways the events may take place is the product $m_1 m_2 \cdots m_k$.

   Returning to the first illustration above, we let E_1 represent the determination of the first-place team, so that $m_1 = 4$. If E_2 denotes the determination of the second-place team, then $m_2 = 3$. Hence by (10.42) the number of outcomes for the sequence E_1, E_2 is $4 \cdot 3 = 12$, which is the same as that found by means of the tree diagram. If we proceed to E_3, the determination of the third-place team, then $m_3 = 2$, and hence $m_1 m_2 m_3 = 24$. Finally, if E_1, E_2, and E_3 have occurred, there is only one possible outcome for E_4. Thus $m_4 = 1$ and $m_1 m_2 m_3 m_4 = 24$.

   Instead of teams, let us now regard a, b, c, and d merely as symbols and consider the various *orderings* or *arrangements* that may be assigned to these symbols, taking them either two at a time, three at a

time, or four at a time. By abstracting in this way we may apply our
methods to other similar situations. For example, the problem of determining the number of two-digit numbers that can be formed from the
digits 1, 2, 3, and 4 so that no digit occurs twice in any number is essen-
tially the same as our illustration of listing first and second place in the
tournament. Incidentally, the arrangements discussed above are called
arrangements without repetitions, since a symbol may not be used
twice in the same arrangement. In Example 1 below we consider
arrangements in which repetitions *are* allowed.

Previously we defined ordered pairs and ordered triples. Simi-
larly, an **ordered 4-tuple** is a set of elements $\{x_1, x_2, x_3, x_4\}$ in which an
ordering has been specified for the elements, so that one of the elements
may be referred to as the *first element,* another as the *second element,*
and so on. The symbol (x_1, x_2, x_3, x_4) is used for the ordered 4-tuple
having first element x_1, second element x_2, third element x_3, and fourth
element x_4. In general, for any positive integer r, we speak of the **ordered r-tuple**

(10.43) $(x_1, x_2, \cdots, x_r)$

as a set of elements in which x_1 is designated as the first element, x_2 as
the second element and so on.

EXAMPLE 1   How many ordered triples can be obtained by using the
letters a, b, c, and d? How many ordered 4-tuples can be obtained?
How many r-tuples?

Solution

To answer the first question we must determine the number
of symbols of the form (x_1, x_2, x_3) that can be obtained by
using only the letters a, b, c, and d. This is not the same as
listing first, second, and third place as in our previous illustration, since we have not ruled out the possibility of repetitions. For example, (a, b, a), (a, a, b), and (b, a, a) are different triples of the desired type. If for $i = 1, 2, 3$, we let E_i
represent the determination of x_i in the triple (x_1, x_2, x_3),
then since repetitions are allowed, there are four possibilities for each of E_1, E_2, and E_3. Hence by (10.42) the total
number of ordered triples is $4 \cdot 4 \cdot 4$, or 64. Similarly, the
number of possible 4-tuples (x_1, x_2, x_3, x_4) is $4 \cdot 4 \cdot
4 \cdot 4 = 256$. Evidently, the number of r-tuples is the product
$4 \cdot 4 \cdots 4$, where 4 appears as a factor r times. That product equals $4r$.

EXAMPLE 2   A class consists of 60 boys and 40 girls. In how many
ways can a president, vice-president, treasurer, and secretary be
chosen if the treasurer must be a boy, the secretary must be a girl,
and a student may not hold more than one office?

Solution

Let E_1 represent the choice of treasurer, E_2 the choice of secretary, and E_3 and E_4 the choice for president and vice-president respectively. As in (10.42), we let m_i denote the number of different ways E_i can occur, for $i = 1, 2, 3$, and 4. It follows that $m_1 = 60$, $m_2 = 40$, $m_3 = 98$, and $m_4 = 97$ (Why?). By (10.42) the total number of possibilities is $60 \cdot 40 \cdot 98 \cdot 97 = 22{,}814{,}400$.

The preceding example illustrates the fact that if some of the events E_i are specialized in any way, then in applying (10.42) those special E_i should be performed first in the sequence.

Let n be a positive integer and let S be a set consisting of n distinct elements. When we speak about sets or subsets of sets, we are generally not concerned about the order or arrangement of the elements. For example, as in Chapter One, the set $\{a, b, c\}$ is the same as the set $\{b, a, c\}$. In the remainder of this section, however, the arrangement of the elements will be our main concern. In the following definition, S denotes a set containing n elements, and r is a positive integer such that $r \leq n$.

(10.44) Definition

A **permutation** of r elements of a set S is an arrangement, without repetitions, of r elements of S.

We shall use the symbol $_nP_r$ to denote the number of different permutations of r elements that can be obtained from a set containing n elements. It is also common to refer to a permutation of r elements of a set containing n elements as a **permutation of n elements taken r at a time.** As a special case, $_nP_n$ denotes the number of arrangements of n elements of S, that is, $_nP_n$ is the number of ways of arranging *all* the elements of S.

In our first illustration involving the four teams $\{a, b, c, d\}$ we had $_4P_2 = 12$, since there were 12 different ways of arranging the four teams in groups of two. It was also shown that the number of ways to arrange all the elements a, b, c, d is 24. In the present notation this would be written $_4P_4 = 24$.

It is not difficult to find a general formula for $_nP_r$. If S denotes a set containing n elements, then the problem of determining $_nP_r$ is equivalent to determining the number of different r-tuples (10.43), where each x_i is an element of S and no element of S appears twice in the same r-tuple. We shall find this number by means of (10.42). For each $i = 1$, $2, \cdots, r$, let E_i represent the determination of the element x_i in (10.43), and let m_i be the number of different ways of performing E_i. We wish to apply the sequence $E_1, E_2, \cdots, E_r$. There are n possible choices for x_1 and consequently we have $m_1 = n$. Since repetitions are not allowed, there are $n - 1$ choices for x_2, so that $m_2 = n - 1$. Continuing in this manner, we successively obtain $m_3 = n - 2$, $m_4 = m - 3$, and ultimately $m_r = n - (r - 1)$, or $m_r = n - r + 1$. Hence by (10.42) we have

(10.45) $\quad {}_nP_r = n(n - 1)(n - 2) \cdots (n - r + 1)$,

where there are r factors on the right side of the equation. As illustrations, if we consider the special cases $r = 1, 2, 3$, and 4, then $n - r + 1$ equals n, $n - 1$, $n - 2$, and $n - 3$, respectively, and using (10.45) gives us

$$
\begin{aligned}
{}_nP_1 &= n, \\
{}_nP_2 &= n(n - 1), \\
{}_nP_3 &= n(n - 1)(n - 2), \\
{}_nP_4 &= n(n - 1)(n - 2)(n - 3).
\end{aligned}
$$

EXAMPLE 3   Find ${}_nP_r$ if

(a) $n = 5, r = 2$;   (b) $n = 6, r = 4$;   (c) $n = 5, r = 5$.

Solutions

Using (10.45) we have

(a) ${}_5P_2 = 5 \cdot 4 = 20$,
(b) ${}_6P_4 = 6 \cdot 5 \cdot 4 \cdot 3 = 360$,
(c) ${}_5P_5 = 5 \cdot 4 \cdot 3 \cdot 2 \cdot 1 = 120$.

EXAMPLE 4   A baseball team consists of nine players. Find the number of ways of arranging the first four positions in the batting order, if the pitcher is excluded.

Solution

Using (10.45) with $n = 8$ and $r = 4$, we obtain

$$
{}_8P_4 = 8 \cdot 7 \cdot 6 \cdot 5 = 1680.
$$

As a special case of (10.45), if $r = n$ we obtain the number of different arrangements of *all* the elements of S. In this case, $n - r + 1 = n - n + 1 = 1$, and (10.45) becomes

(10.46) $\quad {}_nP_n = n(n - 1)(n - 2) \cdots 3 \cdot 2 \cdot 1 = n!$

Consequently ${}_nP_n$ is the product of the first n positive integers. We may also obtain a form for ${}_nP_r$ which involves the factorial notation. If r and n are positive integers with $r \le n$, then

$$
\begin{aligned}
\frac{n!}{(n - r)!} &= \frac{n(n - 1)(n - 2) \cdots (n - r + 1) \cdot [(n - r)!]}{(n - r)!} \\
&= n(n - 1)(n - 2) \cdots (n - r + 1).
\end{aligned}
$$

Comparison with (10.45) gives us

$$(10.47) \quad {}_nP_r = \frac{n!}{(n-r)!}.$$

Substituting $r = n$ in (10.47) we obtain

$${}_nP_n = \frac{n!}{(n-n)!} = \frac{n!}{0!} = \frac{n!}{1} = n!,$$

which is in agreement with (10.46).

EXERCISES

Solve Exercises 1–14 by using tree diagrams or (10.42).

1   How many three-digit numbers can be formed from the digits 1, 2, 3, 4, and 5 if   **a** repetitions are not allowed?   **b** repetitions are allowed?

2   Repeat Exercise 1 for four-digit numbers.

3   How many numbers can be formed from the digits 1, 2, 3, and 4 if repetitions are not allowed? (*Note:* 42 and 231 are examples of such numbers.)

4   Work Exercise 3 if repetitions are allowed.

5   If eight basketball teams take part in a tournament, find the number of different ways that first, second, and third place can be decided, if ties are not allowed.

6   Repeat Exercise 5 for twelve teams.

7   A girl has four skirts and six blouses. How many different skirt-blouse combinations can she wear?

8   If the girl in Exercise 7 also has three sweaters, in how many different ways can she combine the three articles of clothing?

9   In a certain state, automobile license plates start with one letter of the alphabet followed by five numerals, using the digits 0, 1, 2, $\cdots$, 9. Find how many different license plates are possible if   **a** the first digit following the letter cannot be 0,   **b** the first letter cannot be O or I and the first digit cannot be 0.

10   Two dice are tossed, one after the other. In how many different ways can they fall? List the number of different ways the sum of the dots can equal **a** three,   **b** five,   **c** seven,   **d** nine,   **e** eleven.

11   A row of six seats in a classroom is to be filled by selecting individuals from a group of ten students. In how many different ways can the seats be occupied? If there are six boys and four girls in the group and if boys and girls are to be alternated, find the number of different seating arrangements.

12 A student in a certain school may take mathematics at 8, 10, 11, or 2 o'clock; English at 9, 10, 1, or 2; and History at 8, 11, 2, or 3. Find the number of different ways in which the student can schedule the three courses.

13 In how many different ways can a test consisting of ten true-or-false questions be answered?

14 A test consists of six multiple-choice questions, and the number of choices for each question is five. In how many different ways can the test be answered?

In each of Exercises 15–22, find the given number.

15 $_6P_3$ **16** $_7P_4$ **17** $_8P_5$ **18** $_4P_2$

19 $_6P_6$ **20** $_7P_7$ **21** $_4P_1$ **22** $_5P_1$

Use permutations to solve Exercises 23–30.

23 In how many different ways can eight people be seated in a row?

24 In how many different ways can ten books be arranged on a shelf?

25 A signal man has six different flags. How many different signals can be sent by placing three flags, one above the other, on a flag pole?

26 In how many different ways can five books be selected from a set of twelve different books?

27 How many three-digit numbers can be formed from the digits 2, 4, 6, 8, and 9 if   **a**   repetitions are not allowed?   **b**   repetitions are allowed?

28 There are 24 letters in the Greek alphabet. How many fraternities may be specified by choosing three Greek letters if   **a**   repetitions are not allowed?   **b**   repetitions are allowed?

29 How many seven-digit phone numbers can be formed from the digits 0, 1, 2, 3, $\cdots$, 9 if the first digit may not be 0?

30 After selecting nine players for a baseball game, the manager of the team arranges the batting order so that the pitcher bats last and the best hitter bats fourth. In how many different ways can the remainder of the batting order be arranged?

8 DISTINGUISHABLE PERMUTATIONS AND COMBINATIONS

Certain problems involve finding different arrangements of objects, some of which are indistinguishable. For example, suppose we are given five discs of the same size, where three are black, one is white, and one is red. Let us find the number of ways they can be arranged in a row so that different color arrangements are obtained. If the discs were all different, then by (10.46) the number of arrangements would be 5!, or 120. However, since some of the discs have the same appearance, we cannot obtain 120 different arrangements. To clarify this point, let us write

B B B W R

for the arrangement having black discs in the first three positions in the row, the white disc in the fourth position, and the red disc in the fifth position. Now the first three discs can be arranged in 3!, or 6, different ways, but these arrangements cannot be distinguished from one another because the first three discs look alike. We say that those 3! permutations are **nondistinguishable.** Similarly, given any other arrangement, say

B R B W B,

there are 3! different ways of arranging the three black discs, but again each such arrangement is nondistinguishable from the others. Let us call two arrangements of the five objects **distinguishable permutations** if one arrangement cannot be obtained from the other by rearranging the like objects. Thus B B B W R and B R B W B are distinguishable permutations. Let n denote the number of distinguishable permutations. Since with each such arrangement there corresponds 3! nondistinguishable permutations, we must have $3!n = 5!$, the number of permutations of five *different* objects. Hence $n = 5!/3! = 5 \cdot 4 = 20$. By the same type of reasoning we can obtain the following extension of the previous problem: *If r objects in a collection of n objects are alike, and if the remaining objects are different from each other and also from the r objects, then the number of distinguishable permutations of the n objects is n!/r!.*

That result can be generalized to the case in which there are several subcollections of indistinguishable objects. For example, consider eight discs as above, four of which are black, three white, and one red. In this case, with each arrangement, such as

(10.48) B W B W B W B R,

there are 4! arrangements of the black discs and 3! arrangements of the white discs which have no effect on the color arrangement. Hence there are 4!3! arrangements of discs that can be made in (10.48) which will not produce distinguishable permutations. If we let n denote the number of *distinguishable* permutations of the objects, then it follows that $4!3!n = 8!$, since 8! is the number of permutations we would obtain if the objects were all different. This gives us

$$n = \frac{8!}{4!3!} = \frac{8 \cdot 7 \cdot 6 \cdot 5}{3!} \cdot \frac{4!}{4!} = 280.$$

The following general result can be proved.

(10.49) Theorem

Consider a collection of n objects in which n_1 are alike, n_2 are alike of another kind, $\cdots$, n_k are alike of a further kind, and let $n = n_1 + n_2 + \cdots + n_k$. Then the number of distinguishable permutations of the n objects is

$$\frac{n!}{n_1!n_2! \cdots n_k!}.$$

EXAMPLE 1 Find the number of distinguishable permutations of the letters in the word "MISSISSIPPI."

Solution

In this example we are given a collection of eleven objects in which four are of one kind (the letter S), four are of another kind (I), two are of a third kind (P), and one is a fourth kind (M). Hence by (10.49) the number of distinguishable permutations is

$$\frac{11!}{4!4!2!1!},$$

which simplifies to 34,650.

When we work with permutations our concern is with ordered subsets or arrangements of elements in subsets. Let us now ignore the order, or arrangements, and consider only the *elements* in a subset. We would like to answer the following question: Given a set containing n distinct elements, in how many ways can a subset of r elements be chosen, where $r \leq n$? Before answering, let us state a definition.

(10.50) Definition

A **combination** of r elements of a set S is a subset of S which contains r distinct elements.

We shall use the symbol $_nC_r$ to denote the number of combinations of r elements that can be obtained from a set containing n elements. The phrase "a combination of n elements taken r at a time" is often used instead of the phrase in (10.50).

If S contains n elements, then to find $_nC_r$ we must find the total number of subsets of the form

(10.51) $\{x_1, x_2, \cdots, x_r\}$,

where the x_i are different elements of S. Since the elements of the subset (10.51) can be arranged in $r!$ different ways, each such subset produces $r!$ different r-tuples. Hence the total number of different r-tuples is $r!\,_nC_r$.

However, we saw in the previous section that the number of r-tuples is $_nP_r$, which by (10.47) equals $\dfrac{n!}{(n-r)!}$. This implies that

$$r!\,_nC_r = \frac{n!}{(n-r)!}$$

and so

$$(10.52) \quad _nC_r = \frac{n!}{(n-r)!\,r!}.$$

EXAMPLE 2 A baseball squad has six outfielders, seven infielders, five pitchers, and two catchers. In how many different ways can a team of nine players be chosen?

Solution

The number of ways of choosing three outfielders from the six candidates is

$$_6C_3 = \frac{6!}{3!(6-3)!} = \frac{6!}{3!3!} = 20.$$

The number of ways of choosing the four infielders is

$$_7C_4 = \frac{7!}{4!(7-4)!} = \frac{7\cdot 6\cdot 5}{3!} = 35.$$

There are five ways of choosing a pitcher and two choices for the catcher. It follows from (10.42) tht the total number of ways to choose a team is $20\cdot 35\cdot 5\cdot 2 = 7000$.

It is worth noting that if $r = n$, then the right side of (10.52) becomes

$$\frac{n!}{(n-n)!n!} = \frac{n!}{0!n!} = 1.$$

Moreover, $_nC_n$ is the number of subsets consisting of n elements that can be obtained from a set of n elements. This is also equal to 1. Hence (10.52) is true for $r = n$.

It is convenient to assign a meaning to $_nC_r$ when $r = 0$. If (10.52) is to be true in this case, then

$$_nC_0 = \frac{n!}{n!0!} = 1.$$

Hence we *define* $_nC_0 = 1$, which is the same as $_nC_n$. Finally, for consistency we also *define* $_0C_0 = 1$. Thus $_nC_r$ has meaning for all nonnegative integers n and r with $r \leq n$, and formula (10.52) is valid in all those cases.

As a final observation, note that

$$_nC_r = \frac{n!}{(n-r)!r!} = \frac{n(n-1) \cdots (n-r+1)}{r!}.$$

Comparison with (10.40) shows that $_nC_r$ is identical with the binomial coefficient $\binom{n}{r}$. An interesting application of that fact is given in the following example.

EXAMPLE 3 If a set S contains n elements, find the number of distinct subsets of S.

Solution

Let r be any nonnegative integer such that $r \leq n$. From our previous work the number of subsets of S which contain r elements is $_nC_r$, or $\binom{n}{r}$. Hence, to find the total number of subsets we form the sum

$$\binom{n}{0} + \binom{n}{1} + \binom{n}{2} + \binom{n}{3} + \cdots + \binom{n}{n}.$$

By (10.41) this is precisely the binomial expansion of $(1 + 1)^n$. Thus there are 2^n subsets of a set of n elements. In particular, a set of 3 elements has 2^3, or 8 different subsets. A set of 4 elements has 2^4, or 16 subsets. A set of 10 elements has 2^{10}, or 1024 subsets.

EXERCISES

In Exercises 1–8 find the given number.

1 $_6C_4$ **2** $_8C_5$ **3** $_8C_7$ **4** $_6C_3$

5 $_nC_{n-1}$ **6** $_nC_1$ **7** $_6C_0$ **8** $_6C_6$

9 If five black, three red, two white, and two green discs are to be arranged in a row, find the number of possible color arrangements for the discs.

10 Work Exercise 9 if there are three black, three red, three white, and three green discs.

11 Find the number of distinguishable permutations of the letters in the word "BOOKKEEPER."

12 Find the number of distinguishable permutations of the letters in the word "MOON." List all of the permutations.

13 Ten boys wish to play a basketball game. In how many different ways can two teams consisting of five players each be chosen?

14 A student may answer any six of ten questions in an examination. In how many ways can a choice be made? How many choices are possible if the first two questions must be answered?

15 How many straight lines are determined by eight points if no three of the points are collinear? How many triangles are determined?

16 A committee of five persons is to be chosen from a group of twelve men and eight women. If the committee is to consist of three men and two women, determine the number of ways of selecting the committee.

17 A student has five mathematics books, four history books, and eight fiction books. In how many ways can they be arranged on a shelf if books in the same category are kept next to one another?

18 A basketball squad consists of twelve individuals. Disregarding positions, in how many ways can a team of five be selected? If the center of a team must be selected from two specific individuals on the squad and the other four members of the team from the remaining ten players, find the number of different teams possible.

19 A football squad consists of three centers, ten linemen who can play either guard or tackle, three quarterbacks, six halfbacks, four ends, and four full-backs. In how many ways can a team consisting of one center, two guards, two tackles, two ends, two halfbacks, a quarterback, and a fullback be selected?

20 In how many different ways can seven keys be arranged on a key ring if the keys can slide completely around the ring?

9 REVIEW EXERCISES

Oral

Define or discuss each of the following.

1 The axiom of mathematical induction

2 The principle of mathematical induction

3 Infinite sequence

4 Summation notation

5 The nth partial sum of an infinite sequence

6 Arithmetic sequence

7 Geometric sequence

8 Arithmetic mean of two numbers

Written

Prove that Exercises 1–4 are true for every positive integer n.

1 $2 + 5 + 8 + \cdots + (3n - 1) = n(3n + 1)/2$

2 $2^2 + 4^2 + 6^2 + \cdots + (2n)^2 = 2n(2n + 1)(n + 1)/3$

3 $\dfrac{1}{1 \cdot 3} + \dfrac{1}{3 \cdot 5} + \dfrac{1}{5 \cdot 7} + \cdots + \dfrac{1}{(2n - 1)(2n + 1)} = \dfrac{n}{2n + 1}$

4 $1 \cdot 2 + 2 \cdot 3 + 3 \cdot 4 + \cdots + n(n + 1) = n(n + 1)(n + 2)/3$

In each of Exercises 5–8 find the first four terms and the seventh term of the sequence that has the given nth term.

5 $a_n = \dfrac{3n}{1 - 2n^2}$

6 $a_n = (-1)^{n+1} 10 - (0.1)^n$

7 $a_n = 1 + \left(-\tfrac{1}{2}\right)^{n-1}$

8 $a_n = \dfrac{3^n}{(n + 1)(n + 2)}$

Find the first five terms of the infinite sequence defined recursively in each of Exercises 9–12.

9 $a_1 = 2, \ a_{k+1} = 1 + 1/a_k$

10 $a_1 = 3, \ a_{k+1} = a_k!$

11 $a_1 = 4, \ a_{k+1} = \sqrt{a_k}$

12 $a_1 = 1, \ a_{k+1} = (1 + a_k)^{-1}$

Find the number represented by the sum in each of Exercises 13–16.

13 $\displaystyle\sum_{k=1}^{5} (k^2 + 3)$

14 $\displaystyle\sum_{k=2}^{6} \dfrac{k + 4}{k - 1}$

15 $\displaystyle\sum_{k=1}^{25} 8$

16 $\displaystyle\sum_{i=1}^{4} (2^i - 1)$

In Exercises 17–20 use summation notation to express the sums.

17 $2 + 5 + 8 + 11 + 14$

18 $1 + 2 + 4 + 8 + 16 + 32$

19 $42 - 44 + 46 - 48 + 50$

20 $a_0 + a_2 x^2 + a_4 x^4 + \cdots + a_{20} x^{20}$

21 Find the tenth term and the sum of the first ten terms of the arithmetic sequence whose first two terms are $3 + \sqrt{2}$ and 2.

22 Find the sum of the first eight terms of an arithmetic sequence in which the fourth term is 7 and the common difference is -3.

23 The fifth and thirteenth terms of an arithmetic sequence are 8 and 80, respectively. Find the first term and the tenth term.

24 Insert four arithmetic means between 20 and -10.

25 Find the tenth term of the geometric sequence whose first two terms are $\frac{1}{4}$ and $\frac{1}{2}$.

26 If a geometric sequence has 0.05 and -0.005 as its third and fourth terms, find the eighth term.

27 Find the geometric mean of 3 and 50.

28 In a certain geometric sequence the eighth term is 10 and the common ratio is $-\frac{3}{2}$. Find the first term.

Find the sums in Exercises 29–32.

29 $\displaystyle\sum_{k=1}^{12} (2k - \tfrac{1}{2})$
30 $\displaystyle\sum_{k=1}^{10} \left(2 - \frac{k}{2}\right)$

31 $\displaystyle\sum_{k=1}^{8} (2^k - \tfrac{1}{2})$
32 $\displaystyle\sum_{k=1}^{7} (\tfrac{1}{2} - 2^k)$

33 Find the sum of the infinite geometric series

$$1 - \tfrac{2}{3} + \tfrac{4}{9} - \tfrac{8}{27} + \cdots$$

34 Find the rational number whose decimal representation is $3.\overline{145}$.

35 Expand and simplify $(3x - y^2)^6$.

36 Find the first four terms in the binomial expansion of $(x^{1/5} + 3x^{-4/5})^{20}$.

37 Find the sixth term in the expansion of $(2x^3 - y/2)^{10}$.

38 In the expansion of $(4c^3 + 3bc^{-2})^{10}$ find the term which does not contain c.

39 In how many ways can thirteen cards be selected from a deck of 52 cards? In how many ways can thirteen cards be selected if one wishes to obtain five spades, three hearts, three clubs, and two diamonds?

40 How many four-digit numbers can be formed from the digits 1, 2, 3, 4, 5, and 6 if **a** repetitions are not allowed? **b** repetitions are allowed?

41 If a student must answer eight of twelve questions on an examination, in how many ways can a choice be made? How many choices are possible if the first three questions must be answered?

42 If six black, five red, four white, and two green discs are to be arranged in a row, find the number of possible color arrangements.

n	n^2	$\sqrt{n}$	n^3	$\sqrt[3]{n}$	n	n^2	$\sqrt{n}$	n^3	$\sqrt[3]{n}$
1	1	1.000	1	1.000	51	2,601	7.141	132,651	3.708
2	4	1.414	8	1.260	52	2,704	7.211	140,608	3.733
3	9	1.732	27	1.442	53	2,809	7.280	148,877	3.756
4	16	2.000	64	1.587	54	2,916	7.348	157,464	3.780
5	25	2.236	125	1.710	55	3,025	7.416	166,375	3.803
6	36	2.449	216	1.817	56	3,136	7.483	175,616	3.826
7	49	2.646	343	1.913	57	3,249	7.550	185,193	3.849
8	64	2.828	512	2.000	58	3,364	7.616	195,112	3.871
9	81	3.000	729	2.080	59	3,481	7.681	205,379	3.893
10	100	3.162	1,000	2.154	60	3,600	7.746	216,000	3.915
11	121	3.317	1,331	2.224	61	3,721	7.810	226,981	3.936
12	144	3.464	1,728	2.289	62	3,844	7.874	238,328	3.958
13	169	3.606	2,197	2.351	63	3,969	7.937	250,047	3.979
14	196	3.742	2,744	2.410	64	4,096	8.000	262,144	4.000
15	225	3.873	3,375	2.466	65	4,225	8.062	274,625	4.021
16	256	4.000	4,096	2.520	66	4,356	8.124	287,496	4.041
17	289	4.123	4,913	2.571	67	4,489	8.185	300,763	4.062
18	324	4.243	5,832	2.621	68	4,624	8.246	314,432	4.082
19	361	4.359	6,859	2.668	69	4,761	8.307	328,509	4.102
20	400	4.472	8,000	2.714	70	4,900	8.367	343,000	4.121
21	441	4.583	9,261	2.759	71	5,041	8.426	357,911	4.141
22	484	4.690	10,648	2.802	72	5,184	8.485	373,248	4.160
23	529	4.796	12,167	2.844	73	5,329	8.544	389,017	4.179
24	576	4.899	13,824	2.884	74	5,476	8.602	405,224	4.198
25	625	5.000	15,625	2.924	75	5,625	8.660	421,875	4.217
26	676	5.099	17,576	2.962	76	5,776	8.718	438,976	4.236
27	729	5.196	19,683	3.000	77	5,929	8.775	456,533	4.254
28	784	5.292	21,952	3.037	78	6,084	8.832	474,552	4.273
29	841	5.385	24,389	3.072	79	6,241	8.888	493,039	4.291
30	900	5.477	27,000	3.107	80	6,400	8.944	512,000	4.309
31	961	5.568	29,791	3.141	81	6,561	9.000	531,441	4.327
32	1,024	5.657	32,768	3.175	82	6,724	9.055	551,368	4.344
33	1,089	5.745	35,937	3.208	83	6,889	9.110	571,787	4.362
34	1,156	5.831	39,304	3.240	84	7,056	9.165	592,704	4.380
35	1,225	5.916	42,875	3.271	85	7,225	9.220	614,125	4.397
36	1,296	6.000	46,656	3.302	86	7,396	9.274	636,056	4.414
37	1,369	6.083	50,653	3.332	87	7,569	9.327	658,503	4.431
38	1,444	6.164	54,872	3.362	88	7,744	9.381	681,472	4.448
39	1,521	6.245	59,319	3.391	89	7,921	9.434	704,969	4.465
40	1,600	6.325	64,000	3.420	90	8,100	9.487	729,000	4.481
41	1,681	6.403	68,921	3.448	91	8,281	9.539	753,571	4.498
42	1,764	6.481	74,088	3.476	92	8,464	9.592	778,688	4.514
43	1,849	6.557	79,507	3.503	93	8,649	9.644	804,357	4.531
44	1,936	6.633	85,184	3.530	94	8,836	9.695	830,584	4.547
45	2,025	6.708	91,125	3.557	95	9,025	9.747	857,375	4.563
46	2,116	6.782	97,336	3.583	96	9,216	9.798	884,736	4.579
47	2,209	6.856	103,823	3.609	97	9,409	9.849	912,673	4.595
48	2,304	6.928	110,592	3.634	98	9,604	9.899	941,192	4.610
49	2,401	7.000	117,649	3.659	99	9,801	9.950	970,299	4.626
50	2,500	7.071	125,000	3.684	100	10,000	10.000	1,000,000	4.642

Table 1
Powers and
Roots

485

Table 2
Common
Logarithms

N	0	1	2	3	4	5	6	7	8	9
1.0	.0000	.0043	.0086	.0128	.0170	.0212	.0253	.0294	.0334	.0374
1.1	.0414	.0453	.0492	.0531	.0569	.0607	.0645	.0682	.0719	.0755
1.2	.0792	.0828	.0864	.0899	.0934	.0969	.1004	.1038	.1072	.1106
1.3	.1139	.1173	.1206	.1239	.1271	.1303	.1335	.1367	.1399	.1430
1.4	.1461	.1492	.1523	.1553	.1584	.1614	.1644	.1673	.1703	.1732
1.5	.1761	.1790	.1818	.1847	.1875	.1903	.1931	.1959	.1987	.2014
1.6	.2041	.2068	.2095	.2122	.2148	.2175	.2201	.2227	.2253	.2279
1.7	.2304	.2330	.2355	.2380	.2405	.2430	.2455	.2480	.2504	.2529
1.8	.2553	.2577	.2601	.2625	.2648	.2672	.2695	.2718	.2742	.2765
1.9	.2788	.2810	.2833	.2856	.2878	.2900	.2923	.2945	.2967	.2989
2.0	.3010	.3032	.3054	.3075	.3096	.3118	.3139	.3160	.3181	.3201
2.1	.3222	.3243	.3263	.3284	.3304	.3324	.3345	.3365	.3385	.3404
2.2	.3424	.3444	.3464	.3483	.3502	.3522	.3541	.3560	.3579	.3598
2.3	.3617	.3636	.3655	.3674	.3692	.3711	.3729	.3747	.3766	.3784
2.4	.3802	.3820	.3838	.3856	.3874	.3892	.3909	.3927	.3945	.3962
2.5	.3979	.3997	.4014	.4031	.4048	.4065	.4082	.4099	.4116	.4133
2.6	.4150	.4166	.4183	.4200	.4216	.4232	.4249	.4265	.4281	.4298
2.7	.4314	.4330	.4346	.4362	.4378	.4393	.4409	.4425	.4440	.4456
2.8	.4472	.4487	.4502	.4518	.4533	.4548	.4564	.4579	.4594	.4609
2.9	.4624	.4639	.4654	.4669	.4683	.4698	.4713	.4728	.4742	.4757
3.0	.4771	.4786	.4800	.4814	.4829	.4843	.4857	.4871	.4886	.4900
3.1	.4914	.4928	.4942	.4955	.4969	.4983	.4997	.5011	.5024	.5038
3.2	.5051	.5065	.5079	.5092	.5105	.5119	.5132	.5145	.5159	.5172
3.3	.5185	.5198	.5211	.5224	.5237	.5250	.5263	.5276	.5289	.5302
3.4	.5315	.5328	.5340	.5353	.5366	.5378	.5391	.5403	.5416	.5428
3.5	.5441	.5453	.5465	.5478	.5490	.5502	.5514	.5527	.5539	.5551
3.6	.5563	.5575	.5587	.5599	.5611	.5623	.5635	.5647	.5658	.5670
3.7	.5682	.5694	.5705	.5717	.5729	.5740	.5752	.5763	.5775	.5786
3.8	.5798	.5809	.5821	.5832	.5843	.5855	.5866	.5877	.5888	5899
3.9	.5911	.5922	.5933	.5944	.5955	.5966	.5977	.5988	.5999	.6010
4.0	.6021	.6031	.6042	.6053	.6064	.6075	.6085	.6096	.6107	.6117
4.1	.6128	.6138	.6149	.6160	.6170	.6180	.6191	.6201	.6212	.6222
4.2	.6232	.6243	.6253	.6263	.6274	.6284	.6294	.6304	.6314	.6325
4.3	.6335	.6345	.6355	.6365	.6375	.6385	.6395	.6405	.6415	.6425
4.4	.6435	.6444	.6454	.6464	.6474	.6484	.6493	.6503	.6513	.6522
4.5	.6532	.6542	.6551	.6561	.6571	.6580	.6590	.6599	.6609	.6618
4.6	.6628	.6637	.6646	.6656	.6665	.6675	.6684	.6693	.6702	.6712
4.7	.6721	.6730	.6739	.6749	.6758	.6767	.6776	.6785	.6794	.6803
4.8	.6812	.6821	.6830	.6839	.6848	.6857	.6866	.6875	.6884	.6893
4.9	.6902	.6911	.6920	.6928	.6937	.6946	.6955	.6964	.6972	.6981
5.0	.6990	.6998	.7007	.7016	.7024	.7033	.7042	.7050	.7059	.7067
5.1	.7076	.7084	.7093	.7101	.7110	.7118	.7126	.7135	.7143	.7152
5.2	.7160	.7168	.7177	.7185	.7193	.7202	.7210	.7218	.7226	.7235
5.3	.7243	.7251	.7259	.7267	.7275	.7284	.7292	.7300	.7308	.7316
5.4	.7324	.7332	.7340	.7348	.7356	.7364	.7372	.7380	.7388	.7396

Table 2
Common
Logarithms

N	0	1	2	3	4	5	6	7	8	9
5.5	.7404	.7412	.7419	.7427	.7435	.7443	.7451	.7459	.7466	.7474
5.6	.7482	.7490	.7497	.7505	.7513	.7520	.7528	.7536	.7543	.7551
5.7	.7559	.7566	.7574	.7582	.7589	.7597	.7604	.7612	.7619	.7627
5.8	.7634	.7642	.7649	.7657	.7664	.7672	.7679	.7686	.7694	.7701
5.9	.7709	.7716	.7723	.7731	.7738	.7745	.7752	.7760	.7767	.7774
6.0	.7782	.7789	.7796	.7803	.7810	.7818	.7825	.7832	.7839	.7846
6.1	.7853	.7860	.7868	.7875	.7882	.7889	.7896	.7903	.7910	.7917
6.2	.7924	.7931	.7938	.7945	.7952	.7959	.7966	.7973	.7980	.7987
6.3	.7993	.8000	.8007	.8014	.8021	.8028	.8035	.8041	.8048	.8055
6.4	.8062	.8069	.8075	.8082	.8089	.8096	.8102	.8109	.8116	.8122
6.5	.8129	.8136	.8142	.8149	.8156	.8162	.8169	.8176	.8182	.8189
6.6	.8195	.8202	.8209	.8215	.8222	.8228	.8235	.8241	.8248	.8254
6.7	.8261	.8267	.8274	.8280	.8287	.8293	.8299	.8306	.8312	.8319
6.8	.8325	.8331	.8338	.8344	.8351	.8357	.8363	.8370	.8376	.8382
6.9	.8388	.8395	.8401	.8407	.8414	.8420	.8426	.8432	.8439	.8445
7.0	.8451	.8457	.8463	.8470	.8476	.8482	.8488	.8494	.8500	.8506
7.1	.8513	.8519	.8525	.8531	.8537	.8543	.8549	.8555	.8561	.8567
7.2	.8573	.8579	.8585	.8591	.8597	.8603	.8609	.8615	.8621	.8627
7.3	.8633	.8639	.8645	.8651	.8657	.8663	.8669	.8675	.8681	.8686
7.4	.8692	.8698	.8704	.8710	.8716	.8722	.8727	.8733	.8739	.8745
7.5	.8751	.8756	.8762	.8768	.8774	.8779	.8785	.8791	.8797	.8802
7.6	.8808	.8814	.8820	.8825	.8831	.8837	.8842	.8848	.8854	.8859
7.7	.8865	.8871	.8876	.8882	.8887	.8893	.8899	.8904	.8910	.8915
7.8	.8921	.8927	.8932	.8938	.8943	.8949	.8954	.8960	.8965	.8971
7.9	.8976	.8982	.8987	.8993	.8998	.9004	.9009	.9015	.9020	.9025
8.0	.9031	.9036	.9042	.9047	.9053	.9058	.9063	.9069	.9074	.9079
8.1	.9085	.9090	.9096	.9101	.9106	.9112	.9117	.9122	.9128	.9133
8.2	.9138	.9143	.9149	.9154	.9159	.9165	.9170	.9175	.9180	.9186
8.3	.9191	.9196	.9201	.9206	.9212	.9217	.9222	.9227	.9232	.9238
8.4	.9243	.9248	.9253	.9258	.9263	.9269	.9274	.9279	.9284	.9289
8.5	.9294	.9299	.9304	.9309	.9315	.9320	.9325	.9330	.9335	.9340
8.6	.9345	.9350	.9355	.9360	.9365	.9370	.9375	.9380	.9385	.9390
8.7	.9395	.9400	.9405	.9410	.9415	.9420	.9425	.9430	.9435	.9440
8.8	.9445	.9450	.9455	.9460	.9465	.9469	.9474	.9479	.9484	.9489
8.9	.9494	.9499	.9504	.9509	.9513	.9518	.9523	.9528	.9533	.9538
9.0	.9542	.9547	.9552	.9557	.9562	.9566	.9571	.9576	.9581	.9586
9.1	.9590	.9595	.9600	.9605	.9609	.9614	.9619	.9624	.9628	.9633
9.2	.9638	.9643	.9647	.9652	.9657	.9661	.9666	.9671	.9675	.9680
9.3	.9685	.9689	.9694	.9699	.9703	.9708	.9713	.9717	.9722	.9727
9.4	.9731	.9736	.9741	.9745	.9750	.9754	.9759	.9763	.9768	.9773
9.5	.9777	.9782	.9786	.9791	.9795	.9800	.9805	.9809	.9814	.9818
9.6	.9823	.9827	.9832	.9836	.9841	.9845	.9850	.9854	.9859	.9863
9.7	.9868	.9872	.9877	.9881	.9886	.9890	.9894	.9899	.9903	.9908
9.8	.9912	.9917	.9921	.9926	.9930	.9934	.9939	.9943	.9948	.9952
9.9	.9956	.9961	.9965	.9969	.9974	.9978	.9983	.9987	.9991	.9996

Table 3
Four
Place
Trigonometric
Functions

t	t degrees	sin t	tan t	cot t	cos t		
.0000	**0° 00′**	.0000	.0000	—	1.0000	**90° 00′**	1.5708
.0029	10	.0029	.0029	343.77	1.0000	50	1.5679
.0058	20	.0058	.0058	171.89	1.0000	40	1.5650
.0087	30	.0087	.0087	114.59	1.0000	30	1.5621
.0116	40	.0116	.0116	85.940	.9999	20	1.5592
.0145	50	.0145	.0145	68.750	.9999	10	1.5563
.0175	**1° 00′**	.0175	.0175	57.290	.9998	**89° 00′**	1.5533
.0204	10	.0204	.0204	49.104	.9998	50	1.5504
.0233	20	.0233	.0233	42.964	.9997	40	1.5475
.0262	30	.0262	.0262	38.188	.9997	30	1.5446
.0291	40	.0291	.0291	34.368	.9996	20	1.5417
.0320	50	.0320	.0320	31.242	.9995	10	1.5388
.0349	**2° 00′**	.0349	.0349	28.636	.9994	**88° 00′**	1.5359
.0378	10	.0378	.0378	26.432	.9993	50	1.5330
.0407	20	.0407	.0407	24.542	.9992	40	1.5301
.0436	30	.0436	.0437	22.904	.9990	30	1.5272
.0465	40	.0465	.0466	21.470	.9989	20	1.5243
.0495	50	.0494	.0495	20.206	.9988	10	1.5213
.0524	**3° 00′**	.0523	.0524	19.081	.9986	**87° 00′**	1.5184
.0553	10	.0552	.0553	18.075	.9985	50	1.5155
.0582	20	.0581	.0582	17.169	.9983	40	1.5126
.0611	30	.0610	.0612	16.350	.9981	30	1.5097
.0640	40	.0640	.0641	15.605	.9980	20	1.5068
.0669	50	.0669	.0670	14.924	.9978	10	1.5039
.0698	**4° 00′**	.0698	.0699	14.301	.9976	**86° 00′**	1.5010
.0727	10	.0727	.0729	13.727	.9974	50	1.4981
.0756	20	.0756	.0758	13.197	.9971	40	1.4952
.0785	30	.0785	.0787	12.706	.9969	30	1.4923
.0814	40	.0814	.0816	12.251	.9967	20	1.4893
.0844	50	.0843	.0846	11.826	.9964	10	1.4864
.0873	**5° 00′**	.0872	.0875	11.430	.9962	**85° 00′**	1.4835
.0902	10	.0901	.0904	11.059	.9959	50	1.4806
.0931	20	.0929	.0934	10.712	.9957	40	1.4777
.0960	30	.0958	.0963	10.385	.9954	30	1.4748
.0989	40	.0987	.0992	10.078	.9951	20	1.4719
.1018	50	.1016	.1022	9.7882	.9948	10	1.4690
.1047	**6° 00′**	.1045	.1051	9.5144	.9945	**84° 00′**	1.4661
.1076	10	.1074	.1080	9.2553	.9942	50	1.4632
.1105	20	.1103	.1110	9.0098	.9939	40	1.4603
.1134	30	.1132	.1139	8.7769	.9936	30	1.4573
.1164	40	.1161	.1169	8.5555	.9932	20	1.4544
.1193	50	.1190	.1198	8.3450	.9929	10	1.4515
.1222	**7° 00′**	.1219	.1228	8.1443	.9925	**83° 00′**	1.4486
		cos t	cot t	tan t	sin t	t degrees	t

Table 3
Four
Place
Trigonometric
Functions

t	t degrees	sin t	tan t	cot t	cos t		
.1222	**7° 00′**	.1219	.1228	8.1443	.9925	**83° 00′**	1.4486
.1251	10	.1248	.1257	7.9530	.9922	50	1.4457
.1280	20	.1276	.1287	7.7704	.9918	40	1.4428
.1309	30	.1305	.1317	7.5958	.9914	30	1.4399
.1338	40	.1334	.1346	7.4287	.9911	20	1.4370
.1367	50	.1363	.1376	7.2687	.9907	10	1.4341
.1396	**8° 00′**	.1392	.1405	7.1154	.9903	**82° 00′**	1.4312
.1425	10	.1421	.1435	6.9682	.9899	50	1.4283
.1454	20	.1449	.1465	6.8269	.9894	40	1.4254
.1484	30	.1478	.1495	6.6912	.9890	30	1.4224
.1513	40	.1507	.1524	6.5606	.9886	20	1.4195
.1542	50	.1536	.1554	6.4348	.9881	10	1.4166
.1571	**9° 00′**	.1564	.1584	6.3138	.9877	**81° 00′**	1.4137
.1600	10	.1593	.1614	6.1970	.9872	50	1.4108
.1629	20	.1622	.1644	6.0844	.9868	40	1.4079
.1658	30	.1650	.1673	5.9758	.9863	30	1.4050
.1687	40	.1679	.1703	5.8708	.9858	20	1.4021
.1716	50	.1708	.1733	5.7694	.9853	10	1.3992
.1745	**10° 00′**	.1736	.1763	5.6713	.9848	**80° 00′**	1.3963
.1774	10	.1765	.1793	5.5764	.9843	50	1.3934
.1804	20	.1794	.1823	5.4845	.9838	40	1.3904
.1833	30	.1822	.1853	5.3955	.9833	30	1.3875
.1862	40	.1851	.1883	5.3093	.9827	20	1.3846
.1891	50	.1880	.1914	5.2257	.9822	10	1.3817
.1920	**11° 00′**	.1908	.1944	5.1446	.9816	**79° 00′**	1.3788
.1949	10	.1937	.1974	5.0658	.9811	50	1.3759
.1978	20	.1965	.2004	4.9894	.9805	40	1.3730
.2007	30	.1994	.2035	4.9152	.9799	30	1.3701
.2036	40	.2022	.2065	4.8430	.9793	20	1.3672
.2065	50	.2051	.2095	4.7729	.9787	10	1.3643
.2094	**12° 00′**	.2079	.2126	4.7046	.9781	**78° 00′**	1.3614
.2123	10	.2108	.2156	4.6382	.9775	50	1.3584
.2153	20	.2136	.2186	4.5736	.9769	40	1.3555
.2182	30	.2164	.2217	4.5107	.9763	30	1.3526
.2211	40	.2193	.2247	4.4494	.9757	20	1.3497
.2240	50	.2221	.2278	4.3897	.9750	10	1.3468
.2269	**13° 00′**	.2250	.2309	4.3315	.9744	**77° 00′**	1.3439
.2298	10	.2278	.2339	4.2747	.9737	50	1.3410
.2327	20	.2306	.2370	4.2193	.9730	40	1.3381
.2356	30	.2334	.2401	4.1653	.9724	30	1.3352
.2385	40	.2363	.2432	4.1126	.9717	20	1.3323
.2414	50	.2391	.2462	4.0611	.9710	10	1.3294
.2443	**14° 00′**	.2419	.2493	4.0108	.9703	**76° 00′**	1.3265
		cos t	cot t	tan t	sin t	t degrees	t

Table 3
Four
Place
Trigonometric
Functions

t	t degrees	sin t	tan t	cot t	cos t		
.2443	**14° 00′**	.2419	.2493	4.0108	.9703	**76° 00′**	1.3265
.2473	10	.2447	.2524	3.9617	.9696	50	1.3235
.2502	20	.2476	.2555	3.9136	.9689	40	1.3206
.2531	30	.2504	.2586	3.8667	.9681	30	1.3177
.2560	40	.2532	.2617	3.8208	.9674	20	1.3148
.2589	50	.2560	.2648	3.7760	.9667	10	1.3119
.2618	**15° 00′**	.2588	.2679	3.7321	.9659	**75° 00′**	1.3090
.2647	10	.2616	.2711	3.6891	.9652	50	1.3061
.2676	20	.2644	.2742	3.6470	.9644	40	1.3032
.2705	30	.2672	.2773	3.6059	.9636	30	1.3003
.2734	40	.2700	.2805	3.5656	.9628	20	1.2974
.2763	50	.2728	.2836	3.5261	.9621	10	1.2945
.2793	**16° 00′**	.2756	.2867	3.4874	.9613	**74° 00′**	1.2915
.2822	10	.2784	.2899	3.4495	.9605	50	1.2886
.2851	20	.2812	.2931	3.4124	.9596	40	1.2857
.2880	30	.2840	.2962	3.3759	.9588	30	1.2828
.2909	40	.2868	.2994	3.3402	.9580	20	1.2799
.2938	50	.2896	.3026	3.3052	.9572	10	1.2770
.2967	**17° 00′**	.2924	.3057	3.2709	.9563	**73° 00′**	1.2741
.2996	10	.2952	.3089	3.2371	.9555	50	1.2712
.3025	20	.2979	.3121	3.2041	.9546	40	1.2683
.3054	30	.3007	.3153	3.1716	.9537	30	1.2654
.3083	40	.3035	.3185	3.1397	.9528	20	1.2625
.3113	50	.3062	.3217	3.1084	.9520	10	1.2595
.3142	**18° 00′**	.3090	.3249	3.0777	.9511	**72° 00′**	1.2566
.3171	10	.3118	.3281	3.0475	.9502	50	1.2537
.3200	20	.3145	.3314	3.0178	.9492	40	1.2508
.3229	30	.3173	.3346	2.9887	.9483	30	1.2479
.3258	40	.3201	.3378	2.9600	.9474	20	1.2450
.3287	50	.3228	.3411	2.9319	.9465	10	1.2421
.3316	**19° 00′**	.3256	.3443	2.9042	.9455	**71° 00′**	1.2392
.3345	10	.3283	.3476	2.8770	.9446	50	1.2363
.3374	20	.3311	.3508	2.8502	.9436	40	1.2334
.3403	30	.3338	.3541	2.8239	.9426	30	1.2305
.3432	40	.3365	.3574	2.7980	.9417	20	1.2275
.3462	50	.3393	.3607	2.7725	.9407	10	1.2246
.3491	**20° 00′**	.3420	.3640	2.7475	.9397	**70° 00′**	1.2217
.3520	10	.3448	.3673	2.7228	.9387	50	1.2188
.3549	20	.3475	.3706	2.6985	.9377	40	1.2159
.3578	30	.3502	.3739	2.6746	.9367	30	1.2130
.3607	40	.3529	.3772	2.6511	.9356	20	1.2101
.3636	50	.3557	.3805	2.6279	.9346	10	1.2072
.3665	**21° 00′**	.3584	.3839	2.6051	.9336	**69° 00′**	1.2043
		cos t	cot t	tan t	sin t	t degrees	t

Table 3
Four
Place
Trigonometric
Functions

t	t degrees	sin t	tan t	cot t	cos t		
.3665	**21° 00'**	.3584	.3839	2.6051	.9336	**69° 00'**	1.2043
.3694	10	.3611	.3872	2.5826	.9325	50	1.2014
.3723	20	.3638	.3906	2.5605	.9315	40	1.1985
.3752	30	.3665	.3939	2.5386	.9304	30	1.1956
.3782	40	.3692	.3973	2.5172	.9293	20	1.1926
.3811	50	.3719	.4006	2.4960	.9283	10	1.1897
.3840	**22° 00'**	.3746	.4040	2.4751	.9272	**68° 00'**	1.1868
.3869	10	.3773	.4074	2.4545	.9261	50	1.1839
.3898	20	.3800	.4108	2.4342	.9250	40	1.1810
.3927	30	.3827	.4142	2.4142	.9239	30	1.1781
.3956	40	.3854	.4176	2.3945	.9228	20	1.1752
.3985	50	.3881	.4210	2.3750	.9216	10	1.1723
.4014	**23° 00'**	.3907	.4245	2.3559	.9205	**67° 00'**	1.1694
.4043	10	.3934	.4279	2.3369	.9194	50	1.1665
.4072	20	.3961	.4314	2.3183	.9182	40	1.1636
.4102	30	.3987	.4348	2.2998	.9171	30	1.1606
.4131	40	.4014	.4383	2.2817	.9159	20	1.1577
.4160	50	.4041	.4417	2.2637	.9147	10	1.1548
.4189	**24° 00'**	.4067	.4452	2.2460	.9135	**66° 00'**	1.1519
.4218	10	.4094	.4487	2.2286	.9124	50	1.1490
.4247	20	.4120	.4522	2.2113	.9112	40	1.1461
.4276	30	.4147	.4557	2.1943	.9100	30	1.1432
.4305	40	.4173	.4592	2.1775	.9088	20	1.1403
.4334	50	.4200	.4628	2.1609	.9075	10	1.1374
.4363	**25° 00'**	.4226	.4663	2.1445	.9063	**65° 00'**	1.1345
.4392	10	.4253	.4699	2.1283	.9051	50	1.1316
.4422	20	.4279	.4734	2.1123	.9038	40	1.1286
.4451	30	.4305	.4770	2.0965	.9026	30	1.1257
.4480	40	.4331	.4806	2.0809	.9013	20	1.1228
.4509	50	.4358	.4841	2.0655	.9001	10	1.1199
.4538	**26° 00'**	.4384	.4877	2.0503	.8988	**64° 00'**	1.1170
.4567	10	.4410	.4913	2.0353	.8975	50	1.1141
.4596	20	.4436	.4950	2.0204	.8962	40	1.1112
.4625	30	.4462	.4986	2.0057	.8949	30	1.1083
.4654	40	.4488	.5022	1.9912	.8936	20	1.1054
.4683	50	.4514	.5059	1.9768	.8923	10	1.1025
.4712	**27° 00'**	.4540	.5095	1.9626	.8910	**63° 00'**	1.0996
.4741	10	.4566	.5132	1.9486	.8897	50	1.0966
.4771	20	.4592	.5169	1.9347	.8884	40	1.0937
.4800	30	.4617	.5206	1.9210	.8870	30	1.0908
.4829	40	.4643	.5243	1.9074	.8857	20	1.0879
.4858	50	.4669	.5280	1.8940	.8843	10	1.0850
.4887	**28° 00'**	.4695	.5317	1.8807	.8829	**62° 00'**	1.0821
		cos t	cot t	tan t	sin t	t degrees	t

Table 3
Four
Place
Trigonometric
Functions

t	t degrees	sin t	tan t	cot t	cos t		
.4887	**28° 00′**	.4695	.5317	1.8807	.8829	**62° 00′**	1.0821
.4916	10	.4720	.5354	1.8676	.8816	50	1.0792
.4945	20	.4746	.5392	1.8546	.8802	40	1.0763
.4974	30	.4772	.5430	1.8418	.8788	30	1.0734
.5003	40	.4797	.5467	1.8291	.8774	20	1.0705
.5032	50	.4823	.5505	1.8165	.8760	10	1.0676
.5061	**29° 00′**	.4848	.5543	1.8040	.8746	**61° 00′**	1.0647
.5091	10	.4874	.5581	1.7917	.8732	50	1.0617
.5120	20	.4899	.5619	1.7796	.8718	40	1.0588
.5149	30	.4924	.5658	1.7675	.8704	30	1.0559
.5178	40	.4950	.5696	1.7556	.8689	20	1.0530
.5207	50	.4975	.5735	1.7437	.8675	10	1.0501
.5236	**30° 00′**	.5000	.5774	1.7321	.8660	**60° 00′**	1.0472
.5265	10	.5025	.5812	1.7205	.8646	50	1.0443
.5294	20	.5050	.5851	1.7090	.8631	40	1.0414
.5323	30	.5075	.5890	1.6977	.8616	30	1.0385
.5352	40	.5100	.5930	1.6864	.8601	20	1.0356
.5381	50	.5125	.5969	1.6753	.8587	10	1.0327
.5411	**31° 00′**	.5150	.6009	1.6643	.8572	**59° 00′**	1.0297
.5440	10	.5175	.6048	1.6534	.8557	50	1.0268
.5469	20	.5200	.6088	1.6426	.8542	40	1.0239
.5498	30	.5225	.6128	1.6319	.8526	30	1.0210
.5527	40	.5250	.6168	1.6212	.8511	20	1.0181
.5556	50	.5275	.6208	1.6107	.8496	10	1.0152
.5585	**32° 00′**	.5299	.6249	1.6003	.8480	**58° 00′**	1.0123
.5614	10	.5324	.6289	1.5900	.8465	50	1.0094
.5643	20	.5348	.6330	1.5798	.8450	40	1.0065
.5672	30	.5373	.6371	1.5697	.8434	30	1.0036
.5701	40	.5398	.6412	1.5597	.8418	20	1.0007
.5730	50	.5422	.6453	1.5497	.8403	10	.9977
.5760	**33° 00′**	.5446	.6494	1.5399	.8387	**57° 00′**	.9948
.5789	10	.5471	.6536	1.5301	.8371	50	.9919
.5818	20	.5495	.6577	1.5204	.8355	40	.9890
.5847	30	.5519	.6619	1.5108	.8339	30	.9861
.5876	40	.5544	.6661	1.5013	.8323	20	.9832
.5905	50	.5568	.6703	1.4919	.8307	10	.9803
5934	**34° 00′**	.5592	.6745	1.4826	.8290	**56° 00′**	.9774
.5963	10	.5616	.6787	1.4733	.8274	50	.9745
.5992	20	.5640	.6830	1.4641	.8258	40	.9716
.6021	30	.5664	.6873	1.4550	.8241	30	.9687
.6050	40	.5688	.6916	1.4460	.8225	20	.9657
.6080	50	.5712	.6959	1.4370	.8208	10	.9628
.6109	**35° 00′**	.5736	.7002	1.4281	.8192	**55° 00′**	.9599
		cos t	cot t	tan t	sin t	t degrees	t

Table 3
Four
Place
Trigonometric
Functions

t	t degrees	sin t	tan t	cot t	cos t		
.6109	35° 00'	.5736	.7002	1.4281	.8192	55° 00'	.9599
.6138	10	.5760	.7046	1.4193	.8175	50	.9570
.6167	20	.5783	.7089	1.4106	.8158	40	.9541
.6196	30	.5807	.7133	1.4019	.8141	30	.9512
.6225	40	.5831	.7177	1.3934	.8124	20	.9483
.6254	50	.5854	.7221	1.3848	.8107	10	.9454
.6283	36° 00'	.5878	.7265	1.3764	.8090	54° 00'	.9425
.6312	10	.5901	.7310	1.3680	.8073	50	.9396
.6341	20	.5925	.7355	1.3597	.8056	40	.9367
.6370	30	.5948	.7400	1.3514	.8039	30	.9338
.6400	40	.5972	.7445	1.3432	.8021	20	.9308
.6429	50	.5995	.7490	1.3351	.8004	10	.9279
.6458	37° 00'	.6018	.7536	1.3270	.7986	53° 00'	.9250
.6487	10	.6041	.7581	1.3190	.7969	50	.9221
.6516	20	.6065	.7627	1.3111	.7951	40	.9192
.6545	30	.6088	.7673	1.3032	.7934	30	.9163
.6574	40	.6111	.7720	1.2954	.7916	20	.9134
.6603	50	.6134	.7766	1.2876	.7898	10	.9105
.6632	38° 00'	.6157	.7813	1.2799	.7880	52° 00'	.9076
.6661	10	.6180	.7860	1.2723	.7862	50	.9047
.6690	20	.6202	.7907	1.2647	.7844	40	.9018
.6720	30	.6225	.7954	1.2572	.7826	30	.8988
.6749	40	.6248	.8002	1.2497	.7808	20	.8959
.6778	50	.6271	.8050	1.2423	.7790	10	.8930
.6807	39° 00'	.6293	.8098	1.2349	.7771	51° 00'	.8901
.6836	10	.6316	.8146	1.2276	.7753	50	.8872
.6865	20	.6338	.8195	1.2203	.7735	40	.8843
.6894	30	.6361	.8243	1.2131	.7716	30	.8814
.6923	40	.6383	.8292	1.2059	.7698	20	.8785
.6952	50	.6406	.8342	1.1988	.7679	10	.8756
.6981	40° 00'	.6428	.8391	1.1918	.7660	50° 00'	.8727
.7010	10	.6450	.8441	1.1847	.7642	50	.8698
.7039	20	.6472	.8491	1.1778	.7623	40	.8668
.7069	30	.6494	.8541	1.1708	.7604	30	.8639
.7098	40	.6517	.8591	1.1640	.7585	20	.8610
.7127	50	.6539	.8642	1.1571	.7566	10	.8581
.7156	41° 00'	.6561	.8693	1.1504	.7547	49° 00'	.8552
.7185	10	.6583	.8744	1.1436	.7528	50	.8523
.7214	20	.6604	.8796	1.1369	.7509	40	.8494
.7243	30	.6626	.8847	1.1303	.7490	30	.8465
.7272	40	.6648	.8899	1.1237	.7470	20	.8436
.7301	50	.6670	.8952	1.1171	.7451	10	.8407
.7330	42° 00'	.6691	.9004	1.1106	.7431	48° 00'	.8378
		cos t	cot t	tan t	sin t	t degrees	t

493

Table 3
Four
Place
Trigonometric
Functions

t	t degrees	sin t	tan t	cot t	cos t		
.7330	**42° 00′**	.6691	.9004	1.1106	.7431	**48° 00′**	.8378
.7359	10	.6713	.9057	1.1041	.7412	50	.8348
.7389	20	.6734	.9110	1.0977	.7392	40	.8319
.7418	30	.6756	.9163	1.0913	.7373	30	.8290
.7447	40	.6777	.9217	1.0850	.7353	20	.8261
.7476	50	.6799	.9271	1.0786	.7333	10	.8232
.7505	**43° 00′**	.6820	.9325	1.0724	.7314	**47° 00′**	.8203
.7534	10	.6841	.9380	1.0661	.7294	50	.8174
.7563	20	.6862	.9435	1.0599	.7274	40	.8145
.7592	30	.6884	.9490	1.0538	.7254	30	.8116
.7621	40	.6905	.9545	1.0477	.7234	20	.8087
.7650	50	.6926	.9601	1.0416	.7214	10	.8058
.7679	**44° 00′**	.6947	.9657	1.0355	.7193	**46° 00′**	.8029
.7709	10	.6967	.9713	1.0295	.7173	50	.7999
.7738	20	.6988	.9770	1.0235	.7153	40	.7970
.7767	30	.7009	.9827	1.0176	.7133	30	.7941
.7796	40	.7030	.9884	1.0117	.7112	20	.7912
.7825	50	.7050	.9942	1.0058	.7092	10	.7883
.7854	**45° 00′**	.7071	1.0000	1.0000	.7071	**45° 00′**	.7854
		cos t	cot t	tan t	sin t	t degrees	t

Table 4
Logarithms
of
Trigonometric
Functions

angles	log sin	log cos	log tan	log cot	
0° 00′		10.0000			**90° 00′**
10	7.4637	.0000	7.4637	12.5363	50
20	.7648	.0000	.7648	.2352	40
30	7.9408	.0000	7.9409	12.0591	30
40	8.0658	.0000	8.0658	11.9342	20
50	.1627	10.0000	.1627	.8373	10
1° 00′	8.2419	9.9999	8.2419	11.7581	**89° 00′**
10	.3088	.9999	.3089	.6911	50
20	.3668	.9999	.3669	.6331	40
30	.4179	.9999	.4181	.5819	30
40	.4637	.9998	.4638	.5362	20
50	.5050	.9998	.5053	.4947	10
2° 00′	8.5428	9.9997	8.5431	11.4569	**88° 00′**
10	.5776	.9997	.5779	.4221	50
20	.6097	.9996	.6101	.3899	40
30	.6397	.9996	.6401	.3599	30
40	.6677	.9995	.6682	.3318	20
50	.6940	.9995	.6945	.3055	10
3° 00′	8.7188	9.9994	8.7194	11.2806	**87° 00′**
10	.7423	.9993	.7429	.2571	50
20	.7645	.9993	.7652	.2348	40
30	.7857	.9992	.7865	.2135	30
40	.8059	.9991	.8067	.1933	20
50	.8251	.9990	.8261	.1739	10
4° 00′	8.8436	9.9989	8.8446	11.1554	**86° 00′**
10	.8613	.9989	.8624	.1376	50
20	.8783	.9988	.8795	.1205	40
30	.8946	.9987	.8960	.1040	30
40	.9104	.9986	.9118	.0882	20
50	.9256	.9985	.9272	.0728	10
5° 00′	8.9403	9.9983	8.9420	11.0580	**85° 00′**
10	.9545	.9982	.9563	.0437	50
20	.9682	.9981	.9701	.0299	40
30	.9816	.9980	.9836	.0164	30
40	8.9945	.9979	8.9966	11.0034	20
50	9.0070	.9977	9.0093	10.9907	10
6° 00′	9.0192	9.9976	9.0216	10.9784	**84° 00′**
10	.0311	.9975	.0336	.9664	50
20	.0426	.9973	.0453	.9547	40
30	.0539	.9972	.0567	.9433	30
40	.0648	.9971	.0678	.9322	20
50	.0755	.9969	.0786	.9214	10
7° 00′	9.0859	9.9968	9.0891	10.9109	**83° 00′**
	log cos	log sin	log cot	log tan	angles

* Add −10 to all logarithms.

495

Table 4
Logarithms
of
Trigonometric
Functions

angles	log sin	log cos	log tan	log cot	
7° 00′	9.0859	9.9968	9.0891	10.9109	**83° 00′**
10	.0961	.9966	.0995	.9005	50
20	.1060	.9964	.1096	.8904	40
30	.1157	.9963	.1194	.8806	30
40	.1252	.9961	.1291	.8709	20
50	.1345	.9959	.1385	.8615	10
8° 00′	9.1436	9.9958	9.1478	10.8522	**82° 00′**
10	.1525	.9956	.1569	.8431	50
20	.1612	.9954	.1658	.8342	40
30	.1697	.9952	.1745	.8255	30
40	.1781	.9950	.1831	.8169	20
50	.1863	.9948	.1915	.8085	10
9° 00′	9.1943	9.9946	9.1997	10.8003	**81° 00′**
10	.2022	.9944	.2078	.7922	50
20	.2100	.9942	.2158	.7842	40
30	.2176	.9940	.2236	.7764	30
40	.2251	.9938	.2313	.7687	20
50	.2324	.9936	.2389	.7611	10
10° 00′	9.2397	9.9934	9.2463	10.7537	**80° 00′**
10	.2468	.9931	.2536	.7464	50
20	.2538	.9929	.2609	.7391	40
30	.2606	.9927	.2680	.7320	30
40	.2674	.9924	.2750	.7250	20
50	.2740	.9922	.2819	.7181	10
11° 00′	9.2806	9.9919	9.2887	10.7113	**79° 00′**
10	.2870	.9917	.2953	.7047	50
20	.2934	.9914	.3020	.6980	40
30	.2997	.9912	.3085	.6915	30
40	.3058	.9909	.3149	.6851	20
50	.3119	.9907	.3212	.6788	10
12° 00′	9.3179	9.9904	9.3275	10.6725	**78° 00′**
10	.3238	.9901	.3336	.6664	50
20	.3296	.9899	.3397	.6603	40
30	.3353	.9896	.3458	.6542	30
40	.3410	.9893	.3517	.6483	20
50	.3466	.9890	.3576	.6424	10
13° 00′	9.3521	9.9887	9.3634	10.6366	**77° 00′**
10	.3575	.9884	.3691	.6309	50
20	.3629	.9881	.3748	.6252	40
30	.3682	.9878	.3804	.6196	30
40	.3734	.9875	.3859	.6141	20
50	.3786	.9872	.3914	.6086	10
14° 00′	9.3837	9.9869	9.3968	10.6032	**76° 00′**
	log cos	log sin	log cot	log tan	angles

Table 4
Logarithms
of
Trigonometric
Functions

angles	log sin	log cos	log tan	log cot	
14° 00′	9.3837	9.9869	9.3968	10.6032	**76° 00′**
10	.3887	.9866	.4021	.5979	50
20	.3937	.9863	.4074	.5926	40
30	.3986	.9859	.4127	.5873	30
40	.4035	.9856	.4178	.5822	20
50	.4083	.9853	.4230	.5770	10
15° 00′	9.4130	9.9849	9.4281	10.5719	**75° 00′**
10	.4177	.9846	.4331	.5669	50
20	.4223	.9843	.4381	.5619	40
30	.4269	.9839	.4430	.5570	30
40	.4314	.9836	.4479	.5521	20
50	.4359	.9832	.4527	.5473	10
16° 00′	9.4403	9.9828	9.4575	10.5425	**74° 00′**
10	.4447	.9825	.4622	.5378	50
20	.4491	.9821	.4669	.5331	40
30	.4533	.9817	.4716	.5284	30
40	.4576	.9814	.4762	.5238	20
50	.4618	.9810	.4808	.5192	10
17° 00′	9.4659	9.9806	9.4853	10.5147	**73° 00′**
10	.4700	.9802	.4898	.5102	50
20	.4741	.9798	.4943	.5057	40
30	.4781	.9794	.4987	.5013	30
40	.4821	.9790	.5031	.4969	20
50	.4861	.9786	.5075	.4925	10
18° 00′	9.4900	9.9782	9.5118	10.4882	**72° 00′**
10	.4939	.9778	.5161	.4839	50
20	.4977	.9774	.5203	.4797	40
30	.5015	.9770	.5245	.4755	30
40	.5052	.9765	.5287	.4713	20
50	.5090	.9761	.5329	.4671	10
19° 00′	9.5126	9.9757	9.5370	10.4630	**71° 00′**
10	.5163	.9752	.5411	.4589	50
20	.5199	.9748	.5451	.4549	40
30	.5235	.9743	.5491	.4509	30
40	.5270	.9739	.5531	.4469	20
50	.5306	.9734	.5571	.4429	10
20° 00′	9.5341	9.9730	9.5611	10.4389	**70° 00′**
10	.5375	.9725	.5650	.4350	50
20	.5409	.9721	.5689	.4311	40
30	.5443	.9716	.5727	.4273	30
40	.5477	.9711	.5766	.4234	20
50	.5510	.9706	.5804	.4196	10
21° 00′	9.5543	9.9702	9.5842	10.4158	**69° 00′**
	log cos	log sin	log cot	log tan	angles

Table 4
Logarithms
of
Trigonometric
Functions

angles	log sin	log cos	log tan	log cot	
21° 00′	9.5543	9.9702	9.5842	10.4158	**69° 00′**
10	.5576	.9697	.5879	.4121	50
20	.5609	.9692	.5917	.4083	40
30	.5641	.9687	.5954	.4046	30
40	.5673	.9682	.5991	.4009	20
50	.5704	.9677	.6028	.3972	10
22° 00′	9.5736	9.9672	9.6064	10.3936	**68° 00′**
10	.5767	.9667	.6100	.3900	50
20	.5798	.9661	.6136	.3864	40
30	.5828	.9656	.6172	.3828	30
40	.5859	.9651	.6208	.3792	20
50	.5889	.9646	.6243	.3757	10
23° 00′	9.5919	9.9640	9.6279	10.3721	**67° 00′**
10	.5948	.9635	.6314	.3686	50
20	.5978	.9629	.6348	.3652	40
30	.6007	.9624	.6383	.3617	30
40	.6036	.9618	.6417	.3583	20
50	.6065	.9613	.6452	.3548	10
24° 00′	9.6093	9.9607	9.6486	10.3514	**66° 00′**
10	.6121	.9602	.6520	.3480	50
20	.6149	.9596	.6553	.3447	40
30	.6177	.9590	.6587	.3413	30
40	.6205	.9584	.6620	.3380	20
50	.6232	.9579	.6654	.3346	10
25° 00′	9.6259	9.9573	9.6687	10.3313	**65° 00′**
10	.6286	.9567	.6720	.3280	50
20	.6313	.9561	.6752	.3248	40
30	.6340	.9555	.6785	.3215	30
40	.6366	.9549	.6817	.3183	20
50	.6392	.9543	.6850	.3150	10
26° 00′	9.6418	9.9537	9.6882	10.3118	**64° 00′**
10	.6444	.9530	.6914	.3086	50
20	.6470	.9524	.6946	.3054	40
30	.6495	.9518	.6977	.3023	30
40	.6521	.9512	.7009	.2991	20
50	.6546	.9505	.7040	.2960	10
27° 00′	9.6570	9.9499	9.7072	10.2928	**63° 00′**
10	.6595	.9492	.7103	.2897	50
20	.6620	.9486	.7134	.2866	40
30	.6644	.9479	.7165	.2835	30
40	.6668	.9473	.7196	.2804	20
50	.6692	.9466	.7226	.2774	10
28° 00′	9.6716	9.9459	9.7257	10.2743	**62° 00′**
	log cos	log sin	log cot	log tan	angles

Table 4
Logarithms
of
Trigonometric
Functions

angles	log sin	log cos	log tan	log cot	
28° 00′	9.6716	9.9459	9.7257	10.2743	**62° 00′**
10	.6740	.9453	.7287	.2713	50
20	.6763	.9446	.7317	.2683	40
30	.6787	.9439	.7348	.2652	30
40	.6810	.9432	.7378	.2622	20
50	.6833	.9425	.7408	.2592	10
29° 00′	9.6856	9.9418	9.7438	10.2562	**61° 00′**
10	.6878	.9411	.7467	.2533	50
20	.6901	.9404	.7497	.2503	40
30	.6923	.9397	.7526	.2474	30
40	.6946	.9390	.7556	.2444	20
50	.6968	.9383	.7585	.2415	10
30° 00′	9.6990	9.9375	9.7614	10.2386	**60° 00′**
10	.7012	.9368	.7644	.2356	50
20	.7033	.9361	.7673	.2327	40
30	.7055	.9353	.7701	.2299	30
40	.7076	.9346	.7730	.2270	20
50	.7097	.9338	.7759	.2241	10
31° 00′	9.7118	9.9331	9.7788	10.2212	**59° 00′**
10	.7139	.9323	.7816	.2184	50
20	.7160	.9315	.7845	.2155	40
30	.7181	.9308	.7873	.2127	30
40	.7201	.9300	.7902	.2098	20
50	.7222	.9292	.7930	.2070	10
32° 00′	9.7242	9.9284	9.7958	10.2042	**58° 00′**
10	.7262	.9276	.7986	.2014	50
20	.7282	.9268	.8014	.1986	40
30	.7302	.9260	.8042	.1958	30
40	.7322	.9252	.8070	.1930	20
50	.7342	.9244	.8097	.1903	10
33° 00′	9.7361	9.9236	9.8125	10.1875	**57°00′**
10	.7380	.9228	.8153	.1847	50
20	.7400	.9219	.8180	.1820	40
30	.7419	.9211	.8208	.1792	30
40	.7438	.9203	.8235	.1765	20
50	.7457	.9194	.8263	.1737	10
34° 00′	9.7476	9.9186	9.8290	10.1710	**56° 00′**
10	.7494	.9177	.8317	.1683	50
20	.7513	.9169	.8344	.1656	40
30	.7531	.9160	.8371	.1629	30
40	.7550	.9151	.8398	.1602	20
50	.7568	.9142	.8425	.1575	10
35° 00′	9.7586	9.9134	9.8452	10.1548	**55° 00′**
	log cos	log sin	log cot	log tan	angles

Table 4
Logarithms of Trigonometric Functions

angles	log sin	log cos	log tan	log cot	
35° 00′	9.7586	9.9134	9.8452	10.1548	**55° 00′**
10	.7604	.9125	.8479	.1521	50
20	.7622	.9116	.8506	.1494	40
30	.7640	.9107	.8533	.1467	30
40	.7657	.9098	.8559	.1441	20
50	.7675	.9089	.8586	.1414	10
36° 00′	9.7692	9.9080	9.8613	10.1387	**54° 00′**
10	.7710	.9070	.8639	.1361	50
20	.7727	.9061	.8666	.1334	40
30	.7744	.9052	.8692	.1308	30
40	.7761	.9042	.8718	.1282	20
50	.7778	.9033	.8745	.1255	10
37° 00′	9.7795	9.9023	9.8771	10.1229	**53° 00′**
10	.7811	.9014	.8797	.1203	50
20	.7828	.9004	.8824	.1176	40
30	.7844	.8995	.8850	.1150	30
40	.7861	.8985	.8876	.1124	20
50	.7877	.8975	.8902	.1098	10
38° 00′	9.7893	9.8965	9.8928	10.1072	**52° 00′**
10	.7910	.8955	.8954	.1046	50
20	.7926	.8945	.8980	.1020	40
30	.7941	.8935	.9006	.0994	30
40	.7957	.8925	.9032	.0968	20
50	.7973	.8915	.9058	.0942	10
39° 00′	9.7989	9.8905	9.9084	10.0916	**51° 00′**
10	.8004	.8895	.9110	.0890	50
20	.8020	.8884	.9135	.0865	40
30	.8035	.8874	.9161	.0839	30
40	.8050	.8864	.9187	.0813	20
50	.8066	.8853	.9212	.0788	10
40° 00′	9.8081	9.8843	9.9238	10.0762	**50° 00′**
10	.8096	.8832	.9264	.0736	50
20	.8111	.8821	.9289	.0711	40
30	.8125	.8810	.9315	.0685	30
40	.8140	.8800	.9341	.0659	20
50	.8155	.8789	.9366	.0634	10
41° 00′	9.8169	9.8778	9.9392	10.0608	**49° 00′**
10	.8184	.8767	.9417	.0583	50
20	.8198	.8756	.9443	.0557	40
30	.8213	.8745	.9468	.0532	30
40	.8227	.8733	.9494	.0506	20
50	.8241	.8722	.9519	.0481	10
42° 00′	9.8255	9.8711	9.9544	10.0456	**48° 00′**
	log cos	log sin	log cot	log tan	angles

Table 4
Logarithms
of
Trigonometric
Functions

angles	log sin	log cos	log tan	log cot	
42° 00′	9.8255	9.8711	9.9544	10.0456	**48° 00′**
10	.8269	.8699	.9570	.0430	50
20	.8283	.8688	.9595	.0405	40
30	.8297	.8676	.9621	.0379	30
40	.8311	.8665	.9646	.0354	20
50	.8324	.8653	.9671	.0329	10
43° 00′	9.8338	9.8641	9.9697	10.0303	**47° 00′**
10	.8351	.8629	.9722	.0278	50
20	.8365	.8618	.9747	.0253	40
30	.8378	.8606	.9772	.0228	30
40	.8391	.8594	.9798	.0202	20
50	.8405	.8582	.9823	.0177	10
44° 00′	9.8418	9.8569	9.9848	10.0152	**46° 00′**
10	.8431	.8557	.9874	.0126	50
20	.8444	.8545	.9899	.0101	40
30	.8457	.8532	.9924	.0076	30
40	.8469	.8520	.9949	.0051	20
50	.8482	.8507	.9975	.0025	10
45° 00′	9.8495	9.8495	10.0000	10.0000	**45° 00′**
	log cos	log sin	log cot	log tan	angles

Answers to Odd-Numbered Exercises

CHAPTER ONE

Section 1 (page 5)

1 (a) $\{u, v, w, x, y, z\}$. (b) $\varnothing$. (c) $\{10n : n \in \mathbf{N}\}$
3 (a) True. (b) False, $2 \notin \{1, 3\}$. (c) True. (d) True.
   (e) False, $\{2\} \subseteq \{1, 2\}$. (f) True. (g) False, $2 \in \{1, 2\}$.
   (h) False, $\{2\} \subseteq \{2\}$.
5 $\{a, b, c, d\}, \{a, b, c\}, \{a, b, d\}, \{b, c, d\}, \{a, c, d\}, \{a, b\}, \{a, c\}, \{a, d\}, \{b, c\},$
   $\{b, d\}, \{c, d\}, \{a\}, \{b\}, \{c\}, \{d\}, \varnothing$.
7 (a) $\{1, 2, 3, 5, 6\}$; $\{3\}$. (b) $\{a, b, c, d, e\}$; $\{a, d\}$. (c) $\{a, b, c, d, e, f\}$;
   $\varnothing$. (d) $\{1, 2, 3\}$; $\{1, 2\}$.
9 (a) $\{6\}$. (b) $\{6, 2\}$. (c) $\{1, 2, 5, 6\}$. (d) $\{1, 2, 5, 6\}$.
11 (a) $T \subseteq S$. (b) $S \subseteq T$. (c) $S = \varnothing$. (d) Any S. (e) S and T
   disjoint. (f) $S = T = \varnothing$. (g) Any S and T. (h) Any S and T.

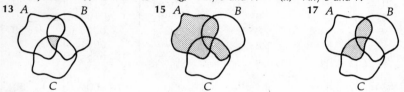

13 A B 15 A B 17 A B
    C C C

19 Same answer as Exercise 17.

Section 2 (page 13)

1 Commutative law. 3 Associative law. 5 Identity element.
7 Inverse element. 9 Identity element. 11 (a) $>$. (b) $<$.
   (c) $=$. (d) $>$. (e) $=$. (f) $<$.
13 $a > 0$. 15 $-3 < a < 4$. 17 $a > -6$. 19 $a \geq 0$.
21 (3) of (1.11). 23 (4) of (1.12). 25 (2) of (1.12).
27 $2 - 3 \neq 3 - 2$; $2 - (3 - 4) \neq (2 - 3) - 4$. 29 Multiply both sides by -1.
31 If $a + c = b + c$, add $-c$ to both sides to obtain $a = b$. If $ac = bc$ and $c \neq 0$, multiply
   both sides by $1/c$ to obtain $a = b$.
33 If $a + c < b + c$, add $-c$ to both sides to obtain $a < b$. If $ac < bc$ and c is positive,
   then $a < b$. If $ac < bc$ and c is negative, then $a > b$.
35 If $a < b$ where a and b are both positive, multiply both sides by $1/ab$ to obtain
   $1/b < 1/a$. Conversely, if $1/a > 1/b$ where a and b are both positive, multiply by
   ab to obtain $b > a$. The result is false if, for example, a is negative and b is positive.
   Thus, $-2 < 3$ and $1/(-2) < 1/3$.
37 If $a > b$ and $c < 0$, then $a - b$ and $-c$ are both positive. Hence the product $(a - b)(-c)$
   is positive, that is, $bc - ac > 0$. Using (1.10) this means that $ac < bc$.

Section 3 (page 20)

1 (a) 3. (b) 7. (c) 7. (d) 3. (e) $\frac{22}{7} - \pi$. (f) -1. (g) 0.
   (h) 9.
3 (a) 4. (b) 8. (c) 8. (d) 12.
5 (a) 4. (b) 8. (c) -8. (d) 12.
7 (a) -3. (b) 3. (c) 1.
9 Let p, q, and r be the coordinates of P, Q, and R, respectively. Then $\overline{PQ} + \overline{QR} =$
   $(q - p) + (r - q) = r - p = \overline{PR}$.
17 $-3 < a < 3$. 19 $-13 < a < 5$.
21 Suppose $|a| = b$. Case 1. If $a > 0$, then $|a| = a$ and hence $a = b$. Case 2. If $a < 0$,

then $|a| = -a$ and hence $-a = b$, or $a = -b$. Conversely, suppose $a = b$ or $a = -b$, where $b > 0$. Case 1. If $a > 0$, then $|a| = a = b$. Case 2. If $a < 0$, then $|a| = -a$. In this case we must have $a = -b$ since $b > 0$. Hence $|a| = -a = -(-b) = b$.

Section 4   (page 25)

1 $16/81$.   **3** $9/8$.   **5** $-71/9$.   **7** $-1/108$.   **9** 1.   **11** $8x^9$.
13 $6/x$.   **15** $-2a^{14}$.   **17** $9/2$.   **19** $12u^{11}/v^2$.   **21** $4/xy$.
23 $9y^6/x^8$.   **25** $81y^6/64$.   **27** $s^6/(4r^8)$.   **29** $20y/x^3$.   **31** $9x^{10}y^{14}$.
33 $-4a^4c^2/b^9$.   **35** $2a^5/5b^6c^{11}$.   **37** 1.   **39** a^4b^8.   **41** 0.
43 $b - a$.   **45** $(a + b)^2/ab$.   **47** $-1/x^2y^2$.   **49** x^{3-n}.

Section 5   (page 31)

1 5.   **3** -4.   **5** 2.   **7** $(\frac{1}{3})\sqrt[3]{9}$.   **9** $2a^2/b^4$.   **11** $2x/y^2$.
13 y^3/x^2.   **15** $(5x^3y^5/z)\sqrt{2}$.   **17** $(1/2xy^2)\sqrt{2xy}$.   **19** $2u^2v^2\sqrt{3}$.
21 $3a^5c^3/b$.   **23** $-2x^2y$.   **25** $-3x\sqrt[3]{y}/y^2$.   **27** $(1/a)\sqrt{a}$.
29 $(2a/3b)\sqrt[3]{18ab^2}$.   **31** x^7z/y^2.   **33** $2x^2y^2\sqrt{3x}$.   **35** $(-5y/x^2)\sqrt[3]{2y}$.
37 $xy\sqrt{y}$.   **39** $\sqrt{x}/x^2$.   **41** $(a - \sqrt{b})/(a^2 - b)$.   **43** $(x - \sqrt{x})/(x - 1)$.
45 $|x - y|$.   **47** $2\sqrt{5}$.   **49** $(2x^2 - x + 4)\sqrt{x}$.

Section 6   (page 35)

1 $x^{4/3}$.   **3** $(a + 1)^{3/2}$.   **5** $(x^2 + y^2)^{1/3}$.   **7** (a) $8\sqrt[3]{a^2}$.   (b) $4\sqrt[3]{a^2}$.
(c) $8 + \sqrt[3]{a^2}$.   **9** $\frac{1}{8}$.   **11** 0.04.   **13** $8a^2$.   **15** $24x^{3/2}$.
17 $1/(9a^4)$.   **19** $8/x^{1/2}$.   **21** $4x^2y^4$.   **23** $3/x^3y^2$.   **25** 1.
27 $a^{5/6}/b^3$.   **29** x.   **31** $4u^4/b^2$.   **33** $x^2 + x + 1$.   **35** $\sqrt[6]{108}$.
37 $\sqrt[12]{x^{11}y^7}$.   **39** $\sqrt[6]{b}/a$.   **41** $\sqrt[4]{x^3}$.   **43** $(1.7)10^{-24}$.
45 (a) $(6.43)10^4$.   (b) $(1.2)10^{-10}$.   (c) $(3.42)10^7$.   **47** (a) $44{,}000$.
(b) $.0000000000000000000000027$.   (c) 76800000000.   **49** $(2.4)10^{39}$.
51 $(\frac{1}{3})10^{-22}$.   **53** $(1.6)10^{10}$ miles; $(5.8)10^{12}$ miles.

Section 7   (page 41)

1 $x^4 + 8x^3 - 2x - 2$; 4.   **3** $-x^3 + x - 8$; 3.   **5** $x^4 - 2x^3 - 2x^2 + 8x + 3$; 4.
7 $5x^2 - 10x$; 2.   **9** $8z^5 + 6z^4 - 25z^3 - 4z^2 + 17z - 10$; 5.
11 $6c^3 - 5c^2d + 3cd^2 - d^3$; 3.   **13** $6y^5 + 4y^4 - 11y^3 + 13y^2 + 14y - 20$; 5.
15 $a^3 + b^3$.   **17** $3s^2 + 2r$; 2.   **19** $5x - 4y^2$; 2.   **21** $5r - (\frac{3}{4})r^2s^2 + (\frac{3}{2})s$; 4.
23 $x^2 - 3x - 28$.   **25** $6x^2 - (1/6)y^2$.   **27** $4x^2 - 12xy + 9y^2$.
29 $6x^2 + 11x - 35$.   **31** $80x^2 - 22xy - 3y^2$.   **33** $36x^2 - 132xy + 121y^2$.
35 $4/x^4 + 2 + x^4/4$.   **37** $9a^2 - 4b^4$.   **39** $u - v$.
41 $8r^3 + 12r^2s + 6rs^2 + s^3$.   **43** $x^6 - 6x^3 + 12 - 8/x^3$.   **45** $4x - 16y^4$.
47 $x + 2x^{3/2} + 3x^2 + 2x^{5/2} + x^3$.

Section 8   (page 46)

1 $a(2b + c)$.   **3** $y(2xz + x + 5z)$.   **5** $(3x - 2)(x + 4)$.
7 $(2x + 5)(6x - 3)$.   **9** $(7m - 6n)(7m + 6n)$.   **11** $x(y + z)$.
13 $u(x + y + 3z)$.   **15** $(2x + 1)(x - 5)$.   **17** $(6x + 1)(2x + 5)$.
19 $(3 - 5y)(3 + 5y)$.   **21** $(4r^2 + 9s^2)(2r - 3s)(2r + 3s)$.
23 $(2a + 3b)(4a^2 - 6ab + 9b^2)$.   **25** $x^2y(3y - x)^2$.   **27** $z(3z^2 + 4)^2$.
29 $(a - b)(2c + 3d)$.   **31** $(x^2 + 5)(3x + 2)$.   **33** $a^2b^2(a^2 + b^2)(a + b)(a - b)$.
35 $(y - 2x)(y^2 + 2xy + 4x^2)$.   **37** $xy^2(2x - 3)^2$.   **39** Prime.
41 $(z + 4)(z + 6)$.   **43** $(y - m)(2k + 3n)$.
45 $(r + s)(r - s)(r^2 - rs + s^2)(r^2 + rs + s^2)$.   **47** $(a - b)(c - d)$.   **49** Prime.
51 $(a^2 + 3b^2 - 2ab)(a^2 + 3b^2 + 2ab)$.   **53** $(x^2 + 6y^2 - 4xy)(x^2 + 6y^2 + 4xy)$.

Section 9   (page 50)

1 $(2x + 3)/(3x + 1)$.   **3** $(x^2 + xy + y^2)/(x + y)$.   **5** $\dfrac{3x + 4}{2x - 1}$.   **7** $\dfrac{a^2 + 3}{3 - a^2}$.

503

9 $\dfrac{17 - 6x}{(2x - 1)(x + 3)}.$ **11** $\dfrac{4(a - 3)}{3a + 4}.$ **13** $(6x - 11)/(3x + 1)(7x - 2).$

15 $(y + 5)/(2y - 3).$ **17** $(9x^2 + 21x + 10)/2(x + 3)(x - 3).$

19 $-3/(3x + 3h - 5)(3x - 5).$ **21** $\dfrac{6x^2 + 9x + 2}{(3x + 1)(3x - 1)}.$ **23** $-x/(x + 1)^2.$

25 $(x^2 - xy + y^2)/(x^2 + xy + y^2).$ **27** $(6a^2 + 9a + 2)/(3a + 1)(3a - 1).$

29 $y^2/(x^2 + y^2).$

Review Exercises   (page 51)

1 $A \cup B = \{u, v, w, x, y, z\},$ $A \cap B = \{y, z\}.$

3 $\{u\}, \{v\}, \{w\}, \{u, v\}, \{u, w\}, \{v, w\}, \{u, v, w\}, \emptyset.$

5 4. **7** 0. **9** $a \le 0.$ **11** $a \ge 1.$ **13** $a > 8/3.$

15 $a > 2.$ **17** $-6 < a < 6.$ **19** $-2 < a < 3.$

23 **(a)** 8. **(b)** $-8.$ **(c)** 8. **(d)** $-12.$ **25** $16z^5x^5.$ **27** $8a^{16}/b^8.$

29 $u^{2/3} - v^{2/3}.$ **31** $1/(a + b).$ **33** $b^2/a^4.$ **35** $(xy/z)\sqrt{xz}.$

37 $a\sqrt{a}/2b.$ **39** $(2a - 1)\sqrt{a}.$ **41** $\sqrt{5}/5.$ **43** $-x^3 - 3x^2 + 6x - 9.$

45 $y^3 - 8y^2 + 19y - 20.$ **47** $-8x^2 + 8x.$ **49** $25a^2 - 20ab + 4b^2.$

51 $8x^3 - 12x^2y + 6xy^2 - y^3.$ **53** $p - q.$ **55** $(4x - 1)(2x + 3).$

57 $x^2y(x + y)(x - y).$ **59** $(x^2 + 1)(x - 1).$ **61** $(4x - 1)/(2x - 3).$

63 $(a - b)/ab.$ **65** $(a - 1)/(a + 1).$

CHAPTER 2

Section 1   (page 58)

1 $\{7/3\}.$ **3** $\{-100/3\}.$ **5** $\{-37/8\}.$ **7** $\{8/5\}.$ **9** $\{-13/6\}.$

11 $\{-1\}.$ **13** $\{-5/4\}.$ **15** $\{-40/9\}.$ **17** $\{5/3\}.$ **19** $\{-1/3\}.$

21 $\{3\}.$ **23** $\emptyset.$ **25** All nonzero real numbers. **27** $\{1/6\}.$

29 $-9.$

31 All numbers a and b such that $3a = -2b.$ $a = 0$ and $b \ne 0.$ $a = 0$ and $b = 0.$

33 **(a)** $x + 1 = x.$ **(b)** $\{1\}.$

Section 2   (page 64)

1 $g = 2s/t^2.$ **3** $b_2 = (2A/h) - b_1.$ **5** $R_2 = RR_1R_3/(R_1R_3 - RR_1 - RR_3).$

7 $r = (nE - IR)/(In);$ $n = IR/(E - Ir).$ **9** $k = 1 + (S - a)/d.$

11 $p = A/(1 + rt);$ $t = (A - p)/pr.$ **13** 76, 78, 80. **15** 53.

17 17 nickels, 8 dimes. **19** 70 miles per hour. **21** 5/8 hour; 6 2/13 miles.

23 150 miles. **25** 1237.5 feet. **27** 40 lbs. **29** $d(b - a)/b$ oz.

31 12/5 liters.

Section 3   (page 71)

1 $\{-2, \frac{5}{3}\}.$ **3** $\{\frac{2}{3}\}.$ **5** $\{0, \frac{3}{2}, -1\}.$ **7** $\{1, -1, 2, -2\}.$

9 $\{5, -5, -\frac{3}{2}\}.$ **11** $\{-\frac{1}{4}, 3\}.$ **13** $\{\frac{4}{3}\}.$ **15** $\{(-1 \pm \sqrt{41})/4\}.$

17 $\{(-9 \pm \sqrt{85})/2\}.$ **19** $\{1, \frac{3}{2}\}.$ **21** $\{-3/2, 5\}.$ **23** $\{1/4\}.$

25 $\{0, -4/3\}.$ **27** 10, 2. **29** $-16, -14$ or 14, 16. **31** $6 + \sqrt{14}, 6 - \sqrt{14}.$

33 At $t = 20 - 10\sqrt{3}$ and $t = 20 + 10\sqrt{3};$ $t = 40.$ **35** $-\frac{1}{2}.$

39 **(a)** $x = (y \pm \sqrt{3y^2 - 8})/2.$ **(b)** $y = -x \pm \sqrt{3x^2 + 4}.$

Section 4   (page 76)   -

1 $\{10\}.$ **3** $\{\pm\sqrt{41}\}.$ **5** $\{2, \frac{3}{2}\}.$ **7** $\{6\}.$ **9** $\{-1\}.$

11 $\{\pm\sqrt{3}/2, \pm3\}.$ **13** $\{1\}.$ **15** $\{16\}.$ **17** $\{16/81\}.$

19 $\{1, -\frac{1}{2}\}.$ **21** $\{2, (9 \pm 3\sqrt{5})/2\}.$ **23** $\{-3, \frac{3}{2}\}.$ **25** $\{2/5, 2\}$

27 $\{5, -\frac{11}{7}\}.$ **29** $\emptyset.$

Section 5   (page 82)

Section 5   (page 82)

Interval Solutions:

1 $(5, \infty)$.   **3** $(-\infty, 1/10)$.   **5** $[-7/3, \infty)$.   **7** $(9/11, \infty)$.
9 $(-4/5, 3]$.   **11** $[-3, 1)$.   **13** $[-1, 2]$.   **15** $(7/2, \infty)$.
17 **R**.   **19** $(-\infty, -3/2)$.   **21** $(9.7, 10.3)$.   **23** $[5/3, 3]$.
25 $(-\infty, 1/25) \cup (3/5, \infty)$.   **27** $\varnothing$.   **29** $(-\infty, -2] \cup [10/3, \infty)$.
31 $(-1, 1/8)$.   **33** $[-2/3, 7/2)$.   **35** $(-\infty, -5/2)$.

Section 6   (page 87)

Interval Solutions:

1 $(-2, 1/3)$.   **3** $[0, 13]$.   **5** $(-\infty, -4) \cup (-1/2, \infty)$.
7 $(-\infty, 5) \cup (5, \infty)$.   **9** $(-\infty, 0) \cup (1, \infty)$.   **11** $[-2/3, 7/2)$.
13 $(-\infty, -3) \cup [0, 4)$.   **15** $(-\infty, 1/2) \cup (7/9, \infty)$.   **17** **R**.
19 $[-1, 0] \cup [2, \infty)$.   **21** $(-\sqrt{6}, -2) \cup (2, \sqrt{6})$.   **23** $(-\infty, -15/2)$.
25 $\varnothing$.   **27** $(-\sqrt{7}, -1) \cup (1, \sqrt{7})$.   **29** $(-\infty, -\sqrt{3}) \cup (\sqrt{3}, \infty) \cup (-1, 1)$.
31 $(-\infty, -2) \cup (-1, 0) \cup (1, 2)$.

Review Exercises   (page 88)

1 $\{-\frac{1}{42}\}$.   **3** $\{11\}$.   **5** $\{-2, 0, \frac{1}{4}\}$.   **7** $\{(-1 \pm \sqrt{29})/2\}$.
9 $\{14\}$.   **11** $\{-27, 1\}$.   **13** $(-\infty, -3/5)$.   **15** $[3.495, 3.505]$.
17 $(1, 3/2)$.   **19** $(-5, 1/3) \cup (7/5, \infty)$.   **21** $\{-1/3, 1\}$.
23 $\{(-3 \pm \sqrt{11})/2\}$.   **25** $(-5/4, 3/4)$.   **27** $(-\infty, 0) \cup (2, \infty)$.
29 $\{2/3, 3\}$.   **31** $\{7\}$.   **33** $\{-4, 2/3\}$.   **35** $1/2$ hour.
37 $19, 28$.   **39** 20 pounds.   **41** $t = 25 \pm 10\sqrt{5}$; $t = (50 + 5\sqrt{102})/2$.
43 $h = V/\pi(R^2 - r^2)$.   **45** $t = (-v_0 \pm \sqrt{v_0^2 + 2gs})/g$.

CHAPTER THREE

Section 1   (page 96)

1

3

The line bisecting
quadrants I and III

5 **(a)** The line parallel to the y-axis which intersects the x-axis at $(5, 0)$.   **(b)** The line parallel to the x-axis which intersects the y-axis at $(0, -2)$.   **(c)** The points in the second and fourth quadrants.   **(d)** The x- and y-axes.
7 **(a)** 5.   **(b)** $(4, -\frac{1}{2})$.   **9** **(a)** $\sqrt{26}$.   **(b)** $(-\frac{1}{2}, -\frac{9}{2})$.
11 **(a)** 5.   **(b)** $(-\frac{11}{2}, -2)$.   **13** 35 square units.   **17** $(-10, 47)$.
21 $x^2 + y^2 = 64$.   **23** $(6, 0)$ and $(-2, 0)$.
25 $(3 + \sqrt{46}/2, 3 + \sqrt{46}/2), (3 - \sqrt{46}/2, 3 - \sqrt{46}/2)$.   **27** $2/5 < a < 4$.

Section 2   (page 104)

1 **(a)** $\{(1, x), (1, y), (2, x), (2, y), (3, x), (3, y)\}$.
   (b) $\{(x, 1), (x, 2), (x, 3), (y, 1), (y, 2), (y, 3)\}$.

(c) {(1, 1), (1, 2), (1, 3), (2, 1), (2, 2), (2, 3), (3, 1), (3, 2), (3, 3)}.

(d) {(x, x), (x, y), (y, x), (y, y)}.

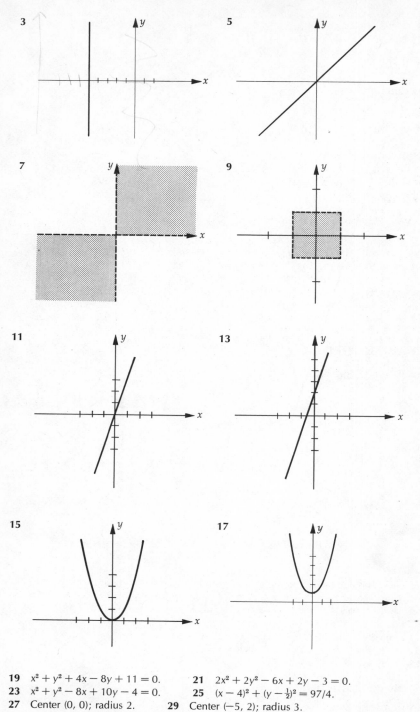

19 $x^2 + y^2 + 4x - 8y + 11 = 0.$ **21** $2x^2 + 2y^2 - 6x + 2y - 3 = 0.$

23 $x^2 + y^2 - 8x + 10y - 4 = 0.$ **25** $(x - 4)^2 + (y - \frac{1}{2})^2 = 97/4.$

27 Center (0, 0); radius 2.   **29** Center (−5, 2); radius 3.

31 Center (−1, −2); radius $\sqrt{2}$.

33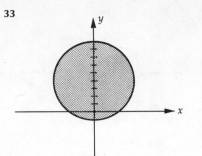

35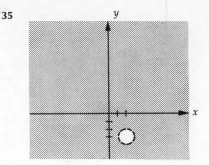

37 $(-2, 3)$; 3. **39** $(-3, 0)$; 3. **41** $(1/4, -1/4)$; $\sqrt{26}/4$.
43 $(0, -3/2)$; 3/2.

Section 3 (page 111)

1 (a) 2. (b) -8. (c) -3. (d) $6\sqrt{2} - 3$.
3 (a) $3a^2 - a + 2$. (b) $3a^2 + a + 2$. (c) $-3a^2 + a - 2$.
   (d) $3a^2 + 6ah + 3h^2 - a - h + 2$. (e) $3a^2 - a + 3h^2 - h + 4$.
   (f) $6a + 3h - 1$.
5 (a) $a^2/(1 + 4a^2)$. (b) $a^2 + 4$. (c) $1/(a^4 + 4)$. (d) $1/(a^2 + 4)^2$.
   (e) $1/(a + 4)$. (f) $1/\sqrt{a^2 + 4}$.
7 11; $a^2 - 5$; the nonnegative real numbers. **9** $\{x : x \geq \frac{3}{2}\}$.
11 $\{x : x \neq 3, x \neq -3\}$. **13** One-to-one. **15** Not one-to-one; for example.
   $f(-1) = f(-\frac{1}{2})$. **17** Yes. **19** No. **21** No. **23** Yes.
25 Odd. **27** Even. **29** Neither. **31** $r = c/2\pi$; $6/\pi \approx 1.9$ inches.
33 $V = 4x(x^2 - 30x + 216)$. **35** $P = 4\sqrt{A}$.

Section 4 (page 118)

1 $\{x : x \neq 0, x \neq 1\}$. **3** $\{x : x^2 \geq 4\}$. **5** $\{x : 5 < x < 7\}$.

7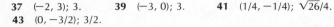

Increasing throughout **R**

9

Decreasing throughout **R**

11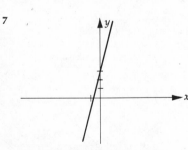

Neither increasing nor decreasing

13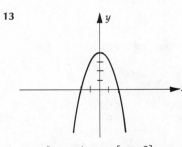

Increasing on $[-\infty, 0]$
Decreasing on $[0, \infty)$

507

15

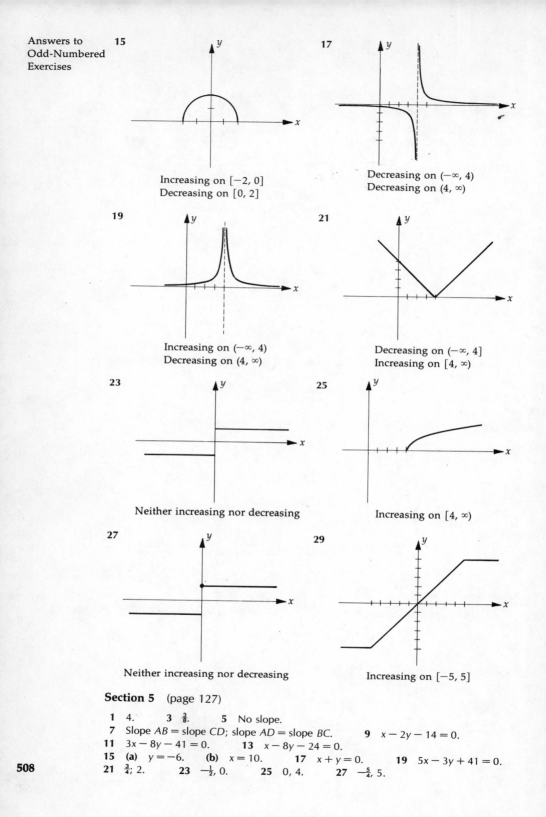

Increasing on $[-2, 0]$
Decreasing on $[0, 2]$

17

Decreasing on $(-\infty, 4)$
Decreasing on $(4, \infty)$

19

Increasing on $(-\infty, 4)$
Decreasing on $(4, \infty)$

21

Decreasing on $(-\infty, 4]$
Increasing on $[4, \infty)$

23

Neither increasing nor decreasing

25

Increasing on $[4, \infty)$

27

Neither increasing nor decreasing

29

Increasing on $[-5, 5]$

Section 5 (page 127)

1 4.   **3** $\frac{3}{8}$.   **5** No slope.
7 Slope AB = slope CD; slope AD = slope BC.   **9** $x - 2y - 14 = 0$.
11 $3x - 8y - 41 = 0$.   **13** $x - 8y - 24 = 0$.
15 **(a)** $y = -6$.   **(b)** $x = 10$.   **17** $x + y = 0$.   **19** $5x - 3y + 41 = 0$.
21 $\frac{3}{4}$; 2.   **23** $-\frac{1}{2}$, 0.   **25** 0, 4.   **27** $-\frac{5}{4}$, 5.

29

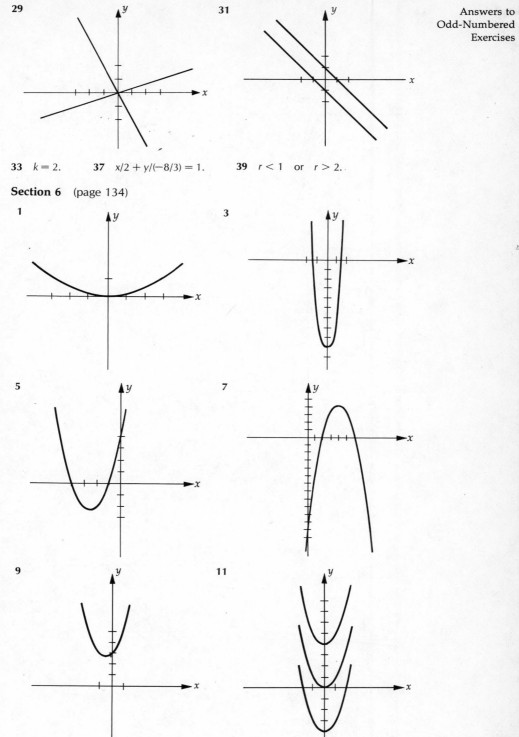

31

33 $k = 2.$ **37** $x/2 + y/(-8/3) = 1.$ **39** $r < 1$ or $r > 2.$

Section 6 (page 134)

1

3

5

7

9

11

13

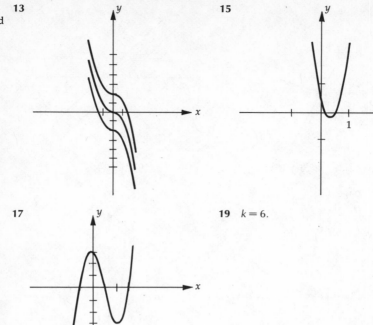

15

17

19 $k = 6$.

Section 7 (page 140)

Answers for Exercises 1–13 are in the order $(f \circ g)(x)$ and $(g \circ f)(x)$.

1 $6x - 9$, $6x - 2$. 3 $18x^2 + 5$, $6x^2 + 15$. 5 $9x^2 - 1$, $3x^2 - 6x + 1$.
7 $1/(3x - 13)$, $-(15x + 9)/(3x + 2)$. 9 $x + 3$, $\sqrt{x^2 + 3}$
11 $3/(3x^2 + 2)^2 + 2$, $1/(27x^4 + 36x^2 + 14)$. 13 $\sqrt{2x^2 + 7}$, $2x + 4$.
19 $f^{-1}(x) = (x - 8)/11$. 21 $f^{-1}(x) = \sqrt{6 - x}$.
23 $f^{-1}(x) = (x^2 + 2)/7$. 25 $f^{-1}(x) = \sqrt[3]{(7 - x)/3}$. 27 $f^{-1}(x) = (x^{1/5} - 8)^{1/3}$.
29

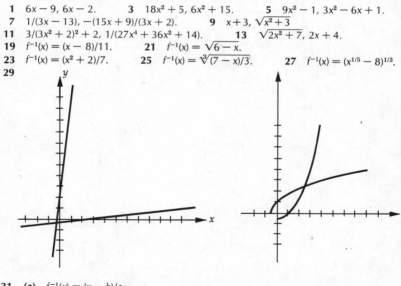

31 (a) $f^{-1}(x) = (x - b)/a$;
    (b) No (not one-to-one);
    (c) No (not one-to-one);
    (d) Yes, it is its own inverse.
33 If g and h are both inverse functions of f then $f(g(x)) = x = f(h(x))$ for all x. Since f is one-to-one this implies that $g(x) = h(x)$ for all x, that is, $h = g$.

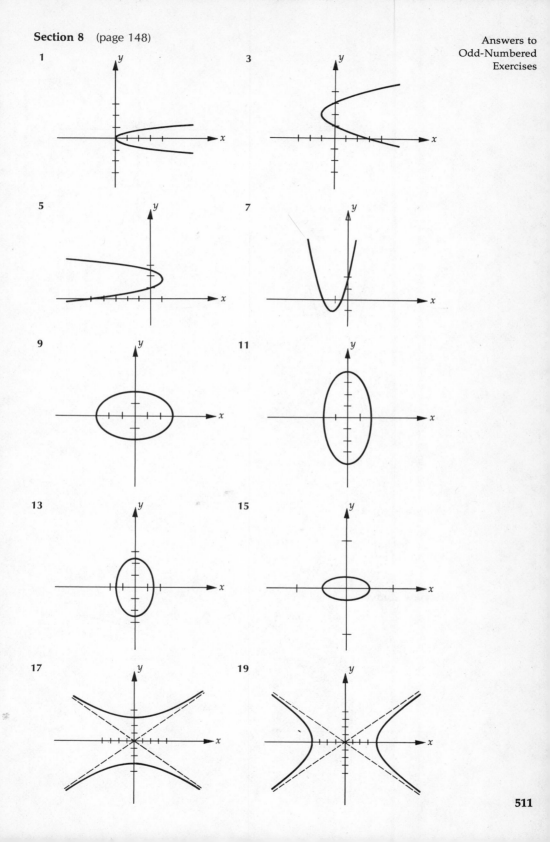

1

3

5

7

9

11

13

15

17

19

21

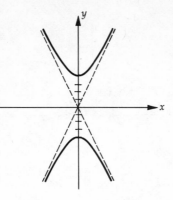

23

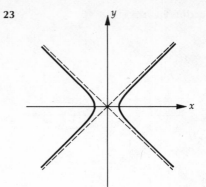

25 $x^2 + 4y^2 = 16$.   **27** $A = 4a^2b^2/(a^2 + b^2)$.

29

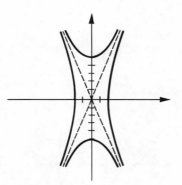

The graphs have the same asymptotes.

Section 9 (page 152)

1 $s = kt$, $k = 2/5$.   **3** $s = kt/d$, $k = 3$.   **5** $w = ku^2v^3$, $k = -1/12$.
7 20 pounds.   **9** $V = 20T/3P$; $\frac{640}{9}$.   **11** $5\frac{5}{9}$ ohms.
13 $3\sqrt{3}/2$ seconds.

Review Exercises (page 153)

1 **(a)** 12 square units.   **(b)** $(\frac{1}{2}, \frac{5}{2})$.   **(c)** 7.   **(d)** $x + y + 5 = 0$.

3 **(a)**

(b)

(c) 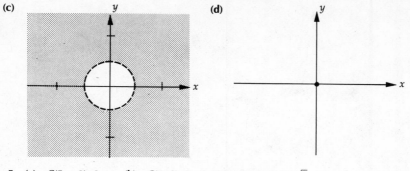 **(d)**

5 **(a)** $C(5, -2)$; 3. **(b)** $C(4, 0)$; 5. **(c)** $C(-3, 3)$; $3\sqrt{2}$.
7 **(a)** $\sqrt{2}/2$. **(b)** $\frac{1}{2}$. **(c)** 1. **(d)** $\sqrt[4]{8}/2$. **(e)** $1/\sqrt{1-x}$.
  (f) $-1/\sqrt{x+1}$. **(g)** $1/\sqrt{x^2+1}$. **(h)** $1/(x+1)$.
9 **(a)** $11x + 3y - 29 = 0$. **(b)** $x = 12$. **(c)** $y = (-3/8)x + 3$.
11 **(a)** $f^{-1}(x) = (5 - x)/7$. **(b)** $f^{-1}(x) = \sqrt{x - 3}/2$. **13** $V = C^3/8\pi^2$.

CHAPTER 4

Section 1 (page 163)

1 **3**

5 **7**

9 **11**

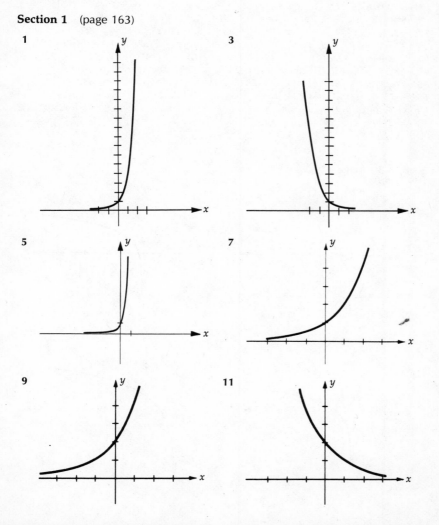

513

13

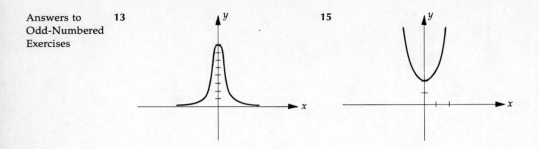

15

17 $-1/1600$.

19 1010.00, 1020.10, 1061.52, 1126.83.

21 a^x is not always real if $a < 0$. **23** Reflection through the x-axis.

Section 2 (page 170)

1 $\log_4 16 = 2$. **3** $\log_2 16 = 4$. **5** $\log_{10}(0.0001) = -4$. **7** $\log_r t = s$.

9 $5^3 = 125$. **11** $4^{-3} = 1/64$. **13** $10^0 = 1$. **15** $p^t = s$.

17 -2. **19** 3. **21** 7. **23** $1/5$. **25** -1. **27** -2.

29 32. **31** $-.2$. **33** 1. **35** $-.3$. **37** $-.1$. **39** $10/3$.

41 $\{19\}$. **43** $\{4\}$. **45** $\{1/10, -1/10\}$. **47** $\{-1, -2\}$.

49 $\{\sqrt{5}\}$. **51** The interval $(-1, 24)$. **53** The interval $(10, 100)$.

55 $\{15/2\}$. **57** $\{4/3\}$. **59** $\{9/8\}$. **61** $\{70\}$. **63** $\{4\}$.

65 $(-33, 101/3)$. **67** $\log_a x + 3 \log_a y - 2 \log_a z$.

69 $(1/3) \log_a x + 2 \log_a z - \log_a y$. **71** $(1/4)(\log_a x - 2 \log_a y - 5 \log_a z)$.

73 $(1/2) \log_a x + (1/4) \log_a y$. **75** $\log_a [x^3 \sqrt{x + 3}/(x - 2)^4]$. **77** $-\log_a x$.

Section 3 (page 176)

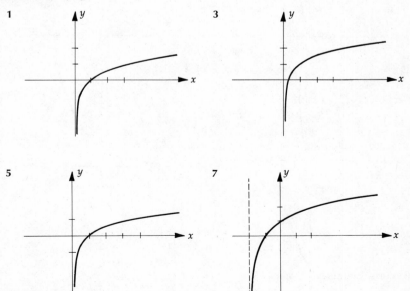

1 **3**

5 **7**

9

11

13

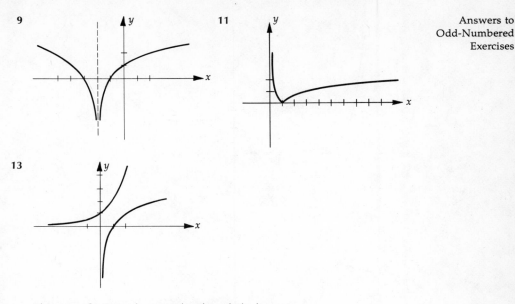

They are reflections of one another through the line $y = x$.

15 $t = (1/k)(\ln Q - \ln Q_0)$.

Section 4 (page 181)

1 2.9689; 7.9689 − 10; 0.9689. **3** 9.5682 − 10; 2.5682; 5.5682.
5 1.7024; 5.7024 − 10; 2.7024. **7** 3.8938; 0.9734; 6.1062 − 10.
9 8.1568 − 10; 1.8432; 9.7952 − 10. **11** 1.2239. **13** 8.3490.
15 6,460. **17** 1.05. **19** 498,000. **21** 0.196. **23** 0.0757.
25 0.0725. **27** 0.957.

Section 5 (page 185)

1 1.4368. **3** 3.8118. **5** 7.2477 − 10. **7** 5.1729.
9 9.9488 − 10. **11** 2.4680. **13** .9276 − 1. **15** 4.7447.
17 .0913 − 2. **19** 27.52. **21** 42,780. **23** 0.1044.
25 0.005994. **27** 275.3. **29** 2.793. **31** .04278.
33 0.02338. **35** .000000002338.

Section 6 (page 189)

1 17.3. **3** 8.08. **5** 715. **7** 0.719. **9** 23.1. **11** 1.97.
13 −0.129. **15** 0.0000238. **17** 1.30. **19** 3.35.
21 88.6 square units. **23** 1.83 seconds.

Section 7 (page 192)

1 $\{\log 5\}$. **3** $\{\log 4/\log 5\}$. **5** $\{\frac{5}{2}\}$. **7** $\{\log (\frac{2}{243})/\log 12\}$.
9 $\{\log 2/\log 6\}$. **11** $\{20\}$. **13** $\{\sqrt{11}/3\}$. **15** $(-\infty, -1)$.
17 $\{1,100\}$. **19** $\{10^{1000}\}$. **21** $\{10,000\}$. **23** $x = \log (y \pm \sqrt{y^2 - 1})$.
25 $x = \ln (y + \sqrt{y^2 + 1})$. **27** $x = (1/2) \ln (1 + y)/(1 - y)$.
29 $t = -(L/R) \ln (1 - Ri/E)$.

Review Exercises (page 193)

1 −3. **3** 7. **5** 5.

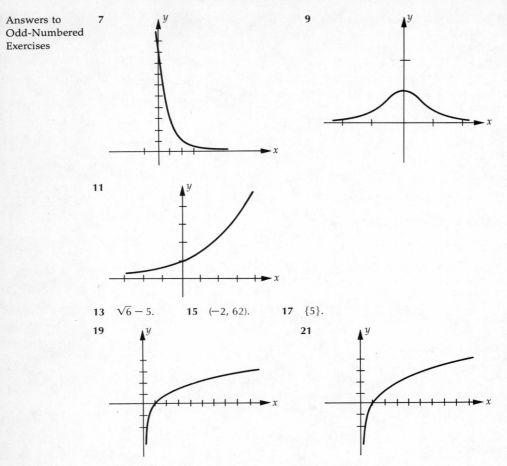

7

9

11

13 $\sqrt{6} - 5$.   **15** $(-2, 62)$.   **17** $\{5\}$.

19

21

23 $(1/5)(2 \log x + \log y - 4 \log z)$.   **25** 1.5532.   **27** 5.3025.

29 425.7.   **31** 2.209.   **33** 76.4.   **35** 8.71.   **37** $\{1\}$.

39 $\log (3/32)/\log 18$.   **41** $(-1, \infty)$.   **43** $x = (1/2) \log (y + 1)/(y - 1)$.

CHAPTER FIVE

Section 1   (page 204)

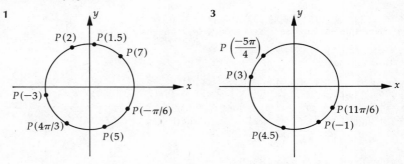

1

3

5 IV.   **7** III.   **9** $(-1, 0); (0, -1); (1, 0); (0, -1), (0, -1), (-1, 0)$.

11 $(-\sqrt{3}/2, -\frac{1}{2})$.   **13** $(-\frac{1}{2}, \sqrt{3}/2)$.   **15** $(\frac{1}{2}, \sqrt{3}/2)$.

17 Any number of the form $\pi/6 + 2n\pi$, where $n \in \mathbf{Z}$.

19 (a) $(-\frac{4}{5}, -\frac{3}{5})$.   (b) $(\frac{4}{5}, -\frac{3}{5})$.   (c) $(-\frac{4}{5}, -\frac{3}{5})$.   (d) $(-\frac{4}{5}, \frac{3}{5})$.

21 (a) $(-\frac{8}{17}, \frac{15}{17})$. (b) $(\frac{8}{17}, \frac{15}{17})$. (c) $(-\frac{8}{17}, \frac{15}{17})$. (d) $(-\frac{8}{17}, -\frac{15}{17})$.

Section 2 (page 212)

3 II. **5** IV. **7** IV. **9** III.

The following are arranged in the order sin t, cos t, tan t, csc t, sec t, cot t:

11 $-\frac{3}{5}, -\frac{4}{5}, \frac{3}{4}, -\frac{5}{3}, -\frac{5}{4}, \frac{4}{3}$. **13** $\frac{12}{13}, -\frac{5}{13}, -\frac{12}{5}, \frac{13}{12}, -\frac{13}{5}, -\frac{5}{12}$.

15 $-\frac{2}{3}, \sqrt{5}/3, -2\sqrt{5}/5, -\frac{3}{2}, 3\sqrt{5}/5, -\sqrt{5}/2$.

17 $1/8, -\sqrt{63}/8, -\sqrt{63}/63, 8, -8\sqrt{63}/63, -\sqrt{63}$.

19 (a) $0, -1, 0, —, -1, —$. (b) $-1, 0, —, -1, —, 0$.

21 (a) $0, -1, 0, —, -1, —$. (b) $-\sqrt{3}/2, -1/2, \sqrt{3}, -2\sqrt{3}/3, -2, \sqrt{3}/3$.

23 (a) $\frac{1}{2}, -\sqrt{3}/2, -\sqrt{3}/3, 2, -2\sqrt{3}/3, -\sqrt{3}$. (b) $\sqrt{2}/2, -\sqrt{2}/2, -1, \sqrt{2}, -\sqrt{2}, -1$.

33 No. $|\sin t| \le 1$. **35** $\varnothing$. **37** $\sqrt{3}/2, -1/6$.

39 No. They are not one-to-one.

Section 3 (page 221)

1 (a) $\pi/6$. (b) $\pi/6$. (c) $\pi/6$. **3** (a) $\pi/4$. (b) $\pi 4$. (c) $\pi/4$.

5 $\pi - 1.9$. **7** $2\pi - 5$.

9 (a) $-\frac{1}{2}, -\sqrt{3}/2, \sqrt{3}/3$. (b) $\frac{1}{2}, -\sqrt{3}/2, -\sqrt{3}/3$. (c) $-\frac{1}{2}, \sqrt{3}/2, -\sqrt{3}/3$.

11 (a) $\sqrt{2}/2, -\sqrt{2}/2, -1$. (b) $\sqrt{2}/2, -\sqrt{2}/2, -1$. (c) $\sqrt{2}/2, \sqrt{2}/2, 1$.

13 0.5446. **15** 0.3827. **17** −3.2041. **19** 0.8744.

21 .9971. **23** .7934. **25** .9271. **27** .6586. **29** 0.5055.

31 −0.6537. **33** 0.5751. **35** 0.3759. **37** .8621.

39 .7272, 2.4144. **41** 1.1286, 4.2702.

Section 4 (page 226)

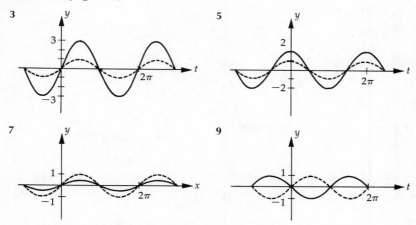

3
5
7
9

11 Same graph as Exercise 9. Reflection through the t-axis.

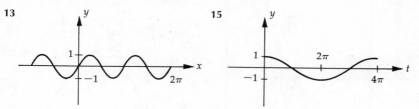

13
15

Section 5 (page 232)

1 The amplitudes and periods are: (a) $4, 2\pi$ (b) $1, \pi/2$. (c) $\frac{1}{4}, 2\pi$.

(d) $1, 8\pi$. (e) $2, 8\pi$. (f) $\frac{1}{2}, \frac{\pi}{2}$. (g) $4, 2\pi$. (h) $1, \frac{\pi}{2}$.

3 The amplitudes and periods are: **(a)** 3, 2π. **(b)** 1, $2\pi/3$. **(c)** 1/3, 2π. **(d)** 1, 6π. **(e)** 2, 6π. **(f)** 1/3, π. **(g)** 3, 2π. **(h)** 1, $2\pi/3$.

5

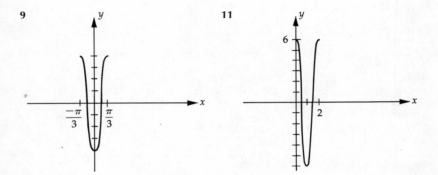

7

9

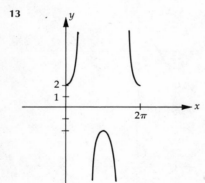

11

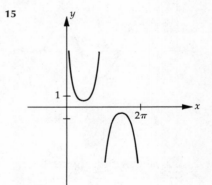

13

15

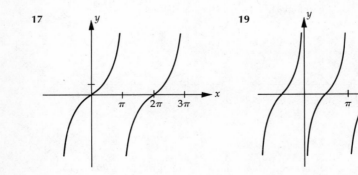

17

19

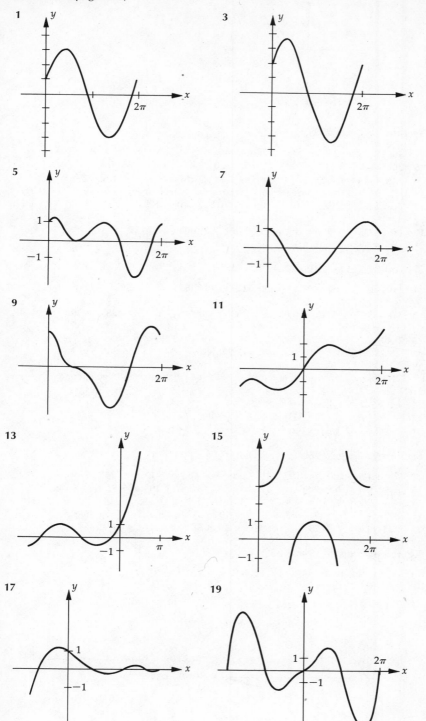

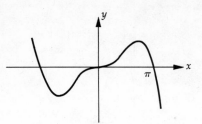

Section 7 (page 242)

1 495°, 855°, −225°, −585° (and others). **3** 570°, 930°, −150°, −510°.
5 300°, 660°, −420°, −780°. **7** 220°, 940°, −140°, −500°.
9 11π/4, 19π/4, −5π/4, −13π/4. **11** 7π/6, 19π/6, −17π/6, −29π/6.
13 (a) π/3. (b) −5π/6. (c) 3π/2. (d) 5π/4. (e) 17π/36. (f) 10π/9.
15 (a) 135°. (b) 120°. (c) −270°. (d) 210°. (e) 900°. (f) 36°.
17 229°11′2″.
19 (a) $\frac{5}{3}$ radians, 300/π degrees. (b) 7.5 square units.
21 (a) 4.19 in. (b) 12.57 sq. in.

Section 8 (page 250)

1 (a) 40°. (b) 25°. (c) 53°. (d) 52°48′. (e) 2°. (f) 60°.
(g) 32°37′42″.
3 (a) 0.8323. (b) 0.9572. (c) 0.2095. (d) −1.4281. (e) −0.4147.
(f) −0.5050.
5 0.4006. **7** 0.6173. **9** 0.2549. **11** −0.1859.
13 28°25′, 151°35′. **15** 49°46′, 229°46′. **17** 155°56′, 335°56′.
19 136°54′, 223°6′.
The following are given in the order sin θ, cos θ, tan θ, csc θ, sec θ, cot θ:
21 $-\frac{4}{5}, -\frac{3}{5}, \frac{4}{3}, -\frac{5}{4}, -\frac{5}{3}, \frac{3}{4}$. **23** $\sqrt{5}/5, -2\sqrt{5}/5, -1/2, \sqrt{5}, -\sqrt{5}/2, -2$.
25 $-2\sqrt{5}/5, \sqrt{5}/5, -2, -\sqrt{5}/2, \sqrt{5}, -\frac{1}{2}$.
27 $-5/\sqrt{29}, -2/\sqrt{29}, 5/2, -\sqrt{29}/5, -\sqrt{29}/2, 2/5$. **29** 1, 0, —, 1, —, 0.
31 $\sqrt{2}/2, -\sqrt{2}/2, -1, \sqrt{2}, -\sqrt{2}, -1$. **33** −1, 0, —, −1, —, 0.

Section 9 (page 257)

1 β = 30°, c = 20, a = 10√3. **3** α ≈ 48°, b ≈ 17, c ≈ 26.
5 β ≈ 70°40′, b ≈ 20.8, c ≈ 22.0. **7** α ≈ 21°36′, a ≈ 52.3, c ≈ 142.
9 α ≈ 20°, β ≈ 70°, c ≈ 41. **11** α ≈ 59°, β ≈ 31°, c ≈ 663. **13** 31°.
15 82.9 feet. **17** 163 feet. **19** 71.5 feet. **21** 55 miles.
23 325 miles. **25** h = d(1 + tan α cot β).

Review Exercises (page 259)

1 (−1, 0), (0, 1), (0, −1), (−√3/2, $\frac{1}{2}$), (−$\frac{1}{2}$, −√3/2), (−√2/2, −√2/2).
3 (a) IV. (b) III. (c) II.
7

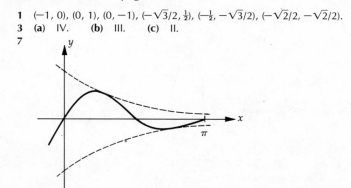

9 (a) $\pi/4$, $\pi/6$, $2\pi/3$.   (b) $70°$, $22°48'$, $31°$.
11 (a) $-\sqrt{2}/2$.   (b) $\sqrt{3}/3$.   (c) $\sqrt{3}/2$.   (d) -1.   (e) $-\sqrt{3}$.   (f) -2.
13 (a) $50°49'$, $309°11'$.   (b) $237°16'$, $302°44'$.   (c) $43°56'$, $223°56'$.
15 (a) $\beta = 30°$, $b = 10\sqrt{3}/3$, $c = 20\sqrt{3}/3$.   (b) $\alpha \approx 37°50'$, $b \approx 41$, $c \approx 52$.
   (c) $\alpha \approx 57°$, $\beta \approx 23°$, $c \approx 76$.

CHAPTER SIX

Section 2   (page 271)

1 $\{7\pi/6 + 2n\pi : n \in \mathbf{Z}\} \cup \{11\pi/6 + 2n\pi : n \in \mathbf{Z}\}$.
3 $\{\pi/2 + 2n\pi : n \in \mathbf{Z}\} \cup \{\pi + 2n\pi : n \in \mathbf{Z}\}$.
5 $\{\pi/6 + n\pi : n \in \mathbf{Z}\} \cup \{5\pi/6 + n\pi : n \in \mathbf{Z}\}$.
7 All $\pi/3 + n\pi$ and $2\pi/3 + n\pi$, where $n \in \mathbf{Z}$.
9 All $2\pi/3 + 2n\pi$, $4\pi/3 + 2n\pi$, $5\pi/4 + 2n\pi$ and $7\pi/4 + 2n\pi$, where $n \in \mathbf{Z}$.
11 All $n(\pi/2)$, $\pi/6 + n\pi$ and $5\pi/6 + n\pi$, where $n \in \mathbf{Z}$.
13 $\{\pi/3, 2\pi/3, 4\pi/3, 5\pi/3\}$. $\{60°, 120°, 240°, 300°\}$.
15 $\{\pi/6, 5\pi/6, 3\pi/2\}$. $\{30°, 150°, 270°\}$.
17 $\{\pi/4, 3\pi/4, 5\pi/4, 7\pi/4, \pi/2, 3\pi/2\}$. $\{45°, 135°, 225°, 315°, 90°, 270°\}$.
19 $\{0, \pi, 7\pi/6, 11\pi/6\}$. $\{0°, 180°, 210°, 330°\}$.   21 $\varnothing$.
23 $\{0, 2\pi/3\}$. $\{0°, 120°\}$.   25 $\{0, \pi/2\}$. $\{0°, 90°\}$.   27 $\{0\}$, $\{0°\}$.
29 $\{0, \pi, 2\pi/3, 4\pi/3\}$. $\{0°, 180°, 120°, 240°\}$.   31 $\{3\pi/4, 7\pi/4\}$. $\{135°, 315°\}$.
33 $\{15°30', 164°30'\}$.   35 $\{41°50', 138°10', 194°30', 345°30'\}$.

Section 3   (page 277)

1 (a) $\cos 56°36'$.   (b) $\sin \pi/6$.   (c) $\cot 22°50'$.
3 (a) $(\sqrt{2} + \sqrt{3})/2$.   (b) $(\sqrt{2} + \sqrt{6})/4$.   5 (a) $1 - \sqrt{3}$.   (b) $\sqrt{3} - 2$.
7 (a) $-(\sqrt{3} + \sqrt{2})/2$.   (b) $(\sqrt{2} - \sqrt{6})/4$.   9 $-\frac{36}{85}, \frac{77}{85}$, II.
11 $\frac{44}{125}, \frac{117}{125}, \frac{44}{117}, -\frac{4}{5}, \frac{3}{5}, -\frac{4}{3}$.
29 $\sin u \cos v \cos w + \cos u \sin v \cos w + \cos u \cos v \sin w - \sin u \sin v \sin w$.

Section 4   (page 284)

1 $\frac{24}{25}, \frac{7}{25}, \frac{24}{7}$.   3 $-\frac{240}{289}, -\frac{161}{289}, \frac{240}{161}$.   5 $\sqrt{5}/5$, $2\sqrt{5}/5$, $\frac{1}{2}$.
7 $-\sqrt{2 + \sqrt{2}}/2$, $\sqrt{2 - \sqrt{2}}/2$, $-1 - \sqrt{2}$.   9 (a) $\sqrt{2 + \sqrt{2}}/2$.
   (b) $\sqrt{2 + \sqrt{3}}/2$.   (c) $\sqrt{2} + 1$.
21 $\{0, 2\pi/3, \pi, 4\pi/3\}$. $\{0°, 120°, 180°, 240°\}$.
23 $\{\pi/3, \pi, 5\pi/3\}$. $\{60°, 180°, 300°\}$.   25 $\{0, \pi\}$. $\{0°, 180°\}$.
27 $\{0, \pi/3, 5\pi/3\}$. $\{0°, 60°, 300°\}$.

Section 5   (page 288)

1 $\sin 11\theta + \sin 3\theta$.   3 $(\frac{1}{2}) \cos 2t - (\frac{1}{2}) \cos 8t$.   5 $(\frac{1}{2}) \cos 2u + (\frac{1}{2}) \cos 14u$.
7 $2 \sin 3x + 2 \sin x$.   9 $2 \cos 2\theta \sin 6\theta$.   11 $-2 \sin 6x \sin x$.
13 $-2 \cos 4t \sin t$.   15 $2 \cos (3x/2) \cos (x/2)$.   23 $\{n\pi/4 : n \in \mathbf{Z}\}$.

Section 7   (page 295)

1 (a) $\pi/3$.   (b) $-\pi/3$.   3 (a) $\pi/4$.   (b) $3\pi/4$.
5 (a) $\pi/3$.   (b) $-\pi/3$.   7 (a) -0.7069.   (b) 1.4573.   9 $\frac{1}{2}$.
11 $\frac{4}{5}$.   13 $\pi - \sqrt{5}$.   15 0.   17 Undefined.   19 $-\frac{24}{25}$.
21 $u\sqrt{1 + u^2}/(1 + u^2)$.   23 $\sqrt{2 + 2u}/2$.
29 $\cot^{-1} u = v$ if and only if $\cot v = u$.

33

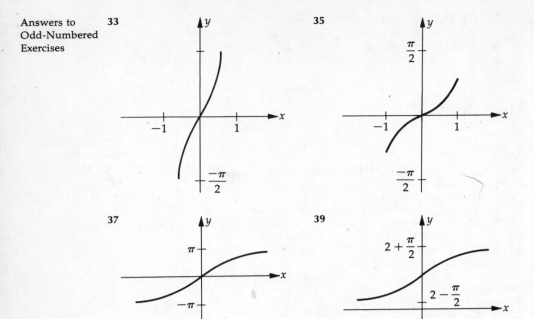

35

37

39

Section 8 (page 304)

1 $\beta \approx 62°$, $b \approx 14.1$, $c \approx 15.6$.   **3** $\gamma \approx 100°10'$, $b \approx 55.1$, $c \approx 68.7$.
5 $\alpha \approx 58°40'$, $a \approx 487$, $b \approx 442$.   **7** $\beta \approx 53°40'$, $\gamma \approx 61°10'$, $c \approx 20.6$.
9 $\alpha \approx 77°30'$, $\beta \approx 49°10'$, $b \approx 108$; $\alpha \approx 102°30'$, $\beta \approx 24°10'$, $b \approx 58.7$.
11 $\alpha \approx 20°30'$, $\gamma \approx 46°20'$, $a \approx 94.5$.   **13** 219.   **15** 50 feet.
17 2.7 miles.   **19** Approximately 3.7 miles from A and 5.4 miles from B.

Section 9 (page 306)

1 $a \approx 26$, $\beta \approx 41°$, $\gamma \approx 79°$.   **3** $b \approx 177$, $\alpha \approx 25°10'$, $\gamma \approx 4°50'$.
5 $c \approx 2.8$, $\alpha \approx 21°10'$, $\beta \approx 43°40'$.   **7** $\alpha \approx 29°$, $\beta \approx 46°30'$, $\gamma \approx 104°30'$.
9 $\alpha \approx 12°30'$, $\beta \approx 136°30'$, $\gamma \approx 31°$.   **11** 63, 87 inches.   **13** 92 feet.
15 24 miles.   **17** 39 miles.

Review Exercises (page 307)

11 $\{0, \pi/4, 3\pi/4, \pi, 5\pi/4, 7\pi/4\}$, $\{0°, 45°, 135°, 180°, 225°, 315°\}$.
13 $\{\pi/2, 3\pi/2\}$, $\{90°, 270°\}$.
15 $\{\pi/2, 7\pi/6, 3\pi/2, 11\pi/6\}$, $\{90°, 210°, 270°, 330°\}$.
17 $\{0, 2\pi/3, 4\pi/3\}$, $\{0°, 120°, 240°\}$.
19 $\{\pi/6, \pi/3, 5\pi/6, 5\pi/3\}$, $\{30°, 60°, 150°, 300°\}$.
21 $\{\pi/3, 5\pi/3\}$, $\{60°, 300°\}$.   **23** $(\sqrt{2} + \sqrt{6})/4$.   **25** $(-\frac{1}{4})(\sqrt{2} + \sqrt{6})$.
27 $-36/85$.   **29** $-36/77$.   **31** $7/25$.   **33** $1/3$.
35 (a) $\frac{1}{2}\cos 3t - \frac{1}{2}\cos 7t$;   (b) $\frac{1}{2}\cos u/12 + \frac{1}{2}\cos 7u/12$;
   (c) $2 \sin 5x - 2 \sin x$.
37 $5\pi/6$.   **39** π.   **41** $\frac{1}{2}$.   **43** $-\frac{7}{23}$.   **45** $\pi/2$.
47 $a = 50\sqrt{6}$, $c = 50(1 + \sqrt{3})$, $\gamma = 75°$.
49 $a = \sqrt{43}$, $\beta = \cos^{-1}(11/16)$, $\gamma = \cos^{-1}(-1/4)$.

CHAPTER SEVEN

Section 1   (page 317)

1 $\{(2, 3), (-1, 0)\}$.   **3** $\{(0, 0), (\frac{1}{16}, \frac{1}{4})\}$.   **5** $\{2, -3)\}$.   **7** $\varnothing$.

9 $\{(0, 5), (3, -4)\}$.   **11** $\{(3, -4), (4, -3)\}$.   **13** $\{(0, 0), (-1, 1)\}$.

15 $\{(1, 0), (-3, 4)\}$.   **17** $\{(1, 6), (-1, 6), (\sqrt{6}, 1), (-\sqrt{6}, 1)\}$.

19 $\{(2\sqrt{3}, \pm\sqrt{2}), (-2\sqrt{3}, \pm\sqrt{2})\}$.   **21** $\{(\pm\sqrt{13/5}, \sqrt{7/5}), (\pm\sqrt{13/5}, -\sqrt{7/5})\}$.

23 $\{(-1, 2, 0)\}$.   **25** $r = 3, s = -4$.   **27** $\{(\log 2/\log 3, 4)\}$.   **29** $13, 9$.

31 57.   **33** 8 inches, 12 inches.   **35** $5 - \sqrt{19}, 5 + \sqrt{19}, 10$.

Section 2   (page 324)

1 Equivalent; multiply $x - 2y = -3$ by -2.

3 Equivalent; add to the first equation -1 times the second equation.

5 Not equivalent; there are solutions of the first equation that are not solutions of the second equation.

7 $\{(3, -2)\}$.   **9** $r = 4, s = -2$.   **11** $\{(7, 5)\}$.   **13** $\varnothing$.

15 $u = 3, v = -6$.   **17** $\{(x, 4x - 2 : x \in \mathbf{R}\}$.   **19** $\{(0, \frac{12}{7})\}$.

21 $\{(\frac{14}{17}, \frac{14}{27})\}$.   **23** $a = \frac{5}{7}, b = \frac{20}{7}$.   **25** $53, 29$.

27 $\frac{7}{4}$ lb. at 75 cents and $\frac{5}{4}$ lb. at 95 cents.   **29** $850 at $6\frac{1}{2}\%$, $425 at 8\%$.

Section 3   (page 331)

1 $\{(3, -1, 2)\}$.   **3** $\{(\frac{3}{4}, \frac{1}{6}, -\frac{11}{12})\}$.   **5** $\varnothing$.   **7** $\{(-c, c, -2c) : c \in \mathbf{R}\}$.

9 $\{(c, 0, -c) : c \in \mathbf{R}\}$.   **11** $\{(-11c - 9, 7c + 5, c) : c \in \mathbf{R}\}$.

13 $\{(3, -1, 2, 1)\}$.   **15** $\varnothing$.   **17** $\{(1, 2, 2)\}$.   **19** $\{(c, 3c, -c) : c \in \mathbf{R}\}$.

21 5.   **23** 2.5 oz. of 10%; 7.5 oz. of 30%; 15 oz. of 50%.

25 $x^2 + y^2 - 4x - 6y - 12 = 0$.

Section 4   (page 337)

19 $\{(2, -3, 1, -4)\}$.   **21** $\{(-c, 1 - 2c, -c, c) : c \in \mathbf{R}\}$.

Section 5   (page 343)

1 $M_{11} = -32, M_{12} = -2, M_{13} = -20, M_{21} = -8, M_{22} = -24, M_{23} = -5, M_{31} = 2$,
$M_{32} = 6, M_{33} = 13, A_{11} = -32, A_{12} = 2, A_{13} = -20, A_{21} = 8, A_{22} = -24, A_{23} = 5$,
$A_{31} = 2, A_{32} = -6, A_{33} = 13$.

3 $M_{11} = 0, M_{12} = -2, M_{21} = 1, M_{22} = 3, A_{11} = 0, A_{12} = 2, A_{21} = -1, A_{22} = 3$.

5 -94.   **7** 2.   **9** -7.   **11** 0.   **13** -72.   **15** 24.

17 266.   **19** $-abcd$.

Section 6   (page 348)

1 7.27(i).   **3** 7.27(iii).   **5** 7.28.   **7** 7.27(i).   **9** 7.27(ii).

11 7.26.   **13** 7.27(iii).   **15** 12.   **17** 318.   **19** 5.   **21** 9.

Section 8   (page 359)

Description of graphs;   **1** The region above the graph of $y = 2x + 1$.

3 The region below and on the graph of $y = 3x - 1$.

5 The region below the graph of $y = x^2 - 3$.

7 The region above and on the graph of $y = x^2 + 2$.

9 The region below the graph of $y = x^3 + 1$.

11 The region above the graph of $3y + 2x = 6$ and also above the graph of $y = x$.

13 The region below the graph of $2x + y = 4$ and also above the graph of $y = 2 - 2x$.

15 The region in the first quadrant bounded by the coordinate axes and the lines with equations $y + 4 = 3x$ and $x + y = 6$.

17 The smaller region bounded by the graphs of $x^2 + y^2 = 1$ and $y - x = 1$.

19 The region bounded by the graphs of $y = 4 - x^2$ and $y + x + 2 = 0$.

21 The region in the first quadrant between the graphs of $y = 2^x$ and $y = 3^x$.

23 The region in the first quadrant bounded by the graphs of $y = x^3 + 1$, $y = 2^{-x}$ and $x = 4$.

25 Let y denote the number of bats and x the number of balls. Then $x \geq 2$, $y \geq 3$ and $5x + 7y \leq 80$.

Section 9 (page 364)

1 Send 25 from W_1 to X and 0 from W_1 to Y. Send 10 from W_2 to X and 60 from W_2 to Y.

3 45 acres of crop A and 55 acres of crop B.

5 Minimum cost: 16 ounces X, 4 ounces Y, 0 ounces Z.
Maximum cost: 0 ounces X, 8 ounces Y, 12 ounces Z.

Section 10 (page 369)

1 $\begin{bmatrix} 3 & 2 \\ 1 & 9 \end{bmatrix}$, $\begin{bmatrix} 1 & -4 \\ 5 & -1 \end{bmatrix}$, $\begin{bmatrix} 6 & -3 \\ 9 & 12 \end{bmatrix}$, $\begin{bmatrix} -2 & -6 \\ 4 & -10 \end{bmatrix}$.

3 $\begin{bmatrix} 5 & -3 \\ -1 & 3 \\ 1 & -3 \end{bmatrix}$, $\begin{bmatrix} 3 & 3 \\ -13 & 1 \\ 1 & -7 \end{bmatrix}$, $\begin{bmatrix} 12 & 0 \\ -21 & 6 \\ 3 & -15 \end{bmatrix}$, $\begin{bmatrix} -2 & 6 \\ -12 & -2 \\ 0 & -4 \end{bmatrix}$.

5 $[-2 \ \ 6 \ \ 7]$, $[4 \ \ -6 \ \ 3]$, $[3 \ \ 0 \ \ 15]$, $[6 \ \ -12 \ \ -4]$.

7 $\begin{bmatrix} 0 & 0 & 0 & 0 \\ 7 & 0 & -3 & -1 \end{bmatrix}$, $\begin{bmatrix} 2 & 0 & 16 & 14 \\ -3 & -2 & -3 & 1 \end{bmatrix}$, $\begin{bmatrix} 3 & 0 & 24 & 21 \\ 6 & -3 & -9 & 0 \end{bmatrix}$, $\begin{bmatrix} 2 & 0 & 16 & 14 \\ -10 & -2 & 0 & 2 \end{bmatrix}$.

9 $\begin{bmatrix} 5 & 21 \\ -8 & 14 \end{bmatrix}$, $\begin{bmatrix} -4 & 22 \\ -15 & 23 \end{bmatrix}$.

11 $\begin{bmatrix} -8 & -5 & 10 \\ 2 & -2 & 8 \\ 7 & 5 & 1 \end{bmatrix}$, $\begin{bmatrix} 10 & -1 & 21 \\ 4 & 1 & 2 \\ -6 & 7 & -20 \end{bmatrix}$.

13 $\begin{bmatrix} 28 & 6 \\ 2 & -30 \end{bmatrix}$, $\begin{bmatrix} -19 & 2 & 19 \\ -5 & 2 & -7 \\ 27 & -8 & 15 \end{bmatrix}$.

15 $[9]$, $\begin{bmatrix} 6 & -3 & 15 \\ 4 & -2 & 10 \\ 2 & -1 & 5 \end{bmatrix}$.

17 $\begin{bmatrix} 1 \\ 3 \\ 7 \end{bmatrix}$.

19 $\begin{bmatrix} -1 & -8 & 16 \\ -12 & 19 & 40 \end{bmatrix}$.

21 $(A + B)(A - B) = \begin{bmatrix} 16 & 32 \\ -8 & -16 \end{bmatrix}$; $A^2 - B^2 = \begin{bmatrix} 8 & 16 \\ -4 & -8 \end{bmatrix}$.

Section 11 (page 375)

1 $\begin{bmatrix} 1 & \frac{1}{2} \\ 2 & \frac{3}{2} \end{bmatrix}$.

3 Inverse does not exist.

5 $\begin{bmatrix} \frac{1}{2} & 0 & 0 \\ 0 & \frac{2}{5} & \frac{3}{10} \\ 0 & -\frac{1}{5} & \frac{1}{10} \end{bmatrix}$.

7 $(\frac{1}{14}) \begin{bmatrix} 12 & 4 & 2 \\ -2 & 4 & 2 \\ 9 & 3 & 5 \end{bmatrix}$.

9 $(\frac{1}{14}) \begin{bmatrix} 6 & 2 & 6 & 4 \\ -2 & 4 & -2 & -6 \\ -4 & 1 & 3 & 2 \\ 6 & 2 & 6 & -10 \end{bmatrix}$.

11 $ad - bc \neq 0$; $1/(ad - bc) \begin{bmatrix} d & -b \\ -c & a \end{bmatrix}$.

15 $\{(\frac{5}{2}, \frac{13}{2})\}$.

17 $\{(\frac{13}{7}, -\frac{1}{7}, \frac{8}{7})\}$.

Review Exercises (page 376)

1 $\{(-3, 10), (1, 2)\}$. **3** $\{(1, 4)\}$. **5** $\{(-2, -9)\}$. **7** $\{(2, -1, 5)\}$. **9** $\emptyset$.

11 The interior of the triangle bounded by the graphs of the given equations.

13 -5. **15** 17. **17** -1.

21 $\begin{bmatrix} -3 & 2 & 0 & 0 \\ 2 & -1 & 0 & 0 \\ 0 & 0 & 1 & -1 \\ 0 & 0 & 1 & -2 \end{bmatrix}$.

23 $\begin{bmatrix} 17 \\ 39 \end{bmatrix}$.

25 $\begin{bmatrix} 9 & -1 \\ -1 & 32 \end{bmatrix}$.

27 $\begin{bmatrix} c & 2c \\ 3c & 4c \end{bmatrix}$.

29 $\begin{bmatrix} -2 & 0 \\ 0 & -2 \end{bmatrix}$.

CHAPTER EIGHT

Section 1 (page 384)

1 $-5 + 8i$. **3** $-3 + i$. **5** $10 - 3i$. **7** $-2 - i$. **9** $4 + 9i$.
11 $-4 - 6i$. **13** $-22 + 7i$. **15** $29 - 11i$. **17** 25.
19 $-56 - 48i$. **21** $24 - 80i$. **23** $3\sqrt{2}$. **25** 7.
27 $-11 + 60i$. **29** $-1 + 17i$. **31** 1. **33** -1. **35** i.
37 $x = -\frac{7}{3}, y = 5$. **39** $x = -10, y = -5$.

Section 2 (page 387)

1 $(\frac{3}{13}) - (\frac{2}{13})i$. **3** $(\frac{12}{61}) + (\frac{10}{61})i$. **5** $(\frac{29}{34}) - (\frac{31}{34})i$. **7** $(-\frac{16}{17}) + (\frac{21}{17})i$.
9 $-8 - 19i$. **11** $(\frac{23}{58}) + (\frac{101}{58})i$. **13** $3 - i$. **15** $(2/125) - (11/125)i$.
17 $3 + i$. **19** $(\frac{1}{64})i$. **21** 5. **23** $\sqrt{106}$. **25** 7. **27** 1.
29 $(7/5) - (1/5)i$. **31** $(1/2)i$.

Section 3 (page 392)

1 -7. **3** $-1 + 5i$. **5** 28. **7** $-3 + 3i$. **9** $-2\sqrt{2}i$.
11 $(11/29) + (13/29)i$. **13** $\{2 \pm 3i\}$. **15** $\{(-3 \pm \sqrt{7}i)/2\}$.
17 $\{(-1 \pm \sqrt{23}i)/6\}$. **19** $\{3, (-3 \pm 3\sqrt{3}i)/2\}$.
21 $\{2, -2, (-2 \pm 2\sqrt{3}i)/2, (2 \pm 2\sqrt{3}i)/2\}$. **23** $\{\pm 2i, \pm(\sqrt{3}/2)i\}$.
25 $\{1, i, -i\}$. **27** $x^2 - 4x + 29 = 0$. **29** $x^2 + 2x + 2 = 0$.
31 $x^2 - 5ix - 6 = 0$. **33** $x^2 - 4x + 13 = 0$. **35** $x^2 + 1 = 0$.
37 $\{\pm(\sqrt{2} + \sqrt{2}i)/2\}$. **39** If $z = a + bi$, then $z + \bar{z} = 2a$ and $z - \bar{z} = 2bi$.

Section 4 (page 396)

Geometric representations: **1** $P(3, 5)$. **3** $P(2, -6)$. **5** $P(-2, 7)$.
7 $P(-9, 4)$. **9** $P(0, -2)$. **11** $P(0, -1)$.
13 $\sqrt{2}(\cos 3\pi/4 + i \sin 3\pi/4)$. **15** $8(\cos 11\pi/6 + i \sin 11\pi/6)$.
17 $100(\cos 3\pi/2 + i \sin 3\pi/2)$. **19** $20(\cos 7\pi/6 + i \sin 7\pi/6)$.
21 $15(\cos \pi + i \sin \pi)$. **23** $\sqrt{5}(\cos \arctan 2 + i \sin \arctan 2)$.
25 $3\sqrt{2}(\cos 5\pi/4 + i \sin 5\pi/4)$. **27** $7(\cos \pi/2 + i \sin \pi/2)$.
29 $8(\cos 2\pi/3 + i \sin 2\pi/3)$. **31** $9(\cos 0 + i \sin 0)$. **33** $2; -i$.
35 $6 - 6\sqrt{3}i; (-\frac{2}{3})(1 - \sqrt{3}i)$. **37** $6; \frac{2}{3}$.

Section 5 (page 402)

1 $32i$. **3** $4 - 4i$. **5** $-8 - 8\sqrt{3}i$. **7** -1. **9** i.
11 $\pm(\sqrt{6} + \sqrt{2}i)/2$.
13 $(\sqrt[4]{2}/2)(\sqrt{3} + i), (\sqrt[4]{2}/2)(-1 + \sqrt{3}i), (\sqrt[4]{2}/2)(-\sqrt{3} - i), (\sqrt[4]{2}/2)(1 - \sqrt{3}i)$.
15 $2i, -\sqrt{3} - i, \sqrt{3} - i$. **17** $\pm 1, \frac{1}{2} \pm \sqrt{3}/2i, -\frac{1}{2} \pm \sqrt{3}/2i$.
19 $i, -i, 1, -1$. **21** $\pm 2i, \sqrt{3} \pm i, -\sqrt{3} \pm i$.
23 $\sqrt[3]{2}[\cos(\pi/4 + n\pi/3) + i \sin(\pi/4 + n\pi/3)]$, where $n = 0, 1, 2, 3, 4, 5$.

Section 6 (page 406)

1 $\langle 3, 1 \rangle, \langle 1, -7 \rangle, \langle 13, 8 \rangle, \langle 3, -32 \rangle, \sqrt{13}, \sqrt{17}$.
3 $\langle -15, 6 \rangle, \langle 1, -2 \rangle, \langle -68, 28 \rangle, \langle 12, -12 \rangle, \sqrt{53}, 4\sqrt{5}$.

Review Exercises (page 408)

1 $-2 + 9i$. **3** $-16 + 11i$. **5** $-20 + 30i$. **7** 10. **9** $5 + 10i$.
11 $(-\frac{22}{25}) + (\frac{29}{25})i$. **13** $(\frac{9}{35}) + (\frac{2}{85})i$. **15** $\sqrt{34}$. **17** $2 + i$.
19 $(-8/17) - (19/17)i$. **21** $\{(1 \pm \sqrt{11}i)/6\}$. **23** $\{\pm\sqrt{3}/2i, \pm 2i\}$.
25 $x^2 - 6x + 34 = 0$. **27** $x^2 + 8x + 17 = 0$.
29 $3\sqrt{2}(\cos 5\pi/4 + i \sin 5\pi/4)$. **31** $8(\cos \pi + i \sin \pi)$.
33 $12(\cos 5\pi/6 + i \sin 5\pi/6)$. **35** $16\sqrt{3} + 16i$. **37** $8i$.
39 $-1, (1/2) \pm (\sqrt{3}/2)i$. **41** $\langle 1, 2 \rangle, \langle 5, -10 \rangle, \langle 13, -24 \rangle, 5, 2\sqrt{10}, \sqrt{5}$.

CHAPTER NINE

Section 1 (page 414)

1 $2x^3 + x^2 + 7$, $2x^3 - x^2 - 2x + 3$, $2x^5 + 2x^4 + 3x^3 + 4x^2 + 3x + 10$.

3 $14x^4 - x^3 + x^2 + 4x - 1$, $x^3 + x^2 - 4x - 1$, $49x^8 - 7x^7 + 7x^6 + 27x^5 - 7x^4 + 5x^3 - 4x$.

5 $x + 3$, $3 - x$, $3x$. **7** Degree of product is $4 + 3$.

13 Not all polynomials have multiplicative inverses. All other field properties are true.

Section 2 (page 418)

1 (a) $(2x + 1)(x - 5)$. (b) $(2x + 1)(x - 5)$.

3 (a) $(x^2 + 4)(x + 2)(x - 2)$. (b) $(x + 2i)(x - 2i)(x + 2)(x - 2)$.

5 (a) $(2x - 3)(4x^2 + 6x + 9)$. (b) $(2x - 3)(4x + 3 - 3\sqrt{3}i)\left(x + \dfrac{3}{4} + \dfrac{3\sqrt{3}}{4}i\right)$.

7 $x^2 - 5x + 20$, $-84x + 104$. **9** $(\frac{5}{2})x$, $(-\frac{29}{2})x$. **11** 0, $9x^3 + 4x - 1$.

13 -37. **15** -9. **17** $\frac{14}{3}$. **19** $f(3) = 0$. **21** $f(c) > 0$.

23 If $f(x) = x^n - y^n$, then $f(y) = 0$. If n is even, then $f(-y) = 0$.

Section 3 (page 422)

1 $4x^2 + 6x + 13$, 19. **3** $x^2 - 4x + 26$, -113.

5 $3x^4 - 6x^3 + 12x^2 - 25x + 50$, -97. **7** $2x^3 + x^2 - (\frac{5}{2})x - \frac{5}{4}$, $\frac{3}{8}$.

9 $x^3 + 3ix^2 - 3x - 4i$, 9. **11** -23, 65. **13** 547.

15 $10 + 9i$, $10 - 9i$.

Section 4 (page 427)

1 $x^3 - 5x^2 + x - 5$. **3** $x^3 - 9x^2 + 31x - 39$.

5 $x^4 - 2x^3 - 23x^2 + 24x + 144$. **7** $x^7 - 3x^6 + 3x^5 - x^4$.

9 -5 (multiplicity 2), 1 (multiplicity 1).

11 0 (multiplicity 3), $(-1 \pm \sqrt{21})/2$ (each of multiplicity 1).

13 3 and -3, (multiplicity 4), $3i$ and $-3i$ (multiplicity 1).

15 -2 (multiplicity 4), -3, 1 (each of multiplicity 1).

17 $f(x) = (x - 3)^2(x - 2i)(x + 2i)$. **19** $f(x) = (x - 1)^3(x + 4)$; -4.

23 $A = -1$, $B = -1$, $C = 0$. **25** $A = \frac{15}{4}$, $B = 2$, $C = \frac{3}{4}$.

Section 5 (page 434)

1 $x^2 - 4x + 13$. **3** $x^4 - 6x^3 + 18x^2 - 30x + 25$. **5** $2, -3, \frac{5}{2}$.

7 $\frac{2}{3}, -2, -\frac{1}{2}$. **9** $\frac{4}{3}, -2 \pm i$. **11** $-2, 3, -1 \pm \sqrt{2}i$. **13** $\frac{1}{3}$.

15 $4, -7, \pm\sqrt{2}$. **21** By (9. 19) complex zeros occur in conjugate pairs.

Review Exercises (page 434)

1 $4x^2 + x - 4$, $2x^2 + x - 6$, $3x^4 + x^3 - 2x^2 + x - 5$.

3 $-x^2 + x + 1$, $-x^2 + x - 1$, $-x^2 + x$. **5** $2, 2, 2, 2, 4$. **7** $2, 0, 2, 2, 2$.

9 $(3 + 2x)(3 - 2x)(9 + 4x^2)$, $(3 + 2x)(3 - 2x)(2x + 3i)(2x - 3i)$.

11 $x(x^2 + 4)(3x - 2)$, $x(3x - 2)(x + 2i)(x - 2i)$. **13** $4x^2 - 10$, $12x^2 + 21x - 23$.

15 $3, -14$. **17** (a) -65. (b) $f(-3) = 0$. **19** $x^2 - 5ix - 3$, $4i$.

21 (a) $x^5 + 6x^4 + 12x^3 + 8x^2$.
(b) $x^5 + 4x^4 + 4x^3$.

23 $5, -1, 2, -3$. **25** No rational solutions.

CHAPTER TEN

Section 2 (page 448)

1 $6, 2, -2, -6, -10$; $a_8 = -22$. **3** $-\frac{1}{3}, \frac{1}{6}, \frac{3}{11}, \frac{5}{18}, \frac{7}{27}$; $a_8 = \frac{13}{66}$.

5 $5, 5, 5, 5, 5$; $a_8 = 5$. **7** $0.9, 1.01, 0.999, 1.0001, 0.99999$; $a_8 = 1.00000001$.

9 $1, -\frac{3}{4}, \frac{2}{3}, -\frac{5}{8}, \frac{3}{5}$; $a_8 = -\frac{9}{16}$. **11** $2, 0, 2, 0, 2$; $a_8 = 0$.

13 $1/5, 4/13, 4/9, 16/25, 16/17$; $a_8 = 256/73$. **15** $4, 9, 25, 49, 121$; $a_8 = 289$.

17 $7, 19, 29, 37, 47$; $a_8 = 79$. **19** $1, 2, 3, 4, 5$; $a_8 = 8$.

21 $3, 13, 63, 313, 1563$. **23** $-2, 4, 16, 256, 65,536$. **25** $1, 1, 2, 6, 24$.

27 $3, 3, 3^{1/2}, 3$ **29** $a_n = 2n + (\frac{1}{24})(n-1)(n-2)(n-3)(n-4)(a-10)$.

Section 3   (page 452)

1 -5.   **3** 34.   **5** 40.   **7** $\frac{123}{12}$.   **9** 43.   **11** 500.
13 $(2n^3 + 12n^2 + 40n)/6$.   **15** $(4n^3 - 12n^2 + 11n)/3$.
17 $\displaystyle\sum_{k=1}^{5}(2k-1)$.   **19** $\displaystyle\sum_{k=1}^{4}k/(3k+1)$.   **21** $1 + \displaystyle\sum_{k=1}^{n}(-1)^k \frac{x^{2k}}{2k}$.
23 $\displaystyle\sum_{k=1}^{7}(-1)^{k-1}/k$.   **25** $\displaystyle\sum_{n=1}^{99}\frac{1}{n(n+1)}$.

Section 4   (page 457)

1 $a_5 = 17, a_{10} = 37, a_n = 4n - 3$.   **3** $a_5 = 0.8, a_{10} = -0.7, a_n = 2.3 - 0.3n$.
5 $a_5 = 8.9, a_{10} = 29.4, a_n = -11.6 + (4.1)n$.
7 $a_5 = \log 32, a_{10} = \log 1024, a_n = \log 2^n$.   **9** -12.4.   **11** -11.
13 -540.   **15** 60.   **17** 477.   **19** 84.   **21** $60; 12{,}780$.
23 24.   **25** $\frac{10}{3}, \frac{14}{3}, 6, \frac{22}{3}, \frac{26}{3}$.   **27** 255.
29 $23\frac{1}{4}, 22\frac{1}{2}, 21\frac{3}{4}, 21, 20\frac{1}{4}, 19\frac{1}{2}, 18\frac{3}{4}$.

Section 5   (page 463)

1 $a_5 = \frac{1}{4}, a_8 = 1/2^5, a_n = 2^{3-n}$.
3 $a_5 = 0.0007, a_8 = -0.0000007, a_n = 7(-0.1)^{n-1}$.
5 $a_5 = x^{12}, a_8 = -x^{21}, a_n = (-x^3)^{n-1}$.   **7** $\frac{243}{4}$.   **9** $a_1 = \frac{1}{81}, S_5 = \frac{211}{1296}$.
11 $4\sqrt{2}, 4, 2\sqrt{2}$.   **13** $2^{11} - 2$.   **15** $(-\frac{1}{3})(1 + 1/2^9)$.   **17** $\frac{25}{256}\%$.
19 $1000(\frac{6}{5})^t; 2^{13}3^{10}/5^7$.   **21** $2/3$.   **23** $\frac{17}{990}$.   **25** No sum.   **27** $\frac{7}{33}$.
29 $\frac{3049}{4995}$.   **31** $\frac{2172}{495}$.   **33** 30 ft.

Section 6   (page 469)

1 $a^6 + 6a^5b + 15a^4b^2 + 20a^3b^3 + 15a^2b^4 + 6ab^5 + b^6$.
3 $a^7 - 7a^6b + 21a^5b^2 - 35a^4b^3 + 35a^3b^4 - 21a^2b^5 + 7ab^6 - b^7$.
5 $625x^4 - 1500x^3y + 1350x^2y^2 - 540xy^3 + 81y^4$.
7 $243u^5 + 810u^4v^2 + 1080u^3v^4 + 720u^2v^6 + 240uv^8 + 32v^{10}$.
9 $x^{-18} - 12x^{-14} + 60x^{-10} - 160x^{-6} + 240x^{-2} - 192x^2 + 64x^6$.
11 $1 + 10x + 45x^2 + 120x^3 + 210x^4 + 252x^5 + 210x^6 + 120x^7 + 45x^8 + 10x^9 + x^{10}$.
13 $a^{15} + 50a^{14} + 1200a^{13} + 18400a^{12}$.   **15** $(15)2^{14}5^{13} - (2s)^{15}$.
17 $(21875/8)x^8y^2$.   **19** $840u^6v^4$.   **21** $924a^2b^2$.   **23** $-258{,}048$.
25 $24x^4y^6$.   **27** $210x^3y^2$.   **29** 1.22.

Section 7   (page 476)

1 (a) 60,   (b) 125.   **3** 64.   **5** 336.   **7** 24.
9 (a) $(2.34)10^6$,   (b) $(2.16)10^6$.   **11** $151{,}200; 2880$.   **13** 1024.
15 120.   **17** 6720.   **19** 720.   **21** 4.   **23** $40{,}320$.
25 120.   **27** (a) 60,   (b) 125.   **29** $9{,}000{,}000$.

Section 8   (page 481)

1 15.   **3** 8.   **5** n.   **7** 1.   **9** $166{,}320$.   **11** $151{,}200$.
13 252.   **15** $28, 56$.   **17** $8!5!4!3!$.   **19** $4{,}082{,}400$.

Review Exercises   (page 482)

5 $-3, -6/7, -9/17, -12/31; -21/97$.   **7** $2, 1/2, 5/4, 7/8; 65/64$.
9 $2, \frac{3}{2}, \frac{5}{3}, \frac{8}{5}, \frac{13}{8}$.   **11** $4, 2, \sqrt{2}, \sqrt[4]{2}, \sqrt[8]{2}$.   **13** 70.   **15** 200.
17 $\displaystyle\sum_{k=1}^{7}(3k-1)$.   **19** $\displaystyle\sum_{k=1}^{5}(-1)^{k-1}2(20+k)$.   **21** $-6 - 8\sqrt{2}, -15 - 35\sqrt{2}$.
23 $-28, 53$.   **25** 128.   **27** $5\sqrt{6}$.   **29** 150.   **31** 506.
33 $3/5$.   **35** $729x^6 - 1458x^5y^2 + 1215x^4y^4 - 540x^3y^6 + 135x^2y^8 - 18xy^{10} + y^{12}$.
37 $-252x^{15}y^5$.   **39** $_{52}P_{13}; {}_{13}P_5 \cdot {}_{13}P_3 \cdot {}_{13}P_3 \cdot {}_{13}P_2$.   **41** $495, 126$.

Index

Abscissa, 92
Absolute value
   of a complex number, 387; of a real
   number, 15
Acute angle, 241
Addition formulas, 271–277
Addition of ordinates, 233
Additive inverse, 8
Adjoint of a matrix, 371
Algebraic equation, 55
Algebraic expression, 37
Algebraic function, 156
Allowable value, 37
Ambiguous case, 301
Amplitude
   of a complex number, 394; of a func-
   tion, 229
Angle(s)
   acute, 241; complementary, 274;
   coterminal, 238; definition of, 237;
   degree measure of, 240; of depres-
   sion, 256; of elevation, 255; nega-
   tive, 237; obtuse, 241; positive, 237;
   quadrantal, 238; radian measure of,
   238; reference, 244; right, 240; stan-
   dard position of, 237; trigonometric
   functions of, 243, 248
Antilogarithm, 185
Arccosine function, 293
Arcsine function, 291
Arctangent function, 294
Area of a circular sector, 240
Argument of a complex number, 394
Arithmetic mean, 457
Arithmetic progression, 454
Arithmetic sequence, 454
Associative laws, 7
Asymptotes
   horizontal, 158; of a hyperbola, 146;
   vertical, 224
Augmented matrix, 334
Axes of an ellipse, 144
Axis
   coordinate, 92; imaginary, 393; real,
   393

Base
   of an exponential function, 157;
   logarithmic, 164
Bearing, 257
Binomial, 38
Binomial coefficients, 468
Binomial Theorem, 465

Cancellation law, 8, 413
Cartesian coordinate system, 92
Characteristic of a logarithm, 177
Circle
   equation of, 101; unit, 102
Circular functions, 205

Closed interval, 79
Closed system, 7
Coefficient, 21, 37
Coefficient matrix, 334
Cofactor, 339, 341
Cofunctions, 274
Column subscript, 333
Combination, 479
Common difference, 454
Common logarithms, 177
Common ratio, 459
Commutative laws, 7
Complementary angles, 274
Completing the square, 46, 68
Complex numbers, 380
   absolute value, 387; amplitude, 394;
   argument, 394; conjugate, 384;
   equality, 380; geometric representa-
   tion, 393; imaginary part, 380; mod-
   ulus, 394; polar form, 394; real part,
   380; trigonometric form, 394; vector
   representation, 403
Complex plane, 393
Components of a vector, 404
Composite function, 135
Compound interest, 161
Conditional equation, 56
Conic section, 147
Conjugate of a complex number, 384
Conjugate hyperbolas, 149
Consistent system of equations, 324
Constant function, 109
Constant of proportionality, 149
Constant of variation, 149
Coordinate
   axes, 92; line, 15; plane, 92; system,
   15
Correspondence, 105
   one-to-one, 109
Cosecant function, 205
   graph of, 225
Cosine function, 205
   graph of, 223
Cotangent function, 205
   graph of, 226
Coterminal angles, 238
Cramer's rule, 350–353
Cube root, 27
Cubic polynomial, 410

Damped sine wave, 236
Decimal fraction, 178
Decreasing function, 113
Degree
   angular measure, 240; of a mono-
   mial, 37, 38; of a polynomial, 38, 120,
   410
De Moivre's Theorem, 398
Denominator, 9
   least common, 49; rationalizing, 30